Y0-CCG-262

Cambridge Studies in Biological and Evolutionary Anthropology 44

Seasonality in Primates

The emergence of the genus *Homo* is linked widely to the colonization of "new" highly seasonal savannah habitats. However, until now, our understanding of the possible impact of seasonality on this shift has been limited because we have little general knowledge of how seasonality affects the lives of primates. This book documents the extent of seasonality in food abundance in tropical woody vegetation, and then presents systematic analyses of the impact of seasonality in food supply on the behavioral ecology of non-human primates. Syntheses in this book then produce for the first time broad generalizations concerning the impact of seasonality on behavioral ecology and reproduction in both human and non-human primates, and apply these insights to primate and human evolution. Written for graduate students and researchers in biological anthropology and behavioral ecology, this is an absorbing account of how seasonality may have affected an important episode in our own evolution.

Diane K. Brockman is Assistant Professor of Anthropology at the University of North Carolina–Charlotte. Her research concerns environmental mechanisms influencing reproduction, development, and life history in human and non-human primates. Current studies involve the hormonal basis of seasonal reproduction, female mate competition, male life history patterns and aging, and the metabolic costs of reproduction in females.

Carel P. van Schaik is now Professor of Biological Anthropology at the University of Zurich, Switzerland. He studies behavioral ecology and the social evolution of primates, and is also interested in the conservation of tropical forests. His previous books include the edited works *Infanticide by Males and Its Implications* (with Charles Janson; 2000; Cambridge: Cambridge University Press) and *Sexual Selection in Primates* (with Peter Kappeler; 2004; Cambridge: Cambridge University Press).

Cambridge Studies in Biological and Evolutionary Anthropology

Series editors

Also available in the series

21 *Bioarchaeology* Clark S. Larsen 0 521 65834 9 (paperback)
22 *Comparative Primate Socioecology* P. C. Lee (ed.) 0 521 59336 0
23 *Patterns of Human Growth*, 2nd edn Barry Bogin 0 521 56438 7 (paperback)
24 *Migration and Colonisation in Human Microevolution* Alan Fix 0 521 59206 2
25 *Human Growth in the Past* Robert D. Hoppa & Charles M. FitzGerald (eds.) 0 521 63153 X
26 *Human Paleobiology* Robert B. Eckhardt 0 521 45160 4
27 *Mountain Gorillas* Martha M. Robbins, Pascale Sicotte & Kelly J. Stewart (eds.) 0 521 76004 7
28 *Evolution and Genetics of Latin American Populations* Francisco M. Salzano & Maria C. Bortolini 0 521 65275 8
29 *Primates Face to Face* Agustin Fuentes & Linda D. Wolfe (eds.) 0 521 79109 X
30 *Human Biology of Pastoral Populations* William R. Leonard & Michael H. Crawford (eds.) 0 521 78016 0
31 *Paleodemography* Robert D. Hoppa & James W. Vaupel (eds.) 0 521 80063 3
32 *Primate Dentition* Davis R. Swindler 0 521 65289 8
33 *The Primate Fossil Record* Walter C. Hartwig (ed.) 0 521 66315 6
34 *Gorilla Biology* Andrea B. Taylor & Michele L. Goldsmith (eds.) 0 521 79281 9
35 *Human Biologists in the Archives* D. Ann Herring & Alan C. Swedlund (eds.) 0 521 80104 4
36 *Human Senescence* Douglas E. Crews 0 521 57173 1
37 *Patterns of Growth and Development in the Genus* Homo. Jennifer L. Thompson, Gaul E. Krovitz & Andrew J. Nelson (eds.) 0 521 82272 6
38 *Neanderthals and Modern Humans – An Ecological and Evolutionary Perspective* Clive Finlayson 0 521 82087 1

39 *Methods in Human Growth Research* Roland C. Hauspie, Noel Cameron & Luciano Molinari (eds.) 0 521 82050 2

40 *Shaping Primate Evolution* Fred Anapol, Rebecca L. German & Nina G. Jablonski (eds.) 0 521 81107 4

41 *Macaque Societies: A Model for the Study of Social Organization* Bernard Thierry, Mewa Singh & Werner Kaumanns (eds.) 0 521 81847 8

42 *Simulating Human Origins and Evolution* Ken Wessen 0 521 84399 5

43 *Bioarchaeology of Southeast Asia* Marc Oxenham & Nancy Tayles (eds.) 0 521 82580 6

Seasonality in Primates

Studies of Living and Extinct Human and Non-Human Primates

EDITED BY

DIANE K. BROCKMAN
University of North Carolina at Charlotte, Charlotte, NC, USA

CAREL P. VAN SCHAIK
University of Zurich, Zurich, Switzerland

CAMBRIDGE UNIVERSITY PRESS
Cambridge, New York, Melbourne, Madrid, Cape Town, Singapore, São Paulo

CAMBRIDGE UNIVERSITY PRESS
The Edinburgh Building, Cambridge CB2 2RU, UK
Published in the United States of America by Cambridge University Press, New York

www.cambridge.org
Information on this title: www.cambridge.org/9780521820691

First published 2005

Printed in the United Kingdom at the University Press, Cambridge

A catalog record for this publication is available from the British Library

Library of Congress Cataloging in Publication data

ISBN-13 978-0-521-82069-1 hardback
ISBN-10 0-521-82069-3 hardback

Contents

Contributors

Susan C. Alberts
Department of Biology, Duke University, Box 90338, Durham NC 27708, USA; and Institute for Primate Research, National Museums of Kenya, Nairobi, Kenya

Jeanne Altmann
Department of Ecology and Evolutionary Biology, Princeton University, Princeton NJ 08544, USA; and Department of Conservation Biology, Brookfield Zoo, Brookfield, IL, USA; and Institute for Primate Research, National Museums of Kenya, Nairobi, Kenya

Douglas W. Bird
Department of Anthropological Sciences, Stanford University, Stanford CA 94305–2117, USA

Rebecca Bliege Bird
Department of Anthropological Sciences, Stanford University, Stanford CA 94305–2117, USA

Diane K. Brockman
Department of Sociology and Anthropology, University of North Carolina at Charlotte, Charlotte NC 28223, USA

Nora Bynum
Center for Biodiversity and Conservation, American Museum of Natural History, Central Park West at 79th Street, New York NY 10024, USA

Peter T. Ellison
Department of Anthropology, Harvard University, Cambridge MA 02138, USA

Jennifer L. Fish
Max Planck Institute of Molecular Cell Biology and Genetics, Pfotenhauerstrasse 108, 01307 Dresden, Germany

Jörg U. Ganzhorn
Department of Animal Ecology and Conservation, Hamburg University, Martin-Luther-King Platz 3, 20146 Hamburg, Germany

Claire A. Hemingway
Botanical Society of America, PO Box 299, St Louis MO 63166–0299, USA

Russell Hill
Evolutionary Anthropology Research Group, Department of Anthropology, University of Durham, 43 Old Elvet, Durham DH1 3HN, UK

Julie A. Hollister-Smith
Department of Biology, Duke University, Box 90338, Durham NC 27708, USA

Nina G. Jablonski
Department of Anthropology, California Academy of Sciences, 875 Howard Street, San Francisco CA 94103–3009, USA

Charles Janson
Department of Ecology and Evolution, State University of New York, Stony Brook NY 11794–5245, USA

Peter M. Kappeler
Deutsches Primatenzentrum, Kellnerweg 4, 37077 Göttingen, Germany

John D. Kingston
Department of Anthropology, Emory University, Atlanta GA 30322, USA

Cheryl D. Knott
Department of Anthropology, Harvard University, Cambridge MA 02138, USA

Richard Madden
Department of Biological Anthropology and Anatomy, Duke University, Box 90383, Durham NC 27708, USA

W. Scott McGraw
Department of Anthropology, Ohio State University, Columbus OH 43210, USA

John C. Mitani
Department of Anthropology, University of Michigan, Ann Arbor MI 48109–1092, USA

Philip M. Muruthi
African Wildlife Foundation, Box 48177, Nairobi, Kenya

Raphael S. Mututua
Amboseli Baboon Research Project, Amboseli National Park, Kenya; and Institute for Primate Research, National Museums of Kenya, Nairobi, Kenya

Kristina R. Pfannes
Center for Tropical Conservation, Duke University, Box 90381, Durham NC 27708–0381, USA

J. Michael Plavcan
Department of Anthropology, University of Arkansas, Fayetteville AR 72701, USA

Michele A. Rasmussen
Department of Biological Anthropology and Anatomy, Duke University, Durham NC 27708, USA

Kaye E. Reed
Department of Anthropology/Institute of Human Origins, Arizona State University, Tempe AZ 85287, USA

Serah N. Sayialel
Amboseli Baboon Research Project, Amboseli National Park, Kenya; and Institute for Primate Research, National Museums of Kenya, Nairobi, Kenya

Jutta Schmid
Department of Experimental Ecology, University of Ulm, Albert Einstein Allee 11, D-89069 Ulm, Germany

Diana S. Sherry
Department of Anthropology, Harvard University, Cambridge MA 02138, USA

Claudia R. Valeggia
Department of Anthropology, Harvard University, Cambridge MA 02138, USA; and Consejo Nacional de Investigaciones Científicas y Tecnológicas (CONICET), Argentina

Carel P. van Schaik
Anthropologisches Institut, University of Zurich, Winterthurerstrasse 190, CH-8057, Zurich, Switzerland

Jennifer Verdolin
Department of Ecology and Evolution, State University of New York, Stony Brook NY 11794–5245, USA

J. Kinyua Warutere
Amboseli Baboon Research Project, Amboseli National Park, Kenya; and Institute for Primate Research, National Museums of Kenya, Nairobi, Kenya

David P. Watts
Department of Anthropology, Yale University, New Haven CT 06520–8277, USA

Preface

Animals everywhere have to cope with the changing seasons. Even those living in the tropics face serious fluctuations in the abundance of their favorite foods. In this book we examine how seasonal variation in food supply affects what primates eat, where they search for it, how active they are, and when during the day they are active, as well as how these responses affect their body sizes, their social lives, the timing of their reproduction, and the composition of their ecological communities. For the first time, we distinguish several general patterns about these seasonal responses. These generalizations are subsequently applied to see whether major transitions during hominin evolution can be ascribed to dramatic changes in seasonality.

The contributors to this book were selected from among internationally recognized primate field biologists and behavioral ecologists who presented synthetic review papers on how seasonality influences intra-specific variation in primate behavior and socioecology at the Primate Seasonality Symposium during the XVIII Congress of the International Primatological Society in Adelaide, Australia. We subsequently added various authors with special expertise so as to broaden our coverage of seasonality in both humans and non-human primates. This book is timely and important because it is unique to the anthropological literature and it presents new ideas about interactions between seasonality and the adaptive responses of living and extinct primate populations to resource variability.

Part I *Introduction*

1 *Seasonality in primate ecology, reproduction, and life history: an overview*

CAREL P. VAN SCHAIK
Anthropologisches Institut, University of Zurich, Winterthurerstrasse 190, CH-8057, Zurich, Switzerland

DIANE K. BROCKMAN
Department of Sociology and Anthropology, University of North Carolina at Charlotte, Charlotte NC 28223, USA

Introduction

Seasonality refers to recurrent fluctuations that tend to have a period of one year. Seasonality in climate is a basic consequence of the tilt of the Earth's axis relative to its orbital plane (e.g. Pianka 1994). As a result, the position of the zenithal Sun (when it is directly overhead) varies through the year. It is directly overhead at 23.5° S on December 22 (winter solstice, in northern hemisphere terminology), then marches north, reaching the Equator on March 21 (equinox), moving on to the summer solstice on June 22 at 23.5° N, where it turns south again, passing the Equator on September 23 (another equinox) toward the winter solstice. The Sun's march affects not only sunshine but also other aspects of climate, such as windiness and rainfall. Hence, seasonality is felt around the globe.

This book is about the impact of seasonality on the lives of primates. Members of the order Primates are confined largely to the tropics, where they occupy a broad range of terrestrial habitats, although 90% of species live in tropical forest (Mittermeier 1988). Hence, our focus will be on seasonality in the tropics. In this region, seasonal variation in temperature is limited: temperature fluctuations over the 24-hour day exceed the range of monthly means and frosts are extremely rare (MacArthur 1972). Likewise, variation in day length, although present everywhere except right on the Equator, is limited. However,

Seasonality in Primates: Studies of Living and Extinct Human and Non-Human Primates, ed. Diane K. Brockman and Carel P. van Schaik. Published by Cambridge University Press.

seasonal variations in rainfall and sunshine characterize all tropical habitats. Tropical climates are usually classified according to the duration and intensity of the dry season. By and large, seasonality increases as one moves from the Equator, i.e. with increasing latitude (see Chapter 2).

Animals respond to changing weather conditions in direct ways such as seeking shelter from rain or avoiding direct sunshine, but these direct responses tend to have only a limited and transient impact on their behavioral ecology. The major impact of climate is indirect. Climate is the fundamental dynamic force that shapes vegetation and in the long run affects soils. The indirect effects of climate come in two fundamentally different kinds: (i) the effects of climate on the phenology of the plant community, and thus on the abundance, nature, and distribution of potential food items, be they plant or animal, and on the amount of cover where plants are deciduous; and (ii) the effects of climate on the structure of the vegetation, from evergreen tall rainforest to sparsely covered open scrubland with virtually no woody vegetation. In this book, we will explore the impacts of these two distinct kinds of seasonality on the lives of primates, albeit with an emphasis on the first.

Humans are primates, and likewise have a tropical origin. Major events in hominin evolution have been ascribed to changes in seasonality (in this book, hominins are defined on the basis of an adaptive shift toward obligate bipedalism). Most hypotheses focus on the permanent effects of increased seasonality, such as the impact of forest fragmentation on the evolution of bipedalism and rates of speciation in the early to mid Pliocene Epoch (Foley 1993; Klein 1999), while others emphasize the role that fluctuating climates may have played in the appearance of stone tools and regular utilization of animal protein in early *Homo* (Potts 1998).

The aim of this book is to examine the impact of seasonality on the behavioral ecology, reproduction, social life, and life history of primates. The results of this analysis can then be used as a background against which to evaluate hypotheses proposed to explain aspects of hominin evolution. There are, as yet, few clear empirical generalizations on how primate range use, foraging, and reproduction respond to seasonality in resource abundance, and hence no clear explanations for the variability. Much of this book will therefore be devoted to documenting the main patterns in nature and developing hypotheses to explain them. In the rest of this introduction, we will develop the framework for this exercise.

Seasonality and behavioral ecology of primates

Activities

The majority of primates are diurnal and, once habituated, easy to observe. As a result, we have more detailed descriptive information on their activity budgets, range use, and reproduction than for any other mammalian order, and increasingly this information is also accumulating for the smaller and non-diurnal species. These studies have shown that primate behavioral ecology varies over time; in most cases, a seasonal signal in these fluctuations is clearly evident. However, although there is enough seasonal variation to warrant a synthetic examination of general patterns, there has until now been no attempt to develop a set of generalizations that can serve as a framework in which to interpret hominin behavioral ecology.

For example, primates are substrate-bound organisms that cannot undertake lengthy seasonal migrations to escape from periods with unfavorable conditions. Thus, some species appear to increase their activity during times of scarcity, whereas others decrease it, some even to the point of hibernation. At present, there is no good explanation for this variation. Likewise, whereas organisms with fast life histories can complete a full reproductive cycle during a seasonal peak in food abundance, primates, along with other long-lived organisms, with their long gestation and lactation periods, respond to seasonality either by coordinating peak demands of mid lactation with seasonal peaks in food abundance or by using those peaks to initiate a new reproductive cycle (van Schaik & van Noordwijk 1985). Thus, empirical patterns in how slow-breeding, Earth-bound tropical mammals respond to predictable seasonality in their natural habitats remain to be elucidated and explained.

Primates cue into and respond to various indicators of seasonality, particularly variations in the type and distribution of food. Most primates prefer ripe fruit with its readily assimilated energy. Fruit comes in discrete patches, as single fruit trees, lianas, or groves of interconnected plants. Abundant evidence indicates that primates do not cruise through their ranges in search of this food, but rather exploit their range based on continuously updated knowledge about the distribution of suitable patches (e.g. Garber 2000). Thus, temporal changes in the abundance of new plant organs ("phenology"), and the invertebrates dependent on them, will affect these consumers.

Optimal foraging theory (Stephens & Krebs 1986) predicts that as preferred food items become scarce, less preferred foods (fallback foods that have lower energy return per unit foraging time) will become included

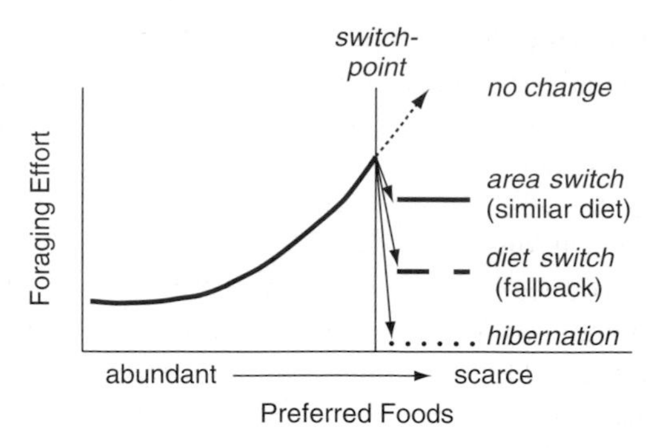

Figure 1.1 The possible responses to a decrease in the abundance of preferred foods, forcing animals to range more widely in order to maintain constant net energy intake. When foraging effort becomes too great to be maintained, animals can switch to different areas or different diets or hibernate.

in the diet and patches of food (e.g. fruit trees) will be more depleted before the forager moves on to the next patch (Charnov 1976). Most primates have a mixed foraging mode in which some foods are encountered as the animals move through their range, but most of the movement is driven by the need to exploit a known set of patches in a near-optimal order (e.g. Janson 2000). In the absence of a formal theory for predicting the response of primates to reductions in the preferred food types in such a foraging system, we can develop some qualitative expectations concerning the responses of consumers (Fig. 1.1). As preferred foods such as succulent fruits become more scarce, either because the patches become smaller or because the distance between them increases, the animals can continue to maintain a constant daily net energy intake by taking more fruit in each patch (perhaps by including less ripe fruit) and by traveling more, either visiting more patches or traveling longer distances between the same number of patches. However, as food abundance continues to decline, full compensation becomes increasingly difficult and net intake will begin to suffer.

Beyond the point of easy compensation for reduced availability of favored foods, animals can respond in a variety of ways (Fig. 1.1). The first, the *area switch*, involves a move into a different habitat that is phenologically out of synchrony with that currently occupied and where fruit abundance is higher. In this way, net energy intake can be maintained. Such a move, of course, requires that different habitats with out-of-phase fruiting peaks are available on a scale that is suitable for the consumers. Usually, this requires topographically complex landscapes, such as floodplains and uplands, valleys and hillsides, or mountain slopes of different exposures. We also expect to find this response in species that can easily travel over longer distances. Flying frugivores such as birds and bats therefore can afford to make the area switch much more easily, and should be more likely to remain specialized frugivores throughout the year, than substrate-bound animals such as primates.

A second response, the *diet switch*, should be more common among primates: the animals will switch to other foods, as predicted by diet breadth models, that produce lower return rates but require less travel to harvest. Examples include young or even mature leaves, unripe fruits, flowers, gum, and insects. Overall net intake on these fallback foods tends to be less because otherwise they would be used at all times (fallback foods also may be available during the period of high fruit abundance; see Chapter 3). Larger animals have lower energy requirements per unit body weight and therefore can afford more easily to switch to foods with lower energy density. Moreover, they have longer gastrointestinal systems, so more time is available to extract nutrients. For these reasons, it is expected that it will generally be easier for large animals, compared with small animals, to switch to a lower-quality diet.

The timing and extent of this diet switch may be affected by the fact that digestive efficiency of the fallback foods may depend on the cumulative intake over longer periods, perhaps because gut morphology and biochemistry become adjusted to them (e.g. Sibly 1981). Because energy intake declines when this switch occurs, we expect animals to reduce energy expenditure as well. This can be achieved because the fallback foods tend to be present at higher densities compared with the preferred foods, thus requiring less expenditure to harvest. Indeed, in some cases, animals may even lower their basal metabolic rates during the period of food scarcity (Pereira 1995). Among primates, this response may be especially common in lemurs, which may explain why many lemur diets are relatively less energy-dense than those of other primates of similar size and even why their reproductive seasonality is more pronounced (see Chapter 11).

Opportunities for diet switching may be more limited where the season of fruit scarcity is also one in which young leaves and organisms depending on them are scarce. There is a trend toward more positive covariation between the production of flush and fruit in areas with longer dry seasons, and independently in Neotropical forests (see Chapter 2). This pattern suggests that diet switches will be observed more commonly among larger species, among Paleotropical species, and among species in less seasonal habitats. Hemingway and Bynum (Chapter 3) demonstrate the importance of both area and diet switches in detail for non-human primates.

The third major response is found where the food situation becomes markedly unfavorable for a predictable period during the year. Animals can then dramatically reduce their requirements by hibernating in a safe place. Small animals consistently need high-quality foods. If they are also not highly mobile, we should therefore expect that smaller animals are more likely to hibernate, especially in the most seasonal environments and especially if they specialize on the most energy-dense foods, such as insects and fruits. Hibernation as a response to seasonal food stress is rare among primates and found only among small strepsirrhines (see Chapter 5), perhaps an extension of the reduced metabolic rate during the lean season.

Where hibernation is not an option, taxa may face a limit to the degree of tolerable seasonality. We should therefore expect to see a relationship between a species' body size and the degree of seasonality that it tolerates. Indeed, in Neotropical forests, smaller species may occur only in less seasonal forests (Cowlishaw & Dunbar 2000: 177), but van Schaik *et al.* (Chapter 15), who examine this more systematically, find little evidence for such a relationship. Within species, however, the response may go in the opposite direction, because smaller individuals need less food overall and therefore are more likely to survive the rigors of food seasonal scarcity. Plavcan *et al.* (Chapter 14) show that this expectation is generally met.

The classification of common responses laid out in Fig. 1.1 does not exhaust the possibilities. First, Alberts *et al.* (Chapter 6), studying baboons, describe a variant of the fallback foods they call *handoff foods*, where net intake remains approximately constant throughout the year in spite of a highly seasonal habitat and major diet switches. Bliege-Bird & Bird (Chapter 9) draw similar conclusions from their review of human foragers living in highly seasonal environments. Both cases suggest that intelligent omnivores might not face a lean season in the same way that regular consumers do, even in the face of severe seasonality in resource availability (see also Chapter 19). Second, some seasonal activities appear to have no direct relationship to seasonal food abundance. One possible reason is that the two sexes show different responses to seasonality, due to the different ways in which males and females maximize their fitness (see also Chapter 9). An example of this is hunting by chimpanzees, in which seasonally varying social conditions affect seasonal variation in hunting activity (see Chapter 8). Students of primate behavioral ecology face the challenge of predicting which of these many responses will be shown by particular species or populations.

Seasonality in cover and day length

Another environmental factor that can show dramatic seasonal variation is cover. In deciduous forest habitats, canopies go from bare to dense, and during the bare phase predation risk for smaller canopy species may be greatly increased. A few primates might respond to seasonal increases in vulnerability to predators due to bare crowns by hibernating (an issue not yet settled; see Chapter 5). Some others respond by changing their activity period. However, the majority will have to make more subtle behavioral adjustments to such dramatic seasonality in predation risk (see Chapter 4).

While it is easy to recognize the central roles that food and cover play in determining seasonal variation in activities, other seasonally entrained factors are also relevant. Especially toward the edge of the tropical region, day length can vary: for instance, at the tropics (23.5° N and S), the difference between maximum and minimum day length is almost three hours (Pianka 1994). Hence, even at constant food abundance, the daily duration of time available for foraging may create an apparent seasonality in food abundance, with its consequences for time budgets (see Chapter 7). Likewise, in open habitats (or in leafless forest canopies during the bare season), the absence of water and high daily maximum temperatures may lead to overheating unless special measures are taken. It turns out, perhaps remarkably, that only the first of these two environmental variables has a dramatic effect on primate range use (see Chapter 7).

Social life and life history

Fluctuating food abundance may affect social life in diverse, and sometimes counterintuitive, ways. Where declining abundance of preferred foods leads to fallback utilization of low-quality food items such as mature leaves, the nature of competition may change from a strong to a weak contest component and, hence, an increased scramble component (cf. Wrangham's [1980] "subsistence diet"). Thus, paradoxically, as food abundance declines, and competitiveness and aggression may be expected to rise, we may actually see an overall decline in rates of aggression. We may also see groups become less spatially cohesive.

Across a seasonality gradient within a species, or among a set of allopatric close relatives, we may see that group size declines, as a result of declining productivity, or increases, as a result of a gradient in diet composition toward lower-quality food but more abundantly available food items. Plavcan *et al.* (Chapter 14) show that both responses occur and no easy generalizations are possible.

Seasonality in climate, food, and reproduction may have subtle effects on the social life of a species that potentially affect demography and hence life history. In particular, seasonal reproduction could affect the operational sex ratio in primate groups, resulting in cascading affects on sexual dimorphism and other aspects of mating competition, male–female relations, and individual career trajectories (see Chapter 14). Overall, however, the impacts of seasonality on social behavior vary enough to prevent clear generalizations.

Seasonality and primate communities

Especially where severe, seasonality may limit the density of a species. Neotropical primates in seasonal forests were found to rely on only a few keystone resources to tide them over the season of fruit scarcity (Terborgh 1983). As a result, seasonality may affect both the species richness and the biomass of thc local primate community. Other work, however, suggests that the main correlate of primate species richness is mean productivity (Ganzhorn *et al.* 1997; Kay *et al.* 1997), as expected by general diversity theory (Rosenzweig 1995). Van Schaik *et al.* (Chapter 15) find little evidence for an effect of seasonality on species richness. However, they do find that aspects of community biomass are predictably linked to seasonality.

Seasonality and reproduction in primates

So far, we have focused on coping with periods of resource scarcity, and animals may avoid reproducing during such periods. However, they may also be able to make use of periods of unusually high abundance of resources to reproduce. It is important to avoid the assumption that animals will attempt to maintain constant net energy intake, even though this is supported by numerous short-term laboratory experiments that varied food rewards per unit effort. In the wild, periods of positive energy balance will alternate with periods when energy losses are minimized. Hence, when food is most abundant, we may see animals work the hardest to gain a major energy surplus, used to support reproductive efforts, or in non-breeders to support growth or the buildup of energy stores in the form of body fat.

That animals such as primates, with slow reproduction involving long gestation periods and often even longer periods of infant dependence (Kappeler *et al.* 2003), respond to seasonal fluctuations in the availability of food with reproductive seasonality is not necessarily a trivial expectation. Seasonal reproduction will entail costs whenever the cycle is not an exact integer number of years, something especially likely to happen when

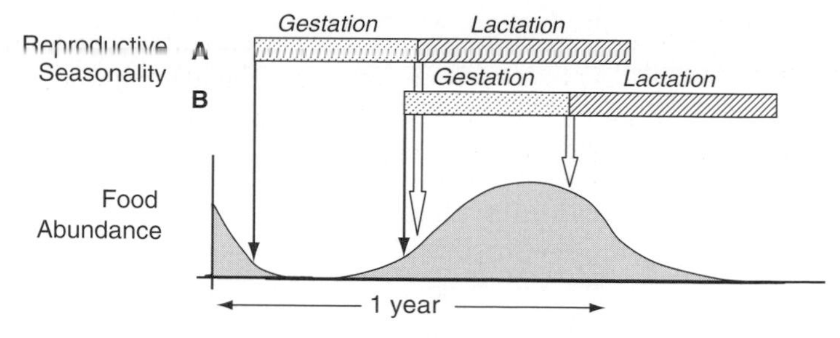

Figure 1.2 Two distinct modes of responding to seasonality in the abundance of preferred foods. In A, females are selected to give birth before the annual fruit peak and therefore have evolved sensitivities to exogenous cues that lead to conception one gestation period before this optimal timing (classic or income breeding). In B, females respond to increases in food abundance as cues for conception, leading to births one gestation period following that event, often coinciding with the end of the fruiting peak (alternative or capital breeding). ↓ approximate timing of conception; ⇩ approximate timing of birth. See Chapter 10 for further details.

infants die. The cost of the delay to resynchronize with the environmental cycle must therefore be outweighed by the benefit of increased prospects for growth of the infant or survival of the mother and/or the infant. Hence, there will be a threshold in resource seasonality below which no reproductive seasonality is expected (see Chapter 11).

In primates, the timing of reproduction relative to predictable seasonal fluctuations in resource abundance is not easy to predict (cf. Kiltie 1988). Figure 1.2 shows two possible reproductive responses. In A, the period of highest food (usually fruit) abundance coincides with the period of greatest energetic burden on the female, mid lactation. In B, increasing food abundance and the positive energy budgets associated with it are used to begin the reproductive cycle (van Schaik & van Noordwijk 1985). These differences are related to the difference between income (A) and capital (B) breeding (Jönsson 1997). Brockman & van Schaik (see Chapter 10) discuss both the proximate and the ultimate aspects of these two main response types, which produce contrasting temporal relationships between peaks in births and food. They show that the different response types vary in many features of their reproductive cycles, ranging from strict income breeding to relaxed income breeding to capital breeding. They propose the *Income–Capital Continuum Model* to accommodate this diversity of primate reproductive responses to seasonality, and test predictions of this model for primates, including humans.

A third response (C), no distinct seasonality at all, is of course also still a possibility. Surprisingly, in the same habitat, some species may show decidedly seasonal reproduction, whereas others do not (e.g. see Chapter 6). We therefore must identify the critical determinants of each response. Janson & Verdolin (Chapter 11) examine the many factors that affect the sharpness of reproductive seasonality in relation to seasonality in food abundance. They also examine the temporal relationship between birth peaks and food peaks to identify the factors affecting the switch from income to capital breeding. Knott (Chapter 12) and Ellison *et al.* (Chapter 13) discuss responses to seasonality and seasonal reproduction in the ultimate capital breeders, great apes and humans in particular.

Seasonality and habitat structure

So far, we have considered how seasonality in climate affects phenology, but its other main impact is on the structure of the vegetation. This effect, while direct, is not necessarily simple. There have been numerous attempts at erecting classificatory schemes for the link between climate and vegetation structure, only to be superseded by others. Nonetheless, some clear trends emerge.

Tropical vegetation varies from evergreen rainforest at the high end of the rainfall spectrum to semi-desert at the low end. The length of the dry season plays a major role in this variation. The wettest evergreen forests experience no predictable dry season or, at most, occasional irregular dry spells lasting a month or less (Walsh 1996). As the dry season becomes longer (Whittaker 1970; Walter 1973; Gentry 1988; Whitmore 1990; Akin 1991; Jacobs 1999):

- the abundance of epiphytes, climbers, and lianas decreases;
- tree size, and thus canopy height, decreases;
- the number of recognizable canopy layers decreases;
- the average size of leaves decreases and entire margins and drip tips become less common;
- canopy connectivity is reduced;
- total plant species richness tends to decline;
- the degree of deciduousness of canopy trees increases (deciduous trees predominate below *c.* 2000 mm, except on the poorest soils);
- the relative importance of the understorey increases, with shrubs and forbs gradually giving way to grasses;
- the density of large terrestrial browsers and the likelihood of natural and anthropogenic fires increases, both potentially producing instability of the forest–savanna boundary over time.

Where dry seasons exceed six to eight months and annual rainfall falls below about 700 mm, the deciduous forest gradually gives way to savanna vegetation dominated by grasses and umbrella-shaped trees and shrubs. At the lowest end of the rainfall spectrum we find semi-desert with thorn scrub. Primates are no longer found at this end of the spectrum.

However, Walter (1973) points out that higher rainfall during the wet season can create buffers ameliorating the effect of the dry season, so that the same vegetation type can be found in areas with longer dry season (see Walter's [1973] Fig. 26; see also Chapter 17). Walsh's (1996) perhumidity index takes this phenomenon into account and therefore is a good indicator of forest structure and species composition within the tropical evergreen forest formation type.

Topographic and edaphic factors also complicate the impact of climatic seasonality. Thus, the permanent availability of water along rivers in generally dry areas can produce deciduous or even evergreen gallery forests in a savanna landscape. Swamp forests, especially those on thick peat, have simpler structures and tend to be poor in species. Regions with highly porous soils, such as karst, usually have vegetation normally associated with climates with longer dry seasons. Poorly drained plains in highly seasonal climates are inundated during the wet season and are bone-dry during the dry season, often having vegetation dominated by palms or other specialists.

In general, then, a primate living in a more seasonal habitat experiences a less connected and lower tree canopy, and a more open understorey, creating the need for at least some terrestrial locomotion and for dealing with a less buffered microclimate. In addition to the generally increased seasonality in the abundance of resources, there are fewer species of plant, especially lianas, which may serve to exacerbate such seasonality.

Altitude modifies this general picture. Temperatures, especially minimum temperatures, decline with increasing altitude and may reach values found more commonly in cold temperate zones, but without the obvious seasonality. Rainfall often peaks at intermediate altitudes along continuous altitudinal gradients. Where the lowlands support evergreen zones, increasing elevation leads to increasingly lower canopy height, more sclerophyllous leaves, fewer lianas, major turnover in species composition, and lower species richness, and, in the cloud forest zone, high loads of epiphytes, especially non-vascular plants. These changes are driven by both changing climate and declining soil fertility. The nature of the local vegetation depends on details of climate and topography. Thus, the "Massenerhebung effect," the observation that transitions to more montane vegetation types occur at higher elevations on higher mountains, is due to higher cloud zones and higher soil nutrient availability on higher mountains. Likewise, local rain shadows produce more deciduous forest, whereas persistent cloudiness produces evergreen forests, sometimes quite close to each other. Some tropical mountains are high enough to reach the tree line. In this book, we do not discuss these high-altitude vegetations or their primate denizens.

Plant productivity is directly dependent on evapotranspiration (Rosenzweig 1968) and thus is a function of insolation and water supply. Hence, it shows a linear increase with annual rainfall up to *c.* 1000 mm, above which the increase is less steep (Whittaker 1970), and may peak at *c.* 2500 mm annual rainfall (Kay *et al.* 1997), declining at higher rainfall values due to persistent cloudiness and soil leaching. At equal rainfall, productivity generally decreases with altitude, but it may actually peak at intermediate altitudes, because high night temperatures cause high rates of transpiration near sea level.

As a rule, the lower the annual rainfall, the greater is the variation in rainfall between years. More important, however, is the tendency for this variation in drier climates to be in the form of runs of drier or wetter years (Jackson 1977). Thus, animals in drier habitats are faced with more dramatic between-year variability than those in closed, evergreen forests (see Chapter 6). In some areas, fires may be common in dry years, requiring consumers to adapt to their occurrence by switching to a diet consisting predominantly of young foliage or else to migrate to unaffected areas.

Implications for primate ecology

Variation in rain and drought produced by predictable changes in the structure and productivity of tropical vegetation is bound to affect features such as group size or body size of the individual species or congeners that occupy different positions in the gradient, as well as the structure and biomass of primate communities.

In general, as dry seasons become more prominent, canopy dwellers will find it increasingly difficult to travel between tree crowns, due to reduced abundance of lianas and increasing crown shyness. The canopy also comes down in height. Group sizes on the savanna are larger than those in the forest, even if we limit the comparison to species habitually traveling on the forest floor (Clutton-Brock & Harvey 1977). These structural changes also have obvious effects on the composition of primate communities in terms of the main locomotor modes and foraging niches allowed in each kind of vegetation (see Chapter 17).

The change in forest structure with decreasing rainfall also may affect primate biomass. However, the available proportion of productivity for consumers, and the relative amount of the various new organs into which this productivity is channeled, varies. Thus, succulent fruits are common in many evergreen forests but are less common and more seasonal in deciduous forests. Moreover, while total leaf biomass may be higher in evergreen forests, the quality of leaves in deciduous forests, with their lower mechanical and chemical protection and higher phosphorus and nitrogen concentrations (Aerts & Chapin 2000), is much higher, thus providing folivores with substantially more edible biomass. Not surprisingly, biomass of primate folivores actually increases as dry seasons become longer (see Chapter 15).

Seasonality and deep time

Climates change over time. A major contributor to long-term changes in climate are the Milankovich cycles (see Chapter 18 for more detail), periodic fluctuations in features of the Earth's orbit around the Sun (eccentricity, obliquity, and precession), each with their own period (Akin 1991). Eccentricity refers to the degree to which the Earth's orbit is circular or elliptical; it determines the total amount of solar energy reaching the Earth, and its period is *c*. 95 000 years. Obliquity refers to the tilt in the Earth's vertical axis. As noted earlier, seasonality is a result of variation in obliquity, and hence varies over time, with a period of approximately 41 000 years. At present, we are close to a maximum tilt of 24.4° (minimum is 21.8°), and hence we experience rather strong seasonality. Precession refers to variation in the season of closest approach to the Sun (at present, the Earth is farthest from the Sun during the northern winter) and has a periodicity of *c*. 22 000 years. Additional natural causes of long-term climate change are changes in the land–sea configuration, providing an important link between continental drift and climate change. For instance, it is surmised that Milankovich cycles can instigate ice ages only when polar regions are not covered by deep sea (Akin 1991).

Particularly over the past ten million years or so, the Earth has experienced numerous, fairly regular, cycles of ice ages alternating with warmer interglacial periods (periodicity over the past one million years has been approximately 125 000 years [Akin 1991]; see Chapters 17 and 18, for more detail). In the tropics, the ice ages were generally characterized by lower sea levels and drier climates, producing greater seasonality. There is now abundant evidence for cyclical contraction and expansion of evergreen forests in what is currently the belt of evergreen forests (Morley 2000). Thus, as climate changes over time, a particular locale may find itself in dense forest at one time and in a savanna at another. Vegetation types gradually move over the landscape until crushed against mountains or water bodies. Animals can move with the vegetation types but, especially on islands (e.g. Madagascar) or in regions surrounded by mountains, opportunities for movements of both plants and animals may be limited. In such cases, the possibility of adjusting to the new seasonality in food abundance or cover depends on their norm of reaction and on the rate of change. At least some primates tend to have fairly broad reaction norms, which allows them to cope with changes in seasonality up to a point (see Chapter 6). However, if reaction norms are exceeded or the climate change is too fast, then plants and animals are likely to become extinct. If the local climate subsequently returns to milder seasonality, then we should expect an imprint of the region's history on the composition of communities through a sorting process.

Such species sorting may have taken place for trees based on their phenological responses: species with a phenology adapted to drier climates survive during wetter periods, but those with a phenology adapted to wet climates will become extinct during the drier periods. As a result, a region that has undergone such climate shifts will tend to have phenological regimes characteristic of drier climates (see Chapter 2).

Sorting due to fluctuating climates is a special case of the more straightforward effects of longer-term and more global changes in climate. Thus, when a forest loses many of its canopy lianas that served as major routes for horizontal and vertical travel, or when the canopy loses its integrity and trees become isolated from their neighbors, basic locomotor modes may become obsolete. Either way, local extinction is often inevitable. Jablonski (Chapter 16) argues that many of the major changes in primate evolution were precipitated by sudden increases in seasonality and the consequent changes in vegetation structure, affecting especially those species with slow life histories and obligatory arboreal locomotion.

Anthropogenic influences have recently been superimposed on these natural cycles. A gradual warming trend is apparent around the globe, affecting phenology and geographic ranges of plants and animals (Walther *et al.* 2002), but the main impacts may well be increases in extreme events in most regions of the earth (Easterling *et al.* 2000). One of the more immediate effects of great relevance for much of the tropics is increased El Niño activity, caused by increases in sea-surface temperatures in the eastern Pacific, which leads to droughts in Southeast Asia (Walsh 1996) and increased rainfall in the western Americas. The patterns in primate seasonality documented in this book can be used as a baseline for assessments of climate change in the near future.

Seasonality and the hominin enigma

In the end, our findings should shed light on major issues in human evolution. The first implication of the work reviewed in this book is that of speciation and direction in hominin evolution. Like Jablonksi (Chapter 16) for primates generally, Reed & Fish (Chapter 17) attribute many major events during hominin evolution to dramatic increases in seasonality (see also Foley 1993), while Kingston (Chapter 18) focuses on the importance of short-term astronomically driven climatic change (Milankovich cycling) in linking shifts in seasonality patterns with hominin evolutionary events. Reed & Fish, in particular, stress the role of seasonality in food availability more than the structural changes in habitat caused by increased seasonality.

But if this is the case, then human evolution poses a major enigma. With the appearance of the genus *Homo*, we see the gradual emergence, completed by the time savanna-dwelling *Homo sapiens* comes around, of traits that reflect both a higher level and greater stability of energy intake relative to the great-ape ancestor. These are: (i) increased mobility, and thus increased energy expenditure; (ii) large brains, and thus significantly increased energy requirements (Leonard & Robertson 1997) and inability to deal with nutrient shortages during development (Levitsky & Strupp 1995); and (iii) faster reproduction than that shown by any of the great apes (Hawkes *et al.* 1998). This increased mean and reduced variance in energy intake is enigmatic given the continuing increase in seasonality (deMenocal 1995) and the reliance by human foragers on a very narrow diet compared with our great-ape relatives (Blurton Jones *et al.* 1999). It suggests that members of the hominin lineage came up with solutions to cope with the increasingly serious seasonality that were radically different from those of the other hominoids (see also Chapter 9). Although several of the contributions to this book touch on this issue, Brockman's final chapter (Chapter 19) will attempt a synthesis of the role of seasonality in hominin evolution.

Acknowledgments

We thank Meredith Bastian for comments.

References

Aerts, R. & Chapin, F. S., III (2000). The mineral nutrition of wild plants revisited: a re-evaluation of processes and patterns. *Advances in Ecological Research*, **30**, 1–61.

Akin, W. E. (1991). *Global Patterns: Climate, Vegetation, and Soils.* Norman, OK: University of Oklahoma Press.

Blurton Jones, N., Hawkes, K., & O'Connell, J. F. (1999). Some current ideas about the evolution of the human life history. In *Comparative Primate Socioecology*, ed. P. C. Lee. Cambridge: University of Cambridge Press, pp. 140–66.

Charnov, E. L. (1976). Optimal foraging: the marginal value theorem. *Theoretical Population Biology*, **9**, 129–36.

Clutton-Brock, T. H. & Harvey, P. H. (1977). Primate ecology and social organization. *Journal of Zoology, London*, **183**, 1–39.

Cowlishaw, G. & Dunbar, R. (2000). *Primate Conservation Biology*. Chicago: Chicago University Press.

DeMenocal, P. B. (1995). Plio-Pleistocene African climate. *Science*, **270**, 53–9.

Easterling, D. R., Meehl, G. A., Parmesan, C., *et al.* (2000). Climate extremes: observations, modeling and impacts. *Science*, **289**, 2068–74.

Foley, R. A. (1993). The influence of seasonality on hominid evolution. In *Seasonality and Human Ecology*, ed. S. J. Ulijaszek & S. Strickland. Cambridge: Cambridge University Press, pp. 17–37.

Ganzhorn, J. U., Malcomber, S., Andrianantoanina, O., & Goodman, S. (1997). Habitat characteristics and lemur species richness in Madagascar. *Biotropica*, **29**, 331–43.

Garber, P. A. (2000). Evidence for the use of spatial, temporal, and social information by primate foragers. In *On the Move: How and Why Animals Travel in Groups*, ed. S. Boinski & P. A. Garber. Chicago: Chicago University Press, pp. 261–98.

Gentry, A. H. (1988). Changes in plant community diversity and floristic composition on environmental and geographical gradients. *Annals of the Missouri Botanical Garden*, **75**, 1–34.

Hawkes, K., O'Connell, J. F., Blurton Jones, N. G., Alvarez, H., & Charnov, E. L. (1998). Grandmothering, menopause, and the evolution of human life histories. *Proceedings of the National Academy of Sciences, USA*, **95**, 1336–9.

Jackson, I. J. (1977). *Climate, Water and Agriculture in the Tropics*. London: Longman.

Jacobs, B. F. (1999). Estimation of rainfall variables from leaf characters in tropical Africa. *Palaeogeography, Palaeoclimatology, Palaeoecology*, **145**, 231–50.

Janson, C. H. (2000). Spatial movement strategies: theory, evidence, and challenges. In *On the Move: How and Why Animals Travel in Groups*, ed. S. Boinski & P. A. Garber. Chicago: Chicago University Press, pp. 165–203.

Jönsson, K. I. (1997). Capital and income breeding as alternative tactics of resource use in reproduction. *Oikos*, **78**, 57–60.

Kappeler, P. M., Pereira, M. E. & van Schaik, C. P. (2003). Primate life histories and socioecology. In *Primate Life Histories and Socioecology*, ed. P. M. Kappeler & M. E. Pereira. Chicago: Chicago University Press, pp. 1–20.

Kay, R. F., Madden, R. H., van Schaik, C. P., & Hendon, D. (1997). Primate species richness is determined by plant productivity: implications for conservation. *Proceedings of the National Academy of Sciences, USA*, **94**, 13 023–7.

Kiltie, R. A. (1988). Gestation as a constraint on the evolution of seasonal breeding in mammals. In *Evolution of Life Histories of Mammals*, ed. M. S. Boyce. New Haven: Yale University Press, pp. 257–89.

Klein, R. G. (1999). *The Human Career: Human Biological and Cultural Origins*, 2nd edn. Chicago: University of Chicago Press.

Leonard, W. R. & Robertson, M. L. (1997). Comparative primate energetics and hominid evolution. *American Journal of Physical Anthropology*, **102**, 265–81.

Levitsky, D. A. & Strupp, B. J. (1995). Malnutrition and the brain: changing concepts, changing concerns. *Journal of Nutrition*, **125**, S2212–20.

MacArthur, R. H. (1972). *Geographical Ecology: Patterns in the Distribution of Species*. New York: Harper & Row.

Mittermeier, R. (1988). Primate diversity and the tropical forest: case studies from Brazil and Madagascar and the importance of megadiversity countries. In *Biodiversity*, ed. E. O. Wilson & F. M. Peter. Washington, DC: National Academy of Sciences, pp. 145–54.

Morley, R. J. (2000). *Origin and Evolution of Tropical Rain Forests*. New York: John Wiley & Sons.

Pereira, M. E. (1995). Development and social dominance among group-living primates. *American Journal of Primatology*, **37**, 143–75.

Pianka, E. R. (1994). *Evolutionary Ecology*, 5th edn. New York: HarperCollins College Publishers.

Potts, R. (1998). Environmental hypotheses of hominin evolution. *Yearbook of Physical Anthropology*, **41**, 93–136.

Rosenzweig, M. L. (1968). Net primary productivity of terrestrial communities: prediction from climatological data. *American Naturalist*, **102**, 67–74.

Rosenzweig, M. L. (1995). *Species Diversity in Space and Time*. Cambridge: Cambridge University Press.

Sibly, R. M. (1981). Strategies of digestion and defecation. In *Physiological Ecology: An Evolutionary Approach to Resource Use*, ed. C. R. Townsend & P. Calow. Oxford: Blackwell Scientific Publications, pp. 109–39.

Stephens, D. W. & Krebs, J. R. (1986). *Foraging Theory*. Princeton: Princeton University Press.

Terborgh, J. (1983). *Five New World Primates: A Study in Comparative Ecology*. Princeton: Princeton University Press.

Van Schaik, C. P. & van Noordwijk, M. A. (1985). Interannual variability in fruit abundance and reproductive seasonality in Sumatran long-tailed macaques (*Macaca fascicularis*). *Journal of Zoology*, **206**, 533–49.

Walsh, R. P. D. (1996). Climate. In *The Tropical Rain Forest: An Ecological Study*, ed. P. W. Richards. Cambridge: Cambridge University Press, pp. 159–205.

Walter, H. (1973). *Vegetation of the Earth in Relation to Climate and the Eco-physiological Conditions*. New York: Springer-Verlag.

Walther, G. R., Post, E., Convey, P., *et al.* (2002). Ecological responses to recent climate change. *Nature*, **416**, 389–95.

Whitmore, T. C. (1990). *An Introduction to Tropical Rain Forests*. Oxford: Clarendon Press.

Whittaker, R. H. (1970). *Communities and Ecosystems*. London: Macmillan.

Wrangham, R. W. (1980). An ecological model of female-bonded primate groups. *Behaviour*, **75**, 262–99.

Part II *Seasonal habitats*

2 *Tropical climates and phenology: a primate perspective*

CAREL P. VAN SCHAIK
Anthropologisches Institut, University of Zurich, Winterthurerstrasse 190, CH-8057, Zurich, Switzerland

KRISTINA R. PFANNES
Center for Tropical Conservation, Duke University, Box 90381, Durham NC 27708–0381, USA

Introduction

The order Primates is one of the few mammalian orders that are confined largely to the tropics (Richard 1985): only a few cercopithecines are found outside the tropics. Thus, the great majority of primate species live in tropical forests and woodlands, with a small minority inhabiting the open savanna.

Our aim here is to explore phenology, the production of young leaves ("flush"), flowers, and fruit, of woody plants in these prime primate habitats to seek useful generalizations for the primate ecologist. Despite the remarkable variability in phenological activity patterns of individual species (e.g. Newstrom *et al.* 1994; Sakai *et al.* 1999), there is enough between-species synchrony to distinguish clear patterns in tropical phenology that should be helpful to predict the responses of non-specialist primate consumers to fluctuations in food availability. This chapter should thus provide a general backdrop for the more detailed studies of the responses of primate consumers to changes in the availability of their various food items presented in subsequent chapters.

Specifically, we present the results of a meta-analysis of studies of phenology of plant communities of tropical forests and woodlands, many of them produced by primatologists in the course of their fieldwork. We explore the extent to which we can distinguish clear relationships between phenology and the timing of climatic events, the extent to which climatic seasonality is translated into phenological seasonality, and the temporal relationship between the fluctuations in availability of flush and ripe fruit. We also explore interannual variation in phenology. Before embarking on these analyses,

Seasonality in Primates: Studies of Living and Extinct Human and Non-Human Primates, ed. Diane K. Brockman and Carel P. van Schaik. Published by Cambridge University Press.

however, we present a brief introduction to tropical climates, illustrated later with climate data from the sites of the phenology studies in our sample.

Seasons in the Sun: tropical climates

Seasonality is found everywhere on Earth, a simple but profound consequence of the 23.5° tilt of the Earth's axis, producing the apparent changes in the solar angle through the year. The tropics are the region in which the Sun is directly overhead in the zenith at least once a year; hence, its edges are the two tropics, the virtual lines at latitudes 23.5° N (Cancer) and 23.5° S (Capricorn).

Tropical climates are defined variously (e.g. Ayoade 1983) as those in which no frost occurs, in which no month occurs with a mean monthly temperature at sea level of less than 18°C, or in which the daily range (maximum–minimum) exceeds the range in monthly means (hence the expression: "night is the winter of the tropics"). They are confined approximately to the region between the tropics, although at the highest altitudes inside the tropics, climates may deviate considerably. Seasonal variation in temperature and in day length is modest, although obviously increasing toward the edges. Thus, most relevant seasonal variation in the tropics is found in rainfall and irradiance.

Although distinguishing seasons is always to some extent arbitrary, traditionally, tropical seasons are based on rainfall (rather than wind, temperature, or insolation). The reason is not only that variation in rainfall is most striking to humans but also that many agricultural crops are highly sensitive to water stress. Whether a plant perceives water stress depends on the species, especially its rooting depth and features of its leaves and canopy, but also on water availability, which is a function of rainfall history as well as soil, slope, and topography. Plant geographers tend to use empirical rules to characterize months as wet or dry. Walter (1971) used the amount of rain in millimeters that is twice the mean temperature in degrees Celsius, which in tropical lowland climates amounts to *c*. 60 mm. Indeed, tropical forests with rainfall exceeding 60 mm in all months are generally considered to have no seasonal water stress (Schulz 1960; Whitmore 1984). However, to take into account between-year variation, a more conservative criterion of a mean of 100 mm as the value below which significant water stress can occur may also be useful.

Seasonality in rainfall and drought is easily appreciated, but that in solar radiation can also be appreciable; this variation is expected not only near the tropics but even near the Equator. Wright & van Schaik (1994) calculated from a set of 24 tropical weather stations with annual rainfall exceeding 1000 mm that the month with highest global radiation received 50% more than the month with the lowest – enough of a range to have a measurable impact on plant productivity. Even though photosynthesis in the top layer of the canopy is often fully saturated, the rapid extinction of photosynthetically active radiation inside the canopy means that stand-level photosynthetic productivity will generally show a roughly linear relationship with irradiance, so long as water is not limiting.

In principle, tropical climates show a predictable seasonal patterning with rainfall peaks following in the wake of sunshine peaks (e.g. MacArthur 1972; Barry & Chorley 1987). The details depend on latitude because of the seasonal march of the Sun. The belt with the zenithal Sun receives intense solar radiation, which will generate rising air, which will cool down and generate rainfall if it contains enough moisture. Moreover, the rising air under the zenithal Sun will be swirled away toward the subtropics. Wherever it comes down, which varies seasonally, it contributes to persistent higher-air-pressure areas, especially over oceans, from which the trade winds rush back (both from the north and the south) to a belt near the low-pressure area under the Sun (known as the intertropical convergence zone).

At the Equator, where the Sun passes overhead twice a year during the equinoxes, there are two rainier and two drier periods every year. Toward the latitudinal limits of the tropics, these two occurrences fall ever closer in time and at the tropics converge on the summer solstice. Hence, as one moves away from the Equator, the two rainfall peaks gradually fuse into one rainy season coinciding with, or following soon after, the period of highest irradiation. In the same direction, dry seasons generally get longer, and hence contrasts between wet and dry seasons increase (cf. Stevens 1989). In the results, we will use the climate data from the phenology sites to illustrate these broad patterns.

In practice, of course, deviations from the idealized pattern are common. On a large scale, trade winds will bring little rain if their area of origin is land or cool sea. Thus, sites in the center of large continents will deviate more from the ideal pattern than those on small oceanic islands. On a small scale, mountain ranges can create rain shadows. These regional and local effects produce variation in rainfall distributions and also deviant relations between rain and sunshine. Thus, sometimes dry seasons are very cloudy (as in eastern central Africa) and/or wet seasons are very sunny (as in much of Sumatra and peninsular Malaysia).

Climate is about averages, but between-year variation can be considerable. Hurricanes (typhoons, cyclones) originate at sea and, apart from causing heavy winds, dump large amounts of rain in short periods of time in the parts of the tropics subject to their influence: Central America, Madagascar, southern Asia, Indochina, and northeast Australia. Not surprisingly, therefore, the hurricane regions tend to experience high interannual variability in rainfall. The Equatorial belt is more or less immune to these effects.

Another source of between-year variation is associated with the El Niño-Southern Oscillation (ENSO) effect, caused by irregular warming of sea temperatures in the eastern Pacific, leading to widespread anomalies in the weather, including periods of long droughts in Southeast Asia and the Australian region and greater rainfall in the western Neotropics.

We do not know of any systematic effects on phenology of hurricanes or ENSO, except that ENSO is implicated in triggering the community-wide mast flowering and mast fruiting in parts of Southeast Asia (Ashton *et al.* 1988; Wich & van Schaik 2000), which has a striking impact on primate reproduction (van Schaik & van Noordwijk 1985; Knott 1998) (see also Chapter 12).

Methods

We compiled 106 studies of tropical forest/woodland phenology, with data collected at the community level, or at least of a large enough set of species that it should reflect the community. We tried to be exhaustive but included only sites within the two tropics (with three exceptions of high-rainfall lowland sites just outside them) and sites at altitudes of less than 1200 m (with one exception). For each site, we recorded latitude and vegetation type (scored as evergreen, semi-deciduous, or deciduous, usually directly from the original sources). Most studies contain mainly canopy trees, but some also include lianas and understorey trees. Canopy trees and lianas were given precedence if results were reported separately for different layers. Studies of single species or a guild of species (for instance, anemochorous trees or lianas) were not included. One site (Prasad & Hegde 1986) was removed from the data set because it produced seriously outlying coefficient of variation (CV) values (see below) that could not be attributed to any biological process.

Phenology data[1]

For each site, we extracted data on community-level phenology, i.e. monthly estimates of flushing, flowering, and fruiting activity of the woody plants in

[1] The full data set is available from the first-named author on request.

the community, although many studies did not report on all three phenophases. Three different kinds of community-level measures are reported in the literature: one qualitative measure (number of species) and two more quantitative measures (percentage of individuals, or some index of abundance). We extracted whichever measures were given for each study. Index methods are least comparable across studies, but the most quantitative methods give estimates of kilograms of production per unit area, ideal for comparisons of overall productivity.

Studies did not apply uniform definitions of flushing, flowering, and fruiting. Often, for instance, a species was recorded as "flushing" if its level of flushing exceeded a certain threshold value. Studies differed in using this criterion and also in how stringent they were in avoiding the scoring of the same leaves as flush twice in consecutive months. Evaluating fruiting created difficulties as well. Because ripeness of fruits is somewhat arbitrary or difficult to determine, many studies avoided defining ripe fruit and report on all visible fruit. Because most primates prefer ripe fruit, we preferred to use ripe fruit where available, but to maximize comparability, we also considered all fruit (though never unripe fruit only) in the analyses. Where average fruit ripening periods are long, the difference between all fruit and ripe fruit only can be substantial.

Variation between years exists, and different studies at the same site may yield different results (e.g. at Makokou, Gabon: Hladik 1973; Hladik 1978; Gautier-Hion *et al.* 1985). Some of the difference is due to the use of different measures, but some of it is real, especially near the Equator, where seasonal variation tends to be less pronounced and less predictable (e.g. Chapman *et al.* 1999). If this variation is strong, then it might obscure any pattern present in long-term data sets.

Seasonality can be described in multiple ways, depending on the aspect that is to be stressed. For phenophases, we calculated two commonly used measures: the coefficient of variation (CV = standard deviation/mean) (cf. Wolda 1978) and the mean vector. The latter is a measure of concentration based on circular statistics considering each month an adjacent 30° segment of a circle (corrected for grouping) (Batschelet 1981). The mean vector estimates the amount of concentration of activity and varies from 0 (even distribution across all months) to 1 (all activity in a single month). If the distribution of a phenophase is neatly unimodal, then these two measures should give very similar results, but if it is not, then discrepancies can arise. Thus, high CV with low concentration suggests dramatic variation that does not display clear seasonality; on the other hand, low CV with high concentration suggests moderate variation that shows strong clumping in the year.

We calculated the various indices of seasonality as the mean of annual indices, whenever available. When studies covered more than one year, but not an integer number, we ignored the first few months to obtain an integer number of years, because early months often show downward bias due to detection and recognition problems.

We assessed flush–fruit covariation in two ways. First, we calculated the correlation between the flush and fruit scores over the 12-month period. Second, we recorded the interval between the flushing and fruiting peaks (in months, positive being flushing peak preceding fruiting peak and maximum value set arbitrarily at eight months). Whereas this measure is probably less precise, it is available for more sites (ten additional sites in our sample) and its value is probably less affected by the measures of phenological activity than the correlation. (We also tried a more direct third measure, the number of months in which both flushing and fruiting reached the lowest n values, where n varied from 3 to 6, but because this measure involved many ties and is subject to more noise, and because it correlated well with the others, we employed only the first two measures in the analyses.)

Climate data

Like phenological seasonality, climatic seasonality can be measured in different ways. A measure that emphasizes the physiological response of plants is the severity of the dry period. For this, we took the number of rather dry (<100 mm rainfall) or clearly dry (<60 mm rainfall) months from mean rainfall data and also estimated the total duration of the dry season by recording the number of rather dry or very dry months that were consecutive. Not surprisingly, these various measures show extremely high correlations. In subsequent analyses, we used the more intuitive measure of dry season length based on fairly dry months.

Rainfall variability was estimated as for phenology through both CV and concentration (mean vector), as above for phenology. We calculated CVs from mean rainfall over a range of years (rather than taking the mean CV over a range of years), because many studies provide only mean rainfall. Note that the mean vector based on circular statistics assumes a unimodal distribution and is therefore not really suitable to capture rainfall distribution in two-peak climates. The various measures show strong correlations among them but contain somewhat different information, as illustrated in the results. We did not have enough data to calculate sunshine variability.

We also estimated the degree to which the rainfall distribution was unimodal or bimodal (i.e. produced one or two peaks in a year) through a measure we call periodicity, defined as the correlation between two consecutive six-month periods, with the first value starting at the (first) driest month of the year. Negative values indicate one peak, and positive values two peaks per year, although in very dry one-peak rainfall areas very occasionally positive periodicities were produced artificially.

To explore how the timing of peaks in phenology is related to climate, we also recorded the months with key climate events. The onset of the wet season was defined as the first month with over 100 mm of rainfall following the most severe dry season or, if no dry season occurred, as the month with the steepest relative increase in rainfall. Wherever possible, we also recorded the month with maximum irradiation from the original study or from climate atlases, such as Müller (1982) or the World Survey of Climatology (especially Schwerdtfeger 1976; Takahashi & Arakawa 1981). Because for most sites we lacked irradiation data, we used sunshine hours instead. Unavoidably, this decision introduced some noise.

Analyses

In the analyses, we had to adjust for the presence of different techniques of scoring phenological activity and of climatic and phenological seasonality. Different analyses used different samples, depending on the criteria for inclusion. However, one site would never be represented by more than one entry.

Seasonality depends on scoring technique. Across sites for which we had both measures, CVs for flowers and fruit were systematically higher for the quantitative measures than for number of species (Wilcoxon matched pairs: $z = 2.20$, $n = 12$, $P < 0.05$; $z = 2.75$, $n = 21$, $P < 0.01$, respectively) but not for flushing ($n = 9$), perhaps reflecting the wide variability of definitions. Hence, we performed separate analyses for number of species and the quantitative measures, although the samples used showed moderate overlap in the set of sites included and therefore are not independent (which is why we avoided analysis of covariance).

We also examined whether estimation of flush-fruit covariation was affected systematically by scoring technique. Peak months were virtually identical in this sample: flushing peaks estimated by number of species and percentage of individuals were on average 0.6 months apart ($n = 10$), flowering peaks were 0.18 months apart ($n = 14$), and fruiting peaks (using ripe fruit where possible) were 0.76 months apart ($n = 21$), with zero being the modal difference in all three cases. Flush–fruit correlations showed somewhat more noise, as expected, but were still fairly close together on average ($n = 10$, mean absolute difference in correlation was 0.167). However, because there was a clear trend for flush–fruit correlations based on ripe fruit to be more positive than those based on all fruit in sites where we had both (Wilcoxon: $z = 1.78$, $n = 12$, $P = 0.08$), we preferentially used values based on ripe fruit. Hence, for analyses of flush–fruit covariation involving peaks or correlations, and for those of peak timing in relation to climate, we used all sites in one sample regardless of technique but gave preference to values based on ripe fruit and averaged where two techniques had been used (unless only one was based on ripe fruit). Mixing techniques in this way may have introduced some noise.

Because our samples were generally quite large, we used standard product–moment correlation, linear regression, and analysis of covariance, but we always checked the results with non-parametric techniques as well and preferred those for analyses in which the homoscedasticity assumption may have been violated. Because of overlap in floristic composition among many sites, they are not statistically independent, so the results should be seen as indicating the robustness of patterns rather than as rigorous tests of hypotheses.

Results

Some climate patterns relevant to tropical phenology

We can illustrate the broad patterns mentioned in the introduction with the climate data from the phenology sites. Even near the Equator, the coldest nights are around January in the northern hemisphere and July in the southern hemisphere, despite the very muted seasonal variation in temperature. Figure 2.1a shows that the month with maximum sunshine (in hours, often coinciding with maximum insolation intensity) tends to precede the path of the overhead (zenithal) Sun, because cloudiness increases once the Sun is directly overhead. It also shows that sunshine tends to be highest before the first of the two annual peaks, e.g. in March–June, rather than June–September, in the northern hemisphere (and September–December, rather than December–March, in the southern hemisphere), probably for the same reason. The exceptions in Fig. 2.1a both are in the Neotropics. In the Guyanas, north of the Equator, sunshine tends to peak in October, probably because the thermal equator

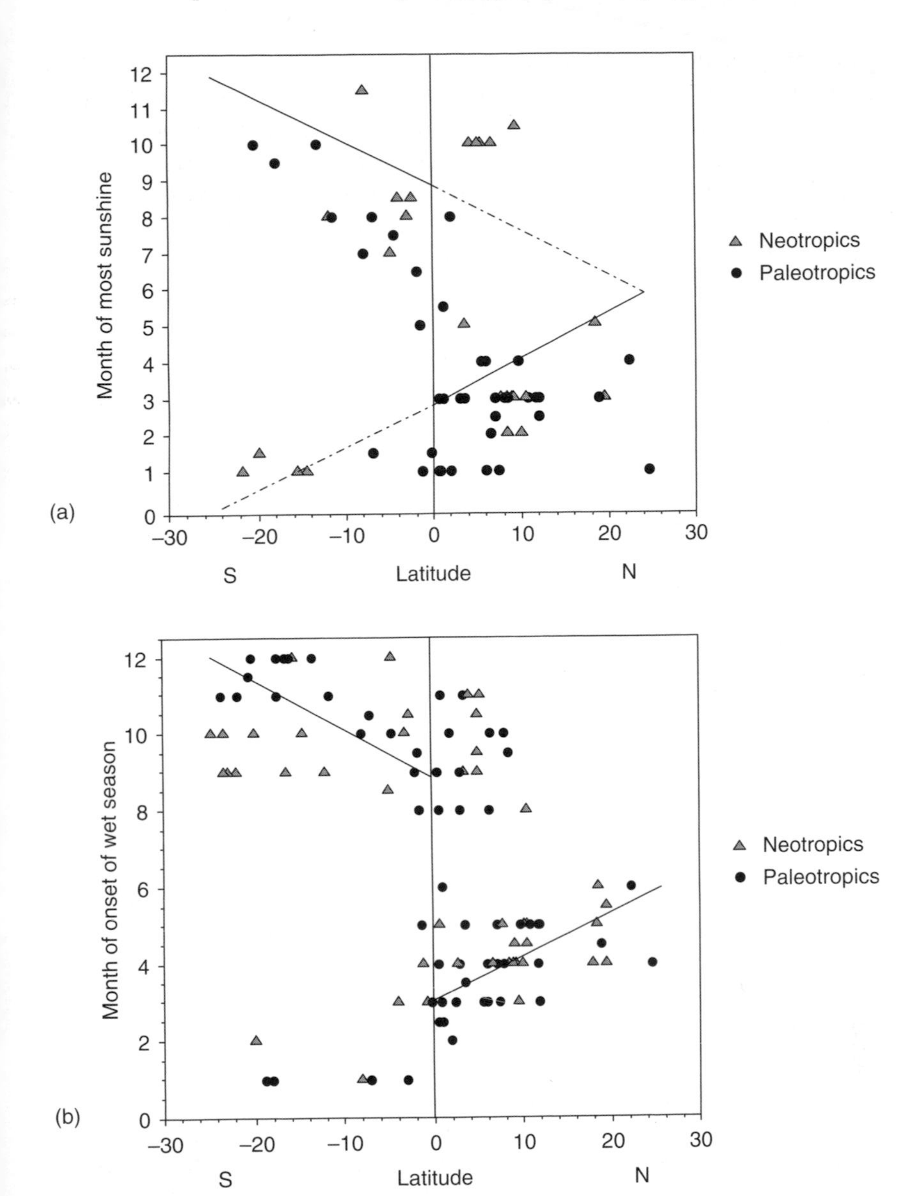

Figure 2.1 Key events in tropical climates, based on the sites included in the phenology data set, as a function of latitude (positive values are for northern hemisphere): (a) month of maximum sunshine (as estimated in hours of sunshine or irradiance); (b) onset of the wet season (first month with >100 mm of rainfall or month with steepest increase in rainfall). Lines indicate the approximate time when the Sun is overhead at a particular latitude.

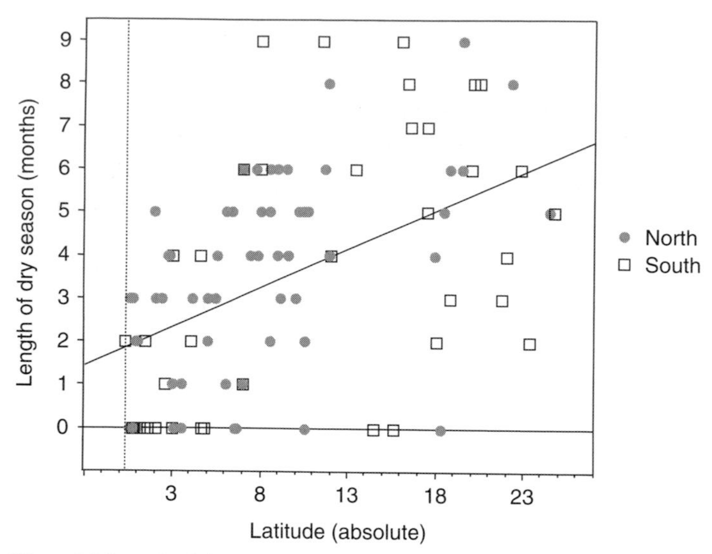

Figure 2.2 Length of the dry season (number of consecutive months with <100 mm of rainfall) in relation to latitude (northern and southern hemisphere combined).

is farther north in this region. Parts of the Atlantic forest region of southeastern Brazil peak around January, probably because the first pass of the Sun can be accompanied by high cloudiness (cf. Ratisbona 1976). We should expect that these regions also show unusual phenologies for their location.

As expected, the wet season shows a clear tendency to follow in the wake of the sunniest period (Fig. 2.1b), usually about one month later, although near the Equator, where two rainy seasons are expected, the strongest rainy season could come with either the first or the second overhead pass of the Sun. We expect climate generally to become more seasonal as distance from the Equator increases. This trend is very distinct for all measures of rainfall seasonality used here. Figure 2.2 illustrates this latitudinal effect, using the length of the longest continuous dry season, but it also serves to stress the great variability. Average rainfall, too, shows a gradual decline in the same direction. The trend for rainfall to become more one-peaked away from the Equator is clear, although even near the Equator many sites show one-peaked rainfall distributions due to topographical effects (continentality, mountain ranges, etc.). Another general trend is that overall cloudiness tends to decrease away from the Equator, producing higher annual irradiance near the tropics than near the Equator (Barry & Chorley 1987).

Does climate affect community-level phenology?

Before we go into the detailed questions, we must assess whether in this data set phenology responds to climate at all rather than, say, to biotic factors (cf. van Schaik *et al.* 1993). Figure 2.3a shows that the latitudinal pattern in the timing of flushing peaks is very similar to that of the onset of the wet season, whereas Fig. 2.3b shows the same thing even more clearly for flowering. Abundant work on the triggering of flushing and flowering at the species level has shown an important role of water stress and its relief in dry climates (e.g. Reich & Borchert 1984) and a role for irradiation or night temperatures in humid climates (e.g. Tutin & Fernandez 1993). Indeed, tropical trees subject to seasonal water stress tend to concentrate flushing and flowering near the onset of the wet season, whereas those that are not tropical generally do so during the sunniest time of the year (ter Steege & Persaud 1991; Wright & van Schaik 1994). There are adaptive benefits to this association. Young leaves are most efficient assimilators, because senescence and herbivory have not yet reduced their performance, and should therefore be produced at the start of a long period of high assimilation potential, i.e. around the first pass of the Sun, when days are sunny and nights are cool (see also Chapman *et al.* 1999). Producing flowers at about the same time is expected since the assimilates of the new crop of leaves can be shunted directly into the production of flowers and ripening fruits, and no storage is needed.

More detailed analyses demonstrate clearly that the similarity also holds at the level of the site (Fig. 2.4). Flushing and flowering peaks are about twice as likely to coincide or nearly coincide (one month before or after) with months of peak sunshine than expected if peaks were reached at random moments (49% and 53%, with $n = 43$ and $n = 57$, respectively; expected value is 25%). The same pattern holds with the month in which the rainy season starts (53% and 49%, with $n = 66$ and $n = 80$, respectively). As shown in Fig. 2.4, the relationship with sunshine peaks is on the whole slightly tighter.

Further analysis shows that the connection of flushing and flowering with the onset of the wet season (but not with the timing of peak insolation) is tighter in climates with a longer dry season, as expected by the impact of water stress: there are strong negative correlations between dry season length and the absolute value of the interval between peak phenological activity and onset of the rainy season (flushing: $n = 66$, $r = -0.457$, $P < 0.0001$; flowering: $n = 80$, $r = -0.326$, $P = 0.003$, respectively). However, where water is not limiting, phenological activity still peaks at approximately the same time of the year

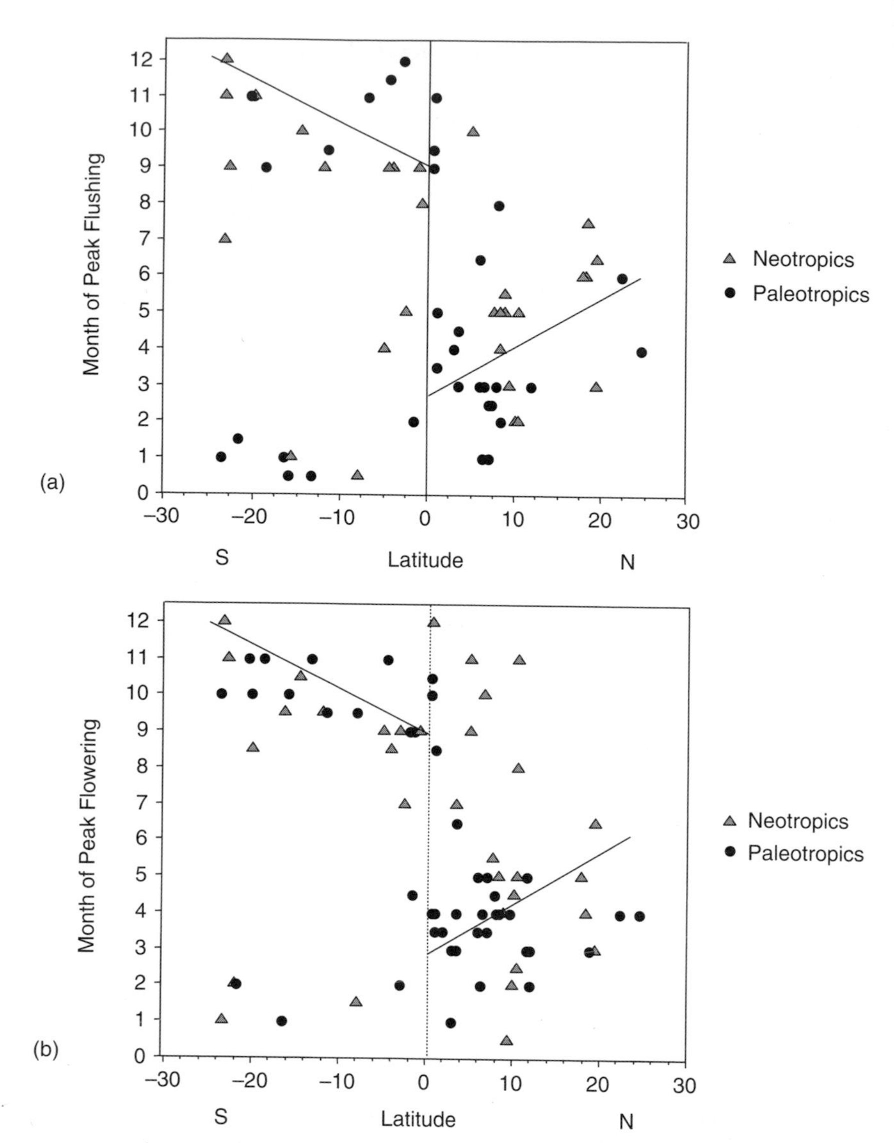

Figure 2.3 Key events in phenology of tropical woody vegetations as a function of latitude (conventions as in Fig. 2.1). (a) Peak month of flushing; (b) peak month of flowering.

because periods of high insolation begin at about the same time as the rainy season (usually a month or so earlier). Thus, even though tropical regions lack a winter, they show clear signs of spring, associated with the return of the high Sun.

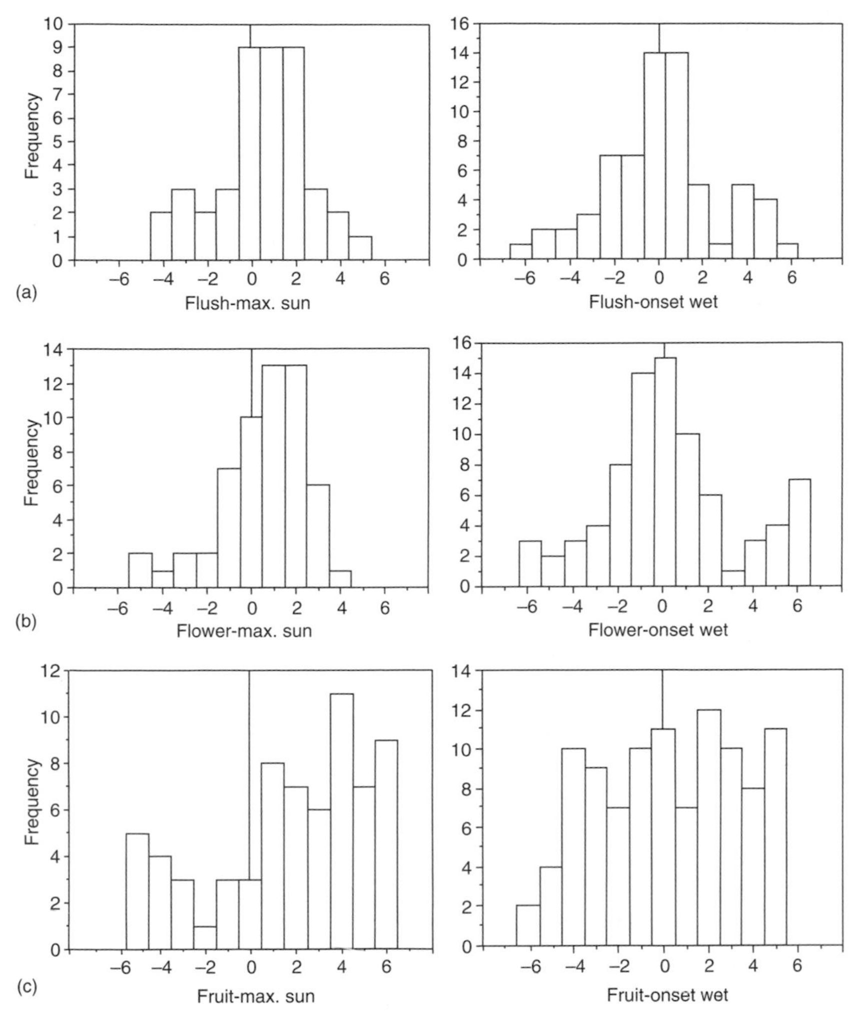

Figure 2.4 Timing of peaks in (a) flushing, (b) flowering, and (c) fruiting of tropical woody-plant communities relative to the timing of maximum sunshine or the onset of the wet season for all sites in the sample. Positive values indicate that the phenology peak followed the climate event. The maximum positive interval was arbitrarily set at 6 months and the most negative interval at – 5.5 months.

Fruiting peaks show a much sloppier latitudinal pattern and likewise, at the level of sites, do not show a relationship with the timing of the sunshine peak or the start of the wet period (Fig. 2.4c). This pattern bolsters the notion that fruiting time in most climates is much less constrained by climate than are flushing and flowering, mainly reflecting the mean fruiting period determined by factors such as fruit size, canopy position, and average weather conditions after flowering. Interestingly, fruiting peaks also do not show a tighter link with the onset of the rainy season in more seasonal climates, and nor do they significantly avoid the start of the dry season (except in the very driest sites). This is unexpected, since the literature has traditionally emphasized that in seasonal climates, fruiting peaks were in the wet season (e.g. Foster 1984).

Upon closer inspection, closeness of fruiting peaks to the start of the rains may still hold in the driest parts. With increasing length of the dry season, the interval from flower to fruit peaks decreases significantly ($r = -0.315$, $n = 76$, $P = 0.006$). Thus, mean flower–fruit intervals decrease from *c.* 3.5 months in ever-wet climates to about one month in the most extreme climates (nine-month dry season); this effect is probably strongest in the Neotropics. Although in the sample as a whole, peak fruiting months fall in the dry season about as often as expected on the basis of its length, this is true no longer for the very driest sites (dry season six to nine months) but only in Neotropics: peak fruiting falls in the dry season in 30% of ($n = 10$) sites there (expected 62.5%), whereas the Paleotropics continued to follow expectation with 57% of peaks in the dry season ($n = 14$ sites). The numbers are too small to reach statistical significance, however, and it is possible that variation in definitions of fruit ripeness confounds this result.

Climatic and phenological seasonality

We now examine the effect of climatic seasonality on the degree of phenological variability at the level of the plant community. Since we do not have good data on variability in insolation, we rely on that in rainfall, which in many cases is correlated due to the association between rain and cloudiness. Table 2.1 shows the correlations between three ways of estimating rainfall seasonality with two ways of estimating phenological seasonality, for two subsets of the total sample: one containing sites with number of species measures of phenology and one with the quantitative measures.

The effects of increased rainfall seasonality on phenological seasonality are generally strongest for flushing (Fig. 2.5, Table 2.1; but note that the samples are not entirely independent), weaker for flowering, and very weak, if present at all, for fruiting, even though half the relationships are statistically significant due to large sample sizes. Other aspects, such as rainfall distribution over the year (periodicity, i.e. regardless of total rainfall) and variation in day length (as indexed by latitude), show weaker correlations. This suggests that it is the wet–dry alternation that is

Table 2.1 *Climate seasonality and phenological seasonality (significant correlations in **bold**)*

(a) Using mean vector as index of seasonality

Rain – mean vector	Flush	Flower	Fruit[a]
Number of species	+0.417 $n=21$, $P<0.10$	**+0.491** $n=40$, $P<0.01$	+0.248 $n=47$, $P<0.10$
Index or % individuals	**+0.536** n = 44, P < 0.001	+0.195 n = 45, n.s.	−0.188 n = 64, n.s.

(b) Using coefficient of variation (CV) as index of seasonality in climate and phenology

Rain – CV	Flush	Flower	Fruit[a]
Number of species	**+0.658** $n=25$, $P<0.001$	**+0.499** $n=41$, $P<0.001$	**+0.281** $n=50$, $P<0.05$
Index or % individuals	**+0.335** $n=43$, $P<0.05$	+0.141 $n=43$, n.s.	−0.106 $n=65$, n.s.

(c) Using length of dry season for climate and CV for phenology

Length of dry season	Flush	Flower	Fruit[a]
Number of species	**+0.685** $n=25$, $P<0.0001$	**+0.415** $n=41$, $P<0.01$	+0.214 $n=50$, n.s.
Index or % individuals	**+0.425** $n=43$, $P<0.01$	+0.031 $n=43$, n.s.	−0.160 $n=65$, n.s.

[a] Ripe fruit where available.

responsible for these phenological responses. The fact that length of the dry season shows very similar relationships with phenological seasonality as do the measures that estimate variability suggests that it is seasonal droughts that are primarily responsible for the seasonality in phenology.

The increased variation in flushing in areas with a stronger dry season is due mainly to deciduous species (at least in the data set based on the number of species). Figure 2.5a shows that evergreen communities in areas with longer dry seasons do not show increased phenological variability, whereas those of semi-deciduous and deciduous forests clearly do. Analysis of covariance confirms this, showing significant effects for both dry season and forest type, as well as their interaction (respectively $F[1,19]=14.57$, $P<0.01$; $F[2,19]=3.70$, $P<0.05$; $F[2,19]=5.79$, $P=0.01$). This indicates that species in evergreen forests increase neither the amount of variability nor the synchrony in their flushing behavior as

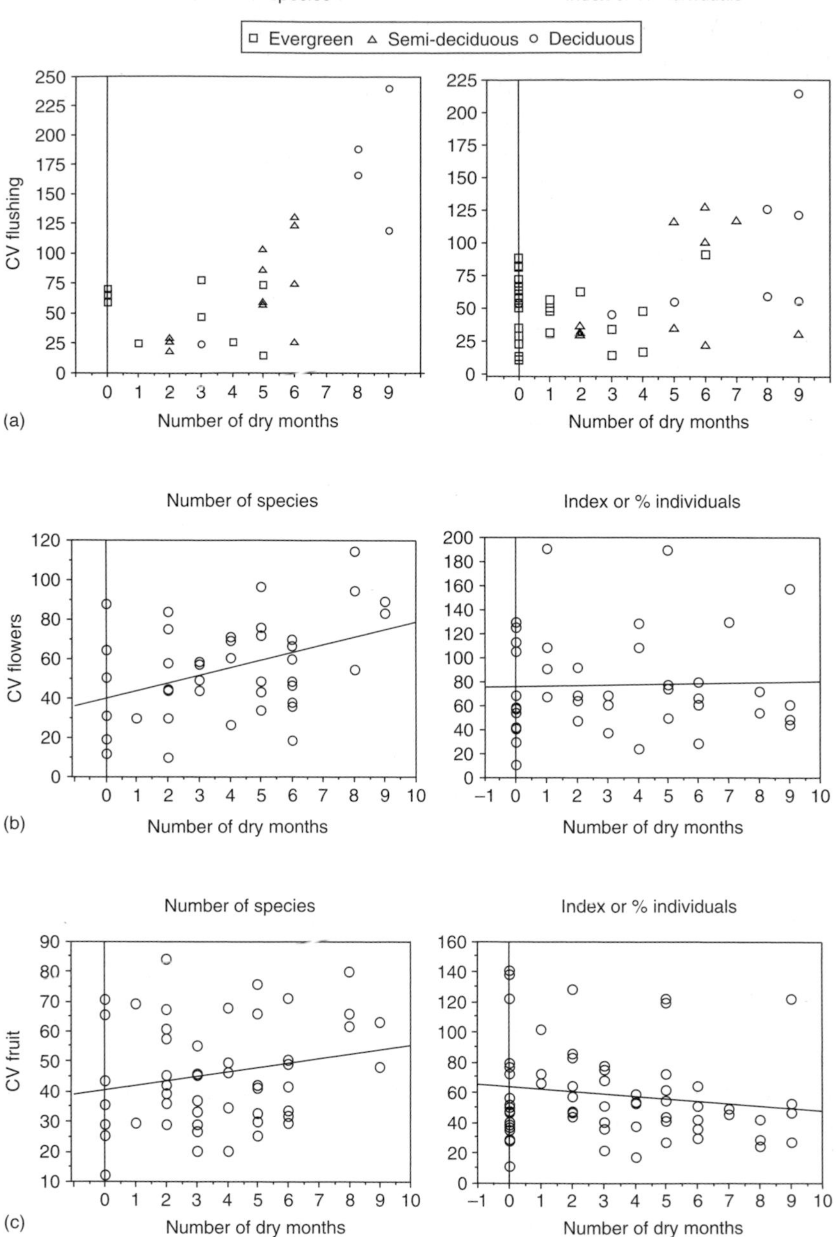

Figure 2.5 Seasonality in climate (estimated by the number of dry months) in relation to seasonality in phenology (measured by the coefficient of variation [CV] in phenological activity), for (a) flushing, (b) flowering, and (c) fruiting activity of tropical woody-plants communities, as measured either qualitatively (number of species) or quantitatively (percentage of individuals or index). Note that samples of different activity measures show overlap. For fruiting, measures involving ripe fruit were selected, wherever possible (see Table 2.1 for statistical significance). In (a), forest type is also indicated.

dry seasons become more pronounced. Probably they are evergreen exactly because they are not affected by water stress. Deciduous species, on the other hand, may become more synchronized as dry season intensity increases since most change their leaves once a year.

Table 2.1 and Fig. 2.5 show that the link of flowering seasonality to rainfall seasonality is weaker than that of flushing, regardless of how we measure seasonality or phenological activity. There is virtually no demonstrable effect of rainfall seasonality on fruiting seasonality. This is not due to the inclusion of all stages of ripeness: strict inclusion of ripe fruit only does not improve the relationship. This result is consistent with the weak links between rainy seasons and fruiting peaks that we found above. It is also consistent with the finding that, at individual sites, seasonality in flowering is greater on average than that in fruiting (for both number of species and quantitative measures, measuring seasonality as CV or mean vector; two of four cases at $P < 0.001$, one at $P < 0.01$, and one not significant). This pattern suggests that the average tree in seasonally dry areas has fewer constraints on when to ripen its fruits than commonly assumed. In other words, overall fruit abundance in a deciduous dry forest is not on average more seasonally variable than in a humid rainforest.

We hasten to add that this conclusion need not be true from a consumer perspective. Many studies have found that the abundance of wind-dispersed fruits peaks in the dry season, whereas that of animal-dispersed, and thus usually pulpy, fruits peaks in the wet season (e.g. Foster 1984; Charles-Dominique *et al.* 1981; Guevara de Lampa *et al.* 1992; Machado *et al.* 1997; Devineau 1999). Thus, it is likely that in seasonal forests, fruits with high water content (those with succulent pulp, for instance) are concentrated in the wet season.

In conclusion, in more seasonal tropical forests, consumers preferentially feeding on flush face a more seasonally variable resource base, as do, to a somewhat lesser extent, those favoring flowers. On the other hand, seasonality in the availability of fruit is at best weakly dependent on climatic seasonality, although the kinds of fruits preferred by primates may well be more seasonal than indicated by this broad survey. Overall, then, tropical primates living in seasonal climates experience seasonality in resource availability.

Relative timing of flushing and fruiting peaks

Climate seasonality could conceivably have another effect on phenology that is relevant to plant consumers. If we focus on frugivores with a preference for ripe fruit, an important question is how do they respond to seasons with low abundance of ripe fruit? In ecological time, there are only two clear options: change their range use to find more fruit elsewhere, perhaps of a special kind (e.g. palm nuts), or switch to alternative foods. We will discuss the phenological basis for these two options in turn.

Switch in food items

Switching to other foods depends strongly on how seasonality in fruit covaries over time with that in young leaves, the major fallback food of many frugivorous primates. Many insects likewise depend on young leaves and show strong dependence of their abundance on the abundance of flush. Thus, we expect that how fruiting and flushing covary over time will be important to frugivores. (This applies much less to folivores, because, unlike frugivores, they can, at least for shorter periods, fall back on mature leaves and on unripe fruit, whereas they usually cannot switch easily to insects.) Hence, temporal correlations between production of young leaves and ripe fruit are potentially of great importance to animals.

An important question is therefore what aspects of climate predict the flush–fruit peak interval and the flush–fruit correlation? Before examining the effect of seasonality, we look for geographic variation. For the inter-peak interval, there is significant geographic variation, with the Neotropical intervals (near one month) being some 2.5 months shorter than those for Asia (Fig. 2.6). This difference becomes even more pronounced when sites reporting "all" fruit (i.e. including unripe) are excluded. Flush–fruit correlations do not show any geographic pattern, however.

As to the effect of seasonality, Figs. 2.7a and 2.7b show that both measures of flush-fruit covariation are correlated significantly with the length of the dry season (interval: $r = -0.351$; $n = 65$, $P = 0.004$; correlation: $r = +0.344$; $n = 56$, $P = 0.009$, respectively). The same is true for other closely related seasonality measures. For intervals, the geographic effect is maintained when we use region as a factor in an analysis of covariance, whereas that of dry season length becomes even stronger. Thus, flushing and fruiting peaks nearly coincide in seasonally very dry climates, but especially so in the Neotropics. This suggests strongly that frugivorous primates in dry forests, especially those in the Neotropics, tend to face a lean season for both fruit and flush. They should have the weakest options for dietary switching to young leaves when fruit is scarce.

One more climate variable warrants discussion. Periodicity shows a significant relationship with the interpeak interval, and this relationship is stronger than the other climate variables for the reduced sample with the

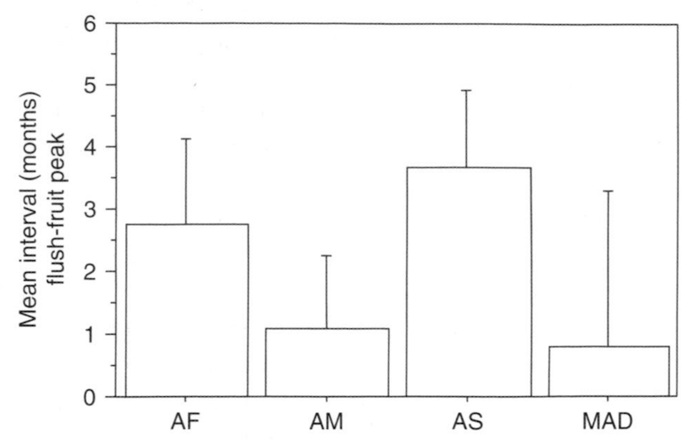

Figure 2.6 Mean intervals between peak flushing and fruiting activity in tropical woody-plant communities (positive values indicating that fruiting follows flushing, with the maximum arbitrarily set at eight months [higher values becoming negative]). Bars show 95% confidence limits. AF, Africa; AM, Americas; AS, Asia; MAD, Madagascar. Pair-wise differences: AM–AS: $P < 0.01$; MAD–AS: $P < 0.05$.

studies reporting all fruits (i.e. explicitly including unripe) removed. It is possible that this strong effect is due to the fact that one-peak rainfall distributions are especially common in the Neotropics and more likely to have a long dry season.

Spatial variation

Phenological cycles can be out of phase over short distances in two different kinds of situation: floodplain–upland contrasts and altitudinal gradients.

The seasonally flooded riverine swamps, called *igapo* and *varzea* in Amazonia, tend to have very different phenologies from the dryland habitats (in both, there is evidence that fruiting peaks are timed to coincide with high water, perhaps because these species rely on fish or water for their dispersal) (e.g. Kubitski & Ziburksi 1994; Defler & Defler 1996). High water levels need not be correlated with high local rainfall. Less extreme flooding, however, can also produce small-scale differences between riverine areas, subject to some flooding during the rainy season, and the uplands (e.g. Terborgh 1983).

Mountain ranges often show steep altitudinal gradients in rainfall and cloudiness, which may produce clear phase differences in phenology over short distances. Thus, at Ketambe in northern Sumatra, fruit peaks are about a month later in a valley north of a ridge compared with those in the area south of the ridge (van Schaik 1986), whereas they are later in the year

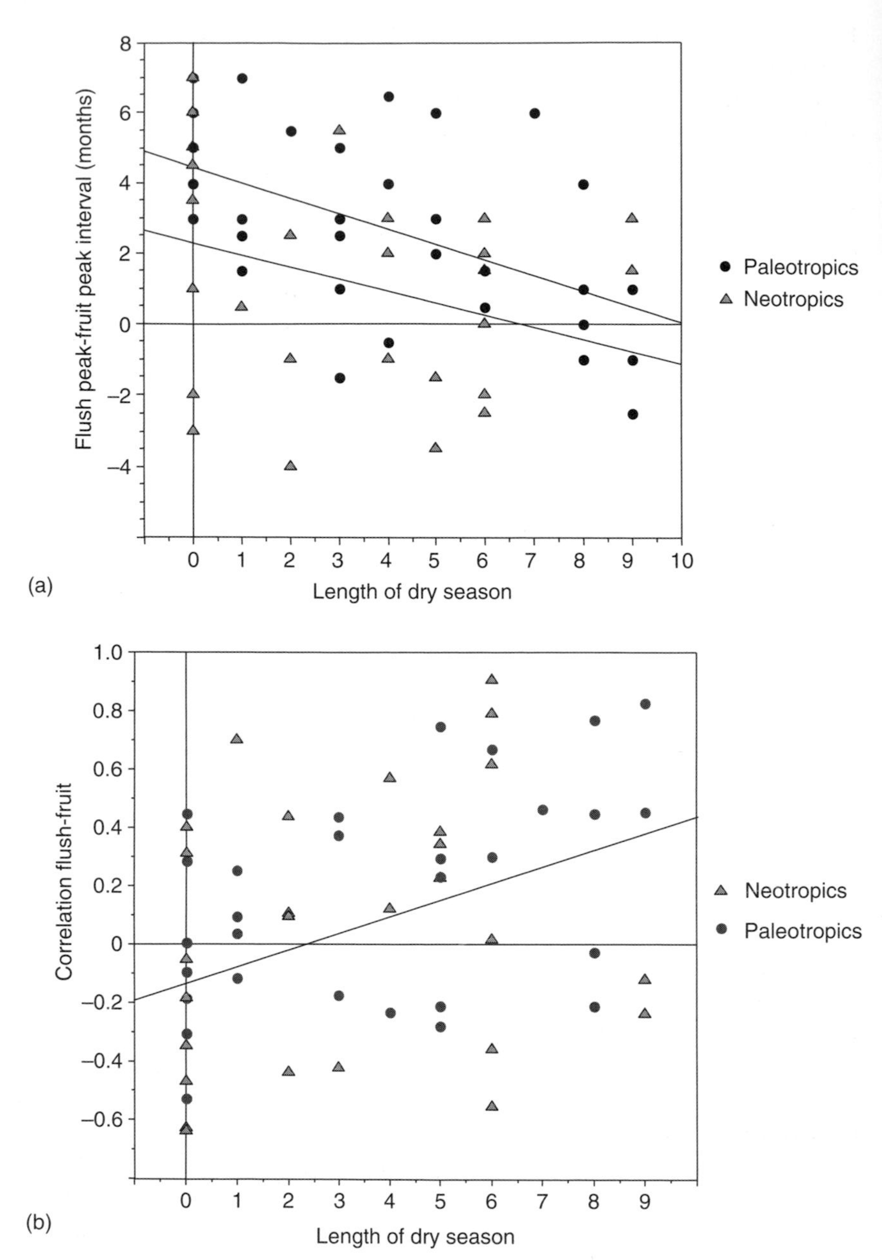

Figure 2.7 Effect of dry season length on flush–fruit covariation, as estimated by (a) the interval between peak flushing and fruiting activity and (b) correlation between flushing and fruiting activity (preferentially using ripe fruit). In (a), the slope for the Neotropical sites is significantly below that of the Paleotropics.

by several months as one moves from the river valley up the mountain slope (Buij *et al.* 2002). Likewise, in peninsular Malaysia, hill sites such as Ulu Gombak (Medway 1972) show later fruit peaks in the year than lowland sites (e.g. Raemaekers & Chivers 1980).

All these major temporal differences are at a scale such that even relatively sedentary arboreal mammals such as primates can exploit phenological differences by moving between habitats. However, they do not work in most landscapes.

Between-year variation

Many individual plant species show pronounced interannual variation in the size of flower and/or fruit crops. This has long been known as masting, although it is not always clear that the interannual variation in crop size is clearly bimodal (Herrera *et al.* 1998). In the tropics, high interannual variability is found especially in the Lecythidaceae in South America (Sabatier 1985; ter Steege & Persaud 1991), the Caesalpiniaceae in Africa (Newbery *et al.* 1998), and the Dipterocarpaceae in Asia (Ng 1977; Ashton *et al.* 1988).

In diverse communities, such masting of individual species need not affect generalist consumers much, because different plant species are on independent schedules. However, if many species show synchrony in their masting, then their behavior could affect plant-consuming animals. Used in this sense, supra-annual peaks in flowering and fruit production at the level of the plant community (or at least trees) is referred to as masting or general flowering and fruiting. It is well known in Southeast Asia, where it is pronounced not only in the Dipterocarpaceae but also among many other large-seeded species (Medway 1972; van Schaik 1986; Sakai *et al.* 1999). During mast peaks, fruit abundance is at least twice, and sometimes up to five times, that in normal years (Medway 1972; van Schaik 1986; Sakai *et al.* 1999). Intervals between masts are irregular, showing links with ENSO in some parts (Wich & van Schaik 2000). Hence, masting is of great importance to animals.

An important question is whether masting occurs outside Southeast Asia and, if not, why it is limited to this region. For the Sahul region (Australia, New Guinea), Crome (1975) and Pratt (1983) suggested a reliable alternation of good and bad years and suggested the general-masting label would be appropriate. However, such an alternation could be an artifact of small sample sizes and supra-annual fruiting cycles (trees that produce fruit in one year in a given sample cannot produce again in

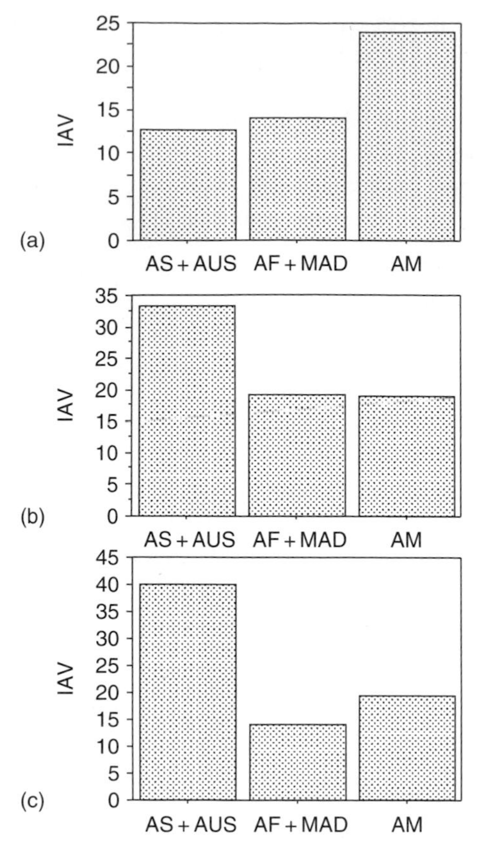

Figure 2.8 Interannual variability (IAV; coefficient of variation over annual mean values) for community-level (a) flushing, (b) flowering, and (c) fruiting activity across three major geographic regions (using quantitative measures, if available, because more sites have this information). Variation is not significant for flushing ($F[2,10] = 1.55$) or flowering ($F[2,10] = 0.90$) but is significant for fruiting ($F[2,14] = 6.31$, $P = 0.01$). AF, Africa; AM, America; AS, Asia; Aus, Australia; MAD, Madagascar.

the next year). More recently, Stocker *et al.* (1995) showed large interannual variation in fruit litterfall. However, their findings may have reflected the contribution of one species more than anything else. Hence, larger samples are needed to establish community-level masting more definitively, but indications are that northern Sahul also follows the masting pattern.

We used all the studies with at least three years of data in the sample and calculated the CV over the means of the individual years, a measure we call interannual variability (IAV). IAVs are available only for a limited number of sites. To examine geographic variation, we distinguished only three regions: Asia and Australia, Africa and Madagascar, and America. As shown in Fig. 2.8, IAV in flushing is not pronounced in any geographic region. IAV in flowering tends to be higher in Asia/Australia but is far from significant. IAV in fruiting, however, is significantly variable and in post-hoc pair-wise comparisons is higher in Asia/Australia than in each of the other two regions (confirmed by Kruskal–Wallis non-parametric one-way analysis of variance followed by pair-wise Mann–Whitney U tests).

The analysis of this very limited data set confirms that Asia/Australia is the only region with community-level mast fruiting. Even within this region, however, there may be variation. Reliable IAV data on South Asian sites are lacking, but the only reports on masting come from Sri Lanka (e.g. Koelmeyer 1958). Thus, there is as yet no evidence to refute the idea that community-level masting is indeed limited to the forests of the weakly seasonal humid Malesian rainforests (but perhaps it also occurs in the more seasonal Sahul rainforests). Clearly, however, more work with better IAV data is needed.

Community-level masting is thought to reflect a shared adaptation of many plant species against generalist seed predators. This hypothesis is supported by various observations. First, seed predation on plants fruiting out of phase with others is high (e.g. Charles-Dominique *et al.* 1981; Curran & Leighton 2000). Second, the species following this pattern are generally large-seeded (van Schaik 1986) and thus vulnerable to large seed predators. Third, the hypothesis explains the absence of masting in habitats without terrestrial rodents, such as peat swamps (van Schaik, unpublished data).

Discussion

We will first summarize the main findings and then examine their validity, discuss some unexpected patterns, and provide a very general discussion of the impacts on primate frugivores.

Phenology–climate relations

Most of the patterns emerging from the meta-analysis are not unexpected and are consistent with those of an earlier analysis of a smaller sample (van Schaik *et al.* 1993). Thus we found the following:

- In very general terms, tropical climates show seasonal patterning, with sunny periods associated with the overhead passage of the Sun followed

soon by rainy seasons (although exceptions are common). Moreover, away from the Equator, the two rainfall peaks come closer and fuse toward the tropics; hence, the dry season gets longer in the same direction.
- Peaks in flushing and flowering follow this climate pattern and thus are linked reliably to latitude, the sunniest period, and the onset of wet seasons, but fruiting peaks do not (except perhaps in the driest areas and then only in the Neotropics).
- As rainfall becomes more seasonal and dry seasons become longer, flushing and flowering become more seasonal, but fruiting may not.
- The interval between flushing and fruiting peaks is shorter in more seasonal climates, especially in the Neotropics, which should make it more difficult there to switch to young leaves during periods of fruit scarcity (flush–fruit correlation also responds to the length of the dry season but does not show the geographic effect).
- Spatial variation in the timing of peak fruiting is found across some floodplain–upland ecotones and along some altitudinal gradients.
- Dramatic between-year variation in fruit production is found only in Southeast Asia and perhaps the adjacent Sahul region, and any effects of this mast fruiting on the timing of primate reproductive patterns are expected only there.

Are these patterns real or are they artifacts? Phenological variables can only be measured with some error, potentially producing low replicability of studies (especially in the Equatorial region). Also, the varying criteria used for including individuals or species and different measures of phenological activity may have introduced serious bias. Although these problems are real, we controlled for them to some extent by repeating analyses with samples composed using different criteria. Moreover, most patterns are robust, emerging whether we analyze qualitative or quantitative measures and using different estimates of seasonality. It is also reassuring that the main correlations between climate on plant phenology found here are entirely consistent with basic plant physiology and numerous detailed studies of single species or communities (e.g. Reich & Borchert 1984; Wright 1996). The only serious ambiguity is whether Neotropical sites show fundamentally different responses to severe seasonality (see below).

Some may object that such general statements will not apply to any primate species in particular because no consumer uses all the plant species in the plant community evenly. Thus, describing what happens at the level of the plant community might not be helpful in understanding each and every primate species. In the end, the quality of the generalizations is judged by their power to predict primate ecology, but two arguments suggest that this concern may be overly cautious. First, quite a few primates prefer to eat fruit when it is abundant and tend to eat it in relation to availability. Likewise, most primates, when they eat leaves, prefer young leaves to mature leaves, which is why we also examined the covariation of the abundance of young leaves and fruit. Second, what is food may to some extent depend on what is available: diets may differ between years in the same group, or between nearby groups in the same year (see Hemingway & Overdorff 1999). Ironically, the qualitative measures of community phenology (number of species) may not give the best quantitative indication of community-level phenological activity or general resource abundance but may provide the best estimates of the average plant's response to seasonality as well as the options for consumers (if they do not simply eat unselectively according to availability). From this perspective, it is interesting that the variation in the qualitative measure often shows a better relationship with climatic variability than the more quantitative measures (e.g. Table 2.1). Thus, qualitative community-level phenology may actually depict the situation facing rather generalized consumers that eat the fruit of many species quite well.

Implications for primate consumers

Phenological changes may affect two different aspects of primate ecology: the behavioral and reproductive responses of individuals (i.e. behavioral ecology) and aspects of the community, such as composition and biomass (i.e. community ecology). Some of our generalizations are supported more strongly than others, but we can assess their validity for primates only by testing their predictive power. The alternative is to "take refuge in nature's complexity as a justification to oppose any search for patterns" (MacArthur 1972). To prepare for chapters that deal with these issues in detail, we develop some predictions.

Behavioral ecology

We found that increased climatic seasonality leads to increased seasonality in flushing and flowering. Although we could not show that fruiting is also more seasonal in more seasonal tropical climates, this is likely for fruits with high water content. Thus, with increased climatic seasonality, we should see longer periods of food scarcity and probably increased fluctuations in at least some of the crude dietary components (e.g. leaves, fruit, etc.). Indeed, in some of the most seasonal habitats, with long periods of overall scarcity, animals can be driven to seasonal hibernation (cf. Chapter 5).

We also found that as climate seasonality increases, the peaks in flushing and fruiting move closer together and the correlation between the amounts of flush and fruit increases. All other things being equal, this should make it more difficult during times of fruit scarcity for primate frugivores in such habitats to switch to flush or the insects that depend on them. On an annual basis, flush is also expected to be a lower component of the diet. These frugivores will have to specialize on some other food item, perhaps nectar or non-seasonal insects, which occur in low biomass, or fruit from highly localized palms or strangling figs, which require very large ranges (Terborgh 1983, 1986; Terborgh & van Schaik 1987). The average primate will either have to be smaller to cope with the lean season or cover a larger area, or both. Detailed tests of these predictions are presented by Hemingway and Bynum (Chapter 3) and Plavcan *et al.* (Chapter 14).

Community ecology

The effects of seasonality on flush-fruit covariation should also affect aspects of the primate assemblage at a site. We have already noted that when switching to young leaves or abundant insects is less common, animals are expected to be smaller. At the level of the assemblage, this should mean lower biomass (see Chapter 15). Obviously, this prediction is not corrected for other systematic trends with seasonality. Thus, in deciduous forests, the nutritional quality of foliage may be higher, because leaves tend to be more nutritious and less tough (Aerts & Chapin 2000), making it possible to switch to mature leaves in some cases and generally improving consumable biomass on an annual basis.

Geographic variation

Primate assemblages show remarkable differences between the four main regions (from a primate phylogenetic perspective): Asia, Africa, Madagascar, and America (Fleagle *et al.* 1999). In general, Asia has lower biomass and species richness than the rest of the Old World, but the Neotropics stand out by having smaller species that tend to live in smaller groups, with overall much less folivory, no terrestrial species, and a lower biomass than much of the Paleotropics (Fleagle *et al.* 1999). An earlier attempt to furnish an ecologically based explanation for these remarkable differences invoked phenological differences. Terborgh & van Schaik (1987) argued that the Neotropics have systematically stronger flush-fruit covariation than the Paleotropics, but in its extreme form this idea is clearly not supported now that more extensive phenological data show that only dry sites show this pattern of covariation (cf. Heymann 2001). Instead, some now believe that historical factors may be more

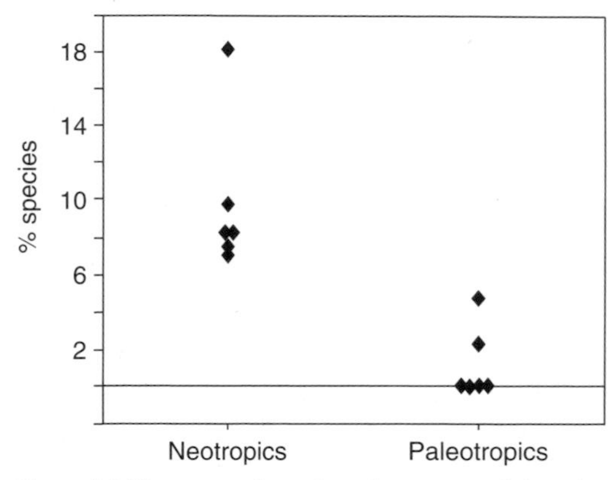

Figure 2.9 The proportion of species measured that show very long flower–fruit intervals (>7 months) for Paleotropics and Neotropics, using all study sites with extensive information.

important, in that competition with other higher taxa of folivores, already present when primates colonized the Neotropics, explains the low folivore biomasses among Neotropical primates (Ganzhorn 1999). We also know that Neotropical primates occupy a much narrower niche space than did their Old World counterparts (Fleagle & Reed 1996), but that could be either cause or effect of their divergent ecology.

Nonetheless, some notable differences between the Neotropics and Paleotropics remain. First, we noted that the effect of strong seasonality on the covariation of flush and fruit was strongest in the Neotropics, although there is much overlap between the continents. Second, there was a trend for seasonally dry Neotropical sites to have fruit peaks coincide with the onset of the rainy season, whereas no such trend was present in the Paleotropics. Third, a comparison of flower–fruit intervals for a range of sites suggests that in the Neotropics, the proportion of species showing very long intervals, approaching one year, is far higher (Fig. 2.9). We should be cautious accepting this result, due to the poor quality of some data sets, differences in how the intervals were measured, and obvious floristic non-independence of sites in the same biogeographic realm. Nonetheless, the pattern is consistent with the suggestion that Neotropical plants are more prone to avoiding shedding seeds during any time but the early wet season (cf. Foster 1984).

Much more systematic work is needed to corroborate this difference, to identify its causes, and to explore its consequences for primates and other consumers. The consequences should be easier to evaluate because primates and other consumers should respond directly to the phenological patterns they are confronted with (see Chapter 3). As to the causes, it is possible that they reflect the imprint of a history of more severe dry seasons in the Neotropics. The impact of such periods with drier climates is best estimated through periodicity. If a climate with a single dry period of a few months (negative periodicity; for definition, see Methods) undergoes a long-term period of drying, then the dry period becomes much longer. If, on the other hand, a climate with two dry periods per year (positive periodicity) is subjected to a period of reduced rainfall, then the effect on the maximum length of the dry period is much less dramatic. In the climates with a single annual rainy season, plants that have phenologies with negative flush–fruit covariation will not survive in the long run because they produce fruits at the wrong time of the year during these dryer periods: the seeds or the seedlings will die. So, even if the present climate seems right for the negative covariation type of phenology, then the species with this kind will have been weeded out over time. The equatorial belt (within $\pm 10^\circ$ of the Equator) has far more sites with one-peak rainfall in the Neotropics than in the Paleotropics (in our sample, 83% of the 24 sites within $\pm 10^\circ$ of the Equator, as compared with 54% of the 39 equivalent sites in the Paleotropics). Thus, it is possible that the current flora of the Equatorial belt in the Neotropics is recruited from a more seasonal stock than those in most of the Paleotropics.

Masting

The data we reported here do agree with the general idea that community-level masting is limited to relatively non-seasonal regions in South and Southeast Asia. It is our impression that high IAV elsewhere is usually the result of normal years alternating with the occasional bust year (Wright *et al.* 1999) rather than the Southeast Asian pattern of generally poor years alternating with the occasional boom year. The pattern of occasional boom years is more likely to produce changes in reproductive timing and seasonality of consumers than that of the occasional, unpredictable bust year.

The implications of mast fruiting should be profound (cf. van Schaik & van Noordwijk 1985). First, regular years should show only moderate seasonal peaks in fruit abundance, and, especially for species with long infant dependency, this means that the reproductive seasonality commonly found elsewhere in the tropics is not expected. Instead, one expects species to wait for signs of increased resource availability before committing to a new pregnancy (see Chapters 10 and 11). Second, during masting periods, behavioral ecology should be dramatically different, with consumers trying to maximize net intake, building up reserves for the impending lean period (cf. Chapter 12). We may also expect sudden migratory movements. For instance, the coastal swamps of northwestern Sumatra do not show any tendencies toward mast fruiting whereas the adjacent hills clearly do. This creates massive movements of swamp inhabitants into the hills during a mast year, though not vice versa during regular years (masting intervals are too long).

As to the effects of masting on community ecology, primate biomass and richness in these habitats is low compared with in other habitats, but it is not easy to prove that they are lower there than expected on the basis of *average* production. Where fruit production of ripe fruit overall is low, we expect consumers, even those that prefer ripe fruit, to evolve adaptations to deal with the more ubiquitous alternative resources of lower quality such as mature leaves or unripe fruit.

Acknowledgments

Many researchers kindly made unpublished data sets or dissertations available for this analysis. CvS's fieldwork was made possible by long-term support from the Wildlife Conservation Society. KP was supported by a Deutscher Akademischer Austauschdienst (DAAD; German Academic Exchange Service) fellowship during this study. We thank John Terborgh and Maria van Noordwijk for constructive comments.

References

Aerts, R. & Chapin, F. S., III (2000). The mineral nutrition of wild plants revisited: a re-evaluation of processes and patterns. *Advances in Ecological Research*, **30**, 1–61.

Ashton, P. S., Givnish, T. J., & Appanah, S. (1988). Staggered flowering in the Dipterocarpaceae: new insights into floral induction and the evolution of mast fruiting in the aseasonal tropics. *American Naturalist*, **132**, 44–66.

Ayoade, J. O. (1983). *Introduction to Climatology for the Tropics*. Chichester: John Wiley & Sons.

Barry, R. G. & Chorley, R. J. (1987). *Atmosphere, Weather and Climate*. London: Methuen.

Batschelet, E. (1981). *Circular Statistics in Biology*. London: Academic Press.

Buij, R., Wich, S.A., Lubis, S. A., A., & Sterck, F. H. M. (2002). Seasonal movements in the Sumatran orangutan (*Pongo Pyymaeus abelii*) and consequences for conservation. *Biological Conservation*,**107**, 83–7.

Chapman, C. A., Wrangham, R. W., Chapman, L. J., Kennard, D. K., & Zanne, A. E. (1999). Fruit and flower phenology at two sites in Kibale National Park, Uganda. *Journal of Tropical Ecology*, **15**, 189–211.

Charles-Dominique, P., Atramentowicz, M., Charles-Dominique, M., *et al.* (1981). Les mammifères frugivores arboricoles nocturnes d'une forêt guyanaise: inter-relations plantes-animaux. *Terre et Vie (Revue Ecologique)*, **35**, 342–435.

Crome, F. H. J. (1975). The ecology of fruit pigeons in tropical Northern Queensland. *Australian Wildlife Research*, **2**, 155–85.

Curran, L. M. & Leighton, M. (2000). Vertebrate responses to spatiotemporal variation in seed production of mast-fruiting Dipterocarpaceae. *Ecological Monographs*, **70**, 101–28.

Defler, T. R. & Defler, S. B. (1996). Diet of a group of *Lagothrix lagothricha lagothricha* in Southeastern Colombia. *International Journal of Primatology*, **17**, 161–90.

Devineau, J.-L. (1999). Seasonal rhythms and phenological plasticity of savanna woody species in a fallow farming system (south-west Burkina Faso). *Journal of Tropical Ecology*, **15**, 497–513.

Fleagle, J. G. & Reed, K. E. (1996). Comparing primate communities: a multi-variate approach. *Journal of Human Evolution*, **30**, 489–510.

Fleagle, J. G., Janson, C. H., & Reed, K. E. (1999). *Primate Communities*. Cambridge: Cambridge University Press.

Foster, R. B. (1984). Plant seasonality in the forests of Panama. In *The Botany and Natural History of Panama*, ed. R. B. Foster. Washington, DC: Smithsonian Institution Press, pp. 255–62.

Ganzhorn, J. U. (1999). Body mass, competition and the structure of primate communities. In *Primate Communities*, ed. J. G. Fleagle, C. H. Janson, & K. E. Reed. Cambridge: Cambridge University Press, pp. 141–57.

Gautier-Hion, A., Duplantier, J.-M., Emmons, L., *et al.* (1985). Coadaptation entre rhythmes de fructification et frugivorie en forêt tropicale humide du Gabon: mythe ou réalité? *Terre et Vie (Revue Ecologique)*, **40**, 405–34.

Guevara de Lampe, M., Bergeron, Y., McNeil, R., & Leduc, A. (1992). Seasonal flowering and fruiting patterns in tropical semi-arid vegetation of Northeastern Venezuela. *Biotropica*, **24**, 64–76.

Hemingway, C. A. & Overdorff, D. J. (1999). Sampling effects on food availability estimates: phenological method, sample size, and species composition. *Biotropica*, **31**, 354–64.

Herrera, C. M., Jordano, P., Guitián, J., & Traveset, A. (1998). Annual variability in seed production by woody plants and the masting concept: reassessment of principles and relationship to pollination and seed dispersal. *American Naturalist*, **154**, 576–94.

Heymann, E. W. (2001). Can phenology explain the scarcity of folivory in New World primates? *American Journal of Primatology*, **55**, 171–5.

Hladik, C. M. (1973). Alimentation et activité d'un groupe de chimpanzés réintroduits en forêt gabonaise. *Terre et Vie, Revue d'Ecologie Appliquée*, **27**, 343–413.

Hladik, A. (1978). Phenology of leaf production in rain forest of Gabon: distribution and composition of food for folivores. In *The Ecology of Arboreal Folivores*, ed. G. G. Montgomery. Washington, D.C.: Smithsonian Institution Press, pp. 51–71.

Knott, C. D. (1998). Changes in orangutan caloric intake, energy balance, and ketones in response to fluctuating fruit availability. *International Journal of Primatology*, **19**, 1061–79.

Koelmeyer, K. O. (1958). The periodicity of leaf change and flowering in the principal forest communities of Ceylon. *Ceylon Forester*, **4**, 157–89.

Kubitzki, K. & Ziburski, A. (1994). Seed dispersal in flood plain forests of Amazonia. *Biotropica*, **26**, 30–43.

MacArthur, R. H. (1972). *Geographical Ecology: Patterns in the Distribution of Species*. New York: Harper & Row.

Machado, I. C. S., Barros, L. M., & Sampaio, E. V. S. B. (1997). Phenology of caatinga species at Serra Talhada, PE, Northeastern Brazil. *Biotropica*, **29**, 57–68.

Medway, L. (1972). Phenology of a tropical rain forest in Malaya. *Biological Journal of the Linnean Society*, **4**, 117–46.

Müller, M. J. (1982). *Selected Climate Data for a Global Set of Standard Stations for Vegetation Science*. The Hague: Dr. W. Junk Publishers.

Newbery, D. M., Songwe, N. C., & Chuyong, G. B. (1998). Phenology and dynamics of an African rainforest at Korup, Cameroon. In *Dynamics of Tropical Communities*, ed. D. M. Newbery, H. H. T. Prins, & N. D. Brown. Oxford: Blackwell Science Publications, pp. 267–308.

Newstrom, L. E., Frankie, G. W., & Baker, H. G. (1994). A new classification for plant phenology based on flowering patterns in lowland tropical rain forest trees at La Selva, Costa Rica. *Biotropica*, **26**, 141–59.

Ng, F. S. P. (1977). Gregarious flowering of dipterocarps in Kepong, 1976. *The Malay Forester*, **40**, 126–37.

Prasad, S. N. & Hegde, M. (1986). Phenology and seasonality in the tropical deciduous forest of Bandipur, South India. *Proceedings of the Indian Academy of Sciences (Plant Sciences)*, **96**, 121–33.

Pratt, T. K. (1983). Seed dispersal in a montane forest in Papua New Guinea. Ph. D. thesis Rutgers University, New Brunswick, NJ.

Raemaekers, J. J. & Chivers, D. J. (1980). Socio-ecology of Malayan forest primates. In *Malayan Forest Primates: 10 Years' Study in Tropical Rain Forest*, ed. D. J. Chivers. New York: Plenum Press, pp. 279–316.

Ratisbona, L. R. (1976). The climate of Brazil. In *Climates of Central and South America (World Survey of Climatology, Vol. 12)*, ed. W. Schwerdtfeger. Amsterdam: Elsevier Scientific, pp. 219–93.

Reich, P. B. & Borchert, R. (1984). Water stress and tree phenology in a tropical dry forest in the lowlands of Costa Rica. *Journal of Ecology*, **72**, 61–74.

Richard, A. F. (1985). *Primates in Nature*. New York: W. H. Freeman and Company.

Sabatier, D. (1985). Saisonalité et déterminisme du pic de fructification en forêt guyanaise. *Terre et Vie (Revue Ecologique)*, **40**, 289–320.

Sakai, S., Momose, K., Yumoto, T., *et al.* (1999). Plant reproductive phenology over four years including an episode of general flowering in a lowland dipterocarp forest, Sarawak, Malaysia. *American Journal of Botany*, **86**, 1414–36.

Schulz, J. P. (1960). *Ecological Studies in Northern Surinam*. Amsterdam: Noord-Hollandse Uitgeversmaatschappij.

Schwerdtfeger, W. (1976). *Climates of Central and South America (World Survey of Climatology, Vol. 12)*. Amsterdam: Elsevier Scientific.

Stevens, G. C. (1989). The latitudinal gradient in geographical range: how so many species coexist in the tropics. *American Naturalist*, **133**, 240–56.

Stocker, G. C., Thompson, W. A., Irvine, A. K., Fitzsimon, J. D., & Thomas, P. R. (1995). Annual patterns of litterfall in a lowland and tableland rainforest in tropical Australia. *Biotropica*, **27**, 412–20.

Takahashi, K. & Arakawa, H. (1981). *Climates of Southern and Western Asia (World Survey of Climatology, Vol. 9)*. Amsterdam: Elsevier Scientific.

Terborgh, J. (1983). *Five New World Primates: A Study in Comparative Ecology*. Princeton: Princeton University Press.

Terborgh, J. (1986). Keystone plant resources in the tropical forest. In *Conservation Biology II*, ed. M. E. Soulé. Sunderland, MA: Sinauer Associates, pp. 330–440.

Terborgh, J. & van Schaik, C. P. (1987). Convergence vs. non-convergence in primate communities. In *Organization of Communities: Past and Present*, ed. J. H. R. Gee & P. S. Giller. Oxford: Blackwell Scientific Publications, pp. 205–26.

Ter Steege, H. & Persaud, C. A. (1991). The phenology of Guyanese timber species: a compilation of a century of observations. *Vegetatio*, **95**, 177–98.

Tutin, C. E. G. & Fernandez, M. (1993). Relationships between minimum temperature and fruit production in some tropical forest trees in Gabon. *Journal of Tropical Ecology*, **9**, 241–8.

Van Schaik, C. P. (1986). Phenological changes in a Sumatran rain forest. *Journal of Tropical Ecology*, **2**, 327–47.

Van Schaik, C. P. & van Noordwijk, M. A. (1985). Interannual variability in fruit abundance and reproductive seasonality in Sumatran long-tailed macaques (*Macaca fascicularis*). *Journal of Zoology*, **206**, 533–49.

Van Schaik, C. P., Terborgh, J. W., & Wright, S. J. (1993). The phenology of tropical forests: adaptive significance and consequences for primary consumers. *Annual Review of Ecology and Systematics*, **24**, 353–77.

Walter, H.(1971). *Ecology of Tropical and Subtropical Vegetation*. Edinburgh: Oliver & Boyd.

Whitmore, T. V. (1984). *Tropical Rain Forests of the Far East*. Oxford: Clarendon Press.

Wich, S. A. & van Schaik, C. P. (2000). The impact of El Niño on mast fruiting in Sumatra and elsewhere in Malesia. *Journal of Tropical Ecology*, **16**, 563–77.

Wolda, H. (1978). Fluctuations in abundance of tropical insects. *American Naturalist*, **112**, 1017–45.

Wright, S. J. (1996). Phenological responses to seasonality in tropical forest plants. In *Tropical Forest Plant Ecophsyiology*, ed. S. S. Mulkey, R. L. Chazdon, & A. P. Smith. New York: Chapman & Hall, pp. 440–60.

Wright, S. J. & van Schaik, C. P. (1994). Light and the phenology of tropical trees. *American Naturalist*, **143**, 192–9.

Wright, S. J., Carrasco, C., Calderón, O., & Paton, S. (1999). The El Niño Southern oscillation, variable fruit production, and famine in a tropical forest. *Ecology*, **80**, 1632–47.

Part III *Seasonality and behavioral ecology*

3 *The influence of seasonality on primate diet and ranging*

CLAIRE A. HEMINGWAY
Botanical Society of America, PO Box 299, St Louis MO 63166–0299, USA

NORA BYNUM
Center for Biodiversity and Conservation, American Museum of Natural History, Central Park West at 79th Street, New York NY 10024, USA

Introduction

To sustain animal populations, an adequate supply of consumable resources is essential. Effects of insufficient resources are well documented in primate populations in the form of reduced rates of fecundity, growth, and survival (Altmann *et al.* 1977; Hamilton 1985; Gould *et al.* 1999). Weight loss (Goldizen *et al.* 1988) and mortality peak during periods of low food availability on an annual (Milton 1980) or interannual basis (Foster 1982; Wright *et al.* 1999). Food availability relative to consumer requirements has been estimated as seasonally deficient in some (Smythe *et al.* 1982; Terborgh 1986; Janson & Emmons 1990) but not all (Coehlo *et al.* 1976) cases. Identifying food-limiting periods generally involves comparisons between estimates of food supply and animal requirements, which in turn require estimates of population density, biomass, energy intake, and metabolic rate. Field techniques measuring doubly labeled water (Nagy & Milton 1979; Williams *et al.* 1997) and products of fat metabolism in urine samples (Knott 1998) (see also Chapter 12) are highly informative in determining whether consumers are operating at a negative energy balance. The great majority of studies, however, rely on phenological monitoring to suggest periods of food scarcity for vertebrate consumers.

Phenological monitoring has revealed spatial and temporal variation in the availability of ripe fruits and young leaves in practically all forests studied (see reviews by van Schaik *et al.* [1993], [Fenner 1998], [Jordano [2000], and van Schaik & Pfannes [Chapter 2 of this book]). Availability of young leaves and flowers appears to be determined in large measure by climatic variables and often concentrated in the months following

Seasonality in Primates: Studies of Living and Extinct Human and Non-Human Primates, ed. Diane K. Brockman and Carel P. van Schaik. Published by Cambridge University Press.

maximum insolation (van Schaik *et al.* 1993) (see also Chapter 2). The timing of peaks in fruit availability across sites is not as strongly correlated with climatic variables. However, with increasing seasonality, the interval between flushing and fruiting peaks decreases (see Chapter 2), and this may have implications for consumers. Fruit availability can vary markedly between adjacent sites in similar forests (Chapman *et al.* 1999) and between years. Insect and prey item availability also vary spatially and temporally; peak insect abundance coincides with leaf flush (Robinson 1986; Boinski & Fowler 1989; Janson & Emmons 1990). In sum, a variable food supply is the norm for consumers in tropical as well as temperate regions, and primates face often marked spatiotemporal changes in the availability of their major food types.

Our objective is to synthesize field data on primate responses to resource scarcity with a focus on behavioral flexibility. After reviewing the theoretical framework, we present a qualitative survey of the nature of responses to resource scarcity among primates. We comment only briefly on physiological responses and do not broach the extensive topic of reproductive responses, which are covered elsewhere in this book. Next, we undertake a quantitative analysis of dietary flexibility to provide context for this common response. Finally, we use both qualitative and quantitative data to address the core question: what is the relationship between seasonality in the environment and changes in ranging and feeding behavior?

Responses to food scarcity

In a seminal review of phenological patterns and their implications to vertebrate consumers, van Schaik *et al.* (1993) identified six responses to food scarcity: occasional famine and mass mortality, dietary switching, seasonal breeding, seasonal movements, altitudinal migration, and hibernation. The range of behavior and physiological responses exhibited by primates is impressive, encompassing all those described for vertebrates. Group- and community-level effects of food scarcity are also of great interest to students of the socially complex primates (Fig. 3.1).

That variation in food availability can elicit numerous, complex responses is well documented in Old and New World primates. Wrangham's (1980) dichotomy of growth and subsistence diets forms the theoretical basis underlying differences in cohesion and competition across primates. Reduced food availability affects interactions among social, diet, and ranging behavior, such as group fissioning (Dittus 1988), reduced feeding party size (Doran 1997), and increased distance

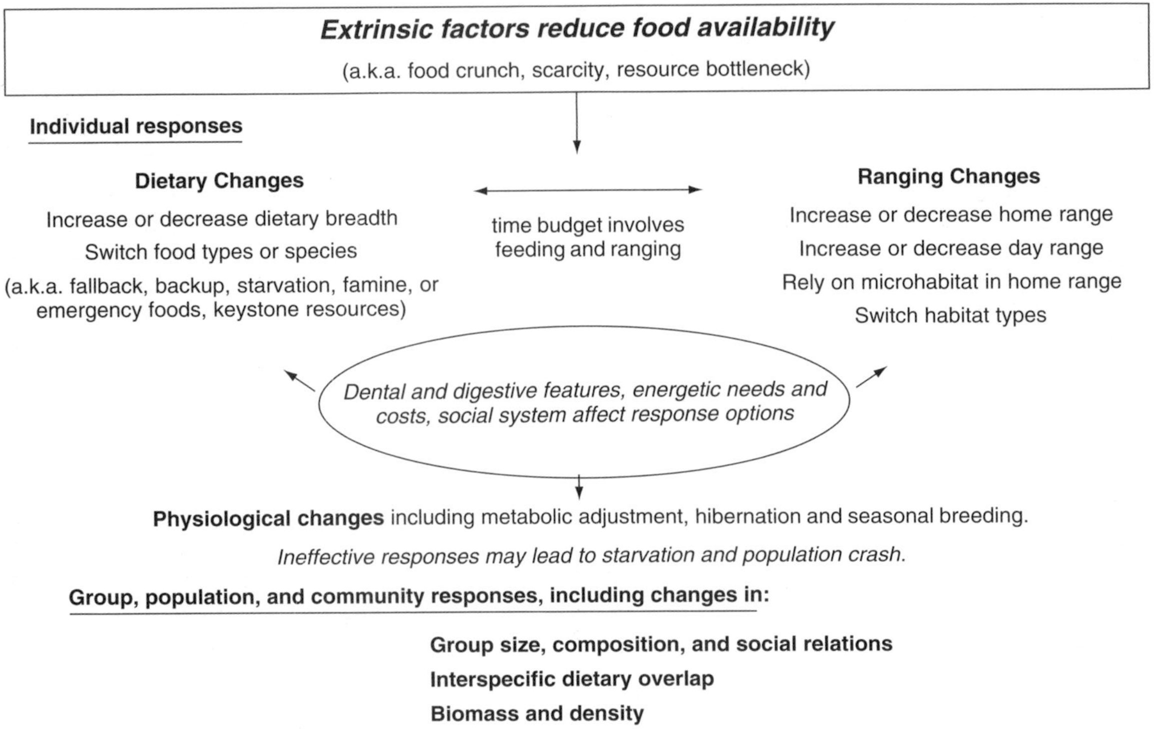

Extrinsic factors reduce food availability

(a.k.a. food crunch, scarcity, resource bottleneck)

Individual responses

Dietary Changes

Increase or decrease dietary breadth

Switch food types or species

(a.k.a. fallback, backup, starvation, famine, or emergency foods, keystone resources)

time budget involves feeding and ranging

Ranging Changes

Increase or decrease home range

Increase or decrease day range

Rely on microhabitat in home range

Switch habitat types

Dental and digestive features, energetic needs and costs, social system affect response options

Physiological changes including metabolic adjustment, hibernation and seasonal breeding.

Ineffective responses may lead to starvation and population crash.

Group, population, and community responses, including changes in:

Group size, composition, and social relations

Interspecific dietary overlap

Biomass and density

Figure 3.1 Schematic diagram of responses to reduced availability in the resource base.

between individuals, feeding conflicts within groups, or territorial disputes (Boinski 1987; Gursky 2000). At the community level, dietary overlap between sympatric species often decreases during food scarcity (Gautier-Hion 1980; Terborgh 1983; Yeager 1989; Overdorff 1993) and the quantity and/or quality of resources, either keystone or staple resources, may limit primate density and biomass and explain population and community differences (Waterman *et al.* 1988; Ganzhorn 1992; Kay *et al.* 1997) (see also contributions in Fleagle *et al.* 1999). In this chapter, our aim is to identify patterns of change in resource acquisition across the order Primates that are related specifically to ranging and dietary flexibility.

In terms of foraging, primates may decrease day range and/or time spent traveling during food shortages (Boinski 1987). Or, primates may increase day range length (Barton *et al.* 1992; Overdorff 1993), traveling and/or foraging time (Milton 1980; Garber 1993; Gursky 2000), and home range size (Clutton-Brock 1977) during times of scarcity. Primates also change locations by moving among habitats and out of their regular home range in search of more resource-rich areas. However, seasonal migration or range shifting in times of resource scarcity are more common and better documented in more mobile taxa such as birds (Karr 1976; Leighton & Leighton 1983; Loiselle & Blake 1991; Kinnaird *et al.* 1996; Curran & Leighton 2000), insects (Janzen 1973; Hunt *et al.* 1999), bats (Fleming & Heithaus 1986; Law 1993), and ungulates (Kiltie & Terborgh 1983; Bodmer 1990; McNaughton 1990; Curran & Leighton 2000).

Dietary switching, i.e. feeding on alternative resources, can take several forms: (i) switching between food categories (e.g. Japanese macaques consume primarily fruit, seeds, leaves, or animal matter, each food type at a different time of year [Hill 1997]); (ii) switching to a different item within a food category (e.g. gibbons select a different set of fruit species when fruit availability is low in central Borneo [McConkey *et al.* 2002]); and (iii) relying on keystone species as defined by authors (e.g. capuchins rely on palm seeds to see them through fruit-scarcity periods in Manu, Peru [Terborgh 1983]). The forms of dietary switching are not mutually exclusive; although chimpanzees in Lope, Gabon, maintain a fruit-dominated diet year-round, during the dry season they rely on a few fruit species as well as leaves and pith as keystone foods (Tutin *et al.* 1997). A change in diet composition is usually accompanied by a change in dietary breadth: the number of species consumed by primates may increase (Oates 1977) or decrease (Struhsaker 1975) during food scarcity.

Framework for interpreting ranging and dietary flexibility

Optimal foraging theory informs practically all aspects of food acquisition: what items to eat, when, where, and with whom. The diet-breadth model predicts that if a highly valuable item becomes too scarce to be exploited profitably and the available items have much lower ratios of net energy value to acquisition time, then dietary diversity is expected to increase when higher-ranked foods are scarce (MacArthur & Pianka 1966; Charnov 1976). Whether a ranging or dietary response is expected depends on how much the travel time would need to increase by to maintain the current food list; how much the foods differ in quality and spatial distribution; and the consumer's general foraging strategy. Animals following an energy-maximizing strategy are expected to increase ranging to meet needs, while time minimizers are expected to reduce energy expenditure (Schoener 1971). If food *x* is substantially more valuable than foods *y* or *z*, then increasing search time for food *x* may be the initial response to a decline in its availability, but switching may follow as it becomes even scarcer.

Items eaten during periods of scarcity are often automatically assumed and occasionally documented to be of lower quality. For example, figs eaten by orangutans, squirrels, and hornbills as fallback foods are less nutritious than more preferred fruit (Leighton 1993). Alternative explanations merit consideration. If foods *x*, *y*, and *z* have equal net energy values, then when *x* becomes scarce, switching to substitutes *y* or *z* is expected, and their relative abundance may determine which is eaten. Guppies presented with two foods differing in quantity preferentially consumed the more abundant item (Murdoch *et al.* 1975). Additionally, foods might be selected on the basis of nutrient, mineral, or chemical content, a complication not accommodated by simple optimal foraging predictions based on energy yields. Mineral content of grasses influences migratory patterns of grazers in the Serengeti (McNaughton 1990). The diet of cotton rats is explained by neither energy-maximizing nor time-minimizing foraging strategies, but rather as a nutrient-balancing strategy (Randolph & Cameron 2001). The benefits of mixing dietary items to reduce the impact of toxins and digestion-inhibiting compounds are particularly important to generalist folivores; their digestive systems may cope better with small quantities of a variety of noxious chemicals than large doses of a few (Freeland & Janzen 1974; Westoby 1978; Cork & Foley 1991). Thus, optimal foraging assumptions, including the lower quality of alternative foods, may be met more easily for frugivores than for folivores (or seed-eaters, discussed below).

We approach the topic of primate foraging from the premise that primates are dietary generalists whose options are set by an interaction between the extrinsic factors influencing spatiotemporal abundance and nutritional quality of foods and the intrinsic factors influencing acquisition and processing of foods (see reviews by Garber 1987; Lambert 1998; Janson & Chapman 1999). In general terms, ripe fruits eaten by primates appear to be designed for vertebrate consumption and dispersal; ripe fruits tend to be relatively rich in sugars and highly digestible, being low in plant secondary compounds and indigestible fiber compared with leaves. Unripe fruit may resemble ripe fruit or seeds in nutritional composition and mechanical and chemical defenses, although the caloric value of a given fruit crop increases as the crop ripens (Schaefer *et al.* 2002). Folivory in primates is hypothesized to have evolved from a seed-eating diet (Chivers 1994; Kay & Davies 1994); seeds, like leaves, often contain secondary or noxious compounds. Seeds also are often mechanically defended and have relatively high amounts of lipids and, sometimes, minerals. Nectar, sap, animal matter – toxic insects excluded – and some fruit provide easily assimilated nutrients, but the complex carbohydrates in leafy matter, gums, and insect exoskeletons require extensive digestive processing. Low-quality foods are considered to be those high in fiber but low in energy. Such generalizations must be tempered by the variation documented in the nutritional quality and chemical composition within food types (Glander 1982; Barton *et al.* 1993; Coley & Barone 1996; Jordano 2000) and the lack of knowledge of specific primate requirements and processing abilities.

Specific predictions regarding extrinsic and intrinsic influences on primate responses

Ranging flexibility

We predict that range shifting will occur in environments that exhibit a habitat grain larger than the species' home range and that range-shifters will have non-defended home ranges that overlap to some extent with conspecifics. Habitat shifting may be constrained by body size and social organization; for example, monogamous species, which are typically relatively small-bodied and defend home ranges, are unlikely to move long distances to exploit a seasonally available resource because of prohibitively high predation risks and costs of territorial interactions.

We expect range use to track resource availability, as demonstrated elegantly in Corbin & Schmid's (1995) field experiment: when their dry-season foods were removed from forest-edge plots, mouse lemurs, *Microcebus murinus*, significantly shifted home range use and rarely visited the once rich area. Alternatively, ranging may be altered by competitive interactions or dictated by the search for mates, the need for sites from which to provision young, and the need for secure sleeping sites or water-holes (Janson 1985; Barton *et al.* 1992; Yeager 1996). Thus, our discussion of habitat shifting in response to seasonality is but one, albeit striking, component of ranging behavior.

Dietary flexibility and body size

We expect that much of the variation in food use between taxa will be determined by their dietary restrictions and preferences and modulated further by resource availability. Energetic, dental, and digestive features constrain or permit dietary choices during scarcity in the same way that they influence overall, annual diets. Dietary components during periods of critical food scarcity may in fact be stronger selective forces on morphology and physiology than overall dietary components (Rosenberger 1992).

Compared with smaller-bodied primates, larger-bodied primates have relatively lower requirements and larger body-fat stores and, thus, a greater ability to withstand short periods of food scarcity (Kleiber 1961; Schmidt-Nielson 1975; Pond 1978; Calder 1984). On a very broad scale, body size explains the non-fruit component of primate diets: smaller primates consume relatively easily digestible, energy-rich items such as nectar, sap, and animal matter; larger primates can tolerate and include less digestible items, such as leaves and other vegetative matter (Kay 1984; Janson & Chapman 1999). Body size alone, however, fails to explain similarities in the macronutrient content of diets of chimpanzee and cercopithecines (Conklin-Brittain *et al.* 1998), ecological features within New World monkeys (Ford & Davis 1992), and gut passage time across all primates (Lambert 1998).

Dietary flexibility and morphology

Taxa with highly specialized morphology are expected to have less flexible diets than those with generalized guts or teeth. Members of the subfamily Colobinae possess a sacculated stomach in which resident microorganisms break down cellulose from the colobine's leafy diet. The pH level required to maintain populations of microorganisms is incompatible with a ripe-fruit-dominated diet (Kay & Davies 1994); therefore, colobines are expected to maintain dietary stability and to avoid large quantities of fleshy, sugary fruit as main or alternative foods, relying instead on leaves and seeds. Folivorous primates relying on an enlarged cecum or colon for the fermentation of structural carbohydrates do not face such a foraging constraint. Facultative folivore-frugivores therefore are expected to consume higher-quality fruit sources whenever possible. In addition to the well-known cases of *Alouatta* and Indridae, some degree of cecocolic fermentation occurs in the relatively well-developed ceca and colons of all New World monkeys, Cercopithecinae, and Pongidae, therefore allowing them to extract nutrients from relatively fibrous vegetation or complex-carbohydrate-rich gums (Chivers 1994; Lambert 1998; Caton *et al.* 1996) and, by extension, to use these as fallback foods.

Dietary flexibility and availability of food types

Primates with diets comprised primarily of temporally reliable foods are not likely to exhibit substantial dietary flexibility. Along a temporal scale, flowers (and, hence, nectar) are highly restricted, seeds are long-persistent, and ripe fruit fall between these extremes. Ripe fruit are available, in an average tropical forest, for less than 1.5 months (Jordano 2000). In contrast, seeds eaten by Malagasy and New World seed-eating primates remain available in the canopy for four months or more (Hemingway 1996; Norconk 1996).

The availability of food items may be modified by the interplay with a species' morphology. New World Pitheciines and *Cebus*, possessing masticatory adaptations for crushing hard seeds, are able to access mechanically protected seeds not accessible to sympatric primates (Terborgh 1983; Kinzey 1992). Dental adaptations of marmosets for gouging trees permit them to stimulate exudate flow on an as-needed basis (Garber 1992). Thus, the consumption of seeds and exudates is expected to be relatively stable compared with that of fruit or flowers.

Generalizations about the temporal availability of leaves make sense only when immature and mature leaves are distinguished. Leaves often are contrasted with fruit and described as a spatially and temporally more available food; that may be true for mature leaves. New leaves, which primates prefer, may be produced year round at low levels or in big bursts, especially in deciduous forests (Fenner 1998) (see also Chapter 2), and even in some tropical evergreen (rain)forests the amplitude of leafing seasonality is greater than that of fruiting (Hemingway 1998).

Behavioral flexibility and biogeographic particularities

We expect feeding and ranging responses to vary as a function of region as the floristic and phenological characteristics particular to each biogeographic region present primates with different foraging options (Fleming *et al.* 1987; Kay *et al.* 1997; Fleagle *et al.* 1999). Southeast Asian forests stand out in several features that encourage dietary flexibility and large ranges of vertebrate consumers (Leighton & Leighton 1983; Lucas & Cortlett 1991): (i) phenological complexity, including masting (van Schaik 1986; Cortlett & LaFrankie 1998); (ii) dominance by the wind-dispersed Dipterocarpaceae (Ashton *et al.* 1988); and (iii) particularly low-quality food in Borneo (Waterman *et al.* 1988). Low diversity and density of palatable fruit may contribute to the dearth of Malagasy frugivores (Goodman & Ganzhorn 1997), and marked seasonality plays an important role in most explanations of Malagasy behavior, physiology, and ecology (Wright 1999). Additionally, differential covariation between leafing and fruiting peaks may have ecological and evolutionary effects on Old World and New World monkeys. Terborgh and van Schaik (1987) proposed that during fruit scarcity, new leaves are available as fallback foods to Old World monkeys, a comparatively larger-bodied and more folivorous clade, but are less commonly available to New World monkeys, who must rely less on young leaves and more on nectar or insects. Van Schaik and Pfannes's larger data set (see Chapter 2) indicates that the situation is more complex than initially hypothesized: leaf and fruit peaks coincide only in seasonally drier habitats but show a stronger trend to do so in the New World.

Behavioral flexibility and resource seasonality

We test in a preliminary fashion whether resource availability determines dietary switching and predict that the resources that primates switch to should be stable in availability or available in off-times relative to preferred resources. Drawing on interspecific comparisons and observations of primate diets in extreme habitats, we predict that populations in more seasonal environments will exhibit greater dietary variability than those in less seasonal environments. Primates inhabiting the margins of their ranges, particularly latitudinal or elevational extremes, consume "unusual" food items, such as bark (*Propithecus verreauxi*: Richard 1978), conifer needles (*Macaca*: Goldstein & Richard 1989; *Rhinopithecus*: Kirkpatrick 1999; Yang & Zhao 2001), lichens (*Rhinopithecus*: Kirkpatrick *et al.* 1999; *Pygathrix*: Li 2001), and bromeliad leaves (*Cebus*: Brown & Zunino 1990). Mountain baboons rely more heavily than other baboons on underground plant matter, especially during winter months (Whiten *et al.* 1987), and savanna-living chimpanzees eat less fruit and more foods that are hard to acquire or process than do forest-living populations (McGrew *et al.* 1988). Similarly, Moraes *et al.* (1998) suggested *Brachyteles* in less seasonal habitats show greater reliance on fruit and greater monthly stability in diet than do populations in more seasonal habitats.

Methods

From the literature, we amassed 234 studies covering 119 species (subspecies not counted separately) at 105 sites from which we extracted data for a survey of qualitative responses and an analysis of quantitative feeding changes. Each species studied at each site represented one record in the database, and thus multiple records may be included from one site and a taxon may be represented by more than one record. From the 234 studies, we gleaned explicit information on primate responses to seasonality as defined by the authors in 157 records, representing 130 studies of 100 species and subspecies as reported by the authors. We recorded home range, day range, and dietary breadth changes in response to seasonal variation; noted whether thc taxon exhibited a physiological response or shifted ranging/habitats; and collected information on the nature of alternative resources in ten food categories.

Just over half of the 234 studies, or 131 studies of 65 primate taxa, presented items eaten on a monthly basis over more than one season as defined by the authors. We extracted from published tables and figures the monthly values to calculate for each food type their coefficient of variation (CV = mean/standard deviation × 100) in use and a grand CV of all dietary items. In addition, van Schaik and Pfannes (see Chapter 2) provided data on the CV for rainfall and the CV for new leaf flushing and fruiting activity in 60 of the sites represented in both of our databases. Primate nomenclature and taxonomy follow Fleagle (1999), and body mass values are from Smith and Jungers (1997).

Despite the proliferation of field studies in the past decades, certain taxa and habitats are under-represented in our study. We obtained no monthly values for food types eaten by the Galagidae and Lorisidae and relatively few responses to scarcity for small-bodied and/or nocturnal taxa. Difficulties in directly identifying foods are sometimes overcome by collecting fecal material; for example, monthly values of percentage of fiber in fecal samples are available for *Microcebus rufus* (Atsalis 1999) and for some of the hard-to-habituate African great apes (Kuroda *et al.* 1996; Remis 1997). Primates that range widely across several habitat types present ideal theoretical conditions for examining ranging responses to seasonal variation in the local food supply. However, they present significant logistic challenges to field studies. Few studies cover altitudinal or other habitat gradients, and even fewer studies simultaneously monitor resource availability and diet across habitats. Marginal, often highly seasonal, habitats and those with low species richness and density of primates are usually not a primatologist's site of first choice. Clearly, our database, while comprehensive, is not exhaustive or bias-free, and this affects our ability to detect and interpret seasonal responses.

Survey of primate responses to scarcity

General patterns

A total of 329 distinct responses to seasonality were recorded in the 157 records where responses were reported. Over 70% of those involved dietary switching; mature leaves, followed by new leaves and other vegetative matter, were the most frequently used alternative resources. Dietary breadth was somewhat more likely to change but could either increase or decrease. Changes in home range area, day range length, and physiological responses each made up less than 10% of the 329 responses. Habitat shifting occurred in 10% of records over all primates. Habitat shifters are scarce among the smallest primates (weight, $n = 4$), most likely due to the confounding effects of social system. Physiological responses included primates hibernating for several months during resource scarcity and changes in body weight. Physiological responses ($n = 22$) and the use of easily digestible items such as gums, nectar, and animals are most common in the smaller primates.

In general, ranging and dietary response patterns fit expectations in light of adult female body mass, except that fruit eating as a response to seasonality is especially important in the largest size, which is made up exclusively of the great apes (Fig. 3.2). This clearly violates body mass and digestive physiology predictions, as the large-bodied apes have the capacity to deal with leafy matter. However, presumably lower-quality items such as mature leaves, other vegetative matter, and seeds are indeed more common in the larger size classes.

We will now review the results of the tests of our predictions in more detail. Because the data are complex at times, we will provide a summary and discussion of our findings in the next main section.

Biogeographic regions and clades

Broad trends mask some important differences in the response to seasonality among the four regions (Fig. 3.3). Home range area appeared to be the most flexible among New World primates and most often increased in size. Day range changes were distributed fairly uniformly among regions,

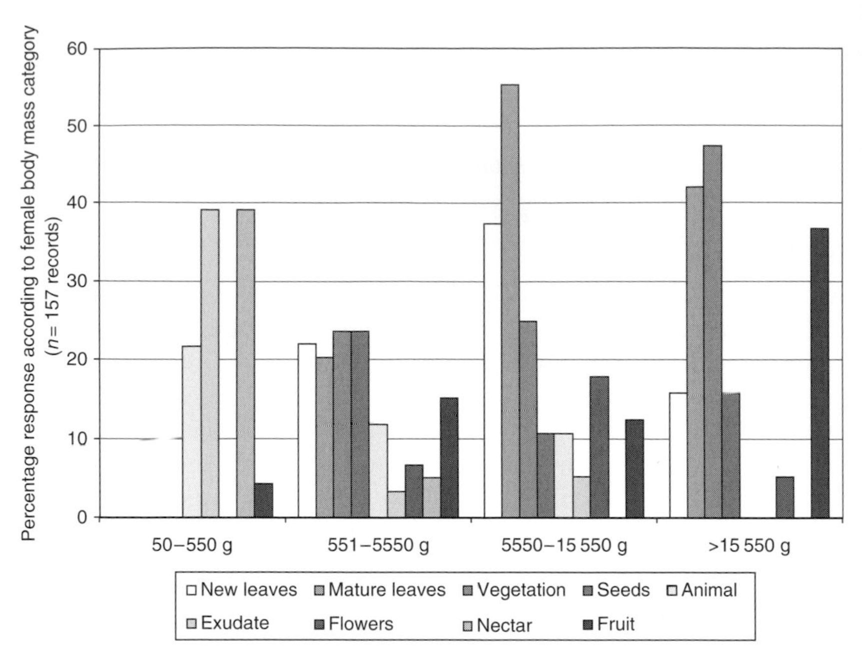

Figure 3.2 Alternative foods used during reduced resource availability, illustrated according to female body mass categories.

except that 25% of primates in Madagascar responded to seasonality by decreasing day range. As expected, primates in Madagascar, specifically the Cheirogalidae, also exhibited the majority of physiological responses to seasonality, such as fattening up and entering periods of torpor. Increases in dietary diversity were common among members of the Indridae, while decreased dietary diversity was a fairly common response across many primates. Home range area changes, day range increases, and habitat shifting were concentrated largely in the New World Atelinae and Cebinae.

Primates in the New World appeared to have little ability to increase their dietary diversity in response to seasonality. In contrast to other regions, New World monkeys did not switch to mature leaves or other vegetative matter such as bark. The Atelinae switched to new leaves, and smaller-bodied taxa switched to exudates, nectar, seeds, and animal matter. Porter (2001) documented that fungus, an uncommon primate food item, was an important food for *Callimico goeldii* both year round and as the main fallback food during the dry season in Bolivia. Dietary switching results in the New World primates are consistent with Terborgh and van Schaik's (1987) expectations based on body size and phenology. Switching

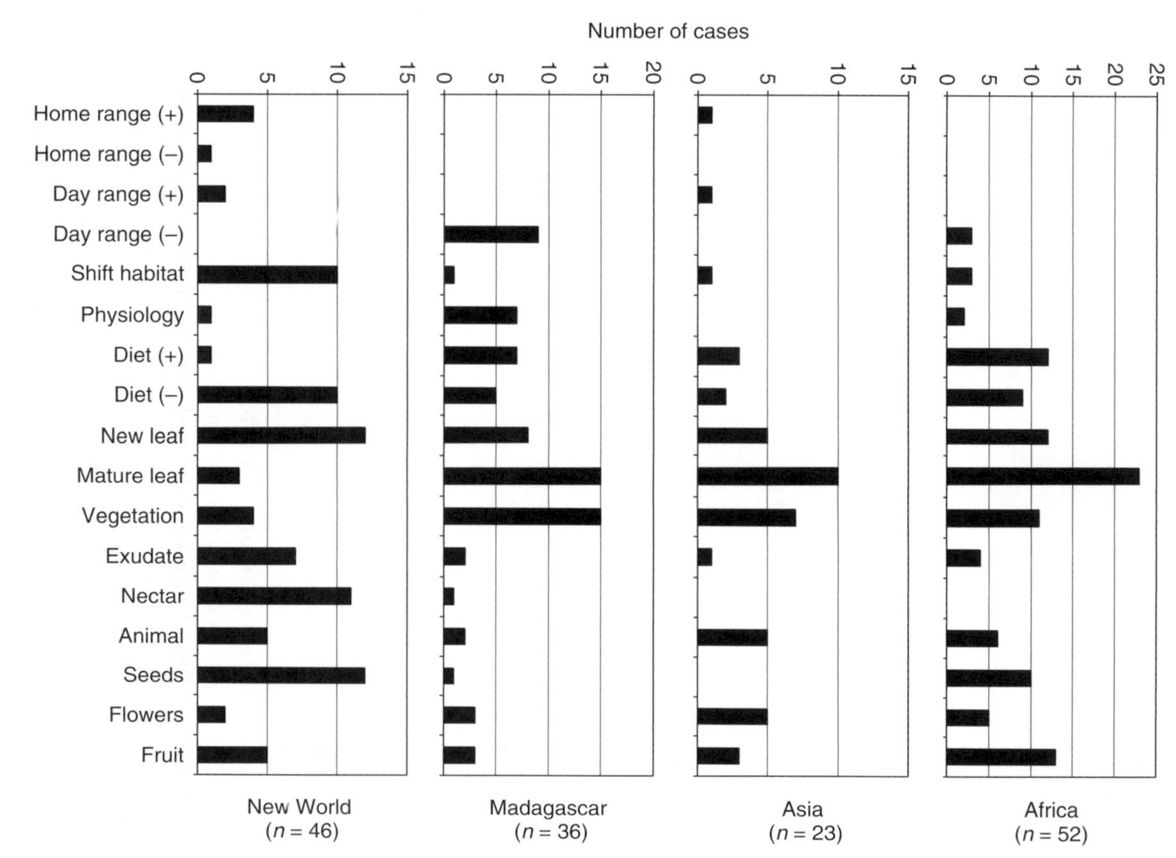

Figure 3.3 Ranging and dietary responses to reduced resource availability by primates in each biogeographic region.

to animal foods was most common in Asia, and switching to seeds was virtually absent from Asia and Madagascar. The main seed-eating primates of Asia and Madagascar, the Colobinae and Indridae, respectively, switch between seed and leaf eating (Davies 1991; Yeager 1989; Hemingway 1998). Seeds are probably not a reliable fallback food in Asia with its supra-annual fruiting peaks (Leighton & Leighton 1983). Phenological peaks on supra-annual or unpredictable cycles have also been suggested for certain Malagasy plants that are favored food sources, including seed sources (Morland 1993; Overdorff 1993; Hemingway 1996).

Subtle and variable dietary switching

Few primates use the same food type (fruit, insects, leaves) year round without switching to another type during a period of scarcity. However, some primates continue to make choices within one food type while varying the specific components in relation to availability. In the Sulawesi wet season, when insect abundance was high, Orthoptera and Lepidoptera were the main foods for *Tarsius spectrum*, whereas in the dry season, Hymenoptera, Isoptera, and Coeloptera formed the bulk of their diet (Gursky 2000). Several Malagasy folivores also consistently consume the same food type year round: *Lepilemur leucopus* (Nash 1998), *Hapalemur alotrensis* (Mutschler *et al.* 1998), and *Hapalemur simus* and *H. aureus* (Tan 1999). Although *H. simus* maintain a bamboo-based diet, they shift between shoots and pith (Tan 1999).

In contrast to most primates, or other gorillas for that matter, *Gorilla gorilla beringei* experience a relatively stable and abundant supply of edible vegetation but little edible fruit; mountain gorillas consume vegetative matter year round but seasonally use bamboo when the shoots emerge in the wettest months (Watts 1998). Thus, annual stability in food type use, which appears common but not exclusive to relatively small-bodied taxa, is accompanied by subtle responses to fluctuations in the resource base.

Dietary switching results for a given species, not surprisingly, vary over space and time. In Kibale, Uganda, *Lophocebus albigenia* is reported to use roughly similar levels of all food types throughout the year (Waser 1977; Wrangham *et al.* 1998; Conklin-Brittain *et al.* 1998) and to increase significantly flower feeding during the dry season (Olupot *et al.* 1997). However, in Dja, Cameroon, the same species shifted from fruit to seeds, flowers, and new leaves (Poulsen *et al.* 2001). In some years, *Cercopithecus mitis doggetti* respond to preferred fruit scarcity in montane rainforest by maintaining some fruit consumption, increasing leaf consumption and dietary diversity, but in others they respond by switching to seeds and decreasing dietary diversity (Kaplin *et al.* 1998). When the food supply is dire, primates may turn to items that were not recorded previously and little resembled their usual foods. For example, during a two-year drought in southwestern Madagascar, *Lemur catta* resorted to feeding on dried leaves and desiccated seed pods (Gould *et al.* 1999).

Critical dietary switching

Alternative, keystone, fallback, or critical foods often are used as synonyms for resources used during periods of low food availability. Keystone resources are more properly restricted to a special case of dietary switching, in which the resources exploited are presumed to be vital to sustaining the consumer population during food scarcity (Howe 1977; Terborgh 1986). Since their addition to the ecological lexicon, the terms "keystone species" and "keystone plant resources" have been critiqued extensively, and efforts to make their study more tractable continue (see Peres 2000). To date, few field researchers have applied rigorous definitions (Gautier-Hion & Michaloud 1989; Lambert & Marshall 1991; Overdorff 1992; Peres 2000), and no primate study has explicitly tested the cascading effects predicted from the keystone resource concept. Circumventing such criticism and seeking to avoid a watered-down concept, authors increasingly opt for theory-neutral terms. Our aim here is not to judge *post factum* the validity of each application of the term but to summarize putative keystone resources, and their characteristics and consumers, as reported by original authors.

Putative keystone resources spanned the array of most primate food types: nectar, exudates, fruit, seeds, leaves, and other vegetative matter (Table 3.1). Most researchers identified resources as important because of their availability year round or only when other foods were scarce. In addition to particular species, sets of species united by certain characteristics and species interactions were reported. Oates (1996) suggested that climbers and colonizer species in forest gaps and terrestrial herbaceous vegetation may be critical for African folivores, thereby functioning as a "keystone habitat" *sensu* Levey (1990). Approaching the topic from the plant's perspective, Baum (1996) suggested that *Adansonia* spp. (Bombacaceae) serve as a keystone mutualist for lemurs, sunbirds, bats, and insects that pollinate the baobabs while consuming flowers and nectar, thereby maintaining ecosystem function in the western dry forests of Madagascar.

Table 3.1 *Putative keystone plant resources for primates*

Plant part eaten and taxa	Consumer	Characteristics	Region	Source
Exudate				
Fabaceae, *Parkia*	Vertebrates	Av-s, Ab, Re	NW	1
Nectar and/or flowers				
Bombacaceae, *Adansonia*	Lemurs, sunbirds, bats, insects	Ab	MAD	3
Bombacaceae, *Quararibea*	Vertebrates	Av-s, Ab, Re	NW	1, 2
Capparidaceae, *Maerua filiformis*	*Lemur catta*	Av-yr	MAD	4
Clusiaceae, *Pentadesma*	*Cercopithecus cephus*	Av-s	AF	5
Combretaceae, *Combretum*	Vertebrates	Av-s, Re	NW	1, 2
Fabaceae, *Erythrina*	Vertebrates	Av-s, Re	NW	1, 2
Five genera	Assorted vertebrates	Av-s, Ab, Re	NW	1, 2
Fruit flesh				
Anacardiaceae, *Pseudospondias*	Primates	Av-s	AF	5
Annonaceae, *Xylopia*	Guenons, *Mandrillus*, *Lophocebus*	Av-s	AF	5
Arecaceae, six genera	Vertebrates	Av-s, Ab, Re	NW	1, 2
Arecaceae, *Elaeis guineensis*	*Pan*, *Mandrillus*, *Lophocebus*, guenons	Av-yr	AF	5
Cecropieaceae, *Musanga cercropoides*	*Pan troglodytes verus*	Av-yr	AF	7
Fabaceae, *Detarium*	*Gorilla*, *Mandrillus*	Av-s	AF	5
Fabaceae, *Tamarindus indica*	*Lemur catta*	Av-yr	MAD	4
Loranthaceae, *Bakerella*	*Microcebus rufus*	Av-yr, Nu	MAD	6
Moraceae, *Ficus*	Primates, except *Gorilla*, *Colobus*	Av-s	AF	5
Moraceae, *Ficus*	Monkeys, marsupials, birds	Av-s, Ab, Re	NW	1, 2
Myristicaceae, *Pycnathus*	*Pan*, guenons	Av-s	AF	5
Rubiaceae, *Enterospermum pruinosum*	*Lemur catta*	Av-rep	MAD	4
Salvadoraceae, *Salvadora augustifolia*	*Lemur catta*	Av-rep, Ab	MAD	4
Tiliaceae, *Duboscia*	*Gorilla*, *Pan*, *Mandrillus*, *Lophocebus*	Av-yr	AF	5
Three to eight genera	Vertebrates	Av-s, Ab, Re	NW	1, 2
Seeds				
Arecaceae, three to six genera	*Cebus*, macaws, rodents, ungulates	Av-s, Ab, Re	NW	1, 2
Arecaceae, *Elaeis guineensis*	*Pan troglodytes verus*	Av-yr	AF	7

Table 3.1 (*cont.*)

Plant part eaten and taxa	Consumer	Characteristics	Region	Source
Clusiaceae, *Pentadesma*	*Colobus satanas*, *Cercopithecus nictitans*	Av-s	AF	5
Fabaceae, *Detarium*	*Gorilla*, *Mandrillus*	Av-s	AF	5
Fabaceae, *Pentaclethra*	*Mandrillus*, *Colobus*, *Lophocebus*	Av-s	AF	5
Fabaceae, *Tetrapleura*	*Mandrillus*, *Colobus*, *Lophocebus*	Av-yr	AF	5
Lecythidaceae, two genera	Monkeys, rodents	Av-s, Ab, Re	NW	2
Olaceae, *Ongokea*	*Mandrillus*, *Colobus*, *Lophocebus*	Av-s	AF	5
Leaves				
Capparidaceae, *Maerua filiformis*	*Lemur catta*	Av-yr	MAD	4
Marantaceae	*Gorilla*, *Mandrillus*, elephants	Av-yr, Ab	AF	5, 9
Moraceae, *Milicia excelsa*	*Gorilla gorilla gorilla*	Av-yr, As, Ab, Nu	AF	9
Climber, colonizer, and ground cover plants	Folivorous primates	Ab, Nu	AF	10
Pith				
Arecaceae, *Elaeis guineensis*	*Pan troglodytes verus*	Av-yr	AF	7
Marantaceae, Zingerberaceae	*Gorilla*, *Pan*, *Mandrillus*	Av-yr, Ab	AF	5, 8
Bark				
Moraceae, *Milicia excelsa*	*Gorilla gorilla gorilla*	Av-yr, As, Ab, Nu	AF	9
ns	*Pongo pygmaeus pygmaeus*	ns	AS	11

Characteristics as identified by authors: Ab, abundant; AF, Africa; AS, Asia; As, asynchronously available; Av-s, available during scarcity; Av-yr, available year round; Av-rep, available during reproductive stress; MAD, Madagascar; Nu, nutritional value; ns, not specified; NW, New World; Re, reliably available.
Source: 1, Peres 2000; 2, Terborgh 1986; 3, Baum 1996; 4, Sauther 1998; 5, Tutin *et al.* 1997; 6, Atsalis 1999; 7, Yamakoshi 1998; 8, White *et al.* 1995; 9, Rogers *et al.* 1994; 10, Oates 1996; 11, Delgado & van Schaik 2000.

The broader ecological literature continues to identify figs as the prime example of keystone resources (Power *et al.* 1996; Fenner 1998; Nason *et al.* 1998), whereas Watson (2001) has described mistletoes as keystone resources worldwide. Figs are undeniably important to many vertebrate consumers, including primates (Kinnaird *et al.* 1992); nevertheless, putative primate keystone resources encompass diverse plant taxa. The prevalence of plant structures that are difficult to access and/or digest (bark, herbaceous vegetation, seeds, and exudates) among primate keystone resources is noteworthy.

Non-primate studies employing the term "keystone resource" frequently stress fruit resources and secondarily flowers or nectar. Focal study taxa are commonly obligate frugivores, particularly hornbills (Kannan & James 1999) and bats (Wendeln *et al.* 2000), whose extensive mobility is linked frequently to their ability to use widely scattered resources (Gautier-Hion & Michaloud 1989) and maintain specialized fig diets (Leighton & Leighton 1983; Lambert & Marshall 1991). Although fruit dominate the diets of monkeys and apes within some communities (Tutin *et al.* 1997), even the most frugivorous of primates are not obligate frugivores.

A closer look at habitat shifting

By our criteria, it was not necessary for a primate community to range exclusively in a different habitat during the time of seasonal food scarcity to be defined as a "habitat shifter." Instead, we looked for temporally and spatially patterned use of the home range over seasons, with home range defined broadly (*sensu* Jolly 1972). Of the 15 instances of habitat shifting in our database, several distinct types of response group together (Table 3.2). In several instances, targeted use of resources in the shifted-into habitat was reported (e.g. *Alouatta seniculus* and *Lagothrix lagotricha* at Caparú [Defler & Defler 1996; Defler 1996] and *Pan troglodytes* at Kalinzu [Furuichi *et al.* 2001]). In contrast, some authors indicated that species become widely nomadic or "vagrant" over large areas (e.g. movement of *Saimiri* and *Cebus albifrons* from flooded blackwater igapo forest to terra firme forest at Urucu (Peres 1994b]). In other cases, authors report an absence or outright emigration from the regular home range but cannot confirm the destination habitat (e.g. *Eulemur fulvus rufus* at Ranomofana [Overdorff 1993], *Cercopithecus mitis* on the Zomba plateau [Beeson 1989], *Lagothrix lagotricha cana* at Caparú [Defler 1996], and *Cebuella pygmaea* at Maniti [Soini 1982, 1993]).

Habitat shifters were scarce among the smallest primates (e.g. only two taxa weigh less than 1 kg), and the majority of documented shifters were in the New World (n = 10), with the majority of these in the Atelinae and the Cebinae (n = 8). At Urucu, the three mobile taxa (*Saimiri*, *Cebus albifrons*, and *Lagothrix*) had the largest group sizes in the primate community. Likewise, at Ranomonfana, Madagascar, of the two congeners similar in body size, the habitat shifter was the species with the larger

Table 3.2 *Habitat shifting from 157 records of response to seasonal scarcity*

Species	Region and site	Body mass (g)	Social system	Responses to food or preferred food reduction	Reference
Alouatta seniculus	NW: Caparu, Colombia	5210	PM	Shift from terrace/transition to igapo; increase day range; switch to young leaves and epiphyte roots	Palacios & Rodriguez (2001)
Lagothrix lagotricha lagotricha	NW: Caparu, Colombia	7020	PM	Shift into igapo; increase dietary diversity; increase new leaves, rely on *Iriartea* fruit and *Micandra* seeds	Defler & Defler (1996), Defler (1996)
Lagothrix lagotricha	NW: Rio Duda, Tinigua, Colombia	7020	PM	Shift into degraded and flooded forest; switch to leaves and unripe fruit; rest more; decrease day length	Stevenson *et al.* (1994)
Lagothrix lagotricha cana	NW: Urucu, Brazil	7650	PM	Travel out of range; switch to young leaves, seeds, and exudate	Peres (1994a, 1994b, 1996)
Callicebus torquatus	NW: Rio Nanay, Peru	1151	M	Shift into streamside vegetation; increase animal matter, seeds, and palm fruit feeding one of two study years	Kinzey (1977)
Cebuella pygmaea	NW: Maniti, Peru	122	M	Emigrate (4/14 study groups) when exudate depleted in home range	Soini (1982, 1993)
Cebus albifrons	NW: Manu, Peru	2290	PM	Shift across habitat mosaic following fruiting peaks; increase home range area; switch to figs and palm fruit	Terborgh (1983, 1986); van Schaik *et al.* (1993)
Cebus albifrons unicolor	NW: Urucu, Brazil	2290	PM	Shift from igapo to terre firme, creekside habitats – 5.5-km movement; dietary switching unknown	Peres (1994b)
Cebus apella	NW: Urucu, Brazil	2520	PM	Rely on creekside and palm swamp habitats; switch to seeds and nectar	Peres (1994b)
Saimiri spp.	NW: Urucu, Brazil	680	PM	Shift from igapo to terre firme – 5.5-km movement – widely vagrant; dietary switching unknown	Peres (1994b)
Lophocebus albigena	AF: Dja, Cameroon	8250	PM	Increase use of swamp habitats; increase dietary diversity; switch to new leaves, flowers, and seeds	Poulson *et al.* (2001)

Table 3.2 (*cont.*)

Species	Region and site	Body mass (g)	Social system	Responses to food or preferred food reduction	Reference
Cercopithecus mitis	AF: Zomba Plateau, Malawai	7930	PM	Travel out of range (one of two groups for one month); switch to leaf petioles of *Ipomoea*	Beeson (1989)
Pan troglodytes	AF: Kalinzu, Uganda	45 800	FF	Shift into secondary habitats; rely on fruit of *Musanga leo-errerae*	Furuichi *et al.* (2001)
Pongo pygmaeus	AS: Suaq Belimbing, Sumatra	35 600	S	In mast years, shift to hill forest; in regular scarcity, non-random use of river, swamp, peat, and hill forest; Shift to vegetation and bark	Singleton & van Schaik (2001); Fox *et al.* (2004)
Eulemur fulvus rufus	MAD: Ranomafana	2147	PM	Travel out of range for six weeks, perhaps 6 km away to fruiting *Psidium*	Overdorff (1993)

AF, Africa; AS, Asia; FF, fission–fusion; PM, multimale-multifemale; M, monogamous; MAD, Madagascar; NW, New World; S, solitary.

group size (*Eulemur fulvus rufus*). At Amazonian sites, multiple species moved between habitat types. For two species at each site (*Alouatta* and *Lagothrix* and Caparú, Colombia [Defler 1996; Palacios & Rodriguez 2001] and *Saimiri* and *Cebus albifrons* at Urucu, Brazil [Peres 1994b]), shifts occurred between igapo and dryland or "terra firme" forests.

Most habitat shifters (n = 10) had polygynous, multimale–multifemale social systems and ranged in relatively large areas that overlapped with conspecifics, consistent with our initial expectations. However, two of the New World habitat shifters were relatively small-bodied primates living in monogamous family groups (*Cebuella pygmaea* [Soini 1982, 1993] and *Callicebus torquatus* [Kinzey 1977]). Interestingly, Gursky (2000) reported an increase in territorial conflicts between neighboring spectral tarsier family groups in the time of insect scarcity, implying groups test home range boundaries during that time. Community-level home ranges of some monogamous primates may also be more dynamic than previously thought.

Habitat shifting was often correlated with dietary switching, most often a switch to seed and fruit eating (n = 6) or a switch to young leaves (n = 4, all New World atelines). Clearly, habitat shifting requires ecological heterogeneity across spatial and temporal scales. In seven cases, the habitats shifted into were secondary or successional forests along creeks and streams. This makes sense, as the magnitude of seasonal differences in production in secondary forest is thought to be lower than that for mature forest (Opler *et al.* 1980). Shifts to riparian areas were often correlated with a dietary switch to palm seeds.

A concentration of the phenomenon of habitat shifting in the New World is not surprising given van Schaik and Pfannes's (Chapter 2) finding that in the Neotropics, the interval between peak leaf flush and peak fruiting within a single habitat is demonstrably shorter than in other regions. Within a single habitat, some New World primates are thus expected to face a lean season for fruits, new leaves, and flowers. Under those conditions, it may be an ecologically attractive option for a primate to shift its ranging into an adjacent habitat with contrasting phenology. The mosaic of terra firme and flooded forests along major rivers of the Amazon basin illustrates this possibility: available evidence suggests fruiting peaks coincide with inundation in both whitewater várzea and blackwater igapó flooded forests (Ayres 1989; Kubitzki & Ziburski 1994), while adjacent terra firme forest does not (Defler 1996; Palacios & Rodriguez 2001; Bynum unpublished data).

In a final twist, ranging and habitat changes coincide with periods of abundance rather than scarcity for both Bornean and Sumatran orangutans (Leighton & Leighton 1983; te Boekhorst *et al.* 1990; Singleton & van Schaik 2001; Fox *et al.* 2004). Shifts across habitat mosaics primarily in response to a food surplus nearby, as opposed to a shortage within the home range, may be tied to a combination of the orangutan's social and locomotor behavior and the uniquely Asian masting phenomenon of exceptionally good years punctuating generally modest availability of preferred food resources.

Primate diet variability: flexibility across food types

General patterns

Primate annual diets are diverse: a mean of 86 plant species in the diet corroborates Oates's (1987) statement that most primates obtain food from over 50 plant species. Across months of the year, primates generally vary the food types they use (mean CV in use of all food types = 81), ranging from 22 for *Gorilla gorilla* (calculated from Tutin *et al.* 1997) to 154 for *Cercopithecus mitis* (calculated from Beeson 1989). Overall diet variability decreases as dietary breadth (defined as the number of food species) increases ($r^2 = 0.08$; $P = 0.045$, $n = 77$). Incorporating many species into the diet on a regular basis appears to mitigate dramatic and frequent switches among food types, particularly for relatively large-bodied primates, as diet breadth increases with body mass ($r^2 = 0.14$; $P = 0.0002$, $n = 93$), although much of the variation is unexplained.

The distribution of coefficients of variation differs significantly across food types (Table 3.3). Temporal variation is highest in flower eating (CV = 131) and lowest in fruit eating (CV = 50). Feeding on flowers regularly throughout a year is typically not an option because of their temporally ephemeral nature. Given the prominence in the literature of fruit-scarcity-mediated changes, the relative consistency in fruit feeding across the primates and for particular taxa is striking and best interpreted in light of the strong preference for fruit as reported in numerous field studies. Items available over long periods, such as mature leaves, animal matter, and exudates, have low CVs in use, as expected. In contrast, seeds have high CVs in use across primates (CV = 103), although variation in seed use is strikingly low in the Callicebinae (CV = 53), Cebinae (CV = 36), and Pitheciinae (CV = 38).

Overall time spent feeding on a food type (fruit, seeds, immature leaves, mature leaves, exudates, and animal matter) greatly impacts the temporal variability in use of both the main and the supplementary items. The more a food contributes to the overall diet, the more consistently across months that food is consumed; not surprisingly, frugivores have low CV fruit-use values and folivores have low CV leaf-use values.

Table 3.3 *Primate diet flexibility*

	Coefficient of variation across months in use of food types										
	Overall	Fruit	Seed	New leaf	Mature leaf	Flower	Exudate	Animal	Diet breadth	% Fruit flesh	Foods used during scarcity
Atelinae	84	56	165	54	101	114	148	43	107	46	nl
Callicebinae	52	18	53	52	–	162	–	105	79	58	sd
Callitrichinae	86	56	–	108	–	207	101	48	54	48	n, ex, a, fu
Cebinae	68	34	36	135	–	146	20	64	86	62	f, sd, a, v
Pitheciinae	112	110	38	183	–	158	–	154	100	42	sd, a, nl
Cheirogalidae	154	39	–	–	–	224	–	63	34	66	ex
Lemuridae	108	47	–	112	88	164	–	100	70	60	nl, ml, v, fl, f, a
Indridae	80	61	81	62	88	106	–	–	82	24	nl, ml, v
Cercopithecinae	73	47	73	84	83	131	35	76	86	50	sd, a, ml, f, nl, v
Colobinae	78	65	116	53	71	105	95	239	53	21	ml, nl, v, fl, f
Hylobatidae	57	25	–	40	–	117	–	62	40	54	fl, ml, nl
Pongidae	69	38	111	53	47	188	–	88	155	57	v, ml, f, nl
Primate-wide	81	51	103	67	82	131	87	76	86		
n	442	114	29	99	43	71	12	45	93		

Sample sizes within clades range from 1 to 21.

Foods used during scarcity are listed in descending number of records: f, fruit; sd, seed; nl, new leaf; ml, mature leaf; v, vegetative matter; fl, flower; n, nectar; ex, exudate; a, animal; fu, fungus.

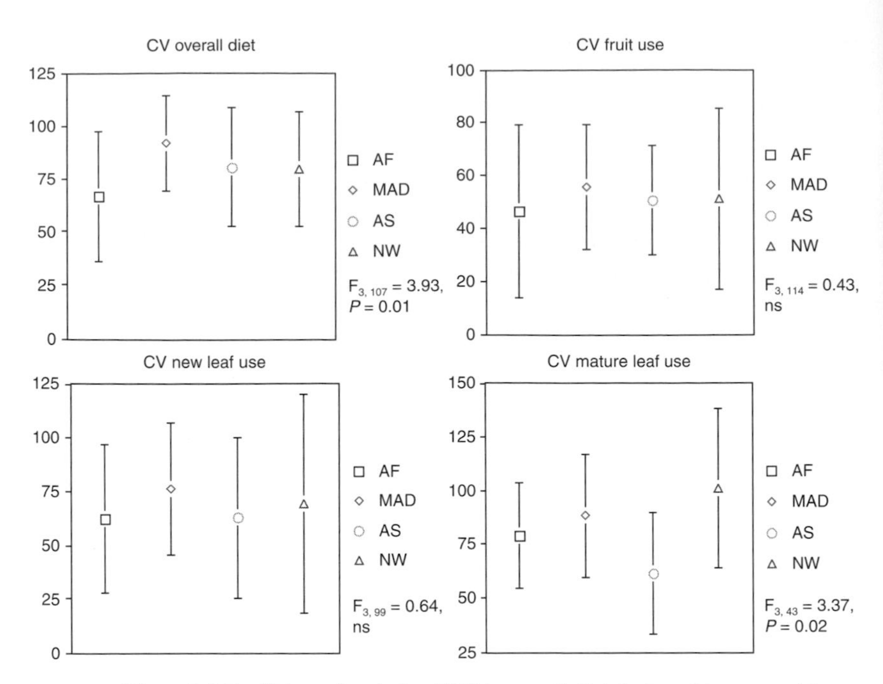

Figure 3.4 Coefficient of variation (CV) in overall diet, fruit, and leaves used by primates in Africa (AF), Madagascar (MAD), Asia (AS), and the New World (NW).

Biogeographic regions and clades

Certain aspects of temporal variation in diet differ across regions and clades (Fig. 3.4). Overall diet values are most variable in Madagascar. This finding is probably due to the extreme diet variability documented in the Cheirogalidae, for which only fruit and flower categories were scored, with flower use extremely variable. Interestingly, variability in fruit and new leaf use is similar across all four regions, but mature leaves are used with greater consistency in Asia. Although one might attribute the low variation in mature leaf use in Asia to a predominance of colobines in the Asian sample, colobines in Asia are less variable in their leaf use than those in Africa (mean CV mature leaf: Asian colobines 57.3; African colobines 87.2; $F_{2,12} = 5.15$, $P = 0.04$). Stability in the use of mature leaves may reflect long periods of fruit scarcity in Asian forests.

Taking a clade perspective rather than a regional view, across the primates overall diet variability is relatively high in the Cheirogalidae, Lemuridae, and Pitheciinae and low in the Callicebinae, Hylobatidae, and Pongidae (Table 3.3). Restricting the scope to contrasts between sister clades, members of the Cercopithecinae are less variable than members of the Colobinae in fruit feeding (t-test−2.48, $P = 0.02$) and more variable in leaf feeding (t-test 3.44, $P = 0.002$). Similarly, the Lemuridae are more variable than the Indridae in leaf feeding (t-test−3.74, $P = 0.002$), although their difference in fruit-feeding variation does not quite reach significance (t-test 2.00, $P = 0.06$). Members of the Lemuridae and Cercopithecinae also consumed the greatest variety of food types as alternative resources. The lemurids of Madagascar generally fill the same ecological niche as the cercopithecines in Africa and Asia, that of medium-sized arboreal frugivores who supplement their fruit-based diet with, often substantial, leaf and insect components (Lemuridae, *Eulemur* [Overdorff 1993; Andrews & Birkinshaw 1998; Rasmussen 1999]; Cercopithecidae, *Cercopithecus* [Harrison 1984; Beeson 1989]; and *Macaca* [O'Brien & Kinnaird 1997; Hill 1997]).

Adequate data also permit contrasts among three "folivore" clades. The main dietary components of both the Malagasy Indridae and the Old World Colobinae span a range of leaves, fruits, and seeds. Adding the Neotropical Atelinae to the comparison allows us to examine the interplay of digestive specialization and dietary flexibility. Like the indrids, atelines use cecocolic fermentation to break down plant matter and combine leaves with other dietary components; however, atelines consume far fewer seeds. Across months, colobines, who use forestomach fermentation, show greater stability only in mature leaf use; however, their range of values for fruit use is much narrower than for other foods (Fig. 3.5). Whereas young leaves figure prominently in the dietary response of atelines, leaves at all stages of maturity and other vegetative matter are important alternative resources to the Indridae and Colobinae, even in the case in which colobines are reported to consume significant quantities of whole fruit when it is abundant (e.g. Fashing 2001).

Our clade contrasts suggest two classic evolutionary patterns: divergence through competition within communities and convergence of independent lineages. Within a community – whether of Old World monkeys or Malagasy lemurs – clades diverge in their feeding flexibility as well as diet composition and morphology. This broad-scale pattern mirrors one of the most comprehensive community studies; comparing responses to scarcity among primates at Lope, Gabon, Tutin *et al.* (1997) reported that cercopithecines and chimpanzees maintained a fruit-dominated diet, whereas colobines increased the leaf component. In contrast, the clades that diversified in separate biogeographic regions and independently derived solutions to food processing generally converge in their monthly variation in food use.

Other tantalizing tests suggested by case studies require larger sample sizes. For example, morphological variation within the Callitrichidae has been related to ecological differentiation: the dental anatomy of

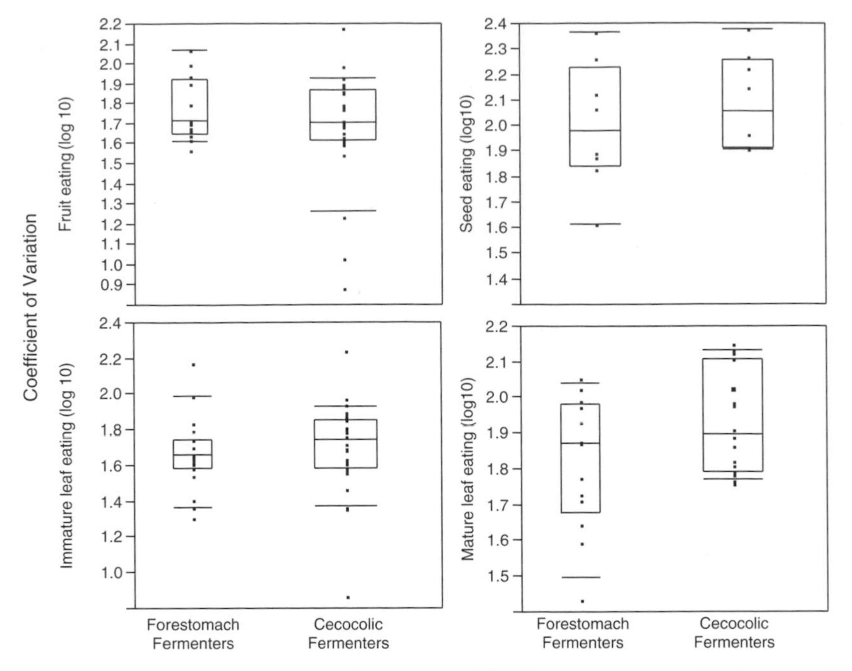

Figure 3.5 Comparison of dietary variability between primates with forestomach fermentation capacities (Colobinae) and cecocolic fermentation capacities (Indridae and Atelinae).

marmosets but not tamarins that permits marmosets to mechanically stimulate food production on an as-need basis is considered important in understanding their tolerance to periods of food scarcity and use of often highly seasonal habitats (Rylands 1993). The savanna-living *Erythrocebus*, however, achieves year-round use of gums without the modified dentition for tree gouging characteristic of specialized exudates feeders (Isbell 1998). At present, we are restricted to documenting broad patterns and pointing out areas of further interest.

Effects of increasing resource scarcity

Response options

We extracted a subset of 83 of the 157 seasonality response records for which the number of dry months (rainfall less than 100 mm) was available. Data were analyzed in four classes: no dry months, one to three dry

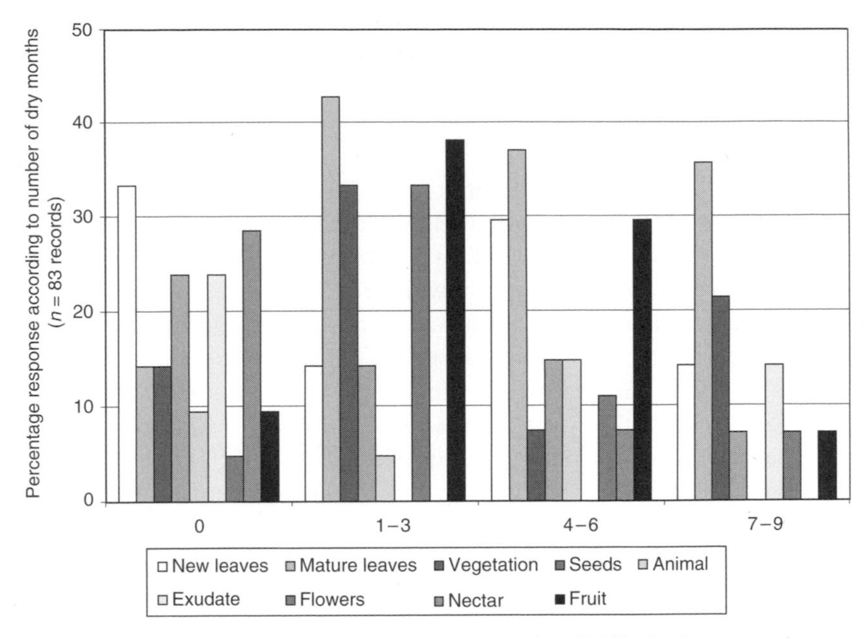

Figure 3.6 Primate dietary responses to reduced availability in the resource base according to seasonality categories, illustrating responses in use of alternative food.

months, four to six dry months, and seven to nine dry months (Fig. 3.6). Habitat shifting appears to be an option that is limited to relatively aseasonal environments as measured by the number of consecutive dry months. In contrast, decreasing day range, increasing dietary breadth, and responding physiologically are concentrated at sites with long dry seasons. New leaves, nectar, exudates, and seeds are the most common fallback foods at sites with no dry season. Although consuming leaves and vegetation are responses found in all four classes of seasonality, mature leaves are particularly important as the length of the dry season increases. Exudate use is bimodal, occurring only in the extreme cases of aseasonal and highly seasonal rainfall. Neither animal matter nor seeds appear to function as fallback foods at sites with long dry seasons.

Dietary flexibility

To address whether greater monthly dietary variability is a response to greater resource seasonality, we began with all primate studies for which we have CV in food type used and then used subsequently smaller, but presumably more cohesive, data sets. We focused on fruits and leaves, the primary foods in many diets, and examined them separately to accommodate the van Schaik and Pfannes finding (see Chapter 2) that leaf flushing is related strongly to climatic variables while fruiting is not.

Seasonality is a complex feature that is difficult to distill into a single variable, and we used surrogate variables commonly encountered in the literature: latitude, rainfall variation (CV rainfall), and dry season length (number of consecutives months in which rainfall was less than 100 mm). Suites of environmental variables are correlated (e.g. length of the dry season is correlated with latitude; see Chapter 2), but certain variables should be more effective with different data sets. Latitude, although available and useful at a primate-wide scale, will not be useful in discriminating between African great ape sites, all of which are Equatorial, and Malagasy primate sites, which cover a relatively short range and do not span the Equator. In addition, the length of dry season may be a more appropriate surrogate variable than latitude or rainfall variation for leaf seasonality, as it may better reflect deciduousness and seasonal leaf flushes across sites.

Coefficient of variation in rainfall was not related to dietary flexibility in any of the data sets. Across the primate-wide sample, dietary flexibility increased with latitude (Table 3.4). Looking at variability of fruit and new leaf use within each of the biogeographic regions, we see slightly different relationships. In Africa and the Neotropics, the relationship between increased latitude and variation in fruit eating held (Table 3.4). Variation in leaf eating at the regional level was explained better by the length of the dry season, at least in Africa and Madagascar (Table 3.4). Latitude held as an important predictor of variation in fruit feeding within clades that span a latitudinal range (Atelinae: $r^2 = 0.58$, $P = 0.003$, $n = 15$; Colobinae: $r^2 = 0.92$, $P = 0.03$, $n = 11$; Cercopithecinae: $r^2 = 0.52$, $P = 0.004$, $n = 17$) but, as expected, not within the Equatorial great apes ($r^2 = 0.09$, not significant, $n = 17$). A conundrum emerges from these results: although variability in fruit production is not related to latitude, variability in fruit use is. Variability in fruit use may be related to some other feature(s) that covaries with latitude, for example plant species diversity or productivity.

Information on variation in food availability for particular sites is likely to be more effective in predicting variation in food use than general surrogates of seasonality. By restricting the analysis to those sites for which we can use van Schaik and Pfannes's data on the CV in monthly food availability (specifically, CV across months in the number of species

Table 3.4 *Relationships between environmental seasonality and dietary flexibility*

Dietary variable	Environmental variable		Primate-wide	Region: Africa	Madagascar	Asia	Neotropics
CV overall diet	Latitude	r^2	0.16[a]	0.34[b]	0.10	0.03	0.02
		n	105	31	23	23	28
	Length of dry season	r^2	0.04	0.23	0.10	0.00	0.01
		n	58	13	9	18	18
CV fruit use	Latitude	r^2	0.17[a]	0.27[a]	0.14	0.06	0.29[c]
		n	111	35	22	25	31
	Length of dry season	r^2	0.02	0.13	0.14	0.19[c]	0.05
		n	63	16	9	20	17
CV new leaf use	Latitude	r^2	0.12[b]	0.11	0.14	0.11	0.06
		n	98	33	22	21	22
	Length of dry season	r^2	0.06	0.54[a]	0.73[b]	0.03	0.11
		n	56	16	8	17	8

Variables were log-10 transformed for least squares regression analyses. Length of dry season: number of consecutive months with less than 100 mm rainfall.
[a] $P < 0.001$.
[b] $P < 0.01$;
[c] $P < 0.05$;
CV, coefficient of variation.

with new leaves or fruit), we aim to test more directly the effect of resource seasonality. Surprisingly, across the entire primate sample, temporal restriction in resource availability and resource use were not related ($r^2 = 0.01$, not significant, $n = 58$). Higher CVs in fruit availability were not reflected by higher CVs in fruit eating, in any subsample of the data set except for the Malagasy primates, where it explained 45% of the variation. As for the effect of new leaf seasonality, the CV of new leaf flush was a relatively good predictor of variation in leaf feeding in Madagascar ($r^2 = 0.63$, $P = 0.003$, $n = 11$) but a relatively poor predictor in Asia ($r^2 = 0.05$, not significant, $n = 14$); sample sizes were too small for tests in Africa and the Neotropics. The ineffectiveness of the most direct resource seasonality measure in explaining resource use variation across most data sets is perplexing.

Tracking resources

Whether food availability is high or low, a primate's daily diet usually includes only a small subset of the resources that it encounters. While acknowledging that a strict definition of preference or selectivity for a food would require that a food is used to a greater extent than expected by its availability in the environment (Chesson 1978), researchers commonly employ positive correlations between representation in diet and habitat to indicate preferred foods. A positive correlation between the use of a resource and its availability in the environment may suggest that a consumer prefers the food item and indicates unquestionably that a consumer tracks the resource.

Correlations have the advantage of being reported widely in the literature and amenable to further analysis. The interpretation of correlations with respect to preferences is not always as straightforward as the case reported by Peres (1994b): feeding time at Urucu by *Cebus apella* on ripe fruit correlated with availability; in contrast, feeding time on young seeds and unripe fruit did not correlate with availability. Peres supported his interpretation that *Cebus* prefer ripe fruit with data on the high annual consumption of fruit and monthly values reaching 90% of diet. However, poor correlations between feeding time and availability do not, by themselves, necessarily indicate that the food is non-preferred. Correlations may be absent because the food type, being eaten in high proportion despite low availability, is highly selected. For example, neither monthly feeding time on whole fruit nor seeds by *Propithecus diadema edwardsi* correlated with overall fruit availability. Yet sifaka clearly preferred certain seeds; they resorted to feeding on fallen fruit for several months after trees of four species were exhausted, and they sought out the few remaining fruiting trees of a top-ranked seed source after the population was generally no longer productive (Hemingway 1996, 1998). In years when preferred seeds were not available, sifaka turned to liana leaves. As neither fallen fruit nor lianas were part of the routine phenological sampling regime, the sifaka study also illustrates that measures may be too crude to detect a relationship between availability and use.

No correlations may also be found if the food type, being too unpredictable, is non-trackable or, being ubiquitous, does not require tracking. Comparing significant correlation values reported in case studies of *Alouatta palliata* conducted in the relatively drier sites of La Pacifica, Costa Rica, and Barro Colorado Island, Panama, with her results at the relatively aseasonal forest of La Selva, Costa Rica, Stoner (1996) attributed the absence of relationships between availability and use of items at La Selva to a more stable and diverse food supply.

Ranging changes discussed earlier as responses to seasonality indicated clearly that some primates track resources spatially. The caveats above aside, we also considered how primates track resources temporally by analyzing correlation values reported by authors and CV food availability for the study sites reported by van Schaik and Pfannes (Chapter 2). The tracking of fruit resources increased as fruiting cycles increased in variability ($r^2 = 0.338$, $n = 26$, $P = 0.02$). Interestingly, although the studies seem to group into a set of low correlation values and high correlation values, diet did not explain the pattern: resource tracking did not increase as fruit and/or seeds dominated the diet ($r^2 = 0.025$, $n = 37$, not significant). Where fruit is highly variable, it is tracked by frugivores and folivore-frugivores alike. In this sample, the degree of new leaf tracking was not related to new leaf seasonality ($r^2 = 0.092$, $n = 9$, not significant). Taken together, these results suggest that many primates, almost regardless of diet, track fruit availability.

Summary, future directions, and implications

The problem of seasonal resource scarcity has numerous possible behavioral and physiological solutions. Rather than following the food source or altering energetic requirements, most primates respond to scarcity by changing their diets. This finding, which essentially uses perhaps the best-researched mammalian order as a test case, reinforces the contention that dietary switching is the most common response to scarcity among vertebrates (van Schaik *et al.* 1993). In addition, our analysis of dietary flexibility provides support for the long-held view that primates are dietary generalists (Harding 1981; Garber 1987; Janson & Chapman 1999). Within a group as diverse as the primates, whose members inhabit tropical to temperate areas in four biogeographic zones and span a range of body sizes from mouse lemur to gorilla, ranging and dietary responses were expected to vary. Our examination of the primate literature yields structure to the data at primate-wide, regional, and clade scales, as discussed earlier and summarized below.

Ranging flexibility

Range shifting, as we expected, involves habitat shifting and occurs where the average home range does not contain all kinds of habitat present in the landscape. It is uncommon among small-bodied primates, but it is recorded in *Saimiri* (Peres 1994b) and *Cebuella* (Soini 1982, 1993), which have a body weight of less than 1 kg. Range-shifters generally have multimale– multifemale social systems and non-defended home ranges. Habitat shifting is correlated with dietary switching in all cases in which the destination habitat is known. Habitat shifting encompasses a variety of ranging patterns but generally may be characterized as temporary shifts across habitats. If vegetation patterns change due to global warming, as is projected, then permanent shifts out of an area are likely to be noted as primates track habitats and the resources contained therein.

Dietary flexibility

The degree to which overall diet changes across months decreases as the number of species in the diet increases. Incorporating many food species into the diet on a regular basis may offer some buffer against needing to switch dramatically between food types in response to seasonality. As dietary breadth increases with body mass, this buffering effect is not available equally to all primates. Large-bodied primates may have more diverse diets than small-bodied primates because their larger oral cavities and digestive systems permit them to extract nutrients from a greater variety of difficult-to-access or -digest foods and/or their larger home ranges mean they encounter more potential foods.

Ignoring the difficulties of identifying cause and effect, consuming a broad set of foods on a regular basis might provide some buffer against periods in which consumer requirements exceed local production; if one element of the food set is scarce, then one of many substitutes might be available. In the face of food scarcity, the alternative resources used by primates run the gamut of possible food types, but difficult-to-access or -digest resources are common. At present, it remains an open question whether dietary flexibility in general or reliance on particular keystone resources is more important in seeing primates through resource bottlenecks.

Several of our findings on dietary flexibility do not meet expectations with respect to body size, morphology, and food type. First, given their large body size and capacity to digest fibrous foods, the apes continue to rely on fruit during periods of scarcity far more than expected, and more than their monkey relatives. Second, although the Colobinae are expected to maintain dietary stability, particularly in fruit consumption, to accommodate pH-level requirements, the colobines do not differ in their mean variability in fruit use across months compared with indrids and atelines. Instead, colobines exhibit lower variability in mature leaf use. Across primates in general, leaf feeding is most variable among primates who feed primarily on fruit and use leaves only as a backup. The final exception of note is that variability in seed use is higher and fruit use is lower than predicted on the basis of their availability. Variability in fruit feeding is surprisingly consistently low across most primates, regardless of their body size or geographic region.

This finding supports the generalization that many primates prefer fruit and that diet differentiation, seen at both ecological and evolutionary timescales, is more pronounced in terms of non-preferred foods. Fruit feeding is particularly consistent in those clades with well-noted fruit preferences (e.g. Cebinae, Hylobatidae, Pongidae). Most primates consume some fruit resources throughout the year; however, few species manage to maintain a fruit-dominated diet during periods of fruit scarcity.

Behavioral flexibility among biogeographic regions

In regional contrasts, responses reported for primates in Madagascar, the New World, and, to a lesser extent, Asia stand out in keeping with the regional particularities. Primates in Madagascar exhibit the majority of physiological responses (see Chapter 5) and often decrease day range area in response to seasonality. Relationships between feeding variability and climatic variables and phenological patterns are evident most clearly in Madagascar, which falls neatly within traditional views that the severe food and thermal stresses that Malagasy lemurs experience require dramatic responses (Morland 1993; Wright 1999; Atsalis 1999).

Primates in the New World appear to have little ability to increase their dietary diversity in response to seasonality and, in contrast to those in other regions, do not switch to mature leaves or other vegetative matter. Balancing their relative inflexibility in diet, New World monkeys appear more flexible in home range area use and shift habitats compared with primates in other regions. The Amazonian waterways play an important role in creating habitat heterogeneity and permitting habitat shifting. Although habitat shifting may appear to be primarily a New World phenomenon, additional reports are likely in other biogeographic regions, perhaps among the temperate-living Asian colobines that form large, socially flexible bands (Kirkpatrick 1999). Finally, Asian primates exhibit two dietary features, absence of seeds as a fallback food and consistency in mature leaf use, that appear related to boom and bust phenological patterns unique to the region.

Response flexibility and resource variability

The relationship between increasing seasonality and primate response is complex. Our initial prediction was that primates in more seasonal environments would exhibit greater dietary variability than taxa in less seasonal environments. However, this prediction does not hold across the board. Overall primate diet variability is related to latitude but not to other standard climatic variables indicating seasonality. Likewise, in the only prior review, Chapman and Chapman (1990) uncovered no significant effect of habitat seasonality, as measured by CV rainfall, on dietary variability. We found, however, that particular dietary features vary with certain measures of seasonality. Across the primates, fruit use is more variable with increasing latitude, and fruit resources are tracked more closely at sites with high variation in fruit availability. Oddly, fruit use is not more flexible across months at sites with high variation in fruit availability; perhaps a strong preference for fruit, no matter its level of availability, explains this pattern. Variability in leaf use, particularly in Africa and Madagascar, appears strongly tied to the length of the dry season.

In relatively aseasonal forests, new leaves, nectar, exudates, and seeds are at their highest levels as fallback foods. In the most extremely seasonal environments, primates decrease day range, exhibit physiological responses, increase the number of species eaten, and include primarily mature leaves and other vegetative matter along with new leaves and exudate in their diets. Suites of behavioral responses are well documented in case studies. In particular, increases in leaf eating, increases in dietary diversity, and decreases in time spent moving often go hand in hand (e.g. *Macaca fuscata yakui* [Hill 1997]).

Where do we go from here?

Exciting directions for future work include greater attention to variation in responses within clades and additional interacting variables not captured by our analysis, such as food quality and physiological flexibility. Within the colobines, members of *Trachypithecus* and *Presbytis* differ in gut size (Chivers 1994) and perhaps in the degree of feeding on mature leaves (Yeager & Kool 2000); dietary flexibility might accompany those differences. Lambert (1998) suggested that researchers should look for the possibility that primates that switch to lower-quality foods also exhibit some physiological adjustment in intestinal morphology or gut transit time, as seen in other taxa. It is tempting to suggest new sorts of data to collect for refined, small-scale comparative studies. However, the data presented in this review, which represent an enormous research effort by numerous fieldworkers, illustrate that several critical gaps in our knowledge of primate dietary and ranging variability and responses to scarcity remain.

From a practical perspective, our review highlights difficulties in choosing appropriate variables to represent seasonality and identifying the appropriate scale for answering the question of whether increased seasonality mandates increased dietary and ranging flexibility. The effects of spatial and temporal resource heterogeneity, the essence of ecology (MacArthur & Pianka 1966; Wiens 1976), are notoriously complicated to study (Naeem & Colwell 1991). Further work is needed to capture the full gradient of habitats and seasonality that primate taxa experience and to uncover the basis of geographic differences in the relationships between seasonality in the environment and variability in feeding.

Students of folivorous primates have long maintained that regardless of dietary components during good times of the year, the ability to subsist on relatively low-quality leaves during food scarcity defines a folivorous foraging strategy (Clutton-Brock 1977; Milton 1980; Stanford 1991). It appears too simplistic, however, to say that regardless of what is eaten during bad times, the ability to benefit substantially during fruit abundance defines a frugivorous foraging strategy. For example, fruits eaten by masked titi monkeys in periods of abundance and periods of scarcity generally are nutritionally similar (Heiduck 1997).

Taking advantage of good times may be particularly characteristic of the more frugivorous great apes, *Pan* and *Pongo*. During periods of fruit abundance, chimpanzees but not cercopithecines at Kibale, Uganda, reduce their fiber and increase carbohydrate content by consuming a relatively higher proportion of ripe fruit (Wrangham *et al.* 1998; Conklin-Brittain *et al.* 1998). During relative fruit scarcity, chimpanzees, gorillas, and mangebeys at Bai Hokou, central Africa, consume nutritionally similar fruits (Remis 2000). Leighton (1993) characterized the orangutan feeding strategy as alternating between feast and famine, building up fat reserves during periods of fruit abundance in order to draw on them during periods of scarcity in which the available resources cannot meet their energetic needs. Knott's (1998) data (see also Chapter 12) support this interpretation for Bornean orangutans. We encountered several comments, primarily with anecdotal support, that primates across various clades put on weight and store fat during times of resource abundance (*Perodicticus*: Oates [1984]; *Pongo*: Leighton [1993]; Knott [1998];

Lagothrix: Di Fiore & Rodman [2001]). Certainly, the behavioral, nutritional, and physiological responses to periods of food abundance and their relation to periods of food scarcity deserve further study.

Perhaps the most notable omission of this review is discussion of the conservation implications of primate responses to scarcity, a thorough consideration of which is beyond the scope of this study. Questions awaiting testing include: (i) does reliance on keystones resources increase in more seasonal environments (Kay *et al.* 1997)?; (ii) relative to species-poor forests, do species-rich forests provide a buffer against periods of food scarcity by containing more alternative foods (Chapman *et al.* 1999)?; and (iii) does lower plant diversity in forest fragments result in greater food shortages than in intact forests (Tutin *et al.* 1997)? Deforestation, forest fragmentation, and global warming all threaten to dramatically alter relationships between consumers and their resources. Our review of behavioral responses to food scarcity focused on short-term, regular environmental changes, i.e. annual resource bottlenecks of average duration and intensity. We know very little about how primates respond to large-scale and long-term alteration to food availability regimes. Wright *et al.* (1999) provided evidence that El Niño-Southern Oscillation cycles are related in a complex manner to famine years for vertebrate consumers, including primates, in Panama. Foods used during annual periods of scarcity often differ from those eaten in dire famine situations. Understanding vertebrate responses to food scarcity ultimately requires detailed information over a long timescale.

Implications for hominin evolution

Environmental change and its effect on resource distribution and availability during the Plio-Pleistocene Epoch figure prominently in hominin evolution (Foley 1994; Aiello & Wheeler 1995; Isbell & Young 1996). One lesson from our study for hominin evolution, regardless of the preferred scenario, is that primates within any community solve the problem of seasonal food scarcity in diverse, multifaceted ways. The various hominin species likely diverged in the details of how they responded to seasonal food scarcity.

Nonetheless, if current primate responses to seasonality are any indication, then a relatively large-bodied hominin might decrease day range and switch to a diet heavy in vegetative matter or alternatively switch to a habitat where more preferred food is available. The continued reliance by some great apes on fruit during scarcity suggests that hominins may have tried to maintain dietary quality during scarcity. Primates with relatively large brains also might be buffered against periods of scarcity, due to efficient (e.g. spatial memory use by *Cebus* [Janson 1998]) or innovative ways of acquiring alternative foods (e.g. extractive foraging and tool use by *Cebus* [Panger 1998]; *Pan* [McGrew 1992; Yamakoshi 1998]; *Pongo* [Fox *et al.* 2004]). This suggests that hominins may have been able to maintain dietary quality during scarcity through these specialized responses. Behavioral flexibility is surely a contributing factor to the "success" of primates, enabling them to adapt to wide ranging and changing conditions.

Acknowledgments

We thank Carlos Peres for constructive comments, Carel van Schaik for generously sharing data and exchanging ideas, and Carel and Diane Brockman for their perseverance. Chris Birkinshaw and Simon Malcomber provided comments on early incarnations of the manuscript.

References

Aiello, L. C. & Wheeler, P. (1995). The expensive-tissue hypothesis: the brain and digestive system in human and primate evolution. *Current Anthropology*, **36**, 199–221.

Altmann, J. A., Altmann, S. A., Hausfater, G., & McCluskey, S. S. (1977). Life history of yellow baboons: infant mortality, physical development, and reproductive parameters. *Primates*, **18**, 315–30.

Andrews, J. R. & Birkinshaw, C. R. (1998). A comparison between the daytime and night-time diet, activity and feeding height of the black lemur, *Eulemur macaco* (Primates:Lemuridae), in Lokobe Forest, Madagascar. *Folia Primatologica*, **69**, 175–82.

Ashton, P. S., Givnish, T. J., & Appanah, S. (1988). Staggered flowering in the Dipterocarpaceae: new insights into floral induction and the evolution of mast fruiting in the aseasonal tropics. *American Naturalist*, **132**, 44–66.

Atsalis, S. (1999). Diet of the brown mouse lemur (*Microcebus rufus*) in Ranomafana National Park, Madagascar. *International Journal of Primatology*, **20**, 193–229.

Ayres, J. M. (1989). Comparative feeding ecology of the uakari and bearded saki, *Cacajao* and *Chiropotes*. *Journal of Human Evolution*, **18**, 697–716.

Barton, R. A., Whiten, A., Strum, S. C., Byrne, R. W., & Simpson, A. J. (1992). Habitat use and resource availability in baboons. *Animal Behaviour*, **43**, 831–44.

Barton, R. A., Whiten, A., Byrne, R. W., & English, M. (1993). Chemical composition of baboon plant foods: implications for the interpretation of intra- and interspecific differences in diet. *Folia Primatologica*, **61**, 1–20.

Baum, D. A. (1996). The ecology and conservation of the baobabs of Madagascar. *Primate Report*, **46**, 311–27.

Beeson, M. (1989). Seasonal dietary stress in a forest monkey (*C. mitis*). *Oecologia*, **78**, 565–70.

Bodmer, R. E. (1990). Responses of ungulates to seasonal inundations in the Amazon floodplains. *Journal of Tropical Ecology*, **6**, 191–201.

Boinski, S. (1987). Habitat use by squirrel monkeys in a tropical lowland forest. *Folia Primatologica*, **49**, 151–67.

Boinski, S. & Fowler, N. (1989). Seasonal patterns in a tropical lowland forest. *Biotropica*, **21**, 223–33.

Brown, A. D. & Zunino, G. E. (1990). Dietary variability in *Cebus apella* in extreme habitats: evidence for adaptability. *Folia Primatologica*, **54**, 187–95.

Calder, W. A., III (1984). *Size, Function and Life History*, Cambridge, MA: Harvard University Press.

Caton, J. M., Hill, D. M., Hume, I. D., & Crook, G. A. (1996). The digestive strategy of the common marmoset, *Callithrix jacchus*. *Comparative Biochemistry and Physiology*, **114A**, 1–8.

Chapman, C. A. & Chapman, L. J. (1990). Dietary variability in primate populations. *Primates*, **31**, 121–8.

Chapman, C. A., Wrangham, R. W., Chapman, L. J., Kennard, D. K., & Zanne, A. E. (1999). Fruit and flower phenology at two sites in Kibale National Park, Uganda. *Journal of Tropical Ecology*, **15**, 189–211.

Charnov, E. L. (1976). Optimal foraging: the marginal value theorem. *Theoretical Population Biology*, **9**, 129–36.

Chesson, J. (1978). Measuring preference in selective predation. *Ecology*, **59**, 211–15.

Chivers, D. J. (1994). Functional anatomy of the gastrointestinal tract. In *Colobine Monkeys: Their Ecology, Behavior and Evolution*, ed. G. Davies & J. Oates. Cambridge: Cambridge University Press, pp. 205–28.

Clutton-Brock, T. H. (1977). Some aspects of intraspecific variation in feeding and ranging behavior in primates. In *Primate Ecology*, ed. T. H. Clutton-Brock. New York: Academic Press, pp. 539–56.

Coehlo, A. M., Bramblett, C. A., Quick, L. B., & Bramblett, S. S. (1976). Resource availability and population density in primates: a sociobioenergetic analysis of the energy budgets of Guatemalan howler and spider monkeys. *Primates*, **17**, 63–80.

Coley, P. D. & Barone, J. A. (1996). Herbivory and plant defenses in tropical forests. *Annual Review of Ecology and Systematics*, **27**, 305–35.

Conklin-Brittain, N. L., Wrangham, R. W., & Hunt, K. D. (1998). Dietary response of chimpanzee and Cercopithecines to seasonal variation in fruit abundance. II. Macronutrients. *International Journal of Primatology*, **19**, 971–98.

Corbin, G. D. & Schmid, J. (1995). Insect secretions determine habitat use patterns by a female lesser mouse lemur (*Microcebus murinus*). *American Journal of Primatology*, **37**, 317–24.

Cork, S. J. & Foley, W. J. (1991). Digestive and metabolic strategies of arboreal mammalian folivores in relation to chemical defenses in temperate and tropical forests. In *Plant Defenses Against Mammalian Herbivory*, ed. D. J. Chivers & P. Langer. Boca Raton, FL: CRC Press, pp. 133–66.

Cortlett, R. T. & LaFrankie, J. V. (1998). Potential impacts of climate change on tropical Asian forests through an influence on phenology. *Climate Change*, **39**, 439–53.

Curran, L. M. & Leighton, M. (2000). Vertebrate responses to spatiotemporal variance in seed production of mast-fruiting Dipterocarpaceae. *Ecological Monographs*, **70**, 101–28.

Davies, G. (1991). Seed-eating by red leaf monkeys (*Presbytis rubicunda*) in dipterocarp forest of northern Borneo. *International Journal of Primatology*, **12**, 119–44.

Defler, T. R. (1996). Aspects of the ranging pattern in a group of wild woolly monkeys (*Lagothrix lagotricha*). *American Journal of Primatology*, **38**, 289–302.

Defler, T. R. & Defler, S. B. (1996). Diet of a group of *Lagothrix lagotricha lagotricha* in southeastern Colombia. *International Journal of Primatology*, **17**, 161–90.

Delgado, R. A. & van Schaik, C. P. (2000). The behavioral ecology and conservation of the orangutan (*Pongo pygmaeus*): a tale of two islands. *Evolutionary Anthropology*, **9**, 201–18.

Di Fiore, A. & Rodman, P. S. (2001). Time allocation patterns of lowland woolly monkeys (*Lagothrix lagotricha poeppigii*) in a neotropical *terra firme* forest. *International Journal of Primatology*, **22**, 449–80.

Dittus, W. P. J. (1988). Group fission among wild toque macaques as a consequence of female resource competition and environmental stress. *Animal Behaviour*, **36**, 1626–45.

Doran, D. (1997). Influence of seasonality on activity patterns, feeding behavior, ranging, and grouping patterns in Tai chimpanzees. *International Journal of Primatology*, **18**, 183–206.

Fashing, P. J. (2001). Feeding ecology of guerezas in the Kakamega forest, Kenya: the importance of Moraceae in their diet. *International Journal of Primatology*, **22**, 579–609.

Fenner, M. (1998). The phenology of growth and reproduction in plants. *Perspectives in Plant Ecology, Evolution and Systematics*, **1**, 78–91.

Fleagle, J. G. (1999). *Primate Adaptation and Evolution*, 2nd edn. San Diego: Academic Press.

Fleagle, J. G., Janson, C., & Reed, K. E. (1999). *Primate Communities*. Cambridge: Cambridge University Press.

Fleming, T. H. & Heithaus, E. R. (1986). Seasonal foraging behavior of the frugivorous bat *Carollia perspicillata*. *Journal of Mammalogy*, **67**, 660–71.

Fleming, T. H., Breitwisch, R., & Whitesides, G. H. (1987). Patterns of tropical vertebrate frugivore diversity. *Annual Review of Ecology and Systematics*, **18**, 91–109.

Foley, R. A. (1994). Speciation, extinction and climate change in hominid evolution. *Journal of Human Evolution*, **26**, 275–89.

Ford, S. M. & Davis, L. C. (1992). Systematics and body size: implications for feeding adaptations in New World monkeys. *American Journal of Physical Anthropology*, **88**, 415–68.

Foster, R. B. (1982). Famine on Barro Colorado Island. In *The Ecology of a Tropical Forest: Seasonal Rhythms and Long-Term Changes*, ed. E. G. J. Leigh, A. S. Rand, & D. M. Windsor. Washington, DC: Smithsonian Institution Press, pp. 151–72.

Fox, E. A., van Schaik, C. P., Sitompul, A., & Wright, D. N. (2004). Intra- and interpopulational differences in orangutan (*Pongo pygmaeus*) activity and diet: implications for the invention of tool use. *American Journal of Physical Anthropology*, **125**, 162–74.

Freeland, W. J. & Janzen, D. H. (1974). Strategies in herbivory by mammals: the role of plant secondary compounds. *American Naturalist*, **108**, 269–89.

Furuichi, T., Hashimoto, C., & Tashiro, Y. (2001). Fruit availability and habitat use by chimpanzees in the Kalinzu Forest, Uganda: examination of fallback foods. *International Journal of Primatology*, **22**, 929–46.

Ganzhorn, J. U. (1992). Leaf chemistry and the biomass of folivorous primates in tropical forests: test of a hypothesis. *Oecologia*, **91**, 540–47.

Garber, P. A. (1987). Foraging strategies of living primates. *Annual Review of Anthropology*, **16**, 339–64.

— (1992). Vertical clinging, small body size, and the evolution of feeding adaptations in the callitrichinae. *American Journal of Physical Anthropology*, **88**, 499–514.

— (1993). Seasonal patterns of diet and range in two species of tamarin monkeys: stability versus variability. *International Journal of Primatology*, **14**, 145–66.

Gautier-Hion, A. (1980). Seasonal variations of diet related to species and sex in a community of *Cercopithecus* monkeys. *Journal of Animal Ecology*, **49**, 237–69.

Gautier-Hion, A. & Michaloud, G. (1989). Are figs always keystone resources for tropical frugivorous vertebrates? A test in Gabon. *Ecology*, **70**, 1826–33.

Glander, K. E. (1982). The impact of plant secondary compounds on primate feeding behavior. *Yearbook of Physical Anthropology*, **25**, 1–18.

Goldizen, A. W., Terborgh, J., Cornejo, F., Porras, D. T., & Evans, R. (1988). Seasonal food shortages, weight loss, and the timing of births in saddle-backed tamarins (*Saguinus fuscicollis*). *Journal of Animal Ecology*, **57**, 893–902.

Goldstein, S. J. & Richard, A. F. (1989). Ecology of rhesus macaques (*Macaca mulatta*) in northwest Pakistan. *International Journal of Primatology*, **10**, 531–67.

Goodman, S. M. & Ganzhorn, J. U. (1997). Rarity of figs (*Ficus*) on Madagascar and its relationship to a depauperate frugivore community. *Revue d'Ecologie*, **52**, 321–9.

Gould, L., Sussman, R. W., & Sauther, M. L. (1999). Natural disasters and primate populations: the effects of a 2-year drought on a naturally occurring population of ring-tailed lemurs (*Lemur catta*) in southwestern Madagascar. *International Journal of Primatology*, **20**, 69–84.

Gursky, S. (2000). Effect of seasonality on the behavior of an insectivorous primate, *Tarsius spectrum*. *International Journal of Primatology*, **21**, 477–95.

Hamilton, W. J. I. (1985). Demographic consequences of a food and water shortage to desert chacma baboons, *Papio ursinus*. *International Journal of Primatology*, **6**, 451–62.

Harding, R. S. O. (1981). An order of omnivores; nonhuman primate diets in the wild. In *Omnivorous Primates: Gathering and Hunting in Human Evolution*, ed. R. S. O. Harding & G. Teleki. New York: Columbia University Press, pp. 191–214.

Harrison, M. J. S. (1984). Optimal foraging strategies in the diet of the green monkey, *Cercopithecus sabaeus*, at Mt. Assirik, Senegal. *International Journal of Primatology*, **5**, 435–72.

Heiduck, S. (1997). Food choice in masked titi monkeys (*Callicebus personatus melanochir*): selectivity or opportunism? *International Journal of Primatology*, **18**, 487–502.

Hemingway, C. A. (1996). Morphology and phenology of seeds and whole fruit eaten by Milne-Edwards' sifaka, *Propithecus diadema edwardsi*, in Ranomafana National Park, Madagascar. *International Journal of Primatology*, **17**, 637–59.

— (1998). Selectivity and variability in the diet of Milne-Edwards' sifaka (*Propithecus diadema edwardsi*): implications for folivory and seed-eating. *International Journal of Primatology*, **19**, 355–77.

Hill, D. A. (1997). Seasonal variation in the feeding behavior and diet of Japanese macaques (*Macaca fuscata yakui*) in lowland forest of Yakushima. *American Journal of Primatology*, **43**, 305–22.

Howe, H. F. (1977). Bird activity and seed dispersal of a tropical wet forest tree. *Ecology*, **58**, 539–50.

Hunt, J. H., Brodie, R. J., Carithers, T. P., Goldstein, P. Z., & Janzen, D. H. (1999). Dry season migration by Costa Rican lowland paper wasps to high elevation cold dormancy sites. *Biotropica*, **31**, 192–6.

Isbell, L. A. (1998). Diet for a small primate: insectivory and gummivory in the (large) patas monkey (*Erythrocebus patas pyrrhonotus*). *American Journal of Primatology*, **45**, 381–98.

Isbell, L. A. & Young, T. P. (1996). The evolution of bipedalism in hominids and reduced groups size in chimpanzees: alternative responses to decreasing resource availability. *Journal of Human Evolution*, **30**, 389–97.

Janson, C. H. (1985). Aggressive competition and individual food consumption in wild brown capuchin monkeys (*Cebus apella*). *Behavioral Ecology and Sociobiology*, **18**, 125–38.

— (1998). Experimental evidence for spatial memory in wild brown capuchin monkeys (*Cebus apella*). *Animal Behaviour*, **55**, 1129–43.

Janson, C. H. & Chapman, C. A. (1999). Resources and primate community structure. In *Primate Communities*, ed. J. G. Feagle, C. H. Janson, & K. E. Reed. Cambridge: Cambridge University Press, pp. 237–67.

Janson, C. H. & Emmons, L. H. (1990). Ecological structure of the non-flying mammal community at the Cocha Cashu Biological Station, Manu National Park, Peru. In *Four Neotropical Rain Forests*, ed. A. Gentry. New Haven: Yale University Press, pp. 314–38.

Janzen, D. H. (1973). Sweep samples of tropical foliage insects: effects of seasons, vegetation types, elevation, time of day, and insularity. *Ecology*, **54**, 687–708.

Jolly, A. (1972). *The Evolution of Primate Behavior*, 2nd edn. New York: Macmillan.

Jordano, P. (2000). Fruits and Frugivory. In *Seeds: The Ecology of Regeneration in Plant Communities*, ed. M Fenner. London: CAB International Publishing, pp. 125–65.

Kannan, R. & James, D. A. (1999). Fruiting phenology and the conservation of the great pied hornbill (*Buceros bicornis*) in the Western Ghats of Southern India. *Biotropica*, **31**, 167–77.

Kaplin, B. A., Munyaligoga, V., & Moermond, T. C. (1998). The influence of temporal changes in fruit availability on diet composition and seed handling in blue monkeys (*Cercopithecus mitis doggetti*). *Biotropica*, **30**, 56–71.

Karr, J. R. (1976). Seasonality, resource availability, and community diversity in tropical bird communities. *American Naturalist*, **110**, 973–94.

Kay, R. F. (1984). On the use of anatomical features to infer foraging behavior in extinct primates. In *Adaptations for Foraging in Nonhuman Primates: Contributions to an Organismal Biology of Prosimians, Monkeys, and Apes*, ed. P. S. Rodman & J. G. H. Cant. New York: Columbia University Press, pp. 21–53.

Kay, R. F., Madden, R. H., van Schaik, C. P., & Higdon, D. (1997). Primate species richness is determined by plant productivity: Implications for conservation. *Proceedings of the National Academy of Sciences, USA*, **94**, 13023–7.

Kay, R. N. & Davies, A. G. (1994). Digestive physiology. In *Colobine Monkeys: Their Ecology Behavior and Evolution*, ed. A. G. Davies & J. F. Oates. Cambridge: Cambridge University Press, pp. 229–59.

Kiltie, R. A. & Terborgh, J. (1983). Observations on the behavior of rain forest peccaries in Peru: why do white-lipped peccaries form herds. *Zeitschrift Fur Tierpsychologie*, **62**, 241–55.

Kinnaird, M. F., O'Brien, T. G., & Suryadi, S. (1992). The importance of figs to Sulawesi's imperiled wildlife. *Tropical Biodiversity*, **6**, 5–18.

— (1996). Population fluctuation in Sulawesi red-knobbed hornbills: tracking figs in space and time. *Auk*, **113**, 431–40.

Kinzey, W. G. (1977). Diet and feeding behavior of *Callicebus torquatus*. In *Primate Ecology: Studies of Feeding and Ranging Behaviour in Lemurs, Monkeys, and Apes*, ed. T. H. Clutton-Brock. London: Academic Press, pp. 127–51.

— (1992). Dietary and dental adaptations in the Pitheciinae. *American Journal of Physical Anthropology*, **88**, 499–514.

Kirkpatrick, R. C. (1999). Ecology and behavior in snub-nosed and douc langurs. In *The Natural History of the Doucs and Snub-nosed Monkeys*, ed. N. Jablonski. Singapore: World Scientific Publishing, pp. 155–98.

Kirkpatrick, R. C., Gu, H. J., & Zhou, X. P. (1999). A preliminary report on Sichuan snub-nosed monkeys (*Rhinopithecus roxellana*) at Baihe Nature Reserve. *Folia Primatologica*, **70**, 117–20.

Kleiber, M. (1961). *The Fire of Life: An Introduction to Animal Energetics*. New York: John Wiley & Sons.

Knott, C. D. (1998). Changes in orangutan caloric intake, energy balance, and ketones in response to fluctuating fruit availability. *International Journal of Primatology*, **19**, 1061–79.

Kubitzki, K. & Ziburski, A. (1994). Seed dispersal in floodplain forests of Amazonia. *Biotropica*, **26**, 30–43.

Kuroda, S., Suzuki, S., & Nishihara, T. (1996). Preliminary report on predatory behavior and meat sharing in Ttschego chimpanzees (*Pan troglodytes troglodytes*) in the Ndoki Forest, northern Congo. *Primates*, **37**, 253–9.

Lambert, J. E. (1998). Primate digestion: interactions among anatomy, physiology, and feeding ecology. *Evolutionary Anthropology*, **7**, 8–20.

Lambert, F. R. & Marshall, A. G. (1991). Keystone characteristics of bird-dispersed *Ficus* in a Malaysian lowland rain forest. *Journal of Ecology*, **79**, 8–20.

Law, B. S. (1993). Roosting and foraging ecology of the Queensland blossom-bat (*Syconycteris australis*) in New South Wales: flexibility in response to seasonal variation. *Wildlife Research*, **20**, 419–31.

Leighton, M. (1993). Modeling dietary selectivity by Bornean orangutans: evidence for integration of multiple criteria in fruit selection. *International Journal of Primatology*, **14**, 257–313.

Leighton, M. & Leighton, D. R. (1983). Vertebrate responses to fruiting seasonality within a Bornean rain forest. In *Tropical Rain Forest: Ecology and Management*, ed. S. L. Sutton, T. C. Whitmore, & A. C. Chadwick. London: Blackwell Scientific, pp. 181–96.

Levey, D. J. (1990). Habitat-dependent fruiting behaviour of an understory tree, *Miconia cetrodesma*, and tropical treefall gaps as keystone habitats for frugivores in Costa Rica. *Journal of Tropical Ecology*, **6**, 409–20.

Li, Y. (2001). The seasonal diet of the Sichuan snub-nosed monkey (*Pygathrix roxellana*) in Shennongjia Nature Reserve, China. *Folia Primatologica*, **72**, 40–43.

Loiselle, B. A. & Blake, J. G. (1991). Temporal variation in birds and fruits along an elevational gradient in Costa Rica. *Ecology*, **72**, 180–93.

Lucas, P. W. & Cortlett, R. W. (1991). Relationships between the diet of *Macaca fascicularis* and forest phenology. *Folia Primatologica*, **57**, 201–15.

MacArthur, R. H. & Pianka, E. R. (1966). On the optimal use of a patchy environment. *American Naturalist*, **100**, 603–9.

McConkey, K. M., Aldy, F., & Chivers, D. J. (2002). Selection of fruit by gibbons (*Hylobates muelleri* × *agilis*) in the rain forests of Central Borneo. *International Journal of Primatology*, **23**, 123–45.

McGrew, W. C. (1992). *Chimpanzee Material Culture: Implications for Human Evolution*. Cambridge: Cambridge University Press.

McGrew, W. C., Baldwin, P. J., & Tutin, C. E. G. (1988). Diet of wild chimpanzees (*Pan troglodytes verus*) at Mt. Assirik, Senegal: I. Composition. *American Journal of Primatology*, **16**, 213–26.

McNaughton, S. J. (1990). Mineral nutrition and seasonal movements of African migratory ungulates. *Nature*, **345**, 613–15.

Milton, K. (1980). *The Foraging Strategy of Howler Monkeys*. New York: Columbia University Press.

— (1990). Annual mortality patterns of a mammal community in central Panama. *Journal of Tropical Ecology*, **6**, 493–9.

Moraes, P. L. R., Carvalho, O., Jr, & Strier, K. B. (1998). Population variation in patch and party size in muriquis (*Brachyteles arachnoides*). *International Journal of Primatology*, **19**, 325–37.

Morland, H. S. (1993). Seasonal behavioral variation and its relationship to thermoregulation in ruffed lemurs (*Varecia variegata variegata*). In *Lemur Social Systems and Their Ecological Basis*, ed. P. M. Kappeler & J. U. Ganzhorn. New York: Plenum Press, pp. 193–203.

Murdoch, W. W., Avery, S., & Smith, M. E. B. (1975). Switching in predatory fish. *Ecology*, **56**, 1094–105.

Mutschler, T., Feistner, A. T. C., & Nievergelt, C. M. (1998). Preliminary field data on group size, diet and activity in the Alaotran gentle lemur, *Hapalemur griseus alaotrensis*. *Folia Primatologica*, **69**, 325–30.

Naeem, S. & Colwell, R. K. (1991). Ecological consequences of heterogeneity of consumable resources. In *Ecological Studies 86: Ecological Heterogeneity*, ed. J. Kolasa & S. T. A. Pickett. New York: Springer-Verlag, pp. 224–55.

Nagy, K. A. & Milton, K. (1979). Energy metabolism and food consumption by wild howler monkeys. *Ecology*, **60**, 475–80.

Nash, L. (1998). Vertical clingers and sleepers: seasonal influences on the activities and substrate use of *Lepilemur leucopus* at Beza Mahafaly Special Reserve, Madagascar. *Folia Primatology*, **69**, 204–17.

Nason, J. D., Herre, E. A., & Hamrick,, J. L. (1998). The breeding structure of a tropical keystone plant resource. *Nature*, **391**, 685–7.

Norconk, M. A. (1996). Seasonal variation in the diets of white-faced and bearded sakis (*Pithecia pithecia* and *Chiropotes satanas*) in Guri Lake, Venezuela. In *Adaptive Radiations of Neotropical Primates*, ed. M. A. Norconk, A. L. Rosenberger, & P. A. Garber. New York: Plenum Press, pp. 403–23.

Oates, J. F. (1977). The guereza and its food. In *Primate Ecology*, ed. T. H. Clutton-Brock, London: Academic Press, pp. 276–321.

— (1984). The niche of the potto, *Perodicticus potto*. *International Journal of Primatology*, **5**, 51–61.

— (1987). Food distribution and foraging behavior. In *Primate Societies*, ed. B. B. Smuts, D. L. Cheney, R. M. Seyfarth, R. W. Wrangham, & T. T. Struhsaker. Chicago: University of Chicago Press, pp. 197–209.

— (1996). Habitat alteration, hunting and the conservation of folivorous primates in African forests. *Australian Journal of Ecology*, **21**, 1–9.

O'Brien, T. G. & Kinnaird, M. F. (1997). Behavior, diet, and movements of the Sulawesi crested black macaque (*Macaca nigra*). *International Journal of Primatology*, **18**, 321–51.

Olupot, W., Chapman, C. A., Waser, P. M., & Isabirye-Basuta, G. (1997). Mangabey (*Cercopithecus albigena*) ranging patterns in relation to fruit availability and the risk of parasite infection in Kibale National Park, Uganda. *American Journal of Primatology*, **43**, 65–78.

Opler, P. A., G. W. Frankie, & H. G. Baker. 1980. Comparative phenological studies of treelet and shrub species in tropical wet and dry forests in the lowlands of Costa Rica. *Journal of Ecology*, **68**, 167–88.

Overdorff, D. J. (1992). Differential patterns in flower feeding by *Eulemur fulvus rufus* and *Eulemur rubriventer* in Madagascar. *American Journal of Primatology*, **28**, 191–203.

— (1993). Similarities, differences and seasonal patterns in the diet of *Eulemur rubriventer* and *Eulemur fulvus rufus* in the Ranomafana National Park, Madagascar. *International Journal of Primatology*, **14**, 721–53.

Palacios, E. & Rodriguez, A. (2001). Ranging pattern and use of space in a group of red howler monkeys (*Alouatta seniculus*) in a southeastern Colombian rainforest. *American Journal of Primatology*, **55**, 233–51.

Panger, M. A. (1998). Object-use in free-ranging white-faced capuchins (*Cebus capuchinus*) in Costa Rica. *American Journal of Physical Anthropology*, **106**, 311–21.

Peres, C. A. (1994a). Diets and feeding ecology of gray woolly monkeys (*Lagothrix lagotricha cana*) in central Amazonia: comparison with other Atelines. *International Journal of Primatology*, **15**, 333–72.

— (1994b). Primate responses to phenological changes in an Amazonian terra firme forest. *Biotropica*, **26**, 98–112.

— (1996). Use of space, spatial group structure, and foraging group size of gray woolly monkeys (*Lagothrix lagotricha cana*) at Urucu, Brazil: a review of the Atelinae. In *Adaptive Radiation of Neotropical Primates*, ed. M. A. Norconk, A. L. Rosenberger, & P. A. Garber. New York: Plenum Press, pp. 467–88.

— (2000). Identifying keystone plant resources in tropical forests: the case of gums from *Parkia* pods. *Journal of Tropical Ecology*, **16**, 287–317.

Pond, C. M. (1978). Morphological aspects and the ecological and mechanical consequences of fat deposition in wild vertebrates. *Annual Review of Ecology and Systematics*, **9**, 519–70.

Porter, L. M. (2001). Dietary differences among sympatric Callitrichinae in northern Bolivia: *Callimico goeldii*, *Saguinus fuscicollis*, and *S. labiatus*. *International Journal of Primatology*, **22**, 961–92.

Poulson, J. R., Clark, C. J., & Smith, T. B. (2001). Seasonal variation in the feeding ecology of the grey-cheeked mangabey (*Lophocebus albigena*) in Cameroon. *American Journal of Primatology*, **54**, 91–105.

Power, M. E., Tilman, D., Estes, J., *et al.* (1996). Challenges in the quest of keystones. *Bioscience*, **46**, 609–20.

Randolph, J. C. & Cameron, G. N. (2001). Consequences of diet choice by a small generalist herbivore. *Ecological Monographs*, **71**, 117–36.

Rasmussen, M. A. (1999). Ecological influence on activity cycle in two cathemeral primates, the mongoose lemur (*Eulemur mongoz*) and the common brown lemur (*Eulemur fulvus fulvus*). Ph. D. thesis, Duke University.

Remis, M. J. (1997). Western lowland gorillas (*Gorilla gorilla gorilla*) as seasonal frugivores: use of variable resources. *American Journal of Primatology*, **43**, 87–109.

— (2000). Initial studies on the contributions of body size and gastrointestinal passage rates to dietary flexibility among gorillas. *American Journal of Physical Anthropology*, **112**, 171–80.

Richard, A. F. (1978). Variability in the feeding behavior of a Malagasy prosimian, *Propithecus verreauxi*: Lemuriformes. In *The Ecology of Arboreal Folivores*, ed. G. G. Montgomery. Washington, DC: Smithsonian Institution Press, pp. 519–33.

Robinson, J. T. (1986). Seasonal variation in use of time and space by the wedge-capped capuchin monkey, *Cebus olivaceus*: implications for foraging theory. *Smithsonian Contributions to Zoology*, **431**, 1–60.

Rogers, M. E., Tutin, C. E. G., Williamson, E. A., Parnell, R. J., & Voysey, B. C. (1994). Seasonal feeding on bark by gorillas, an unexpected keystone food? In *Current Primatology I: Ecology and Evolution*, ed. B. Thierry, J. E. Anderson, J. J. Roeder, & N. Herrera Schmidt. Strasbourg, France: Universitie Louis Pasteur, pp. 34–43.

Rosenberger, A. L. (1992). Evolution of feeding niches in the New World Monkeys. *American Journal of Physical Anthropology*, **88**, 525–62.

Rylands, A. B. (1993). The ecology of the lion tamarins, *Leontopithecus*: some intrageneric differences and comparisons with other callitrichids. In *Marmosets and Tamarins: Systematics, Behaviour, and Ecology*, ed. A. B. Rylands. Oxford: Oxford University Press, pp. 296–312.

Sauther, M. L. (1998). Interplay of phenology and reproduction in ring-tailed lemurs, implications for ring-tailed lemur conservation. *Folia Primatologica*, **69**, 309–20.

Schaefer, H. M., Schmidt, V., & Wesenberg, J. (2002). Vertical stratification and caloric content of the standing fruit crop in a tropical lowland forest. *Biotropica*, **34**, 244–53.

Schmidt-Nielson, K. (1975). Scaling in biology: the consequences of size. *Journal of Experimental Zoology*, **194**, 287–307.

Schoener, T. W. (1971). Theory of feeding strategies. *Annual Review of Ecology and Systematics*, **2**, 369–404.

Singleton, I. & van Schaik, C. P. (2001). Orangutan home range size and its determinants in a Sumatran swamp forest. *International Journal of Primatology*, **22**, 877–911.

Smith, R. J. & Jungers, W. L. (1997). Body mass in comparative primatology. *Journal of Human Evolution*, **32**, 523–59.

Smythe, N., Glanz, W. E., & Leigh, E. G., Jr (1982). Population regulation in some terrestrial frugivores. In *The Ecology of a Tropical Forest*, ed. E. G. Leigh, Jr, A. S. Rand, & D. M. Windsor. Washington, DC: Smithsonian Institution Press, pp. 227–38.

Soini, P. (1982). Ecology and population dynamics of the pygmy marmoset, *Cebulla pygmaea*. *Folia Primatologica*, **39**, 1–21.

— (1993). The ecology of the pygmy marmoset, *Cebuella pygmaea*: some comparisons with two sympatric tamarins. In *Marmosets and Tamarins: Systematics, Behaviour, and Ecology*, ed. A. B. Rylards. Oxford: Oxford University Press, pp. 257–61.

Stanford, C. B. (1991). The diet of the capped lemur (*Presbytis pileata*) in a moist deciduous forest in Bangladesh. *International Journal of Primatology*, **12**, 199–216.

Stevenson, P. R., Quiñones, M. J., & Ahumada, J. A. (1994). Ecological strategies of woolly monkeys (*Lagothrix lagotricha*) at Tinigua National Park, Colombia. *American Journal of Primatology*, **32**, 123–40.

Stoner, K. E. (1996). Habitat selection and seasonal patterns of activity and foraging of mantled howling monkeys (*Alouatta palliata*) in northeastern Costa Rica. *International Journal of Primatology*, **17**, 1–30.

Struhsaker, T. T. (1975). *The Red Colobus*. Chicago: University of Chicago Press.

Tan, C. (1999). Group composition, home range size, and diet of three sympatric bamboo lemur species (Genus *Hapalemur*) in Ranomafana National Park, Madagascar. *International Journal of Primatology*, **20**, 547–66.

Te Boekhorst, I. J. A., Schurmann, C. L., & Sugardjito, J. (1990). Residential status and seasonal movements of wild orang-utans in the Gunung Leuser Reserve (Sumatra, Indonesia). *Animal Behaviour*, **39**, 1098–109.

Terborgh, J. (1983). *Five New World Primates: A Study in Comparative Ecology*. Princeton, NJ: Princeton University Press.

— (1986). Keystone plant resources in the tropical forest. In *Conservation Biology: The Sciences of Scarcity and Diversity*, ed. M. E. Soule. Sunderland, MA: Sinauer Associates, pp. 330–44.

Terborgh, J. & van Schaik, C. P. (1987). Convergence and non-convergence in primate communities. In *Organization of Communities*, ed. J. H. R. Gee & P. S. Giller. Oxford: Blackwell Scientific Publications, pp. 205–36.

Tutin, C. E. G., Ham, R. M., White, L. T., & Harrison, M. J. S. (1997). The primate community of the Lope Reserve, Gabon: diets, responses to fruit scarcity, and effects on biomass. *American Journal of Primatology*, **42**, 1–24.

Tutin, C. E. G., White, L. J. T., & Mackanga-Missandzou, A. (1997). The use by rain forest mammals of natural forest fragments in an equatorial African savanna. *Conservation Biology*, **11**, 1190–203.

Van Schaik, C. P. (1986). Phenological changes in a Sumatran rain forest. *Journal of Tropical Ecology*, **2**, 327–47.

Van Schaik, C. P., Terborgh, J. W., & Wright, S. J. (1993). The phenology of tropical forests: adaptive significance and consequences for primary consumers. *Annual Review of Ecology and Systematics*, **24**, 353–77.

Waser, P. (1977). Feeding, ranging and group size in the mangabey *Cercocebus albigena*. In *Primate Ecology*, ed. T. H. Clutton-Brock. London: Academic Press, pp. 183–222.

Waterman, P. G., Ross, J. A. M., Bennett, E. L., & Davies, A. G. (1988). A comparison of the floristics and leaf chemistry of the tree flora in two Malaysian rain forests and the influence of leaf chemistry on populations of colobine monkeys in the Old World. *Biological Journal of Linnean Society*, **34**, 1–32.

Watson, D. M. (2001). Mistletoe: a keystone resource in forests and woodlands worldwide. *Annual Review of Ecology and Systematics*, **32**, 219–49.

Watts, D. P. (1998). Seasonality in the ecology and life histories of mountain gorillas (*Gorilla gorilla beringei*). *International Journal of Primatology*, **19**, 929–49.

Wendeln, M. C., Runkle, J. R., & Kalko, E. K. V. (2000). Nutritional values of 14 fig species and bat feeding preferences in Panama. *Biotropica*, **32**, 489–501.

Westoby, M. (1978). What are the biological bases of variable diets? *American Naturalist*, **112**, 627–31.

White, L. J. T., Rogers, M. E., Tutin, C. E. G., Williamson, E. A., & Fernandez, M. (1995). Herbaceous vegetation in different types of forest in the Lope Reserve, Gabon: implications for keystone food availability. *African Journal of Ecology*, **33**, 124–41.

Whiten, A., Byrne, R. W., & Henzi, S. P. (1987). The behavioral ecology of mountain baboons. *International Journal of Primatology*, **8**, 367–88.

Wiens, J. A. (1976). Population responses to patchy environments. *Annual Review of Ecology and Systematics*, **7**, 81–120.

Williams, J. B., Anderson, M. D., & Richardson, P. R. K. (1997). Seasonal differences in field metabolism, water requirements, and foraging behavior of free-living aardwolves. *Ecology*, **78**, 2588–602.

Wrangham, R. W. (1980). An ecological model of female-bonded primate groups. *Behaviour*, **75**, 262–300.

Wrangham, R. W., Conklin-Brittain, N. L., & Hunt, K. D. (1998). Dietary response of chimpanzees and cercopithecines to seasonal variation in fruit abundance. I. Antifeedants. *International Journal of Primatology*, **19**, 949–70.

Wright, P. C. (1999). Lemur traits and Madagascar ecology: coping with an island environment. *Yearbook of Physical Anthropology*, **42**, 31–72.

Wright, S. J., Carrasco, C., Calderon, C., & Paton, S. (1999). The El Niño Southern Oscillation, variable fruit production, and famine in a tropical forest. *Ecology*, **80**, 1632–47.

Yamakoshi, G. (1998). Dietary responses to fruit scarcity of wild chimpanzees at Bossou, Guinea: possible implications for ecological importance of tool use. *American Journal of Physical Anthropology*, **106**, 283–95.

Yang, S. & Zhao, Q.-K. (2001). Bamboo leaf-based diet of *Rhinopithecus bieti* at Lijiang, China. *Folia Primatologica*, **72**, 792–5.

Yeager, C. P. (1989). Feeding ecology of the proboscis monkey (*Nasalis larvatus*). *International Journal of Primatology*, **72**, 497–530.

— (1996). Feeding ecology of the long-tailed macaque (*Macaca fascicularis*) in Kalimantan Tengah, Indonesia. *International Journal of Primatology*, **17**, 51–62.

Yeager, C. P. and Kool, K. (2000). The behavioral ecology of Asian primates. In *Old World Monkeys*, ed. P. F. Whitehead and C. J. Jolly. Cambridge: Cambridge University Press, pp. 496–521.

4 *Seasonality in predation risk: varying activity periods in lemurs and other primates*

MICHELE A. RASMUSSEN
Department of Biological Anthropology and Anatomy, Duke University, Durham NC 27708, USA

Introduction

A primate's activity cycle is one of its most fundamental characteristics as it determines when an individual engages in the basic behaviors required for survival. The allocation of activity over a single 24-hour period varies substantially across and even within species. Although the biological mechanisms that regulate vertebrate activity cycles and the abiotic cues associated with them have been the subject of much research, there has been relatively less emphasis on the adaptive and evolutionary correlates of the different ways in which animals distribute activity and rest across the 24-hour day.

Most primates display either a nocturnal or a diurnal activity cycle. Although diurnality frequently is associated with haplorhine primates, along with their superior visual acuity and trichromatic vision (at least in the catarrhines), several strepsirrhine species are also habitually day-active (Tattersall 1982; Mittermeier *et al.* 1994), and at least three taxa display an X-linked polymorphism that permits trichromacy in females (Tan & Li 1999). Furthermore, not all haplorhines are diurnal; exceptions include the tarsier and the owl monkey, taxa that secondarily evolved a nocturnal activity cycle from diurnal ancestors (Martin 1990).

The morphological and behavioral dichotomy that exists between nocturnal and diurnal species is striking, especially within the visually adapted primate order, and it would seem difficult to shift from day activity to night activity, or vice versa, once adaptations for either of these activity periods have evolved. Nonetheless, a third activity cycle observed in primates requires such transitions between day and night activity. Cathemeral species display significant levels of activity across the 24-hour cycle, either year

Seasonality in Primates: Studies of Living and Extinct Human and Non-Human Primates, ed. Diane K. Brockman and Carel P. van Schaik. Published by Cambridge University Press.

Table 4.1 *Extant cathemeral primates*

Taxon	References
Strepsirrhini	
Eulemur coronatus (crowned lemur)	Wilson *et al.* [1989]; Freed [1996]
E. fulvus (brown lemur)	Harrington [1975]; Tattersall [1979]; Andriatsarafara [1988]; Wilson *et al.* [1989]; Overdorff & Rasmussen [1995]; Donati *et al.* [2001]; Rasmussen [1999]
E. macacao macaco (black lemur)	Colquhoun [1993, 1998]; Andrews & Birkinshaw [1998]
E. mongoz (mongoose lemur)	Tattersall & Sussman [1975]; Sussman & Tattersall [1976]; Tattersall [1976]; Harrington [1978]; Andriatsarafara [1988]; Curtis *et al.* [1999]; Rasmussen [1999]
E. rubriventer (red-bellied lemur)	Overdorff [1988]; Overdorff & Rasmussen [1995]
Hapalemur griseus alaotrensis (Alaotran gentle lemur)	Mutschler *et al.* [1998]; Mutschler & Martin [1999]
Haplorhini	
Aotus azarai (owl monkey)	Rathbun & Gache [1980]; Wright [1985, 1989, 1994]; Fernandez-Duque *et al.* [2001]

round or seasonally (Tattersall 1987). Cathemerality is relatively common in carnivores, ungulates, and rodents (Charles-Dominique 1975; Nowak 1991; van Schaik & Griffiths 1996) and has been described for a diverse array of animals living in a variety of habitats (e.g. invertebrates [Bradley 1988], fish [Tonn & Paszkowski 1987; Valdimarsson *et al.* 1997], amphibians [Griffiths 1985], owls [Reynolds & Gorman 1999], platypuses [Serena 1994], armadillos [Greegor 1980, 1985; Layne & Glover 1985], coatis [Valenzuela & Ceballos 2000], moose [Risenhoover 1986], otters [Perrin & D'Inzillo Carranza 2000], possums [Winter 1996], seals [Kooyman 1975], shrews [Woodall *et al.* 1989], and voles [Reynolds & Gorman 1999]). In contrast, this activity cycle is rare in primates, limited to Malagasy strepsirrhines in the family Lemuridae and a few populations of the otherwise nocturnal owl monkey, *Aotus azarai* (Rathbun & Gache 1980; Wright 1985, 1989; Fernandez-Duque *et al.* 2001) (Table 4.1)

Because nocturnal and diurnal activity patterns are associated so strongly with critical aspects of primate lives, including body size, sensory systems, habitat use, sociality, and anti-predator strategies, the morphological and behavioral correlates of cathemerality in primates are of great interest.

Despite extensive research in recent years, the functional underpinnings of primate cathemerality are not understood fully (Martin 1972; Tattersall 1976; Overdorff 1988; Engqvist & Richard 1991; Overdorff & Rasmussen 1995; Andrews & Birkinshaw 1998; Curtis *et al.* 1999; Mutschler & Martin 1999) (see also van Schaik & Kappeler 1996; Wright 1999). However, among the ecological explanations for cathemerality, seasonality in predation risk appears to provide one of the strongest links to this activity cycle in lemurids (Overdorff 1988; Andrews & Birkinshaw 1998; Curtis *et al.* 1999; Rasmussen 1999). Therefore, in this chapter we explore the interplay of cathemerality, seasonality, and predation risk by presenting results from a 14-month study of a pair of closely related cathemeral lemurids, *Eulemur mongoz* (mongoose lemur) and *Eulemur fulvus fulvus* (common brown lemur) inhabiting the seasonal dry forest of Ampijoroa in northwestern Madagascar. Data on activity cycle, seasonal changes in the forest environment, and anti-predator behavior are used to construct a simple model for optimal activity patterns in these two lemur species.

Primates, predation risk, and seasonality

When considering the influence of predation on primate behavior, predation risk provides the best insights on how the prey perceive the level of threat facing them and how they might modify their behavior to reduce that threat (Hill & Dunbar 1998). Cowlishaw (1997) separated predation risk into two components: the risk of attack and the risk of capture after an attack has occurred. In many primate species, the risk of attack is dependent mostly on whether the predator detects the prey and whether the prey detects the predator in time to either give an alarm or engage in evasive behavior. Predation risk is highly dependent on habitat-specific conditions, including habitat type, the number and density of both predator and prey species, and the availability and distribution of escape routes and refuges (Hill & Lee 1998). It is also best assessed when there is knowledge of the predator's hunting methods and likelihood for success in a given set of circumstances (Cowlishaw 1997, 1998).

Primates can offset predation risk in numerous ways, including modifying group size (Alexander 1974; van Schaik 1983), being selective over travel routes and sleeping sites, displaying vigilance behaviors, and mobbing (see Isbell 1994 for review). Janson (1998) identifies crypsis as an additional anti-predator strategy not only in nocturnal, non-gregarious species but also in some diurnal, group-living taxa such as titi monkeys, sifaka, and chimpanzees (Wright 1984, 1998; Boesch 1991). Of particular interest to Janson is determining why some primates remain silent and inconspicuous when predators are detected while others emit alarms. He shows that giving alarm calls upon sighting a predator has the immediate effect of alerting the predator to the group's presence and, most probably, its location. Therefore, by using an alarm system, the group "ruins its crypsis" (Janson 1998: 405), and individuals maximize their chances of escaping attack by detecting and advertising the presence of predators as soon as possible. Another way in which groups can enhance this effect is through increasing their size. Janson labels this combination of strategies "early warning large group" (EWLG) behavior. He predicts that primates that are noisy foragers and travelers (e.g. insect-eating species) are more likely to display EWLG anti-predator behavior than are similarly-sized herbivores, and that when group size is equal, larger-bodied primates are more likely than smaller-bodied primates to be EWLG.

Alternatively, some primates minimize predation risk by avoiding all conspicuous behavior, thereby precluding the use of early warning systems. Janson calls this the "cryptic small group" (CSG) strategy and predicts that it is most likely to be observed in species that are more capable of quiet foraging and traveling and are smaller in body size than their EWLG counterparts. CSG is also expected to predominate in nocturnal species and is predicted for solitary individuals in EWLG species following such events as natal group dispersal and group transfer.

Janson (1998) recognizes that EWLG and CSG strategies are not mutually exclusive and that any anti-predator strategy is subject to the potential constraints associated with foraging, reproduction, and habitat type. These constraints, in turn, are all likely to be affected by seasonality. For example, the benefits of large group size (and the EWLG strategy) may be outweighed by the costs associated with increased feeding competition, resulting in smaller foraging parties during periods of food scarcity and seasonal shifts from EWLG to CSG. Conversely, when resources are spatially and temporally clumped, primate groups may temporarily increase in size as individuals converge on common food patches and anti-predator behavior may shift from CSG to EWLG as the potential for vigilance and the benefits of an early warning system also increase. In addition to foraging constraints, seasonal changes in primate anti-predator strategies also may be associated with varying levels of vulnerability due to the presence of infants, the shifting hunting strategies of predators (depending on the availability of prey or increased nutritional needs during critical reproductive periods), and reduced canopy cover in deciduous forest habitats.

Cathemerality as an anti-predator strategy in seasonal habitats

Unlike nocturnality and diurnality, cathemerality can potentially take many forms, and implicit in Tattersall's (1987) definition is the possibility that cathemeral activity can vary by time of year or even by habitat type. Cathemeral activity may afford primates enough flexibility to reduce the possibility of detection by predators while engaging in potentially high-risk activities such as foraging and travel. Moreover, cathemerality could be especially advantageous in highly seasonal forest habitats such as those found in Madagascar, where predation risk may vary substantially with changes in the canopy cover of deciduous trees.

The flexibility in predator response afforded by cathemerality is shown in the various forms it takes (Rasmussen 1999), including (i) the complete shift from daytime to nighttime activity (mongoose lemurs [Tattersall & Sussman 1975; Sussman & Tattersall 1976; Curtis *et al.* 1999; Rasmussen 1999]); (ii) habitual daytime (or nighttime) activity with a seasonal extension of feeding and travel into nighttime (or daytime) hours (brown lemurs [Rasmussen 1999; Gerson 2000]; owl monkeys [Wright 1985, 1989]); and (iii) activity that occurs by day and night, year round (some lemurids and owl monkeys [Rathbun & Gache 1980; Freed 1996; Overdorff 1996; Andrews & Birkinshaw 1998; Mutschler & Martin 1999; Fernandez-Duque *et al.* 2001]) (Table 4.2).

In the absence of direct observations and long-term data on predator–prey interactions and lemur demography, it is difficult to ascertain which predators have the most significant potential impact on mortality (and, hence, behavior) in cathemeral primates in Madagascar. Viverrids, raptors, and constricting snakes all have been observed or strongly inferred to prey upon medium- to large-bodied day active lemurs (Goodman *et al.* 1993, 1997; Overdorff & Strait 1995; Wright 1998; Karpanty & Goodman 1999; Burney 2002; Brockman 2003). The fossa, *Cryptoprocta ferox* (Carnivora: Viverridae), is Madagascar's largest predator, weighing 6–6.5 kg (Hawkins 1998). It is distributed widely in many habitat types and appears to hunt opportunistically, taking advantage of both its climbing ability and the local availability of different terrestrial and arboreal prey species (Goodman *et al.* 1993, 1997; Sommer 2000). Its activity cycle is described as nocturnal or cathemeral (Dollar *et al.* 1997); in the deciduous dry forests of the northwest, it was observed to be active during the day as well as during the night (Rasmussen 1999). Available life history data indicate that the fossa breeds in September and October (which is the late dry season in western Madagascar) and gives birth to litters of two to four offspring in December or January. Offspring are weaned at approximately

Table 4.2 *Patterns of cathemerality*

Cathemeral pattern	Species	Locality	Annual rainfall (mm)	Forest type
1 Alternating day and night activity	*Eulemur mongoz*	Anjamena, Madagascar	1189	Deciduous dry forest[a]
		Ampijoroa, Madagascar	1755	Deciduous dry forest[b,c]
2a Continuous day activity, seasonal night activity	*E. fulvus fulvus*	Ampijoroa, Madagascar	1755	Deciduous dry forest[b,c]
	E. f. rufus	Anjamena, Madagascar	1189	Deciduous dry forest[a]
2b Continuous night activity, seasonal day activity	*Aotus azarai*	Chaco region, Paraguay	<1000	Subtropical dry forest[d]
3 Continuous day and night activity	*E. f. rufus*	Ranomafana, Madagascar	2300–4000	Montane evergreen rainforest[e]
	E. rubriventer			
	E. macaco macaco	Nosy Bé, Madagascar	2356	Lowland evergreen rainforest[f]
		Ambato Massif, Madagascar	2000–2300	Seasonally moist semi-deciduous forest[g]
	E. coronatus	Mt D'Ambre, Madagascar	1959	Predominantly evergreen humid forest[h]
	E. f. sanfordi			
	Hapalemur griseus alaotrensis	Lac Alaotra, Madagascar	905	Reed and marsh beds[i]

[a] Curtis [1997];
[b] Thalmann and Müller, unpublished data;
[c] this study;
[d] Wright [1985, 1989];
[e] Overdorff [1996]; Wright *et al.* [1997]; Atsalis [1998];
[f] Andrews and Birkinshaw [1998];
[g] Colquhoun [1993];
[h] Freed [1996];
[i] Mutschler *et al.* [1998].

Table 4.3 *Characteristics of* Eulemur mongoz *and* E. fulvus fulvus *at Ampijoroa*

	Eulemur mongoz	*Eulemur mongoz fulvus*
Mean body mass (g)	995 ($n = 2$)	1779 ($\pm$ 129 g) ($n = 7$)
Group size	2–4 ($n = 4$)	6–15 ($n = 6$)
Presumed mating system	Monogamy	Polygyny or promiscuity
Diet	Frugivore–folivore; seasonal specializations on flowers and nectar	Frugivore–folivore
Range size (ha)	5.47 for primary study group; territorial	16.24 for primary study group; range overlap with other *Eff* groups

4.5 months, which coincides with the beginning of the dry season in western Madagascar and is the time of year when many small mammals and reptile species are entering torpor (Nowak 1991).

In addition to the fossa, several raptors are present in Madagascar, but the Madagascar harrier hawk, *Polyboroides radiatus* (Falconiformes: Accipitridae), is probably the most dangerous aerial predator on lemurids, given its size (68 cm in length) (Langrand 1990) and ability to prey successfully on large lemurs such as Verreaux's sifaka (Karpanty & Goodman 1999; Brockman 2003). *P. radiatus* displays a variety of hunting methods, including foraging terrestrially for insects, sitting and waiting for extended periods of time in dense foliage, and searching for vertebrate prey while flying above the canopy (Langrand 1990). Like other diurnal birds of prey, *P. radiatus* relies primarily on superior visual acuity and speed to catch vertebrates, using a "gliding attack" or "diving attack" (Brockman 2003: 72).

There are few reports of predation on larger-bodied lemurs by snakes (Goodman *et al.* 1993; Rakotondravony *et al.* 1998; Burney 2002), although constricting snakes do appear to provoke increased vigilance, vocalizations, and/or mobbing in *Eulemur* (Bayart & Anthouard 1992; Goodman *et al.* 1993).

One might argue that the three distinct forms of cathemerality displayed by lemurids are alternative strategies used to offset the risk of predation, especially by terrestrial carnivores and diurnal raptors, and thus are extensions of the CSG and EWLG behaviors identified by Janson (1998). Based on their body mass and social structure, Janson's model would predict that mongoose lemurs and common brown lemurs display CSG and EWLG behaviors, respectively (Table 4.3). In the analysis that follows, we will explore the degree to which *E. mongoz* and *E.f. fulvus* activity patterns are concordant with the predictions of these anti-predator strategies and assess the relationship between these activity patterns and seasonal changes in habitat and the level of risk posed by predators.

Methods

Study site and subjects

The activity patterns of *E. mongoz* and *E.f. fulvus* were studied from May 1996 through June 1997 at the Classified Forest of Ampijoroa in northwestern Madagascar. Activity, behavioral, and phenological data were used to test a series of hypotheses on the influence of ecological correlates to cathemeral activity in lemurs, including predation risk (Rasmussen 1999).

Ampijoroa (16°19′S, 46°49′E) is located within the 135 000-ha system of protected areas comprising the Ankarafantsika Nature Reserve. Ampijoroa's climate is typical for western Madagascar, with a distinct dry season (May through October) and wet season (December through March). The beginning and end of the annual rains vary from year to year, and November and April are considered to be interseasonal months. The forest at Ampijoroa is typical of those seen throughout the western domain of Madagascar (Martin 1972; Tattersall 1982; Sussman & Rakotozafy 1994). Compared with the humid, evergreen forests of the east (e.g. Hemingway & Overdorff 1999), western forests have a lower canopy, are less structurally complex, and contain many semi-deciduous species (Sussman & Rakotozafy 1994).

Mongoose lemurs (*E. mongoz)* and common brown lemurs (*E.f. fulvus)* are medium-sized lemurs with a mixed frugivore–folivore diet. (Table 4.3). The slightly smaller mongoose lemurs display a pair-bonded social structure (Andriatsarafara 1988; Curtis & Zaramody 1998; Curtis 1999), while common brown lemurs display social similarities to the red-fronted lemur, *E.f. rufus*, living in multimale groups of 4–17 individuals with roughly equal adult sex ratios (*E.f. rufus*: Sussman [1974]; Overdorff *et al.* [1999]; Gerson [2000]; *E.f. fulvus*: Harrington [1975]; Andriatsarafara [1988]; *E.f. mayottensis* [= *E.f. fulvus*]: Tattersall [1977]). Preliminary data from short-term studies at Ampijoroa indicated that *E. mongoz* and *E.f. fulvus* displayed different forms of cathemerality (Tattersall & Sussman 1975; Sussman & Tattersall 1976; Harrington 1978; Andriatsarafara 1988).

Potential non-human predators of lemurs at Ampijoroa include three falconiform raptors, one viverrid carnivore, and at least one boid reptile. Sympatric lemur species include *Propithecus verreauxi coquereli*, *Avahi occidentalis*, *Lepilemur edwardsi*, *Cheirogaleus medius*, *Microcebus murinus*, and *Microcebus ravelobensis*.

Behavioral and phenological data collection

Four main lemur study groups (two of each species) were followed at Ampijoroa during day and night observation sessions between May 1996 and July 1997. Follows yielded 418.8 hours of activity data for *E. mongoz* and 692.3 hours of activity data for *E.f. fulvus*. Activity profiles were generated from individual activity records derived from group scan sampling (Altmann 1974). Scan samples were made at five-minute intervals, for a total of 12 scans/hour (detailed methods given in Rasmussen 1999).

All occurrences of predator alarms were recorded during observations of *E. mongoz* and *E.f. fulvus* study groups. Distinct vocalizations, movements, and postures were associated with alarms and were easily recognized and distinguished from other behaviors (for review of lemurid alarm vocalizations, see Macedonia and Stanger [1994]). The date and time of the alarm were recorded along with spatial information for the group immediately before and after the alarm. These data included minimum and maximum height in the forest (of the lowest and highest visible individuals) and horizontal group spread (HGS). The cause of the alarm was assigned to one of three categories: terrestrial, aerial, or unknown. The alarm type was recorded as being "vocalization only," "movement only," or "vocalization and movement." Five categories describing changes in group spatial distribution were used: (i) up: the majority of visible group members increased vertical height; (ii) down: the majority of visible group members decreased vertical height; (iii) increase in the maximum visible HGS; (iv) decrease in the maximum visible HGS; and (v) no change. Relationships between alarm causes and lemur responses were tested for independence using the G-test with Williams' correction, and the significance level was set at $P \leq 0.05$.

To examine whether changes in the lemurs' habitat affected activity patterns, two phenological measures were used to determine canopy cover in the study area: the number of stems with 50–100% crown cover (new and mature leaves) and the number of completely bare stems (no leaves). Phenological data were derived from biweekly checks of the ten botanical plots distributed randomly throughout the study site.

Measures of nighttime activity in the lemurs were derived from activity data collected during nighttime follows; the percentages of monthly group scans scored as "active" were used in the analyses. Associations between canopy cover and lemur activity were tested for using Spearman rank correlation coefficients, with the significance level set at $P \leq 0.05$. Daily temperatures and precipitation were recorded daily throughout the course of the study (for methods, see Müller [1999]).

Results

Anti-predator behavior of E. mongoz *and* E. f. fulvus *at Ampijoroa*

Fifty alarm events were recorded for *E.f. fulvus* groups between July 1996 and July 1997. For many alarms ($n = 23$), the cause was unknown; for the remaining alarms, the most common cause was aerial ($n = 18$) followed by terrestrial ($n = 9$). Only five alarm events were recorded for *E. mongoz* groups between December 1996 and April 1997. Perceived aerial threats caused three of these alarms, and two resulted from unknown causes. Causes of lemur alarms that were identified by observers included various bird species, domesticated dogs and cats, a cow, a tree boa (*Boa manditra*) and the fossa, *Cryptoprocta ferox*. The alarm caused by the latter preceded an actual predation on an adult female common brown lemur.

Because so few alarm events were recorded for *E. mongoz*, the following results are for *E.f. fulvus* only. The type of alarm response displayed by lemurs was dependent on alarm cause (G [Williams] = 9.32, df = 4, $P = 0.05$). Aerial threats were most associated with movements while terrestrial threats were associated with vocalizations. A second analysis showed even greater association between aerial threats and the combined response of vocalizations and movements (G [Williams] = 9.34, df = 2, $P < 0.01$).

A significant dependent relationship was found between alarm cause and the movement response (G [Williams] = 15.59, df = 8, $P < 0.05$). Changes in vertical position (either increased or decreased vertical height) and group cohesiveness (i.e. increased or decreased HGS) were most associated with aerial threats. Thus, *E.f. fulvus*' response to perceived aerial and terrestrial threats were unambiguously non-cryptic, with the most conspicuous alarm responses (vocalizations combined with vertical and horizontal movements) associated with threats from the air. This species appears to combine an early warning system with rapid evasive movements when a raptor threat is detected.

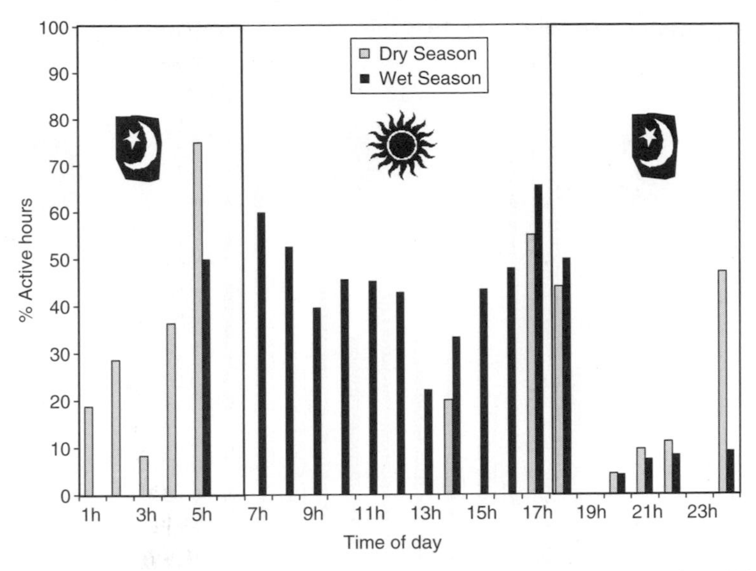

Figure 4.1 *Eulemur mongoz* dry season activity profile, May–November 1996 and May–June 1997, and wet season activity profile, December 1996–April 1997. Bars show the percentage of "active" hours out of total sample of hours. The seasonal activity profiles presented in this chapter were generated for each species by determining the percentage of "active" hours for the total sample of each of the 24 hours contributing to the day and night cycle (00.00–01.00, 01.00–02.00, etc.). An hour was scored as "active" when seven or more group scans (out of a total of 12 for the hour) saw the majority of visible individuals engaged in non-rest activities.

Seasonal activity rhythms in **E. mongoz** *and* **E. f. fulvus**

E. mongoz altered their activity patterns dramatically throughout the year at Ampijoroa. During the dry season months in 1996 (May–November) and 1997 (May–June), activity was restricted largely to the dark portion of the 24-hour cycle. However, during the 1996–97 wet season (December–April), mongoose lemurs were almost exclusively active by day (Fig. 4.1). In contrast, *E.f. fulvus* at Ampijoroa were day-active throughout the year, but like mongoose lemurs, *E.f. fulvus* engaged in night activity during the dry season months (Fig. 4.2). Climate data collected at Ampijoroa during the study period suggest that precipitation was not the proximate cause for changes in activity pattern in either species (Rasmussen 1999). The timing of both species' transition from night activity to exclusive day activity did not coincide with the onset of rainfall, and nor did nighttime activity resume

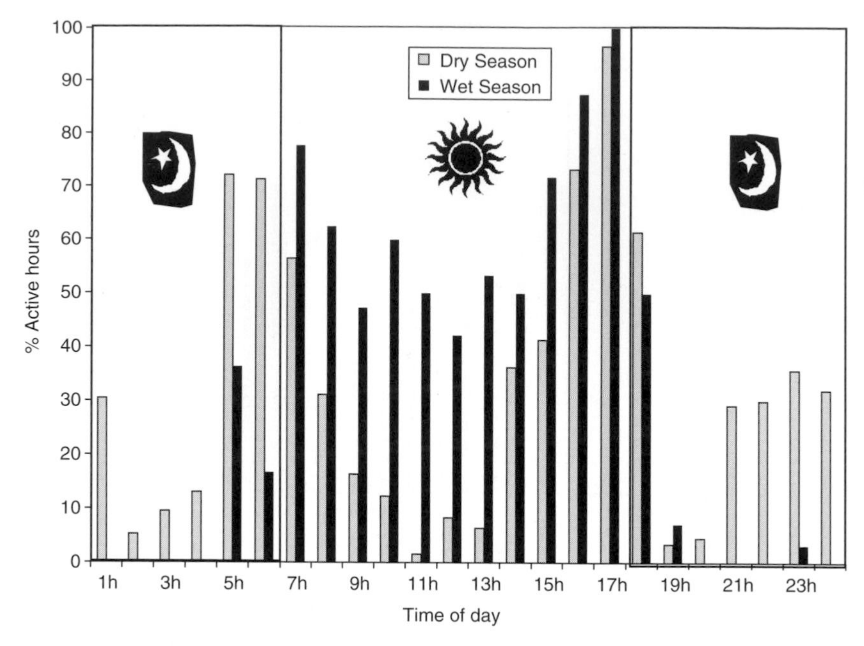

Figure 4.2 *Eulemur fulvus fulvus* dry season activity profile, May–November 1996 and May–June 1997, and wet season activity profile, December 1996–April 1997, showing the percentage of "active" hours out of the total sample.

immediately once the wet season ended in late April. Mongoose lemurs entered a period of predominantly crepuscular activity at this time, while common brown lemurs persisted with exclusive day activity until the middle of June (Rasmussen 1999).

Seasonal changes in canopy cover and lemur behavior

Changes in canopy cover play a role in cueing transitions in activity patterns in lemurids at Ampijoroa. In both *E. mongoz* and *E.f. fulvus*, the shift between day and night activity is associated more tightly with canopy cover than with rainfall (Fig. 4.3). In both lemur species, nighttime activity was correlated negatively with the number of trees with crown coverage between 50% and 100% (*E. mongoz*: $r = -0.706$, $P < 0.025$, $n = 9$; *E.f. fulvus*: $r = -0.674$, $P < 0.01$, $n = 12$) but was correlated positively with the number of bare stems (*E. mongoz*: $r = 0.866$, $P < 0.05$, $n = 9$; *E.f. fulvus*: $r = 0.600$, $P < 0.025$, $n = 12$).

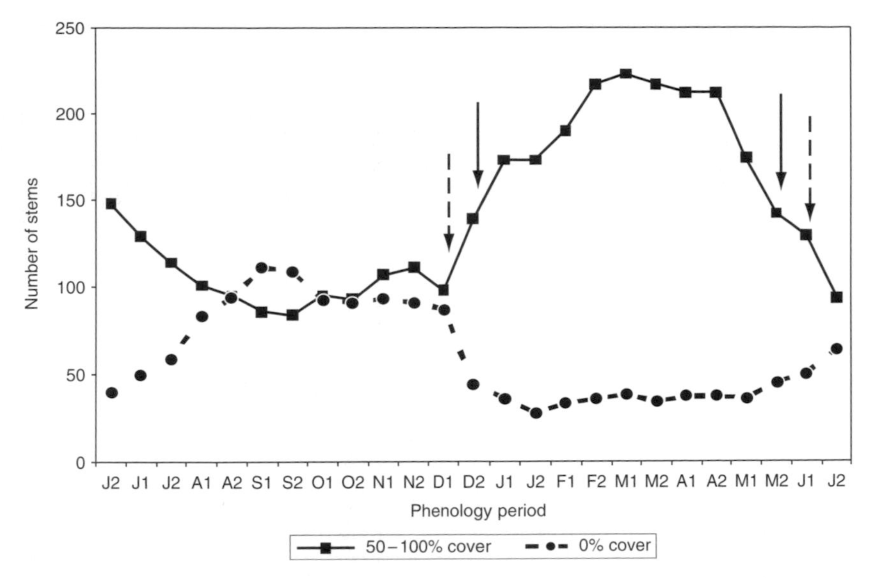

Figure 4.3 Timing of *Eulemur mongoz* and *E. fulvus fulvus* activity shifts in 1996–97 and their relationship with forest canopy cover at Ampijoroa. Dashed arrows indicate the end of *E. f. fulvus* nighttime activity in December and the onset of nighttime activity in late May–early June. Solid arrows indicate the onset of *E. mongoz* daytime activity in December and the onset of nighttime activity in May. Significant negative correlations exist between both species' nighttime activity and the number of stems with 50–100% leafy crown coverage. Conversely, significant positive correlations exist between nighttime activity and the number of stems with completely bare crowns.

Discussion

Cathemerality as a response to seasonal changes in predation risk

E. mongoz and *E. f. fulvus* in northwestern Madagascar display variable patterns of day and night activity that are associated with seasonal changes in forest canopy cover. Potential aerial and terrestrial predators of lemurid-sized primates are present, and lemurs display differentiated alarm behaviors in response to perceived threats from the air and ground; *E. f. fulvus* is more conspicuous than *E. mongoz* in its alarm response to both types of predation threat and is most likely to combine vocalizations with movement when perceiving risk from raptors. In contrast, only five alarm events were recorded for *E. mongoz* in the five-month period when antipredator data were collected for this species (compared with 24 recorded for sympatric *E. f. fulvus* groups during the same period). This small number of alarm events suggests a more cryptic strategy for minimizing predation risk in mongoose lemurs. Similar crypticity in moving and foraging behavior has been reported for the small group-living and cathemeral red-bellied lemur, *Eulemur rubriventer*, inhabiting evergreen rainforest in southeastern Madagascar (Overdorff *et al.* 2002). These results suggest that, concordant with the Janson (1998) model, *E. mongoz* and *E.f. fulvus* do in fact display CSG and EWLG strategies, respectively (Janson 1998). More importantly, these strategies appear to track seasonal changes in activity cycle, which, in turn, offsets varying levels of predation risk at Ampijoroa.

Although there are as yet insufficient data to assess the seasonality of hunting habits of diurnal raptors (but see Karpanty & Goodman [1999] and Brockman [2003]), their excellent photopic vision (Walls 1967; Wolin & Massopust 1970) (see also Jacobs [1993]) and their strategic preference for hunting at and above canopy level suggest that they are most dangerous to *Eulemur* species during daylight hours in the dry season when leafy crown cover protection is most reduced. Of the two lemur species, *E. mongoz* is probably more vulnerable to predation, given its smaller body size and reduced capacity for vigilance due to small group size. However, during the rainy season, these same characteristics should prove advantageous to mongoose lemurs employing a CSG strategy, which is enhanced by extensive canopy cover while the lemurs forage, feed, and travel. During the rainy season, *E.f. fulvus* may also be less vulnerable to detection by diurnal raptors such as *Polyboroides radiatus* when they are flying over the forest canopy; however, the same leafy cover that protects mongoose lemurs also provides cover for raptors using a "sit-and-wait" hunting strategy. Thus, the more conspicuous brown lemurs remain vulnerable to aerial predators during the wet season when forest cover increases, thereby resulting in the maintenance of EWLG behavior in this species.

These differences in predation risk for *E. mongoz* and *E.f. fulvus* might explain partly their divergent cathemeral activity rhythms (Table 4.4). In the dry season, the CSG strategy of *E. mongoz* is probably hindered during the day by the substantial decrease in leafy crown cover. Therefore, mongoose lemurs avoid the risk associated with visually superior diurnal raptors by reserving all activity for nighttime hours during the dry season. Although *Cryptoprocta ferox* is potentially active during the night at this time of the year, CSG behavior and the small body size of mongoose lemurs make them a less obtainable prey than larger-bodied or noisier species. During the wet season, when crown cover is extensive, CSG behavior in mongoose lemurs should be adequate protection against diurnal raptors, and nighttime activity is no longer necessary. *C. ferox* may pose some threat to *E. mongoz*, although it is expected to be minimal given

Table 4.4 *Summary of predator risk facing cathemeral lemurs at Ampijoroa*

Predator	Season and portion of 24-hour cycle when risk is highest (given overall antipredator strategy, CSG or EWLG)	
	Eulemur mongoz (CSG)	*E.f. fulvus* (EWLG)
Polyboroides radiatus and other raptors	Dry season, day	Both seasons, day
Cryptoprocta ferox	Wet season, day	Dry season, day and night

CSG, cryptic small group; EWLG, early warning large group.

the abundance of other, more easily caught prey species (e.g. the insectivore *Tenrec ecaudatus*) at this time of the year.

In contrast, a combination of vigilance and diffused activity may work better for EWLG species in highly seasonal habitats like Ampijoroa. The larger body size and group size of *E.f. fulvus* make it less likely that this species will "escape" predation by shifting activity entirely away from one part of the 24-hour cycle and relying exclusively on crypsis to minimize risk. Brown lemurs are equally vulnerable to day- and night-active predators during the dry season, so lower levels of activity across the entire 24-hour cycle may be more strategic than engaging in more intense and conspicuous activity during shorter time periods and extending sleep bouts (thereby increasing detection by predators) (Reichard 1998; Wright 1998). Furthermore, daytime activity during the dry season is less risky for EWLG species such as *E.f. fulvus* than it is for CSG species such as *E. mongoz* because the former are more likely to detect predators early enough to thwart an attack. In the wet season, EWLG behavior is advantageous during the day as defense against both raptors and *C. ferox*. The lack of nighttime activity in common brown lemurs suggests that, as for mongoose lemurs, predation pressure from both classes of predators is relaxed substantially during this time of the year.

Although this study found no specific association between cathemerality and food resource availability, interspecific competition, or temperature at Ampijoroa, hypotheses citing these variables as important proximate and/or ultimate influences on cathemeral activity for other lemur populations cannot be rejected (e.g. Andrews & Birkinshaw 1998; Curtis *et al.* 1999; Mutschler & Martin 1999). Nonetheless, predation does appear to be an important factor in the cathemerality of some populations of *Aotus*. In the strongly seasonal, subtropical chaco region of Paraguay, Wright (1985,

1989) found that *Aotus* groups spent up to three hours traveling, foraging, and feeding during the day. Unlike their rainforest counterparts in Peru, the chaco owl monkeys have no potential competitors for food resources (during the day or night) and face minimal threat from diurnal eagles, which are rare in the region. Wright (1989) argues that the chaco owl monkeys are released from exclusive nighttime activity by these favorable daytime conditions; furthermore, daytime activity in this population may be a diffusion strategy to counteract the risk posed by the nocturnal great-horned owl (which is not found in Peru). According to Wright (1989), during nighttime travel the chaco *Aotus* were more cryptic than the Peruvian rainforest population and also refrained from making contact calls. Although the absence of diurnal eagles makes daytime activity by the Paraguay *Aotus* possible, cathemerality in this population may actually be driven by the need to minimize exposure to a nocturnal predator.

Predation risk and canopy cover in diurnal primates

The adaptations for diurnality in arboreal haplorhines preclude cathemerality as a viable strategy to offset predation risk, either year round or seasonally. Nonetheless, we expect that these primates should modify their behavior to minimize the risk posed by predators when they occupy exposed substrates resulting from the seasonal loss of leafy cover or the structural characteristics of forest microhabitats. Numerous reports from studies of forest-living monkeys have noted their vulnerability to raptors and even terrestrial carnivores in areas where foliage cover is at a minimum – at forest edges, in tree gaps, in the uppermost strata of the canopy, and in leafless crowns (Eason 1989; Peetz *et al.* 1992; Isbell 1994). However, recent studies of foraging behavior in arboreal African and Neotropical monkeys suggest that many species may actually perceive the highest predation risk in microhabitats where leaf cover is greatest, because individuals there are less likely to detect an approaching raptor or carnivore than when their view is unobstructed by leaves (Cords 2002; Treves 2002; Di Fiore 2002). The "better detection in the open" hypothesis (Di Fiore 2002: 262) predicts increased vigilance whenever a primate's view is impeded, i.e. when it is near the forest floor where ambient light levels are low and the understorey vegetation is dense, or when it is near a tree trunk and surrounded by large leafy branches. This hypothesis also explains why some primates engage in seemingly risky behavior by utilizing leafless crowns or the uppermost levels of the canopy at higher rates than expected (Di Fiore 2002).

Tests of this intriguing hypothesis in strepsirrhines could provide additional new insights into the roles that seasonality and predation risk play in the evolution of cathemeral activity patterns in lemurids, particularly if they focus on the utilization of exposed versus covered substrates in CSG species such as *E. mongoz*. Monkeys that have been shown to exhibit high rates of vigilance when obscured by foliage and that use "risky" locations for foraging, moving, and resting are all species that live in large troops; one would predict that these species use the EWLG anti-predator strategy and also, presumably, benefit from the vigilance of many other group members (Janson 1998; Cords 2002; Treves 2002; Di Fiore 2002). If small group-living species such as mongoose lemurs display similar patterns of microhabitat use and vigilance as those of monkeys, then the hypothesis that they are using cathemerality to avoid daytime activity in seasonally leafless crowns would be weakened.

In this regard, future research on activity patterns in primates should focus on comparisons between microhabitat usage and vigilance rates in cathemeral versus diurnal species, in EWLG versus CSG species, and in species living in evergreen versus deciduous forests. Results from these studies would greatly enhance our understanding of the tradeoffs that primates must make when balancing predation risk against basic maintenance activities as well as further elucidate the ecological basis of cathemeral activity in forest-dwelling primates.

Conclusion

The results from Ampijoroa reported here suggest that a flexible activity cycle may serve a similar function in some primate populations. *E. mongoz* and *E.f. fulvus* augment their CSG and EWLG behaviors, respectively, with seasonally distinct activity rhythms that potentially minimize predation risk. When small group size and crypticity are inadequate defenses against diurnal aerial predation risk during the period of least canopy cover, mongoose lemurs at Ampijoroa shift all activity to the relatively safe dark portion of the 24-hour cycle. Common brown lemurs are unable to completely avoid either day- or night-active predators during the period of least canopy cover, so they diffuse activity over the entire 24-hour day. They maintain EWLG behavior but minimize the likelihood of detection by diurnal raptors and cathemeral carnivores by engaging in low levels of hourly activity. However, reports of higher-than-expected use of exposed substrates by some forest-dwelling diurnal monkeys suggest that arboreal haplorhines may perceive areas with dense foliage cover as a dangerous microhabitat, requiring high rates of vigilance and proximity to nearest neighbors. This hypothesis warrants testing in cathemeral species to further elucidate the relationship between activity cycle, predation risk, and seasonal effects on habitat in these primates.

Acknowledgments

The study of *E. mongoz* and *E. f. fulvus* at Ampijoroa was made possible by funding provided by the National Science Foundation (dissertation improvement grant SBR-9526111) and the Wenner-Gren Foundation for Anthropological Research (pre-doctoral grant 5953). I would also like to thank the Graduate School and Center for International Studies at Duke University and Noel Rowe and Primate Conservation, Inc. for funding preliminary studies in the Ankarafantsika Reserve, which greatly facilitated my long-term research in 1996 and 1997. I am grateful to the Direction des Eaux et Forêts, ANGAP and the government of Madagascar for giving me permission to work at Ampijoroa. I am particularly indebted to Urs Thalmann, Alexandra Müller, Debbie Curtis, and Jayne Gerson for their help and encouragement while I was in Madagascar. Finally, I would like to thank Diane Brockman, Carel van Schaik, and an anonymous reviewer for their helpful comments on this manuscript.

References

Alexander, R. D. (1974). The evolution of social behaviour. *Annual Review of Ecology and Systematics*, **5**, 325–83.

Altmann, J. (1974). Observational study of behaviour: sampling methods. *Behaviour*, **49**, 227–65.

Andrews, J. R. & Birkinshaw, C. R. (1998). A comparison between the daytime and night-time diet, activity and feeding height of the black lemur, *Eulemur macaco* (Primates: Lemuridae), in Lokobe Forest, Madagascar. *Folia Primatologica*, **69 (Suppl. 1)**, 175–82.

Andriatsarafara, R. (1988). Etude eco-éthologique de deux lémuriens sympatriques de la forêt sèche caducifoliée d'Ampijoroa (*Lemur fulvus fulvus, Lemur mongoz*). Ph. D. thesis, Université de Madagascar.

Atsalis, S. (1998). Feeding ecology and aspects of life history in *Microcebus rufus* (family Cheirogaleidae, order Primates). Ph. D. thesis, City University of New York.

Bayart, F. & Anthouard, M. (1992). Responses to a live snake by *Lemur macaco macaco* and *Lemur fulvus mayottensis* in captivity. *Folia Primatologica*, **58**, 41–6.

Boesch, C. (1991). The effects of leopard predation on grouping patterns in forest chimpanzees. *Behaviour*, **117**, 220–42.

Bradley, R. A. (1988). The influence of weather and biotic factors on the behaviour of the scorpion (*Paruroctonus utahensis*). *Journal of Animal Ecology*, **57**, 533–51.

Brockman, D. K. (2003). *Polyboroides radiatus* predation attempts on *Propithecus verreauxi*. *Folia Primatologica*, **74**, 71–4.

Burney, D. A. (2002). Sifaka predation by a large boa. *Folia Primatologica*, **73**, 144–5.

Charles-Dominique, P. (1975). Nocturnality and diurnality: an ecological interpretation of these two modes of life by an analysis of the higher vertebrate fauna in tropical forest ecosystems. In *Phylogeny of the Primates: An Interdisciplinary Approach*, ed. W. P. Luckett & F. S. Szalay. New York: Plenum Press, pp. 69–88.

Colquhoun, I. C. (1993). The socioecology of *Eulemur macaco*: a preliminary report. In *Lemur Social Systems and Their Ecological Basis*, ed. P. M. Kappeler & J. U. Ganzhorn. New York: Plenum Press, pp. 11–23.

— (1998). Cathemeral behavior of *Eulemur macaco macaco* at Ambato Massif, Madagascar. *Folia Primatologica*, **69**, 22–34.

Cords, M. (2002). Foraging and safety in adult female blue monkeys in the Kakamega Forest, Kenya. In *Eat or Be Eaten: Predator Sensitive Foraging Among Primates*, ed. L. E. Miller. Cambridge: Cambridge University Press, pp. 205–21.

Cowlishaw, G. (1997). Trade-offs between foraging and predation risk determine habitat use in a desert baboon population. *Animal Behaviour*, **53**, 667–86.

— (1998). The role of vigilance in the survival and reproductive strategies of desert baboons. *Behaviour*, **135**, 431–52.

Curtis, D. J. (1997). The mongoose lemur (*Eulemur mongoz*): a study in behaviour and ecology. Ph. D. thesis, University of Zurich.

— (1999). Social structure and seasonal variation in the behaviour of *Eulemur mongoz*. *Folia Primatologica*, **70**, 79–96.

Curtis, D. J. & Zaramody, A. (1998). Group size, home range use, and seasonal variation in the ecology of *Eulemur mongoz*. *International Journal of Primatology*, **19**, 811–35.

Curtis, D. J., Zaramody, A., & Martin, R. D. (1999). Cathemerality in the mongoose lemur, *Eulemur mongoz*. *American Journal of Primatology*, **47**, 279–98.

Di Fiore, A. (2002). Predator sensitive foraging in ateline primates. In *Eat or Be Eaten: Predator Sensitive Foraging Among Primates*, ed. L. E. Miller. Cambridge: Cambridge University Press, pp. 242–67.

Dollar, L., Forward, Z., & Wright, P. C. (1997). First study of *Cryptoprocta ferox* in the rainforests of Madagascar. *American Journal of Physical Anthropology*, Suppl. 24, 102–3.

Donati, G., Lunardini, A., Kappeler, P. M., & Borgognini Tarli, S. M. (2001). Nocturnal activity in the cathemeral red-fronted lemur (*Eulemur fulvus rufus*), with observations during a lunar eclipse. *American Journal of Primatology*, **53**, 69–78.

Eason, P. (1989). Harpy eagle attempts predation on adult howler monkey. *Condor*, **91**, 469–70.

Engqvist, A. & Richard, A. (1991). Diet as a possible determinant of cathemeral activity patterns in primates. *Folia Primatologica*, **57**, 169–72.

Fernandez-Duque, E., Rotundo, M., & Sloan, C. (2001). Density and population structure of owl monkeys (*Aotus azarai*) in the Argentinian chaco. *American Journal of Primatology*, **53**, 99–108.

Freed, B. Z. (1996). Co-occurrence among crowned lemurs (*Lemur coronatus*) and Sanford's lemurs (*Lemur fulvus sanfordi*) of Madagascar. Ph. D. thesis, Washington University.

Gerson, J. S. (2000). Social relationships in wild red-fronted brown lemurs (*Eulemur fulvus rufus*). Ph. D. thesis, Duke University.

Goodman, S. M., O'Connor, S., & Langrand, O. (1993). A review of predation on lemurs: implications for the evolution of social behavior in small, nocturnal primates. In *Lemur Social Systems and their Ecological Basis*, ed. P. M. Kappeler & J. U. Ganzhorn. New York: Plenum Press, pp. 51–66.

Goodman, S. M., Langrand, O., & Rasolonandrasana, B. P. N. (1997). The food habits of *Cryptoprocta ferox* in the high mountain zone of the Andringitra Massif, Madagascar (Carnivora: Viverridae). *Mammalia*, **61**, 185–92.

Greegor, D. H. (1980). Diet of the little hairy armadillo, *Chaetophractus vellerosus*, of northwestern Argentina. *Journal of Mammalogy*, **61**, 331–4.

— (1985). Ecology of the little hairy armadillo *Chaetophractus vellerosus*. In *The Evolution and Ecology of Armadillos, Sloths, and Vermilinguas*, ed. G. G. Montgomery. Washington, DC: Smithsonian Institution Press, pp. 397–405.

Griffiths, R. A. (1985). Diel profile of behaviour in the smooth newt, *Triturus vulgaris* (L.): an analysis of environmental cues and endogenous timing. *Animal Behaviour*, **33**, 573–82.

Harrington, J. E. (1975). Field observations of social behavior of *Lemur fulvus fulvus* E. Geoffroy 1812. In *Lemur Biology*, ed. I. Tattersall & R. W. Sussman. New York: Plenum Press, pp. 259–79.

— (1978). Diurnal behavior of *Lemur mongoz* at Ampijoroa, Madagascar. *Primatologica*, **29**, 291–302.

Hawkins, C. E. (1998). Behaviour and ecology of the fossa, *Cryptoprocta ferox* (*Carnivora: Viverridae*) in a dry deciduous forest, western Madagascar. Ph. D. thesis, University of Aberdeen.

Hemingway, C. A. & Overdorff, D. J. (1999). Sampling effects on food availability estimates: phenological method, sample size, and species composition. *Biotropica*, **31**, 354–64.

Hill, R. A. & Dunbar, R. I. M. (1998). An evaluation of the roles of predation rate and predation risk as selective pressures on primate grouping behaviour. *Behaviour*, **135**, 411–30.

Hill, R. A. & Lee, P. C. (1998). Predation risk as an influence on group size in cercopithecoid primates: implications for social structure. *Journal of Zoology, London*, **245**, 447–56.

Isbell, L. A. (1994). Predation on primates: ecological patterns and evolutionary consequences. *Evolutionary Anthropology*, **3**, 61–71.

Jacobs, G. H. (1993). The distribution and nature of colour vision among the mammals. *Biological Review*, **68**, 413–71.

Janson, C. H. (1998). Testing the predation hypothesis for vertebrate sociality: prospects and pitfalls. *Behaviour*, **135**, 389–410.

Karpanty, S. M. & Goodman, S. M. (1999). Diet of the Madagascar harrier-hawk, *Polyboroides radiatus*, in southeastern Madagascar. *Journal of Raptor Research*, **33**, 313–16.

Kooyman, G. L. (1975). A comparison between day and night diving in the Weddell seal. *Journal of Mammalogy*, **56**, 563–74.

Langrand, O. (1990). *Guide to the Birds of Madagascar*. New Haven: Yale University Press.

Layne, J. N. & Glover, D. (1985). Activity patterns of the common long-nosed armadillo *Dasypus novemcinctus* in south-central Florida. In *The Evolution and Ecology of Armadillos, Sloths, and Vermilinguas*, ed. G. G. Montgomery. Washington, DC: Smithsonian Institution Press, pp. 407–17.

Macedonia, J. M. & Stanger, K. F. (1994). Phylogeny of the Lemuridae revisited: evidence from communication signals. *Folia Primatologica*, **63**, 1–43.

Martin, R. D. (1972). Adaptive radiation and behaviour of the Malagasy lemurs. *Philosophical Transactions of the Royal Society of London*, **264**, 295–352.

— (1990). *Primate Origins and Evolution*. Princeton: Princeton University Press.

Mittermeier, R. A., Tattersall, I., Konstant, W. R., Meyers, D. M., & Mast, R. B. (1994). *Lemurs of Madagascar*. Washington, DC: Conservation International.

Müller, A. E. (1999). The social organisation of the fat-tailed dwarf lemur, *Cheirogaleus medius* (Lemuriformes: Primates). Ph. D. thesis, University of Zurich.

Mutschler, T. & Martin, R. D. (1999). Cathemerality in the Alaotran gentle lemur (*Hapalemur griseus alaotrensis*). In *The Alaotran Gentle Lemur (Hapalemur griseus alaotrensis): A Study in Behavioural Ecology*, ed. T. Mutschler. Ph. D. thesis, University of Zurich, pp. 65–87.

Mutschler, T., Feistner, A. T. C., & Nievergelt, C. M. (1998). Preliminary field data on group size, diet and activity in the Alaotran gentle lemur, *Hapalemur griseus alaotrensis*. *Folia Primatologica*, **69**, 325–30.

Nowak, R. M. (1991). *Walker's Mammals of the World*. Baltimore: Johns Hopkins University Press.

Overdorff, D. J. (1988). Preliminary report on the activity cycle and diet of the red-bellied lemur (*Lemur rubriventer*) in Madagascar. *American Journal of Primatology*, **16**, 143–53.

— (1996). Ecological determinants of social structure in *Eulemur fulvus rufus* and *Eulemur rubriventer* in Madagascar. *American Journal of Physical Anthropology*, **100**, 487–506.

Overdorff, D. J. & Rasmussen, M. A. (1995). Determinants of nighttime activity in “diurnal” lemurid primates. In *Creatures of the Dark: The Nocturnal Prosimians*, ed. L. Alterman, G. A. Doyle, & M. K. Izard. New York: Plenum Press, pp. 61–74.

Overdorff, D. J. & Strait, S. G. (1995). Life-history and predation in *Eulemur rubriventer* in Madagascar. *American Journal of Physical Anthropology*, **20** (Suppl.), 164.

Overdorff, D. J., Merenlender, A. M., Talata, P., Telo, A., & Forward, Z. A. (1999). Life history of *Eulemur fulvus rufus* from 1988–1998 in southeastern Madagascar. *American Journal of Physical Anthropology*, **108**, 295–310.

Overdorff, D. J., Strait, S. G., & Seltzer, R. G. (2002). Species differences in feeding in Milne-Edward's sifakas (*Propithecus diadema edwardsi*), rufus lemurs (*Eulemur fulvus rufus*), and red-bellied lemurs (*Eulemur rubriventer*) in southern Madagascar: implications for predator avoidance. In *Eat or Be Eaten: Predator Sensitive Foraging Among Primates*, ed. L. E. Miller. Cambridge: Cambridge University Press, pp. 126–37.

Peetz, A., Norconk, M. A., & Kinzey, W. G. (1992). Predation by jaguar on howler monkeys (*Alouatta seniculus*) in Venezuela. *American Journal of Primatology*, **28**, 223–8.

Perrin, M. R. & D'Inzillo Carranza, I. (2000). Activity patterns of the spotted-necked otter in the Natal Drakensberg, South Africa. *South African Journal of Wildlife Research*, **30**, 1–7.

Rakotondravony, D., Goodman, S. M., & Soarimalala, V. (1998). Predation on *Hapalemur griseus griseus* by *Boa manditra* (Boidae) in the littoral forest of eastern Madagascar. *Folia Primatologica*, **69**, 405–8.

Rasmussen, M. A. (1999). Ecological influences on activity cycle in two cathemeral primates, the mongoose lemur (*Eulemur mongoz*) and the common brown lemur (*Eulemur fulvus fulvus*). Ph. D. thesis, Duke University.

Rathbun, G. .B. & Gache, M. (1980). Ecological survey of the night monkey, *Aotus trivirgatus*, in Formosa Province, Argentina. *Primates*, **21**, 211–19.

Reichard, U. (1998). Sleeping sites, sleeping places, and pre-sleep behavior of gibbons (*Hylobates lar*). *American Journal of Primatology*, **46**, 35–62.

Reynolds, P. & Gorman, M. L. (1999). The timing of hunting in short-eared owls (*Asio flammeus*) in relation to the activity patterns of Orkney voles (*Microtus arvalis orcadensis*). *Journal of Zoology, London*, **247**, 371–9.

Risenhoover, K. L. (1986). Winter activity patterns of moose in interior Alaska. *Journal of Wildlife Management*, **50**, 727–34.

Serena, M. (1994). Use of time and space by platypus (*Ornithorhynchus anatinus*: Monotremata) along a Victorian stream. *Journal of Zoology, London*, **232**, 117–31.

Sommer, S. (2000). Sex-specific predation on a monogamous rat, *Hyogeomys antimena* (Muridae: Nesomyinae). *Animal Behaviour*, **59**, 1087–94.

Sussman, R. W. (1974). Ecological distinctions in sympatric species of *Lemur*. In *Prosimian Biology*, ed. R. D. Martin, G. A. Doyle, & A. C. Walker. Pittsburgh: University of Pittsburgh Press, pp. 75–108.

Sussman, R. W. & Rakotozafy, A. (1994). Plant diversity and structural analysis of a tropical dry forest in southwestern Madagascar. *Biotropica*, **26**, 241–54.

Sussman, R. W. & Tattersall, I. (1976). Cycles of activity, group composition, and diet of *Lemur mongoz* in Madagascar. *Folia Primatologica*, **26**, 270–83.

Tan, Y. & Li, W. -H. (1999). Trichromatic vision in prosimians. *Nature*, **402**, 36.

Tattersall, I. (1976). Group structure and activity rhythm in *Lemur mongoz* (Primates, Lemuriformes). *Anthropological Papers of the American Museum of Natural History*, **53**, 369–80.

— (1977). Ecology and behavior of *Lemur fulvus mayottensis*. *Anthropological Papers of the American Museum of Natural History*, **54**, 421–82.

— (1979). Patterns of activity in the Mayotte lemur, *Lemur fulvus mayottensis*. *Journal of Mammalogy*, **60**, 314–23.

— (1982). *The Primates of Madagascar*. New York: Columbia University Press.

— (1987). Cathemeral activity in primates: a definition. *Folia Primatologica*, **49**, 200–202.

Tattersall, I. & Sussman, R. W. (1975). Observations on the ecology and behavior of the mongoose lemur, *Lemur mongoz mongoz* Linnaeus (Primates, Lemuriformes), at Ampijoroa, Madagascar. *Anthropological Papers of the American Museum of Natural History*, **52**, 193–216.

Tonn, W. M. & Paszkowski, C. A. (1987). Habitat use of the central mudminnow (*Umbra limi*) and yellow perch (*Perca flavescens*) in *Umbra-Perca* assemblages: the roles of competition, predation and the abiotic environment. *Canadian Journal of Zoology*, **65**, 862–70.

Treves, A. (2002). Predicting predation risk for foraging, arboreal monkeys. In *Eat or Be Eaten: Predator Sensitive Foraging Among Primates*, ed. L. E. Miller. Cambridge: Cambridge University Press, pp. 222–41.

Valdimarsson, S. K., Metcalfe, N. B., Thorpe, J. E., & Huntingford, F. A. (1997). Seasonal changes in sheltering: effect of light and temperature on diel activity in juvenile salmon. *Animal Behaviour*, **54**, 1405–12.

Valenzuela, D. & Ceballos, G. (2000). Habitat selection, home range, and activity of the white-nosed coati (*Nasua narica*) in a Mexican tropical dry forest. *Journal of Mammalogy*, **81**, 810–19.

Van Schaik, C. P. (1983). Why are diurnal primates living in groups? *Behaviour*, **87**, 120–44.

Van Schaik, C. P. & Griffiths, M. (1996). Activity periods of Indonesian rain forest mammals. *Biotropica*, **28**, 105–12.

Van Schaik, C. P. & Kappeler, P. M. (1996). The social systems of gregarious lemurs: lack of convergence with anthropoids due to evolutionary disequilibrium? *Ethology*, **102**, 915–41.

Walls, G. L. (1967). *The Vertebrate Eye and its Adaptive Radiation*. New York: Hafner Publishing.

Wilson, J. M., Stewart, P. D., Ramangason, G. -S., Denning, A. M., & Hutchings, M. S. (1989). Ecology and conservation of the crowned lemur, *Lemur coronatus*, at Ankarana, N. Madagascar. *Folia Primatologica*, **52**, 1–26.

Winter, J. W. (1996). Australasian possums and Madagascan lemurs: behavioural comparison of ecological equivalents. In *Comparisons of Marsupial and Placental Behaviour*, ed. D. B. Croft & U. Gansloβer. Fürth: Filander Verlag, pp. 262–92.

Wolin, L. R. & Massopust, L. C. (1970). Morphology of the primate retina. In *The Primate Brain*, ed. C. R. Noback & W. Montagna. New York: Appleton-Century-Crofts, pp. 1–27.

Woodall, P. F., Woodall, L. B., & Bodero, D. A. V. (1989). Daily activity patterns in captive elephant shrews (Macroscelididae). *African Journal of Ecology*, **27**, 63–76.

Wright, P. C. (1984). Biparental care in *Aotus trivirgatus* and *Callicebus moloch*. In *Female Primates: Studies by Women Primatologists*, ed. M. F. Small. New York: Alan R. Liss, pp. 59–75.

— (1985). The costs and benefits of nocturnality for *Aotus trivirgatus* (the night monkey). Ph. D. thesis, City University of New York.

— (1989). The nocturnal primate niche in the New World. *Journal of Human Evolution*, **18**, 635–58.

— (1994). The behavior and ecology of the owl monkey. In *Aotus: The Owl Monkey*, ed. J. F. Baer, R. E. Weller, & I. Kakoma. New York: Academic Press, pp. 97–112.

— (1998). Impact of predation risk on the behaviour of *Propithecus diadema edwardsi* in the rain forest of Madagascar. *Behaviour*, **135**, 483–512.

— (1999). Lemur traits and Madagascar ecology: coping with an island environment. *Yearbook of Physical Anthropology*, **42** (**Suppl. 29**), 31–72.

Wright, P. C., Heckscher, S. K., & Dunham, A. E. (1997). Predation on Milne-Edward's sifaka (*Propithecus diadema edwardsi*) by the fossa (*Cryptoprocta ferox*) in the rain forest of southeastern Madagascar. *Folia Primatologica*, **68**, 34–43.

5 *Physiological adaptations to seasonality in nocturnal primates*

JUTTA SCHMID
Department of Experimental Ecology, University of Ulm, Albert Einstein Allee 11, D-89069 Ulm, Germany

PETER M. KAPPELER
Deutsches Primatenzentrum, Kellnerweg 4, 37077 Göttingen, Germany

Introduction

The current geographic distribution of primates is confined largely to tropical and subtropical regions, where they have colonized a variety of habitats. The majority of primate taxa inhabit tropical forests with little annual fluctuation in environmental conditions. Some species, however, live in habitats characterized by pronounced seasonal fluctuations in climate and or resource availability. These primates tend to live at relatively high latitudes or altitudes, or both. Primates in such seasonal habitats provide an opportunity to identify behavioral and physiological adaptations that enable them to cope with fluctuating environmental conditions. Furthermore, it is interesting to ask whether and how schedules of growth and reproduction are adapted to maximize individual reproductive success under such seasonal conditions, because they may have to be traded off against maintenance requirements during the lean part of the year.

Primates living in seasonal environments exhibit a number of specific behavioral, ecological, and physiological adaptations. For example, during the climatically and or energetically most stressful time of year, they may reduce energy expenditure, e.g. by reducing overall activity, and many have scheduled periods of growth and infant weaning to coincide with seasons of relative abundance. Behavioral and physiological mechanisms of thermoregulation play especially important roles in maintaining homeostasis in seasonally stressed primates. These mechanisms are importantly influenced by circadian activity patterns because diurnal and nocturnal animals are exposed to fundamentally different constraints and options in this respect. The guild of strictly nocturnal primates, which includes members of independent major primate radiations inhabiting parts of

Seasonality in Primates: Studies of Living and Extinct Human and Non-Human Primates, ed. Diane K. Brockman and Carel P. van Schaik. Published by Cambridge University Press.

5 *Physiological adaptations to seasonality in nocturnal primates*

JUTTA SCHMID
Department of Experimental Ecology, University of Ulm, Albert Einstein Allee 11, D-89069 Ulm, Germany

PETER M. KAPPELER
Deutsches Primatenzentrum, Kellnerweg 4, 37077 Göttingen, Germany

Introduction

The current geographic distribution of primates is confined largely to tropical and subtropical regions, where they have colonized a variety of habitats. The majority of primate taxa inhabit tropical forests with little annual fluctuation in environmental conditions. Some species, however, live in habitats characterized by pronounced seasonal fluctuations in climate and or resource availability. These primates tend to live at relatively high latitudes or altitudes, or both. Primates in such seasonal habitats provide an opportunity to identify behavioral and physiological adaptations that enable them to cope with fluctuating environmental conditions. Furthermore, it is interesting to ask whether and how schedules of growth and reproduction are adapted to maximize individual reproductive success under such seasonal conditions, because they may have to be traded off against maintenance requirements during the lean part of the year.

Primates living in seasonal environments exhibit a number of specific behavioral, ecological, and physiological adaptations. For example, during the climatically and or energetically most stressful time of year, they may reduce energy expenditure, e.g. by reducing overall activity, and many have scheduled periods of growth and infant weaning to coincide with seasons of relative abundance. Behavioral and physiological mechanisms of thermoregulation play especially important roles in maintaining homeostasis in seasonally stressed primates. These mechanisms are importantly influenced by circadian activity patterns because diurnal and nocturnal animals are exposed to fundamentally different constraints and options in this respect. The guild of strictly nocturnal primates, which includes members of independent major primate radiations inhabiting parts of

Seasonality in Primates: Studies of Living and Extinct Human and Non-Human Primates, ed. Diane K. Brockman and Carel P. van Schaik. Published by Cambridge University Press.

Asia, Madagascar, Africa, and the New World, therefore provide an opportunity to study convergent solutions to challenges imposed by seasonal habitats in a natural group of animals. Nocturnal strepsirrhine primates are especially interesting in this context because they have additional behavioral and physiological mechanisms at their disposal to cope with adverse environmental conditions, such as torpor, not available to haplorhines. Furthermore, they tend to have faster life histories (Harvey & Clutton-Brock 1985; Harvey *et al.* 1987; Kappeler 1996), which may provide them with more phenotypic plasticity in their life history responses to changing environmental conditions (Lee & Kappeler 2003).

In this chapter, we review information on adaptations of nocturnal primates to seasonal environments. Specifically, we relate the occurrence of seasonal changes in body mass, metabolic rate, body temperature, and energy expenditure in lemurs, galagos, lorises, tarsiers, and owl monkeys to changes in external conditions. The most extreme adaptation to cope with seasonal energy shortages and low temperatures is daily and prolonged torpor, which is found only among Malagasy lemurs. We use detailed field measurements from two species of mouse lemurs to illustrate this most extreme adaptation in more detail. Finally, we summarize available information on life history variation of nocturnal primates that may constitute adaptations to seasonal habitats.

Nocturnal primates in seasonal environments

Nocturnal primates are found today in four superfamilies in as many biogeographic regions. First, they include the majority of Malagasy Lemuroidea from four families (i.e. all Cheirogaleidae, Lepilemuridae, *Daubentonia*, and *Avahi*). Cathemeral lemurs with regular substantial amounts of nocturnal activity (*Eulemur*, *Hapalemur*) are not considered here because their ecophysiology remains poorly studied. All of Madagascar's 29 currently recognized nocturnal lemur species can be considered to inhabit environments with more or less pronounced seasonal variation in climate and resource abundance (Richard & Dewar 1991; Wright 1999). Figure 5.1 illustrates these seasonal changes in two relevant environmental factors, rainfall and ambient temperature, over the year at different sites in Madagascar.

Second, all bushbabies (Galagidae) and lorises (Lorisidae) are strictly nocturnal. Bushbabies are distributed widely in sub-Saharan Africa with at least 17 species in four genera (*Galago*, *Galagoides*, *Otolemur*, and *Euoticus*), of which at least seven species inhabit some seasonal environments in southern and eastern Africa. Lorises are found with a total of at

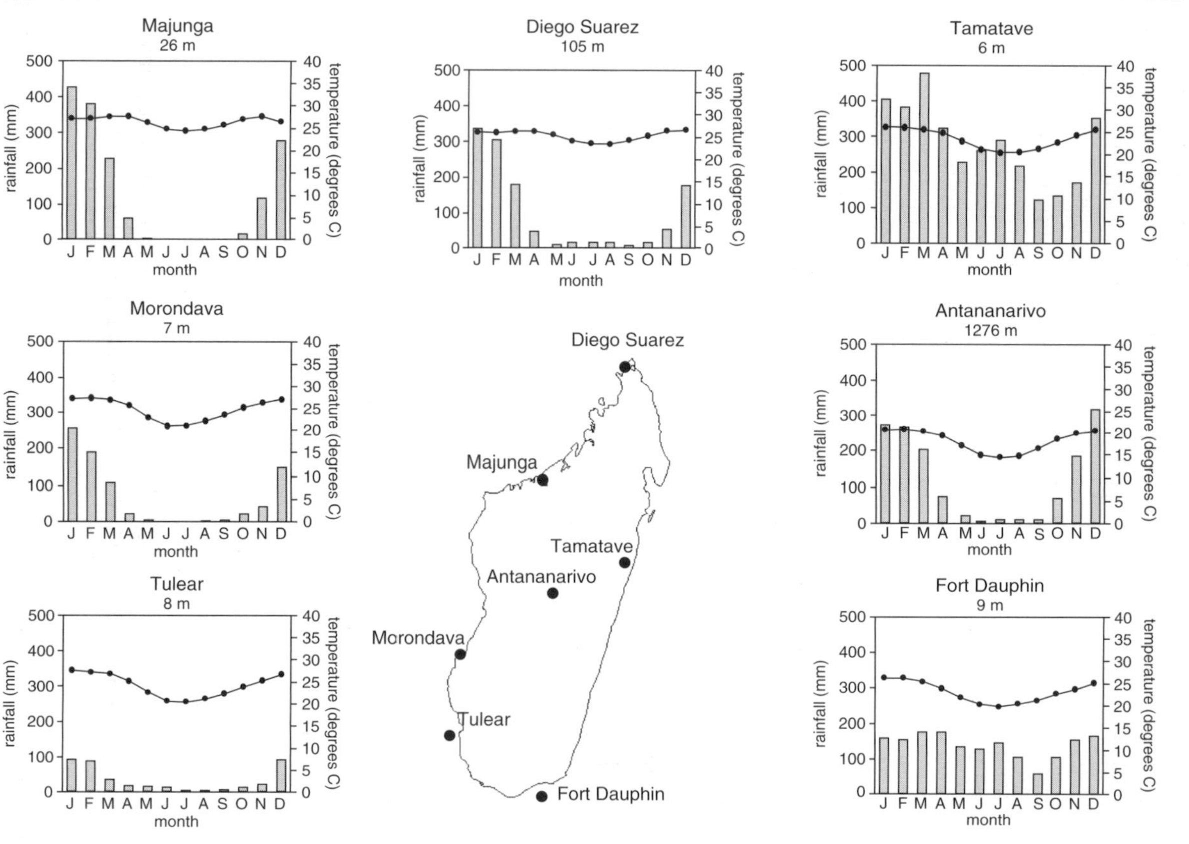

Figure 5.1 Seasonal changes of rainfall (hatched bars) and ambient temperature (lines) over the year at different sites in Madagascar. Values are monthly means; location and altitude of each site is given.

least 18 species from five genera in Equatorial western and central Africa (*Perodicticus*, *Arctocebus*, and *Pseudopotto*), the Indian subcontinent (*Loris*), and southeast Asia (*Nycticebus*), where especially *Perodicticus* and *Loris* inhabit seasonal environments. Finally, among haplorhines, nocturnal activity evolved twice independently from diurnal ancestors: once in Asian tarsiers (*Tarsius*) and once in New World owl monkeys (*Aotus*). The five species of *Tarsius* are restricted to Indonesian and Philippine islands near the Equator with little seasonality. The eight species of owl monkey *Aotus* inhabit a wide range of habitats throughout South America, including seasonally deciduous scrub forest to high-altitude cloud forest up to the timberline at 3200 m (Wright 1981).

Nocturnal primates inhabiting seasonal environments have to cope with three main types of stress – low temperatures, long dry seasons, and seasonal food scarcity – or any combination thereof. First, in western and southern Madagascar, the austal winter is characterized by up to seven months without rain and minimum temperatures below 10 °C (Hladik *et al.* 1980; Sorg & Rohner 1996). Many trees shed their leaves and exhibit marked annual cycles in the presence of new leaves, flowers, and fruit. In eastern and northern Malagasy rainforests, these effects are less marked but nevertheless noticeable (Albrecht *et al.* 1990; Atsalis 1999; Overdorff 1993). Similar conditions prevail in southern and eastern Africa and in southern India (Bearder & Doyle 1974; Harcourt & Nash 1986; Nekaris 2000). Second, nocturnal primates utilize a variety of different foods. The majority is omnivorous, feeding mainly on fruit and animal matter. Some (*Lepilemur*, *Avahi*) are specialized folivores, a few others (*Phaner*, *Euoticus*) are gum specialists, and tarsiers are exclusive carnivores. The seasonal availability of these dictary items is variable, but especially high-energy fruit and animal prey may become rare during the dry season. Finally, nocturnal primates in most of these areas are faced with low temperatures during their periods of activity, combined with relatively high temperatures during their inactivity, which poses particular challenges for their thermoregulation and metabolic homeostasis not faced by diurnal animals.

How do nocturnal primates cope with seasonal stress?

Heat production

Conservation of energy relies on both physiological and behavioral thermoregulatory mechanisms. One important strategy employed by prosimians to optimize their energy balance is to lower resting metabolic rate (RMR), typically by 20–50% below the mammalian mass-specific standard (reviewed in Müller *et al.* 1983; Müller 1985; Genoud *et al.* 1997; Schmid & Ganzhorn 1996) (see Table 5.1). This depressed level of heat production generally is accompanied by a reduction of body temperature (T_b). Several species of lorises (*Nyticebus coucang* [Müller 1979]; *Perodicticus potto* [Hildwein & Goffart 1975]), bushbabies (*Galago senegalensis, Galagoides demidovii* [Dobler 1978]; *Otolemur crassicaudatus* [Müller & Jaschke 1980]), and cheirogaleids (*Cheirogaleus medius* [Russel 1975; Dausmann *et al.* 2000]; *Microcebus murinus* [Ortmann *et al.* 1997; Schmid 2000]; *M. berthae* [formerly *M. myoxinus*; see Rasoloarison *et al.* 2000]; Schmid *et al.* [2000]) exhibit circadian T_b rhythms. Instead of maintaining constant T_b day round, they are able to lower body temperature to around 32°C during their daily periods of inactivity, thereby saving energy for heat production. However, other New World primates exhibit similar T_b fluctuation as similar-sized strepsirrhines (Morrison & Middleton 1967; Dawson *et al.* 1979). Thus, the combination of hypometabolism and reduced body temperature can contribute considerably to energy savings in some nocturnal primates.

An additional strategy for saving energy relies on seasonal adjustments of resting metabolic rate. This mechanism has been reported for the gray mouse lemur (*M. murinus*), a species in which resting metabolic rate of both males and females greatly decreased during the cold and dry season (Perret 1998) (but see Génin & Perret [2000]). Perret (1998) described a mean seasonal reduction in metabolic rate of 20%, which is likely based on seasonal changes in thyroid, corticoadrenal, and gonadal functions (Perret 1992). This energy-conservation strategy has also been described in other Malagasy mammals, such as tenrecids (Stephenson & Racey 1994), but it remains unknown whether it occurs in other nocturnal primates. *Aotus trivirgatus*, the only nocturnal haplorhine for which data on resting metabolic rate are available, has only a slightly lower basal rate than expected (mean 95%) (Table 5.1). Low relative metabolic rates in some lemurs and tenrecs are thought to be adaptations that have evolved in parallel with ecological specialization in response to the seasonality faced by these species (Schmid & Stephenson 2004).

Torpor

The most extreme adaptations to seasonal food scarcity and low ambient temperatures are daily and prolonged torpor. Daily torpor is characterized by a dormancy bout duration of less than 24 hours, whereas hibernation is characterized by a sequence of prolonged torpor bouts with an

Table 5.1 *Summary of life history traits and physiological parameters of nocturnal primates*

Species	Body mass (g)	Diet	Litter size	AFR (months)	IBI (months)	Torpor	BMR as % of expected	Reference
Aotus azarai	1230	–	–	–	–	–	–	
Aotus hershkovitzi	–	–	–	–	–	–	–	
Aotus lemurinus	874	–	–	–	12	–	–	
Aotus miconax	–	–	–	–	–	–	–	
Aotus nancymaae	780	–	–	–	–	–	–	
Aotus nigriceps	1040	–	–	–	–	–	–	
Aotus trivirgatus	1030	Om	1	28	6–14	–	95	Le Maho *et al.* (1981)
Aotus vociferans	698	–	–	–	–	–	–	
Allocebus trichotis	85	Om	–	–	–	–	–	
Cheirogaleus major	356	Om	2–3	–	–	Yes	–	
Cheirogaleus adipicaudatus	–	–	–	–	–	–	–	
Cheirogaleus crossleyi	–	–	–	–	–	–	–	
Cheirogaleus sibreei	–	–	–	–	–	–	–	
Cheirogaleus ravus	–	–	–	–	–	–	–	
Cheirogaleus minusculus	–	–	–	–	–	–	–	
Cheirogaleus medius	139	Om	1–4	12	12	Yes	91	McCormick (1981)
Microcebus berthae	31	Om	–	–	–	Yes	152	Schmid *et al.* (2000)
Microcebus griseorufus	63	–	–	–	–	–	–	
Microcebus murinus	63	Om	1–3	12	12	Yes	80	Perret *et al.* (1998)
Microcebus myoxinus	49	–	–	–	–	–	–	
Microcebus ravelobensis	72	Om	–	–	–	–	–	
Microcebus rufus	43	Om	1–3	–	–	Yes	–	
Microcebus sambiranensis	44	–	–	–	–	–	–	
Microcebus tavaratra	61	–	–	–	–	–	–	
Mirza coquereli	297	Om	2	23	12	No	–	
Phaner furcifer	328	Gu	1	–	–	No	–	

Daubentonia madagascariensis	2572	Om	1	–	20	No	–	
Lepilemur dorsalis		Fo	–	–	–	No	–	
Lepilemur edwardsi	934	Fo	–	–	–	No	–	
Lepilemur leucopus	594	Fo	–	–	–	No	–	
Lepilemur microdon	970	Fo	–	–	–	No	–	
Lepilemur mustelinus	600	Fo	–	22	–	No	–	
Lepilemur ruficaudatus	779	Fo	1	–	–	No	57	Schmid & Ganzhorn (1996)
Lepilemur septentrionalis	700	Fo	–	–	–	No	–	
Avahi laniger	1316	Fo	–	–	12	No	–	
Avahi occidentalis	777	Fo	–	–	–	No	–	
Avahi unicolor	–	–	–	–	–	–	–	
Arctocebus aureus	210	Ca	–	–	–	No	–	
Arctocebus calabarensis	298	Ca	–	14	6	No	–	
Loris grandis	–	–	–	–	–	No	–	
Loris lydekkerianus	269	–	–	–	–	No	–	
Loris malabaricus	193	–	1	–	–	No	–	
Loris nordicus	–	–	–	–	–	No	–	
Loris nycticeboides	–	–	–	–	–	No	–	
Loris tardigradus	255[a]	Om	–	18	6	No	62	Müller (1985)
Nycticebus bengalensis	1020	–	–	–	–	No	–	
Nycticebus coucang	626	Om	1	–	12	No	50	Müller (1979)
Nycticebus intermedius	–	–	–	–	–	No	–	
Nycticebus javanicus	798	–	–	–	–	No	–	
Nycticebus menagensis	511	–	–	–	–	No	–	
Nycticebus pygmaeus	376[a]	–	2	–	–	No	–	
Perodicticus edwardsi	1210	–	–	–	–	No	–	
Perodicticus ibeanus	–	–	–	–	–	No	–	
Perodicticus potto	836	Om	1	25	12	No	71	Hildwein & Goffart (1975)
Pseudopotto martini	–	–	–	–	–	No	–	
Euoticus elegantulus	293	Om	1	–	–	No	–	
Euoticus pallidus	–	–	–	–	–	No	–	
Galago alleni	269	Om	–	10	12	No	–	
Galago gabonensis	–	–	–	–	–	No	–	

Table 5.1 (*cont.*)

Species	Body mass (g)	Diet	Litter size	AFR (months)	IBI (months)	Torpor	BMR as % of expected	Reference
Galago gallarum	200	Om	–	–	–	No	–	
Galago matschiei	212	Om	1	–	–	No	–	
Galago moholi	188	Ca	1	12	6	No	90	Dobler (1978)
Galago senegalensis	199	Om	1–2	10	6	No	98	Dobler (1978)
Galagoides demidoff	60	Om	2	12	12	No	86	Dobler (1978)
Galagoides granti	–	–	–	–	–	No	–	
Galagoides orinus	–	–	–	–	–	No	–	
Galagoides rondoensis	–	–	–	–	–	No	–	
Galagoides thomasi	130	–	–	–	–	No	–	
Galagoides udzungwensis	–	–	–	–	–	No	–	
Galagoides zanzibaricus	137	Om	1	12	6	No	–	
Otolemur crassicaudatus	1110	Om	2	26	12	No	84	Müller & Jaschke (1980)
Otolemur garnettii	800	Om	1–2	19	12	No	–	
Tarsius bancanus	117	Ca	1	24	8	No	–	
Tarsius dianae	107	Ca	1	–	–	No	–	
Tarsius pumilis	–	Ca	–	–	–	No	–	
Tarsius spectrum	108	Ca	1	17	5	No	–	
Tarsius syrichta	117[a]	Ca	1	–	–	No	72	McNab & Wright (1987)

[a] Body mass from captive animals. AFR, age of first reproduction; BMR, basal metabolic rate; Ca, carnivores; Fo, folivores; Gu, gum-eating; In, insectivores; IBI, interbirth interval; Om, omnivores; – = information uncomplete or not available.

Percentage of expected is (BMR) expressed as a precentage of the value expected using the allometric equation $BMR = 3.53M^{0.72}$, where M is mass in grams (McNab 1988). References are given for data on BMR. Unless indicated otherwise, data from Appendix in Kappeler and Pereira (2003).

average bout duration of two weeks (see Hudson [1973], Heldmaier & Ruf [1992], Geiser & Ruf [1995], and Wang [1989] for reviews). Generally, torpor is a regulated state of physiological dormancy during which body temperature can be lowered to levels close to ambient temperatures and metabolic rate can be reduced to as little as 5% of its normothermic level. Among primates, torpor is found only in the small cheirogaleids *Microcebus* and *Cheirogaleus* (Bourlière & Petter-Rousseaux 1966; Martin 1972, 1973; Russel 1975; Chévillard 1976; Petter-Rousseaux 1980; McCormick 1981; Schmid 1996, 1997; Ortmann *et al.* 1997; Dausmann *et al.* 2000) and possibly in *Allocebus* (Wright & Martin 1995). For many years, it was believed that the ability of *Microcebus* and *Cheirogaleus* to become torpid was the result of a primitive thermoregulatory system, but it is now clear that it is part of these lemurs' energy-saving strategy (Müller 1985).

Numerous studies on the thermoregulatory capacities of mouse lemurs, both in the laboratory (Russel 1975; Chévillard 1976; Aujard *et al.* 1998; Perret *et al.* 1998) and in the field (Ortmann *et al.* 1996, 1997; Schmid 1996, 1997, 2000; Schmid *et al.* 2000), have contributed importantly to our understanding of the utilization of torpor and its energetic consequences. In wild *M. murinus*, spontaneous torpor occurred on a daily basis during the dry season (Fig. 5.2), with torpor bouts lasting from 3 to 18 hours and extending into the nocturnal activity phase. Body temperatures during daily torpor dropped to values of less than 10 °C and metabolic rates were reduced to about 20–30% of normothermic level. Arousal from torpor was a two-step process, with an initial passive heating phase, during which the T_b of animals increased to a mean value of 27 °C, carried by the daily increase of ambient temperature (T_a) without a noticeable increase in metabolic rate, followed by endogenous heat production to further raise T_b to normothermic values (Fig. 5.3).

Berthe's mouse lemur, *M. berthae*, is found only in the Forêt de Kirindy in western Madagascar, where it occurs sympatrically with *M. murinus* and consequently is exposed to the same environmental conditions (Schmid & Kappeler 1994). Studies on the thermal and metabolic physiology of *M. berthae* revealed that this species is capable of entering into, and spontaneously arousing from, daily torpor during the dry season when T_a is low (Ortmann *et al.* 1996; Schmid 1996; Schmid *et al.* 2000). Torpid *M. berthae* also arose from lethargy using the rising T_a to rewarm T_b. Thus, the energy-conserving mechanisms of these two mouse lemur species, with passive exogenous heating during arousal from torpor, low minimum torpor T_bs, and extended torpor bouts into the activity phase, comprise

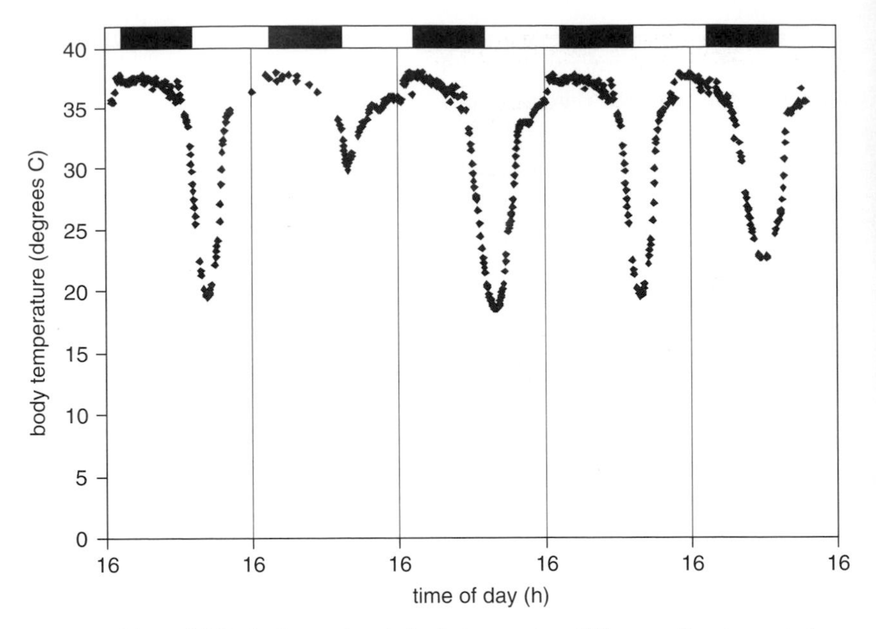

Figure 5.2 Daily fluctuations in body temperature (T_b) over a five-day period between May 4 and 8, 1994, for a *Microcebus murinus* individual entering torpor on a daily basis (implanted radio transmitter). Vertical lines indicate 16 hours in the afternoon; black horizontal bars show the dark phase.

an important and highly adapted mechanism to minimize energetic costs in response to harsh environmental conditions.

Dwarf lemurs (*Cheirogaleus* spp.) are exceptional among primates because in some species, both males and females show extended periods of torpor during the dry season in Madagascar (Petter-Rousseaux 1980). Measurements of body temperature and metabolic rate in hibernating *C. medius* in their natural habitat revealed that the animals' T_b fluctuated over more than 10 °C (between 18.3 and 31.9 °C), following passively the diurnal rise and nocturnal fall of ambient temperatures (Dausmann *et al.* 2000). These lemurs continued to hibernate even when ambient temperatures rose above 30 °C. Complete arousals during hibernation were never recorded in this species, although periodic arousals have been proposed to be a crucial factor for the maintenance of body functions in hibernators (Willis 1982). This lack of periodic interruptions in hibernating *Cheirogaleus* is unique and may represent an adaptation to further save energy during long-term hibernation because periodic arousals are energetically expensive (Dausmann *et al.* 2000). The ability of several newly described species of *Cheirogaleus* (Groves 2000) to enter

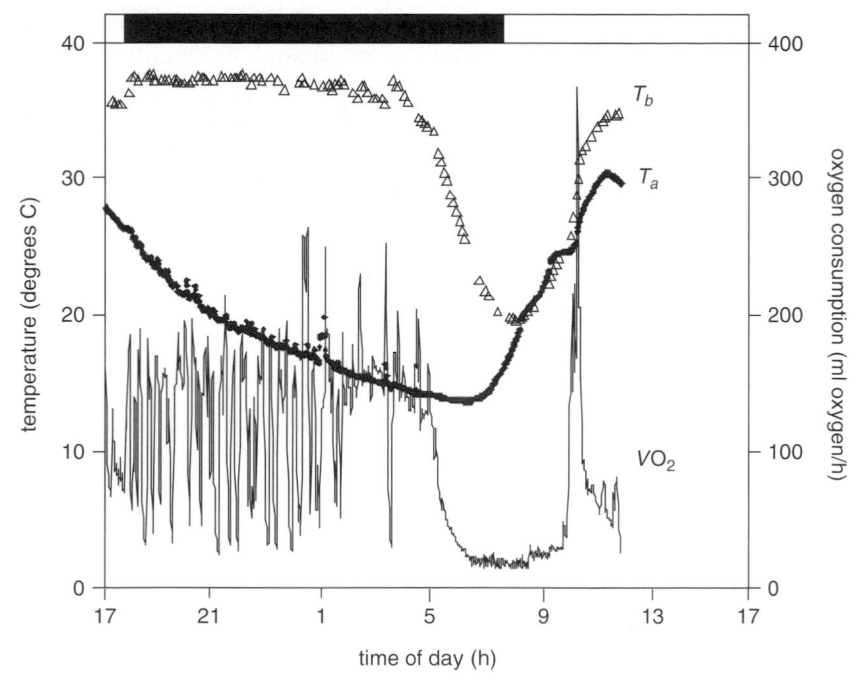

Figure 5.3 Oxygen consumption (VO_2), body temperature (T_b), and ambient temperature (T_a) over a 24-hour period of a torpid *Microcebus murinus* individual measured in Kirindy forest (near Morondava, Madagascar). The dark horizontal bar indicates period of darkness. Redrawn from Ortmann *et al.* (1997).

torpor remains to be determined. Recent studies on the energy-saving strategies of free-ranging *M. murinus* revealed that apart from the use of daily torpor, this species is also a "classical" hibernator with an inactivity phase of several weeks throughout the cold and dry season (Schmid & Kappeler 1998; Schmid 1999, 2000). In contrast to *C. medius*, however, hibernation in *M. murinus* is sexually skewed, with significantly more females than males entering prolonged periods of torpor. During hibernation, individual T_b regularly fluctuated over more than 20 °C from torpid to normothermic levels during the course of the day (Schmid 2000). Thus, even during hibernation, mouse lemurs reached normothermic T_bs everyday but decreased their T_bs when ambient temperatures dropped again. Such fluctuations of body temperature found in hibernating *C. medius* and *M. murinus* are exceptional and previously were unknown for any hibernating mammal species.

The thermoregulatory mechanisms employed by other cheirogaleids are still poorly known. Prolonged periods of torpor are found in neither

M. ravelobensis (Schmelting *et al.* 2000) nor *M. berthae* (Schmid *et al.* 2000), and nothing is known about the physiology of the other mouse lemur species *M. sambiranensis*, *M. tavaratra*, *M. griseorufus*, and *M. myoxinus*. Similarly, there is very little information from closely related inhabitants of Madagascar's rainforests, some of which also experience cool months with reduced precipitation and scarcity of resources. First, hairy-eared dwarf lemur (*Allocebus trichotis*) may be able to enter torpor and remain inactive for several months (Wright & Martin 1995; Rakotoarison *et al.* 1997), but this suggestion is based on reports from local people alone. Second, short-term observations of the greater dwarf lemur (*Cheirgaleus major*) indicated the presence of prolonged periods of inactivity (Wright & Martin 1995), but neither year-round behavioral studies nor any physiological studies of this species have been carried out so far. Finally, in rufous mouse lemurs (*Microcebus rufus*), a high proportion of individuals of both sexes entered torpor during the cooler and drier season suggested by their absence from traps for more than one month during the dry season (Atsalis 1999). However, unlike for *M. murinus* (Schmid & Kappeler 1998; Schmid 1999), there was no clear evidence to suggest that for *M. rufus*, the highly male-biased sex ratio found between June and September may also indicate different sexual strategies for coping with food scarcity. Measurements of T_b and metabolic rate clearly are needed now to clarify the thermoregulatory strategies of *M. rufus*, and further ecophysiological research is required urgently to examine potential latitudinal and climatic effects on the occurrence of torpor in other cheirogaleids.

None of the other nocturnal primates are known to enter daily or extended torpor (Table 5.1), although it has been reported that some species of bushbabies become inactive on cold nights in their natural environment to reduce energy expenditure (Martin & Bearder 1979). However, it remains unclear whether reduced activity in free-ranging bushbabies is associated with torpor, since T_b and metabolic rate have not been measured. According to laboratory studies, galagos strictly maintained T_b at non-torpid levels (Hiley 1976; Dobler 1978; Müller & Jaschke 1980; Knox & Wright 1989). The only field study of galago physiology revealed that T_b of *Euticus elegantulus*, but not of sympatric *Galago alleni*, decreased significantly with lowered T_a, but only to about 34 °C (Vincent 1978). Body temperatures of slender lorises (*Loris tardigradus*) (Müller 1985), slow lorises (*Nycticebus coucang*) (Müller 1979), and pottos (*Perodicticus potto*) (Hildwein & Goffart 1975) were also regulated above 30 °C during the resting phase when exposed to ambient temperatures between 25 and 32 °C.

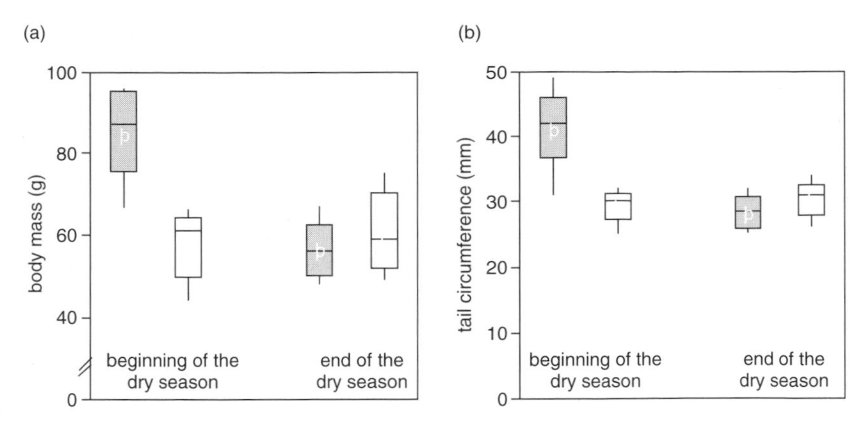

Figure 5.4 Medians (10, 25, 75, and 90% quartiles) of (a) body mass and (b) tail circumference of hibernating *Microcebus murinus* females ($n = 19$; hatched bars) and non-hibernating *M. murinus* of both sexes ($n = 37$; open bars) before and after the dry season, measured in Kirindy forest, Madagascar. Data from Schmid (1999).

Body reserves

Driven by these physiological adaptations to seasonality, mouse lemurs show pronounced annual cycles of body mass changes (*M. rufus*: Atsalis [1999]; *M. murinus*: Bourlière & Petter-Rousseaux [1966]; Petter-Rousseaux [1980]; Fietz [1998]). In particular, mouse lemurs that entered hibernation increased body mass by depositing fat reserves in body and tail, which they burnt up during subsequent months of inactivity (Schmid & Kappeler 1998; Schmid 1999). In Kirindy, adult *M. murinus* females exhibited prolonged torpor for up to five months throughout the cool dry season (Schmid & Kappeler 1998; Schmid 1999). Before the onset of the inactive phase, females increased body mass and tail circumference by about 30%, compared with their pre-hibernation values, which they then lost during the months of hibernation (Fig. 5.4). Adult males, in contrast, did not show any significant changes in body mass, they hardly ever remained inactive for more than a few days, and they were all continuously active weeks before females re-emerged from hibernation (Fietz 1998; Schmid & Kappeler 1998; Schmid 1999; Rasoazanabary 2001). In Ranomafana National Park, some male and female *M. rufus* exhibited annual body mass fluctuations, with the highest values found just around the onset of the cool season (Atsalis 1999). However, other individuals exhibited either no change in body fat or increased body fat over the course of the dry cool season. Thus, age, social status, and probably individual response to seasonality may influence the energy saving strategies of *M. rufus* (Atsalis 1999).

Seasonal accumulation of fat in body and tail during the periods of seasonal activity, followed by torpor, characterizes both *Cheirogaleus major* (Petter *et al.* 1977; Wright & Martin 1995) and *C. medius* (Hladik *et al.* 1980; McCormick 1981). The smaller *C. medius* were able to double their body mass within a few weeks by reducing locomotor activity and consuming fruit species with a high sugar content (Fietz & Ganzhorn 1999). These fat deposits support the energetic costs of overwintering and enable these lemurs to survive the lean dry season without additional food. At Ranomafana, body masses of adult *C. major* varied between 235 and 470 g (Wright & Martin 1995) but, so far, no data on annual body mass fluctuations are available for this species.

Behavioral mechanisms

Behavioral mechanisms to conserve energy include selection of micro-habitats, changes in body posture, and changes in the degree of sociality during resting periods (see also Morland [1993] and Pereira *et al.* [1999]). These behaviors reduce heat loss due to a reduction of exposed surface and local heating. Nocturnal primates rest either alone or in sleeping groups, and they either use shelters or do not. Self-constructed nests and tree holes are used by most of the lemurs, galagos, and owl monkeys, whereas lorises and tarsiers prefer to rest in dense vegetation (summarized by Kappeler [1998]). The energetic consequences of these combinations of resting behavior remain poorly documented. In captive *M. murinus*, individual energetic expenditure was reduced by 20–40% when animals nested together (Perret 1998). Furthermore, a study on the quality and insulation capacity of tree holes used by *M. murinus* revealed that utilization of tree holes prolonged the duration of torpor bouts and led to an increase in energy saving of 5% (Schmid 1998). Nest-sharing and availability of optimal microclimates therefore may be crucial for maintaining viable populations of mouse lemurs (e.g. Ganzhorn & Schmid [1998]; Radespiel *et al.* [1998]). Furthermore, circadian homeothermy in *Galago alleni* has been attributed to their use of insulated tree holes, whereas sympatric *Euoticus elegantulus*, which sleep on open branches, saved energy by lowering T_b with falling T_a (Vincent 1978). Whether taxonomic differences in shelter use are related to the number and developmental state of neonates (Kappeler 1998), and whether thermoregulatory constraints on adults influence daytime shelter choice (as well), remain to be determined in detail by surveying species with large geographical ranges or by comparing sympatric species with different resting behavior.

A second set of behavioral mechanisms to cope with shortages in resource base employed by primates that live in seasonal environments includes modification of foraging behavior and habitat use. For example, during periods of low resource availability, wild spectral tarsiers (*Tarsus spectrum*) increased time spent traveling and foraging compared with their activity budget in the wet season at the expense of time allocated to social behavior and resting (Gursky 2000). They also changed the type of insects that they consumed, and the location where they hunted them, between seasons. In Ampijoroa (northwestern Madagascar), *M. murinus* had smaller home ranges during the months of food shortages compared with those during the rainy season (Schmelting *et al.* 2000). In contrast, seasonal differences in mean home ranges occurred neither in *M. murinus* at Kirindy in western Madagascar (Schmid & Speakman 2000) nor in sympatric *M. berthae* (Schwab 2000). *Lepilemur leucopus* at Beza Mahafaly maintained low costs of locomotion during the cool season by resting significantly more and traveling significantly less than in the hot rainy season (Nash 1998). Thus, nocturnal primates do not differ from their diurnal relatives (see Chapter 3) in that they follow one of two energy-saving strategies at this level: either a reduction in energy expenditure or an increase in foraging effort.

Seasonality and life history strategies

Species-specific life history schedules describe modal adaptive strategies of maintenance, growth, and reproduction under local ecological conditions (Stearns 1976, 1992). Principal specific life history traits determine the size of an organism at birth, how fast and how long it will grow, the age and size at which it will mature, the number, size, and sex of its offspring, and its reproductive and total lifespan. Because the vast majority of data on life history traits of nocturnal primates come from captive studies (Kappeler 1995), few meaningful comparisons between seasonal and non-seasonal taxa are currently possible.

Because of their relatively small size, most nocturnal primates are, at least theoretically, able to reproduce more than once per year. Species inhabiting seasonal habitats therefore may experience a reduction in potential reproductive rates by ecological factors that constrain them to a single annual reproductive event. Seasonal breeding therefore may constitute the main life history cost of inhabiting seasonal habitats. Some galagos and lorises from Equatorial west Africa reproduce throughout the year or have two breeding peaks per year (Charles-Dominique 1977). Similar-sized lemurs, in contrast, typically produce only a single litter during a sharply defined breeding season (Petter-Rousseaux 1962; Rasmussen 1985). Only reproduction in *Daubentonia* is not seasonally constrained (Sterling 1994), but even with *ad libitum* access to high-quality food do aye-ayes fail to produce more than one young within a year (Glander 1994). However, some captive mouse and dwarf lemurs were found to produce more than one litter per year (Glatston 1979; Stanger *et al.* 1995; Schmelting *et al.* 2000), and there is some evidence suggesting that wild females may produce a second litter upon losing the first one (Schwab 2000; M. Eberle, personal communication, 2001). Thus, seasonal reproduction may be an important fitness cost paid by species inhabiting seasonal environments. Field surveys focusing on reproductive biology of species with large geographic ranges may contribute to a quantification of this cost.

Available information on specific key life history traits does not suggest qualitative differences between seasonal and non-seasonal nocturnal primates. Quantitative analyses of interspecific variation in life history traits revealed significant differences in pre- and postnatal litter growth rates among lemurs, lorises, galagos, and tarsiers after controlling for differences in body size (Kappeler 1995), suggesting that phylogenetic effects are more pervasive than those of ecological factors. Moreover, there is little variation among nocturnal primates with respect to their age of first reproduction. This is reached within the first or second year, a difference explained by differences in body size and not by the seasonality of their respective habitats. Juvenile growth and development of nocturnal prosimians therefore may not be tuned firmly to varying ecological conditions as is the case in some larger lemurs, in which photoperiodic cues control periods of seasonal growth (Pereira *et al.* 1999).

Finally, variation in litter size may be expected to be used to fine-tune reproductive effort to current maternal and ambient conditions. Again, phylogenetic effects on the occurrence of litters versus singletons (Kappeler 1998) are more pronounced than potential effects of seasonality, because singletons and larger litters are found across nocturnal primates, independent of the seasonality of their habitat (e.g. Charles-Dominique [1972]; Martin [1972]). Litter size and composition of captive mouse lemurs did vary as a function of the social environment (Perret 1990, 1996), but we do not know yet whether such adjustments also occur in response to ecological conditions. It has been documented that all females in a population of *C. medius* forgo reproduction altogether during a particular year (Fietz 1999), but the causes of this collective decision remained obscure. The cryptic nature of early maternal care and the lack of long-term data from known individuals, especially on age-specific schedules of mortality (see Promislow & Harvey [1990] and Harvey *et al.* [1991]) of wild nocturnal primates, will continue to hamper profound functional analyses of life history variation for some time.

Discussion

The majority of nocturnal primates live in habitats characterized by seasonal variation in climate and resource availability. Our review has revealed that they counter these ecological challenges with various physiological and behavioral adaptations, some of which are not available to other primates. The details and underlying mechanisms of these adaptations have begun to be explored under natural conditions in only a handful of species, however, so it is quite likely that we do not yet know the full range of adaptations that exist. We therefore stress the need for long-term field studies of a much broader range of taxa. Moreover, some suggestive observations, as well as theoretical considerations, indicate that physiological and behavioral responses to seasonality are embedded within general life history adaptations that affect schedules of development and reproduction as well. A major challenge with respect to understanding life history adaptations to seasonality will be to disentangle responses to ecological factors from phylogenetic invariants. In particular, studies of phenotypic plasticity in physiological and life history traits in several populations of species with large geographical ranges are indicated to solve this problem.

Our review has revealed that nocturnal primates exhibit three main physiological responses to cope with the energetic challenges of a seasonal environment. The reduction of basal heat production is an effective strategy to save considerable amounts of energy, and not just a reflection of a primitive thermoregulatory system (Müller 1985). It has been demonstrated in all major taxa of nocturnal strepsirrhines and therefore appears to be the most widespread mechanism. The two other main physiological mechanisms, in contrast, appear to be confined to the Cheirogaleidae. The ability to sustain longer periods of torpor or even hibernation may in fact be contingent upon the ability to accumulate and store large fat reserves (Lyman *et al.* 1982; Wang 1989), so that the phylogenetic non-independence of these traits comes as no surprise.

However, why apparently only a few taxa of basal prosimians evolved (or retained?) these abilities remains puzzling. An analysis of potential costs and additional benefits of torpor and hibernation may hint at possible answers to this key question. First, long-term studies of marked individuals of hibernating cheirogaleid species are beginning to reveal demographic and long-term fitness consequences of individual variation in the utilization of torpor and hibernation. Mouse lemurs in the deciduous Kirindy forest, for example, face a high predation risk during the dry season that is due mainly to the open canopies (Goodman *et al.* 1993; Rasoloarison *et al.* 1995) (see also Chapter 4). Female gray mouse lemurs that had accumulated enough fat reserves to hibernate for several months during the dry season reduced exposure to predators and had higher-than-average annual survival rates (30 versus 50%) (Schmid, unpublished data, 1997). Thus, the energy-saving strategies of *M. murinus* not only have a strong influence on the lemurs' energy budget but also affect their survival – clear evidence for its adaptive significance.

Second, although torpor is employed by animals primarily as an adaptation to conserve energy when food supplies are short, it is generally assumed that the retention of water and metabolites may also constitute important selective advantages of torpor (Schmidt-Nielsen 1964; Speakman & Racey 1989; Thomas & Cloutier 1992). Schmid and Speakman (2000) measured daily energy expenditure (DEE) and water turnover of free-living *M. murinus* during the dry season. They showed that female mouse lemurs save about 25% of energy when they enter daily torpor, whereas the DEE of torpid males was 5% higher than that of normothermic males. However, the mean water turnover was significantly lower in torpid males and females compared with normothermic animals. Thus, a common benefit of torpor for both male and female mouse lemurs was conservation of water, thereby possibly reducing water requirements. Because there is virtually no precipitation during the dry season in western Madagascar (Sorg & Rohner 1996), the effect on water budgets may be a significant factor influencing the utilization of torpor in this species. Again, comparative studies of other nocturnal primates as well as of sympatric mammals are needed to evaluate the relative importance of this effect.

It is also possible that torpor may have other, less beneficial, consequences that may limit its occurrence and explain its absence in some taxa. For example, torpid animals are unable to perform coordinated movements or respond to sensory stimuli, and consequently they are incapable of defending territories or escaping when detected by predators (French 1988, 1992). Furthermore, hypothermia may reduce T_b below the optimal temperature for enzymatic activity, thereby decelerating protein and cellular turnover (Yacoe 1983; Deerenberg *et al.* 1997). Nevertheless, why the result of such a cost/benefit analysis is different in *Microcebus*, *Cheirogaleus*, and, perhaps, *Allocebus* remains unclear. Future studies of physiological adaptations to seasonality in other nocturnal primates may contribute parts of an answer, or at least reveal potential alternative mechanisms.

It therefore remains mysterious why some of the lemurs go into torpor or hibernation to cope with environmental stress, whereas bushbabies, lorises, and other nocturnal primates do not. Madagascar's unusually unpredictable climate may provide a partial answer to this question. Dewar and Wallis (1999) demonstrated that southwest Madagascar, in particular, is characterized by extremely high unpredictability of annual rainfall totals and may experience major droughts more frequently compared with other regions in the tropics with similar average rainfall in which primates occur. Ganzhorn (1995) also pointed out that the high number of cyclones striking Madagascar may have direct as well as indirect effects on the islands' fauna. It is not surprising, therefore, that Malagasy primates have evolved distinctively different adaptation forms to deal with the unpredictability of environmental conditions than did primate species that occur in areas with a more predictable climate (see also Wright [1999]). One prediction of this hypothesis is that similar adaptations should be found in sympatric Malagasy mammals, but this aspect of their ecology remains virtually unstudied (see Garbutt [1999]).

An adaptive explanation based on the unusual climate cannot be complete, however, because there is no obvious reason why the capacity for torpor does not exist in *Mirza coquereli* and *Phaner furcifer*, both cheirogaleids that live sympatrically with hibernating mouse and dwarf lemurs over much of their range. One possible explanation is that daily torpor and hibernation are plesiomorphic, although not necessarily functionally primitive (Malan 1996). Furthermore, the energetics of endothermic animals is affected by body size (Kleiber 1932; McNab 1983) and food habits (McNab 1980) and, therefore, also influence the strategies that the animals utilize to cope with seasonal energy shortages. *Phaner* and *Mirza* are larger than the other sympatric cheirogaleids, and at least *Phaner* has a highly specialized diet (Kappeler 2004, Schülke 2004). Another explanation is that seasonal food shortages and cold must not necessarily mean that all taxa use the same energy-saving mechanisms (see also Morland [1993]). Thus, similar organisms may show different adaptations that represent different solutions to the same problem, and it appears that most of these adaptations of nocturnal primates are still in the dark.

Acknowledgments

We are grateful to Diane K. Brockman and Carel P. van Schaik for the invitation to contribute to this volume and for their constructive comments on earlier drafts of the manuscript. The participation of J. Schmid at the International Primatological Society conference was funded by a grant from the German Research Council (Deutsche Forschungsgemeinschaft, grant SCHM 1391/2–1).

References

Albrecht, G. H., Jenkins, P. D., & Godfrey, L. R. (1990). Ecogeographic size variation among the living and subfossil prosimians of Madagascar. *American Journal of Primatology*, **22**, 1–50.

Atsalis, S. (1999). Seasonal fluctuations in body fat and activity levels in a rain-forest species of mouse lemur, *Microcebus rufus*. *International Journal of Primatology*, **20**, 883–910.

Aujard, F., Perret, M., & Vannier, G. (1998). Thermoregulatory responses to variations of photoperiod and ambient temperature in the male lesser mouse lemur: a primitive or an advanced adaptive character? *Journal of Comparative Physiology B*, **168**, 540–48.

Bearder, S. K. & Doyle G. A. (1974). Ecology of bushbabies *Galago senegalensis* and *Galago crassicaudatus*, with some notes on their behaviour in the field. In *Prosimian Biology*, ed. R. D. Martin, G. A. Doyle, & A. C. Walker. London: Duckworth, pp. 109–30.

Bourlière, F. & Petter-Rousseaux, A. (1966). Existence probable d'un rythme métabolique saisonnier chez les cheirogaleinae (Lemuroidea). *Folia Primatologica*, **4**, 249–56.

Charles-Dominique, P. (1972). Ecologie et vie sociale de *Galago demidovii* (Fischer 1808; Prosimii). *Fortschritte der Verhaltensforschung*, **9**, 7–41.

— (1977). *Ecology and Behaviour of Nocturnal Primates*. New York: Columbia University Press.

Chévillard, M.-C. (1976). Capacités thermorégulatrices d'un lémurien malgache, *Microcebus murinus* (Miller, 1777). Ph. D. thesis, University of Paris.

Dausmann, K. H., Ganzhorn, J. U., & Heldmaier, G. (2000). Body temperature and metabolic rate of a hibernating primate in Madagascar: preliminary results from a field study. In *Life in the Cold. Eleventh International Hibernation Symposium*, ed. G. Heldmaier & M. Klingenspor. New York: Springer-Verlag, pp. 41–7.

Dawson, T. J., Grant, T. R., & Fanning, D. (1979). Standard metabolism of monotremes and the evolution of homeothermy. *Australian Journal of Zoology*, **27**, 511–15.

Deerenberg, C., Apanius, V. A., Daan, S., & Bos, N. (1997). Reproductive effort decreases antibody responsiveness. *Proceedings of the Royal Society of London, B*, **264**, 1021–9.

Dewar, R. E. & Wallis, J. R. (1999). Geographical patterning of interannual rain-fall variability in the tropics and near tropics: an L-moments approach. *Journal of Climate*, **12**, 3457–66.

Dobler, H.-J. (1978). Untersuchungen über die Temperatur- und Stoffwechselregulation von Galagos (Lorisiformes: Galagidae). Ph. D. thesis, University of Tübingen.

Fietz, J. (1998). Body mass in wild *Microcebus murinus* over the dry season. *Folia Primatologica*, **69**, 183–90.

— (1999). Monogamy as a rule rather than exception in nocturnal lemurs: the case of the fat-tailed dwarf lemur, *Cheirogaleus medius*. *Ethology*, **105**, 259–72.

Fietz, J. & Ganzhorn, J. U. (1999). Feeding ecology of the hibernating primate *Cheirogaleus medius*: how does it get so fat? *Oecologia*, **121**, 157–64.

French, A. R. (1988). The patterns of mammalian hibernation. *American Scientist*, **76**, 568–75.

— (1992). Mammalian dormancy. In *Mammalian Energetics: Interdisciplinary Views of Metabolism and Reproduction*, ed. T. E. Tomasi & T. H. Horton. Ithaca: Cornell University Press, pp. 105–121.

Ganzhorn, J. U. (1995). Cyclones over Madagascar: fate or fortune? *Ambio*, **24**, 124–5.

Ganzhorn, J. U. & Schmid, J. (1998). Different population dynamics of *Microcebus murinus* in primary and secondary deciduous dry forests of Madagascar. *International Journal of Primatology*, **19**, 785–96.

Garbutt, N. (1999). *Mammals of Madagascar*. Tonbridge, UK: Pica Press.

Geiser, F. & Ruf, T. (1995). Hibernation versus daily torpor in mammals and birds: physiological variables and classification of torpor patterns. *Physiological Zoology*, **68**, 935–66.

Génin, F. & Perret, M. (2000). Photoperiod-induced changes in energy balance in gray mouse lemurs. *Physiology and Behavior*, **71**, 315–21.

Genoud, M., Martin, R. D., & Glaser, D. (1997). Rate of metabolism in the smallest simian primate, the pygmy marmoset (*Cebuella pygmaea*). *American Journal of Primatology*, **41**, 229–45.

Glander, K. (1994). Morphometrics and growth in captive aye-ayes (*Daubentonia madagascariensis*). *Folia Primatologica*, **62**, 108–14.

Glatston, A. R. H. (1979). *Reproduction and Behaviour of the Lesser Mouse Lemur (*Microcebus murinus*) in Captivity*. London: University of London.

Goodman, S. M., Langrand, O., & Raxworthy, C. J. (1993). Food habits of the Madagascar long-eared owl *Asio madagascariensis* in two habitats in southern Madagascar. *Ostrich*, **64**, 79–85.

Groves, C. (2000). The genus *Cheirogaleus*: unrecognized biodiversity in dwarf lemurs. *International Journal of Primatology*, **21**, 943–62.

Gursky, S. (2000). Effect of seasonality on the behavior of an insectivorous primate, *Tarsus spectrum*. *International Journal of Primatology*, **20**, 69–84.

Harcourt, C. S. & Nash, L. T. (1986). Social organization of galagos in Kenyan coastal forest: I. *Galago zanzibaricus*. *American Journal of Primatology*, **10**, 339–55.

Harvey, P. H. & Clutton-Brock, T. H. (1985). Life history variation in primates. *Evolution*, **39**, 559–81.

Harvey, P. H., Martin, R. D., & Clutton-Brock, T. H. (1987). Life histories in comparative perspective. In *Primate Societies*, ed. B. B. Smuts, D. L. Cheney,

R. M. Seyfarth, R. W. Wrangham, & T. T. Struhsaker. Chicago: University of Chicago Press, pp. 181–96.

Harvey, P., Pagel, M., & Rees, J. (1991). Mammalian metabolism and life histories. *American Naturalist*, **137**, 556–66.

Heldmaier, G. & Ruf, T. (1992). Body temperature and metabolic rate during natural hypothermia in endotherms. *Journal of Comparative Physiology B*, **162**, 696–706.

Hildwein, G. & Goffart, M. (1975). Standard metabolism and thermoregulation in a prosimian, *Perodicticus potto*. *Comparative Biochemistry and Physiology A*, **50**, 201–13.

Hiley, P. G. (1976). The thermoregulatory responses of the galago (*G. crassicaudatus*), the baboon (*Papio ursinus*) and the chimpanzee (*Pan satyrus*) to heat stress. *Journal of Physiology, London*, **254**, 657–71.

Hladik C. M., Charles-Dominique, P., & Petter, J. J. (1980). Feeding strategies of nocturnal prosimians. In *Nocturnal Malagasy Primates: Ecology, Physiology and Behaviour*, ed. P. Charles-Dominique, H. M. Cooper, C. M. Hladik, *et al.* New York: Academic Press, pp. 41–72.

Hudson, J. W. (1973). Torpidity in mammals. In *Comparative Physiology of Thermoregulation*, ed. G. C. Whittow. London: Academic Press, pp. 97–165.

Kappeler, P. M. (1995). Life history variation among nocturnal prosimians. In *Creatures of the Dark: The Nocturnal Prosimians*, ed. L. Alterman, M. K. Izard, & G. A. Doyle. New York: Plenum Press, pp. 75–92.

Kappeler, P. (1996). Causes and consequences of life history variation among strepsirhine primates. *American Naturalist*, **148**, 868–91.

Kappeler, P. M. (1998). Nests, tree holes, and the evolution of primate life histories. *American Journal of Primatology*, **46**, 7–33.

— (2004). The natural history of *Mirza coquereli*. In *The Natural History of Madagascar*, ed. S. M. Goodman & J. P. Benstead. Chicago: University of Chicago Press, pp. 1316–18.

Kappeler, P. M. & Pereira, M. E. (2003). *Primate Life History and Socioecology*. Chicago: University of Chicago Press.

Kleiber, M. (1932). Body size and metabolism. *Hilgardia*, **6**, 315–53.

Knox, C. M. & Wright, P. G. (1989). Thermoregulation and energy metabolism in the lesser bushbaby, *Galago senegalensis moholi*. *South African Journal of Zoology* **24**, 89–94.

Lee, P. C. & Kappeler, P. M. (2003). Socio-ecological correlates of phenotypic plasticity in primate life histories. In *Primate Life History and Socioecology*, ed. P. M. Kappeler & M. E. Pereira. Chicago: University of Chicago Press, pp. 41–65.

Le Maho, Y., Goffart, M., Rochas, A., Felbalbel, H., & Chatonnet, C. (1981). Thermoregulation in the only nocturnal simian: the night monkey *Aotus trivirgatus*. *American Journal of Primatology*, **240**, 156–65.

Lyman, C. P., Willis, J. S., Malan, A., & Wang, L. C. H. (1982). *Hibernation and Torpor in Mammals and Birds*. London: Academic Press.

Malan, A. (1996). The origins of hibernation: a reappraisal. In *Adaptations to the Cold: Tenth International Hibernation Symposium*, ed. F. Geiser,

A. J. Hulbert, & S. C. Nichol. Hanover, NH: University of New England Press, pp. 1–66.

Martin, R. D. (1972). A preliminary field-study of the lesser mouse lemur (*Microcebus murinus* J. F. Miller 1777). *Zeitschrift für Tierpsychologic Supplement*, **9**, 43–89.

— (1973). A review of the behaviour and ecology of the lesser mouse lemur (*Microcebus murinus*). In *Ecology and Behaviour of Primates*, ed. M. Crook. London: Academic Press, pp. 1–68.

Martin, R. D. & Bearder, S. K. (1979). Radio bushbaby. *Natural History* **88**, 77–81.

McCormick, S. A. (1981). Oxygen consumption and torpor in the fat-tailed dwarf lemur (*Cheirogaleus medius*): rethinking prosimian metabolism. *Comparative Biochemistry and Physiology A*, **68**, 605–10.

McNab, B. K. (1980). Food habits, energetics, and the population biology of mammals. *American Naturalist*, **116**, 106–24.

— (1983). Energetics, body size, and the limits to endothermy. *Journal of Zoology, London*, **199**, 1–29.

— (1988). Complications inherent in scaling the basal rate of metabolism in mammals. *Quarterly Review of Biology*, **63**, 25–54.

McNab, B. K. & Wright, P. C. (1987). Temperature regulation and oxygen consumption in the Philippine tarsier *Tarsius syrichta*. *Physiological Zoology*, **60**, 596–600.

Morland, H. S. (1993). Determinants of seasonal behavioral variation in ruffed lemurs (*Varecia variegata variegata*). In *Lemur Social Systems and Their Ecological Basis*, ed. P. M. Kappeler & J. U. Ganzhorn. New York: Plenum Press, pp. 193–204.

Morrison, P. & Middleton, E. H. (1967). Body temperature and metabolism in the pygmy marmoset. *Folia Primatologica*, **6**, 70–82.

Müller, E. F. (1979). Energy metabolism, thermoregulation and water budget in the slow loris (*Nycticebus coucang*, Boddaert 1785). *Comparative Biochemistry and Physiology A*, **64**, 109–19.

Müller, E. (1985). Basal metabolic rates in primates: the possible role of phylogenetic and ecological factors. *Comparative Biochemistry and Physiology A*, **81**, 707–11.

Müller, E. F. & Jaschke, H. (1980). Thermoregulation, oxygen consumption, heart rate and evaporative water loss in the thick-tailed bushbaby (*Galago crassicaudatus* Geoffroy, 1812). *Zeitschrift fur Säugetierk*, **45**, 269–78.

Müller, E. F., Kamau, J. M. Z., & Maloiy, G. M. O. (1983). A comparative study of basal metabolism and thermoregulation in a folivorous (*Colobus guereza*) and an ominivorous (*Cercopithecus mitis*) primate species. *Comparative Biochemistry and Physiology A*, **74**, 319–22.

Nash, L. T. (1998). Vertical clingers and sleepers: seasonal influences on the activities and substrate use of *Lepilemur leucopus* at Beza Mahafaly Special Reserve, Madagascar. *Folia Primatologica*, **69** (Suppl. 1), 204–17.

Nekaris, K. (2000). Socioecology of the slender loris (*Loris tardigradus lydykkerianus*) in Dindigul (DT), Tamil Nadu, South India. Ph. D. thesis, Washington University.

Ortmann, S., Schmid, J., Ganzhorn, J. U., & Heldmaier, G. (1996). Body temperature and torpor in a Malagasy small primate, the mouse lemur. In *Adaptations to the Cold: The Tenth International Hibernation Symposium*, ed. F. Geiser, A. J. Hulbert, & S. C. Nicol. Armidale: University of New England Press, pp. 55–61.

Ortmann, S., Heldmaier, G., Schmid, J., & Ganzhorn, J. U. (1997). Spontaneous daily torpor in Malagasy mouse lemurs. *Naturwissenschaften*, **84**, 28–32.

Overdorff, D. J. (1993). Similarities, differences, and seasonal patterns in the diets of *Eulemur rubriventer* and *Eulemur fulvus rufus* in the Ranomafana National Park, Madagascar. *International Journal of Primatology*, **14**, 721–53.

Pereira, M. E., Strohecker, R., Cavigelli, S., Hughes, C., & Pearson, D. (1999). Metabolic strategy and social behavior in Lemuridae. In *New Directions in Lemur Studies*, ed. H. Rasamimanana, B. Rakotosamimanana, J. Ganzhorn, & S. Goodman. New York: Plenum Press, pp. 93–118.

Perret, M. (1990). Influence of social factors on sex ratio at birth, maternal investment and young survival in a prosimian primate. *Behavioral Ecology and Sociobiology*, **27**, 447–54.

— (1992). Environmental and social determinants of sexual function in the male lesser mouse lemur (*Microcebus murinus*). *Folia Primatologica*, **59**, 1–25.

— (1996). Manipulation of sex ratio at birth by urinary cues in a prosimian primate. *Behavioral Ecology and Sociobiology*, **38**, 259–66.

— (1998). Energetic advantage of nest-sharing in a solitary primate, the lesser mouse lemur (*Microcebus murinus*). *Journal of Mammalogy*, **79**, 1093–102.

Perret, M., Aujard, F., & Vannier, G. (1998). Influence of daylength on metabolic rate and daily water loss in the male prosimian primate *Microcebus murinus*. *Comparative Biochemistry and Physiology A*, **119**, 981–9.

Petter, J.-J., Albignac, R., & Rumpler, Y. (1977). *Mammifères lémuriens (*Primates prosimiens*)*. Paris: ORSTOM-CNRS.

Petter-Rousseaux, A. (1962). Recherche sur la biologie de la reproduction des primates inférieurs. *Mammalia 26 Supplement*, **1**, 1–88.

Petter-Rousseaux, A. (1980). Seasonal activity rhythms, reproduction, and body weight variations in five sympatric nocturnal prosimians, in simulated light and climatic conditions. In *Nocturnal Malagasy Primates: Ecology, Physiology and Behaviour*, ed. P. Charles-Dominique, H. M. Cooper, A. Hladik. New York: Academic Press, pp. 137–51.

Promislow, D. E. L. & Harvey, P. H. (1990). Living fast and dying young: a comparative analysis of life-history variation among mammals. *Journal of Zoology, London*, **220**, 417–37.

Radespiel, U., Cepok, S., Zietemann, V., & Zimmermann, E. (1998). Sex-specific usage patterns of sleeping sites in grey mouse lemurs (*Microcebus murinus*) in Northwestern Madagascar. *American Journal of Primatology*, **46**, 77–84.

Rakotoarison, N., Zimmermann, H., & Zimmermann, E. (1997). First discovery of the hairy-eared dwarf lemur (*Allocebus trichotis*) in a highland rain forest of eastern Madagascar. *Folia Primatologica*, **68**, 86–94.

Rasmussen, D. T. (1985). A comparative study of breeding seasonality and litter size in eleven taxa of captive lemurs (*Lemur* and *Varecia*). *International Journal of Primatology*, **6**, 501–17.

Rasoazanabary, E. (2001). Stratégie adaptive chez les males de *Microcebus murinus* pendant la saison sèche, dans la forêt de Kirindy, Morondava. DEA thesis, Universite d'Antananarivo.

Rasoloarison, R. M., Rasolonadrasana, B. P. N., Ganzhorn, J. U., & Goodman, S. M. (1995). Predation on vertebrates in the Kirindy Forest, Western Madagascar. *Ecotropica*, **1**, 59–65.

Rasoloarison, R. M., Goodman, S. M., & Ganzhorn, J. U. (2000). A taxonomic revision of mouse lemurs (*Microcebus*) occurring in the western portions of Madagascar. *International Journal of Primatology*, **21**, 963–1019.

Richard, A. F. & Dewar, R. E. (1991). Lemur ecology. *Annual Review of Ecology and Systematics*, **22**, 145–75.

Russel, R. J. (1975). Body temperature and behavior of captive cheirogaleids. In *Lemur Biology*, ed. I. Tattersall & R. W. Sussman. New York: Plenum Press, pp. 193–206.

Schmelting, B., Ehresmann, P., Lutermann, H., Randrianambinina, B., & Zimmermann, E. (2000). Reproduction of two sympatric mouse lemur species (*Microcebus murinus* and *M. ravelobensis*) in north-west Madagascar: first results of a long term study. In *Mémoires de la Société de Biogéographie*, ed. W. R. Lourenco & S. M. Goodman. Paris: ORSTOM, pp. 165–75.

Schmid, J. (1996). Oxygen consumption and torpor in mouse lemurs (*Microcebus murinus* and *Microcebus myoxinus*): preliminary results of a study in western Madagascar. In *Adaptations to the Cold: The Tenth Hibernation Symposium*, ed. F. Geiser, A. J. Hulbert, & S. C. Nicol. Armidale: University of New England Press, pp. 47–54.

— (1997). Torpor beim Grauen Mausmaki (*Microcebus murinus*) in Madagascar: Energetische Konsequenzen und ökologische Bedeutung. Ph. D. thesis, University of Tübingen.

— (1998). Tree holes used for resting by gray mouse lemur (*Microcebus murinus*) in Madagascar: insulation capacities and energetic consequences. *International Journal of Primatology*, **19**, 797–809.

— (1999). Sex-specific differences in activity patterns and fattening in the gray mouse lemur (*Microcebus murinus*) in Madagascar. *Journal of Mammalogy*, **80**, 749–57.

— (2000). Daily torpor in the gray mouse lemur (*Microcebus murinus*) in Madagascar: energetical consequences and biological significance. *Oecologia*, **123**, 175–83.

Schmid, J. & Ganzhorn, J. U. (1996). Resting metabolic rates of *Lepilemur ruficaudatus*. *American Journal of Primatology*, **38**, 169–74.

Schmid, J. & Kappeler, P. M. (1994). Sympatric mouse lemurs (*Microcebus* spp.) in western Madagascar. *Folia Primatologica*, **63**, 162–70.

— (1998). Fluctuating sexual dimorphism and differential hibernation by sex in a primate, the gray mouse lemur (*Microcebus murnius*). *Behavioral Ecology and Sociobiology*, **43**, 125–32.

Schmid, J. & Speakman, J. R. (2000). Daily energy expenditure of the gray mouse lemur (*Microcebus murinus*): a small primate that uses torpor. *Journal of Comparative Physiology B*, **170**, 633–41.

Schmid, J. & Stephenson, P. J. (2004). Physiological adaptations of Malagasy mammals: lemurs and tenrecs compared. In *The Natural History of Madagascar*, ed. S. M. Goodman & J. P. Benstead. Chicago: University of Chicago Press, pp. 1198–203.

Schmid, J., Ruf, T., & Heldmaier, G. (2000). Metabolism and temperature regulation during daily torpor in the smallest primate, the pygmy mouse lemur (*Microcebus myoxinus*) in Madagascar. *Journal of Comparative Physiology B*, **170**, 59–68.

Schmidt-Nielsen, K. (1964). *Desert Animals: Physiological Problems of Heat and Water*. Oxford: Clarendon Press.

Schülke, O. (2004). *Phaner furcifer*. In *The Natural History of Madagascar*, ed. S. M. Goodman & J. P. Benstead. Chicago: University of Chicago Press, pp. 1318–80.

Schwab, D. (2000). A preliminary study of spatial distribution and mating system of pygmy mouse lemurs (*Microcebus cf myoxinus*). *American Journal of Primatology*, **51**, 41–60.

Sorg, J.-P. & Rohner, U. (1996). Climate and tree phenology of the dry deciduous forest of the Kirindy forest. In *Primate Report*, ed. J. U. Ganzhorn & J.-P. Sorg. Göttingen, Germany: Kinze, pp. 57–80.

Speakman, J. R. & Racey, P. A. (1989). Hibernal ecology of the pipistrelle bat: energy expenditure, water requirements and mass loss, implications for survival and the function of winter emergence flights. *Journal of Animal Ecology*, **58**, 797–813.

Stanger, K., Coffman, B., & Izard, M. (1995). Reproduction in Coquerel's dwarf lemur (*Mirza coquereli*). *American Journal of Primatology*, **36**, 223–37.

Stearns, S. C. (1976). Life-history tactics: a review of the ideas. *Quarterly Review of Biology*, **51**, 3–47.

Stearns, S. C. (1992). *The Evolution of Life Histories*. Oxford: Oxford University Press.

Stephenson, P. J. & Racey, P. A. (1994). Seasonal variation in resting metabolic rate and body temperature of streaked tenrecs, *Hemizentetes nigriceps* and *H . semispinosus* (Insectivora: Tenrecidae). *Journal of Zoology, London*, **232**, 285–94.

Sterling, E. J. (1994). Evidence for nonseasonal reproduction in wild aye-ayes (*Daubentonia madagascariensis*). *Folia Primatologica* **62**, 46–53.

Thomas, D. W. & Cloutier, D. (1992). Evaporative water loss by hibernating little brown bats, *Myotis lucifugus*. *Physiological Zoology*, **65**, 443–56.

Vincent, F. (1978). Thermoregulation and behaviour in two sympatric galagos: an evolutionary factor. In *Recent Advances in Primatology*, Vol 3, ed. D. A. Chivers & K. A. Joysey. London: Academic Press, pp. 181–7.

Wang, L. C. H. (1989). Ecological, physiological, and biochemical aspects of torpor in mammals and birds. In *Advances in Comparative and Environmental Physiology*, ed. L. C. H. Wang & J. A. Boulant. New York: Springer-Verlag, pp. 361–93.

Willis, J. S. (1982). The mystery of periodic arousal. In *Hibernation and Torpor in Mammals and Birds*, ed. C. P. Lyman, J. S. Willis, A. Malan, & L. C. H. Wang. New York: Academic Press, pp. 92–101.

Wright, P. C. (1981). The night monkeys, genus *Aotus*. In *Ecology and Behavior of Neotropical Primates*, Vol. 1, ed. A. F. Coimbra-Filho & R. A. Mittermeier. Rio de Janeiro: Academia Brasileira de Ciências, pp. 211–40.

— (1999). Lemur traits and Madagascar ecology: coping with an island environment. *Yearbook of Physical Anthropology*, **42**, 31–72.

Wright, P. C. & Martin, L. B. (1995). Predation, pollination and torpor in two nocturnal prosimians: *Cheirogaleus major* and *Microcebus rufus* in the rain forest of Madagascar. In *Creatures of the Dark: The Nocturnal Prosimians*, ed. L. Alterman, G. A. Doyle, & M. K. Izard, New York: Plenum Press, pp. 45–60.

Yacoe, M. E. (1983). Protein metabolism in the pectoralis muscle and liver of hibernating bats, *Eptesicus fuscus*. *Journal of Comparative Physiology B*, **152**, 137–44.

6 *Seasonality and long-term change in a savanna environment*

SUSAN C. ALBERTS
Department of Biology, Duke University, Box 90338,
Durham NC 27708, USA;
and Institute for Primate Research, National Museums
of Kenya, Nairobi, Kenya

JULIE A. HOLLISTER-SMITH
Department of Biology, Duke University, Box 90338,
Durham NC 27708, USA

RAPHAEL S. MUTUTUA
Amboseli Baboon Research Project, Amboseli National Park, Kenya;
and Institute for Primate Research, National Museums
of Kenya, Nairobi, Kenya

SERAH N. SAYIALEL
Amboseli Baboon Research Project, Amboseli National Park, Kenya;
and Institute for Primate Research, National Museums
of Kenya, Nairobi, Kenya

PHILIP M. MURUTHI
African Wildlife Foundation, Box 48177, Nairobi, Kenya

J. KINYUA WARUTERE
Amboseli Baboon Research Project, Amboseli National Park, Kenya;
and Institute for Primate Research, National Museums
of Kenya, Nairobi, Kenya

JEANNE ALTMANN
Department of Ecology and Evolutionary Biology, Princeton University,
Princeton NJ 08544, USA;
and Department of Conservation Biology,
Brookfield Zoo, Brookfield IL, USA;
and Institute for Primate Research, National Museums
of Kenya, Nairobi, Kenya

Seasonality in Primates: Studies of Living and Extinct Human and Non-Human Primates, ed.
Diane K. Brockman and Carel P. van Schaik. Published by Cambridge University Press.

Introduction

The emergence and spread of savannas in Africa during the past five million years is often cited as a major factor in hominid evolution. Tropical savannas are different from forests in having less rainfall, which is strongly seasonal and often very unpredictable, even within seasons (Bourliere & Hadley 1983; Solbrig 1996). Human ancestors are thought to have moved into savannas as a response to cooling and drying climates, and the exigencies of the savanna environment – including the marked seasonal changes in plant food availability – are often cited as key selective pressures shaping the hominid lineage (see reviews and references in Foley [1987, 1993], Potts [1998a, 1998b], Klein [1999], and Chapters 4, 5, and 17). This scenario invites a careful examination of responses to seasonality in extant savanna-dwelling primates.

Like most vertebrates, the large majority of primate species exhibit reproductive seasonality that reflects the seasonality of their habitats (see review in Chapter 11). Indeed, among savanna-dwelling primates, there are only two exceptions to the rule of seasonal reproduction: humans and baboons (genus *Papio*). This shared characteristic – the ability to reproduce throughout the year in seasonal environments – may be related to the extraordinary success of these two genera. While only humans (and their commensals) have spread across the globe, baboons have achieved a nearly continental distribution in Africa. Indeed, the genus *Papio*, increasingly treated as a single species with multiple subspecies (Jolly 1993), occupies habitats ranging from desert to semi-arid tropical savanna to temperate montane grasslands to moist evergreen forest (Estes 1991; Jolly 1993; Kingdon 1997).

These shared features of *Papio* and *Homo* – a wide geographic distribution, success in but not restriction to savanna environments, and non-seasonal reproduction – make analyses of seasonality in baboon behavior especially valuable in light of the role that seasonality is proposed to have played in selecting for unique human traits (Foley 1987, 1993). In this chapter, we examine seasonal patterns of behavior in the well-studied population of savanna baboons in the Amboseli basin at the foot of Mt Kilimanjaro. Amboseli is a semi-arid habitat and one of the drier habitats in which baboons have been studied (see review in Dunbar [1992]). We employ 16 years of behavioral data on adult females, combined with demographic and meteorological data, to test hypotheses about the impacts of the seasonal environment on baboon behavior. Our results are preceded first by a description of savanna seasonality and an outline of the hypotheses that we will test with these data and then by background on baboon ecology and on the Amboseli ecosystem.

The challenges of savanna seasonality

Savannas are tropical habitats dominated by grasses, with scattered drought-resistant trees and shrubs (Bourliere & Hadley 1983; Solbrig 1996; Lincoln *et al.* 1998). The term "savanna" sometimes is used to include subtropical grasslands as well. Savannas show great variability in rainfall across the year but little variation in mean daily temperature, and temperature fluctuations over the course of any given day are larger in magnitude than those in mean temperatures over the course of a year (Bourliere & Hadley 1983; Solbrig 1996). Accordingly, savannas do not exhibit the extreme seasonal shutdown of plant productivity exhibited by temperate-zone grasslands. However, considerable seasonality of plant productivity still occurs in savannas, driven by rainfall seasonality. Hence, a primary challenge for animals living in highly seasonal savanna environments is finding enough food and water during the dry season. For highly social species (most primates), social behavior will also be affected by seasonal changes in time spent foraging. Indeed, discussions about seasonal behavior in primates have centered on two issues: (i) the manner in which foraging behavior changes with season and (ii) the manner in which social behavior changes with season. For each of these issues, contrasting predictions have been made.

Hypotheses about seasonality of foraging behavior

The onset of the dry season in savanna habitats marks the beginning of a long period during which plant productivity is highly constrained. Grasses and many shrubs limit or cease their production of new leaves, grass seed heads vanish, the above-ground parts of many forbs disappear entirely, and fruits and flowers of shrubs and forbs become limited in abundance. How might primates in seasonal environments respond to dry-season food scarcity? Two alternatives have been proposed (see review in Foley [1987] and Chapter 8).

On the one hand, primates might respond to dry-season food scarcities by shifting to foods that are abundant but have low profitability (low ratio of nutrient to harvesting time) (Foley 1987) (see also discussions in Altmann [1998], Dunbar [1983], and Wrangham *et al.* [1991, 1998]). Such a seasonal shift to "fallback foods" typically will result in increased time spent foraging during the dry season relative to the wet season. Indeed, time spent foraging increases during the dry season for a number of primate species (baboons [Post 1981]; muriqis [Strier 1991]; two *Eulemur* species [Overdorff 1996]; tarsiers [Gursky 2000]). Studies of chimpanzees have also provided substantial direct evidence for the fallback foods hypothesis. Chimps increase their intake of herbaceous vegetation during times of fruit scarcity, and work on chimps has formed a model for our understanding of the importance of fallback foods (Wrangham *et al.* 1991, 1998; Malenky & Wrangham 1994).

Alternatively, during times of scarcity, primates may seek out novel foods that are highly profitable but difficult to acquire. This is sometimes proposed as the major strategy of early hominids, and the novel, highly profitable food in question is meat (Blumenschine 1987; Foley 1987, 1993; Potts 1998a, 1998b; Klein 1999) (see also Chapters 4, 5, and 17). There is only limited evidence for this "high-return foods" strategy among tropical human foragers. While many human foragers show marked effects of season on their foraging behavior, this is not commonly manifested as an increase in hunting time or in meat consumption during the dry season (see review in Chapter 9; see also Bunn *et al.* [1988] and Hawkes *et al.* [1991]). Among non-human primates, too, there is limited evidence that hunting increases during the dry season (baboons [Dunbar 1983]; chimpanzees [Stanford 1996; Stanford 1998]), but in general dry-season food scarcity appears to be a poor explanation for primate hunting (Stanford 1996; Stanford 1998; Mitani & Watts 2001) (See also Chapter 8).

These fallback foods and high-return foods hypotheses are not mutually exclusive. While some species or populations may pursue a relatively pure strategy of either type, any given population may pursue both strategies to some extent, shifting to fallback foods but supplementing periodically with high-return foods.

An additional question regarding dry-season foraging is whether primates diversify their diets during times of food scarcity. For species that specialize on one or a few classes of foods (e.g. fruits in the case of chimps), periods of scarcity may prompt the animals both to shift to fallback foods and to diversify their diets to include species and plant parts that are bypassed when fruit is abundant (Foley 1987; Wrangham *et al.* 1991, 1998) (See also Chapter 8). In contrast, for generalists such as baboons, food scarcity will result in reduced diet diversity during the dry season (Post 1982; Norton *et al.* 1987).

Hypotheses about seasonality in social behavior

How might social behavior be affected by the dry season? One hypothesis predicts that the nutritional stresses associated with food scarcity will lead to reduced social activity, while increased competition for food will lead to an expanded spatial distribution within groups. As a consequence, various measures of sociality – grooming rate, time spent in other social interactions, and time spent in proximity to other animals – should decrease in the dry season (Foley 1987) (See also Chapter 8). The underlying assumption here is that social activities are non-essential and will be sacrificed to meet the physiological demands of the dry season; resting time must remain fixed (or not fall below a minimum) as foraging time increases; and the seasonal difference will be taken out of social time. We refer to this as the "dispensable social time" hypothesis.

The alternative hypothesis assumes that social time is functionally important for maintaining social relationships, which in turn are critical in mitigating the effects of both inter- and intragroup competition (e.g. Seyfarth [1977]; Dunbar & Dunbar [1988]; Dunbar [1991]). Under this hypothesis, Dunbar and Dunbar (1988) and Dunbar (1992) propose that animals will conserve social time during food scarcity because social activities (primarily grooming) service relationships and hence represent "social glue" that maintains cohesion of social groups (Dunbar 1992). This hypothesis acknowledges that animals' time budgets must accommodate changes in foraging time but predicts that animals will reduce resting time rather than social time in order to accommodate the increased foraging demands of the dry season. We term this the "social glue" hypothesis.

Baboon ecology

Baboons (genus *Papio*) are large semi-terrestrial monkeys that occupy a wide range of habitats across the continent of Africa (Altmann & Altmann 1970; Jolly 1993; Kingdon 1997). Baboon populations typically are divided into stable social groups, most of which have between 20 and 100 members, including multiple adults and juveniles of both sexes (Altmann & Altmann 1970; Estes 1991). Hamadryas baboons in the horn of Africa deviate markedly from this basic social pattern (Kummer 1968; Stammbach 1987), and we exclude them from consideration here because of their unique socioecological adaptations. We use the term "savanna baboon" to refer to all members of the genus other than hamadryas baboons.

Savanna baboons are eclectic and omnivorous feeders, but this omnivory is combined with great discrimination. They feed very selectively, often choosing a small component of a plant and forgoing the remainder, or focusing on a single species within a genus (Hamilton *et al.* 1978; Post

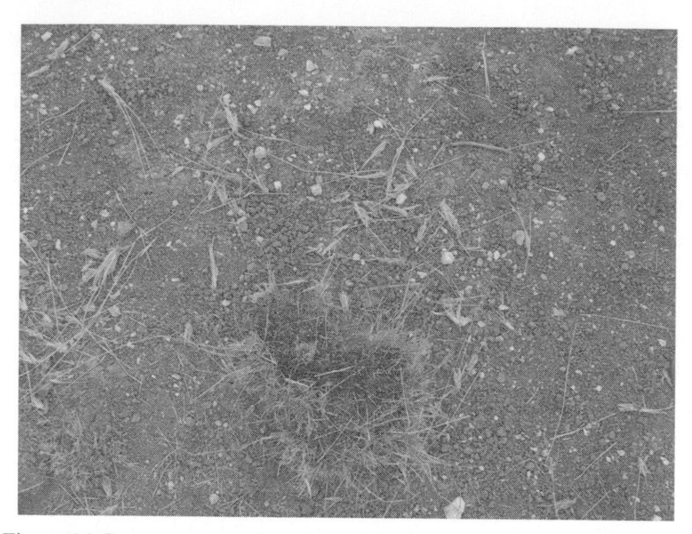

Figure 6.1 Grass corms are the principal food resource for baboons during the dry season at Amboseli.

1982; Norton *et al.* 1987; Muruthi *et al.* 1991; Whiten *et al.* 1991; Byrne *et al.* 1993; Altmann 1998). Plants are the most important source of nutrients; invertebrate and vertebrate animals are eaten but contribute relatively little in calories and protein. In the habitats in which they are best studied (savannas), baboons rely heavily on grasses, consuming both the underground storage organs (corms; Fig. 6.1) and the leaves (Post 1982; Norton *et al.* 1987; Muruthi *et al.* 1991; Whiten *et al.* 1991; Byrne *et al.* 1993; Altmann 1998). Many of the foods consumed by baboons are available and consumed year round; in Amboseli, these include grass corms (Figs 6.2 and 6.3), tree gum, material gleaned from the dung of ungulates and elephants, and the blade bases of grasses (containing the meristem). However, some preferred foods, including the fruits of most species, flowers, green *Acacia* seeds, grass seedheads, and green grass blades (consumed in quantity only when they are new and low in fiber), are highly seasonal (Hamilton *et al.* 1978; Post 1982; Byrne *et al.* 1993; Altmann 1998).

Savanna baboons do not exhibit seasonal patterns of mating or birth; females may conceive and give birth in any month (Fig 6.4) (Melnick & Pearl 1987; Altmann 1980; Bercovitch & Harding 1993; Bentley-Condit & Smith 1997). However, Amboseli births do show a modest peak in August through October, corresponding to conceptions occurring most often from February through May. In fact, 242 of 495 (49%) live births occurred in the five months of the long dry season, June through October, and this is

Figure 6.2 Adult baboon digging grass corms (underground storage organs) to consume during the dry period of food scarcity at Amboseli.

Figure 6.3 Baboons gleaning food from elephant dung during the dry season at Amboseli.

significantly greater than the expected number of 206 (42%) in this season (G test of goodness of fit: $G = 10.64$, $P < 0.005$). An analysis using circular statistics (see Chapter 11) indicates significant but weak clustering of births during the year (corrected vector length $r = 0.12$, $P < 0.001$).

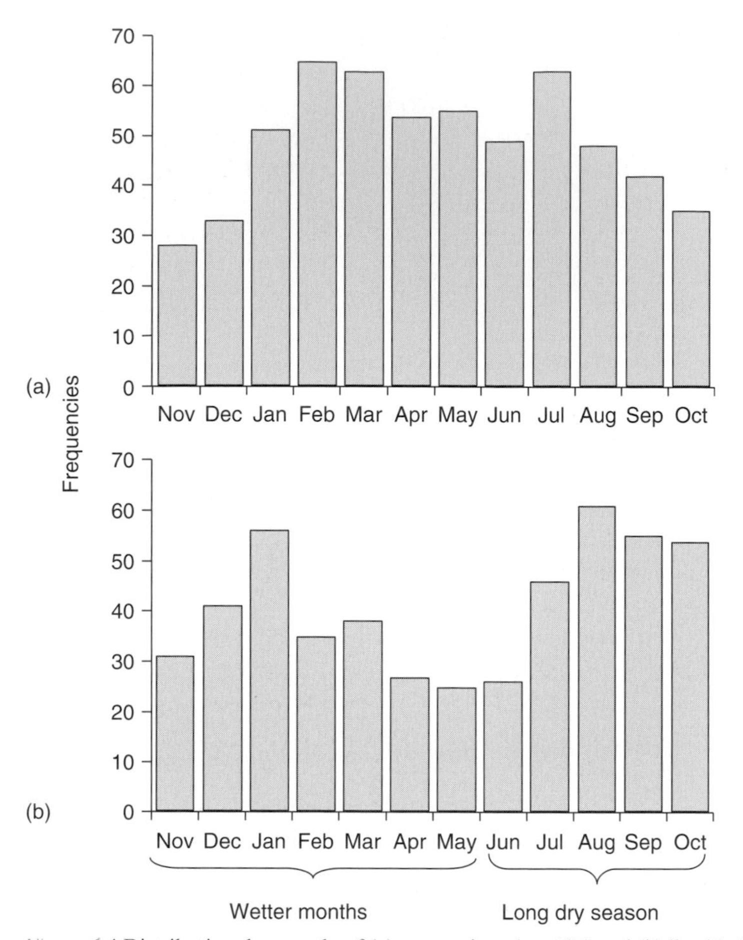

Figure 6.4 Distribution, by month, of (a) conceptions ($n = 586$) and (b) live births ($n = 495$) in Amboseli baboons. The x-axis represents the hydrological year in Amboseli, beginning with the onset of the rains in November (see Fig. 6.2). The difference between the total number of conceptions and the total number of live births represents a combination of miscarriages, stillbirths, and females that died during pregnancy.

Because gestation duration is the least variable life history stage, dates of live births are determined almost entirely by conception dates. Conception dates, in turn, are determined by the timing of the onset of cycling (menarche or postpartum) and the probability of conception. In baboons, these are both highly variable as a result of both stochastic and deterministic processes. For instance, onset of menarche shows modest seasonality, occurring significantly less often than expected during the long dry season

(41 of 138 observed menarches occurred in these months, compared with 58 expected; $G = 10.83$, $P < 0.001$; using circular statistics, corrected vector length $r = 0.18$, $P < 0.01$). In contrast, resumption of cycling after a previous pregnancy shows almost no seasonal effect (205 of 499 resumptions occurred in the long dry season, compared with 208 expected, $G = 0.74$, not significant (NS); corrected vector length $r = 0.08$, $P < 0.05$). Hence, two somewhat similar processes, both of which contribute to the timing of births, are very different in the extent to which they show seasonality. We refer the reader to Chapter 11 for a more detailed discussion of this complex topic.

Amboseli ecology

The Amboseli basin (2°40′ S latitude, 1100 m altitude) is a semi-arid short-grass savanna ecosystem located in an ancient lake basin at the base of Mt Kilimanjaro in east Africa (Williams 1972; Western & van Praet 1973; Behrensmeyer & Boaz 1981; Behrensmeyer 1993; Hay *et al.* 1995). Mean annual rainfall is 348 mm, but the range of annual rainfall is quite large, from less than 150 mm to more than 550 mm (Fig. 6.5a) (Altmann *et al.* 2002).

In the pattern typically described for the area, rainfall occurs in two seasons centered in November–December (the "short rains") and in March–May (the "long rains"), with a "short dry season" in January and February and a "long dry season" during June through October. However, the only component of this pattern that does not vary from year to year is the long dry season. The short rains or the long rains, or both, may fail, or substantial rain may fall during the short dry season. This variability contrasts sharply with the predictability of the long dry season: between the end of May and the last few days of October, virtually no rain falls (Fig. 6.5b) Altmann *et al.* 2002). Mean daily maximum and minimum temperatures exhibit small but predictable seasonal changes; diurnal changes are much larger in magnitude than those that occur seasonally (Altmann *et al.* 2002).

In addition to experiencing year-to-year variability in rainfall, Amboseli has undergone dramatic long-term habitat change over the past four decades (Struhsaker 1973, 1976; Western & van Praet 1973; Hauser *et al.* 1986; Isbell *et al.* 1991; Behrensmeyer 1993; Koch *et al.* 1995; Altmann 1998: 15–19; Cutler *et al.* 1999). In the central part of the Amboseli basin, the dominant tree species, *Acacia xanthophloea* (the fever tree), and various plant species associated with it, have experienced dramatic decline, with complete die-off in some places within Amboseli. At the same time, the number and size of freshwater swamps and ponds has increased

which the baboons relied on for both food and sleeping sites (Altmann & Altmann 1970; Altmann *et al.* 1985; Altmann 1998). Between 1964 and 1969, the population underwent a precipitous decline, probably as a consequence of the die-off of the fever tree woodland in the central part of the basin (Western & van Praet 1973; Altmann *et al.* 1985; Altmann 1998). Over the next decade, the population stabilized, but at a much smaller size than in the early 1960s, and during the late 1980s and 1990s the population grew moderately in size (Altmann *et al.* 1985; Alberts & Altmann 2003). This growth occurred after several social groups abandoned the home range that they had occupied in the central basin during the 1970s and 1980s (Bronikowski & Altmann 1996; Alberts & Altmann 2001; Altmann & Alberts 2003). In each case, the groups moved approximately 8 km west (Alto's in 1987–88, Hook's in 1991–92), to an area still within the Amboseli basin but with a relatively high density of *A. xanthophloea* trees and a relatively low density of baboons. This "western basin" is slightly elevated relative to the central basin and was rarely used by elephants or other browsers when the baboon study groups first moved there, perhaps because of poaching or other human disturbance in that area. These two factors may have contributed to the relative health of the western fever tree woodlands at that time – the elevation resulting in slower effects of changes in the water table and the low elephant density resulting in reduced pressure on the fever tree population. However, during the 1990s, the fever trees in the western basin gradually began to show signs of decline, perhaps partly because elephants and other browsers began using the area more heavily during that period. The impact of this new die-off on the baboons, and their response to it, remains to be seen.

Methods

Study groups, data collection, and subjects

Data were collected on wild-feeding adult female baboons between January 1984 and December 1999 (we exclude data from members of Lodge Group, who augmented their diet with human refuse; see Altmann & Muruthi [1988] and Muruthi *et al.* [1991]). Our two original study groups, Alto's and Hook's groups, fissioned in 1989–91 and 1994–95, respectively (Fig. 6.6). In our analyses, we treated Alto's group and its fission products as one subpopulation and Hook's group and its fission products as a second subpopulation.

Data were collected by R. S. Mututua, S. N. Sayialel, J. K. Warutere, and P. M. Muruthi, who have a cumulative 47 person-years of experience observing baboons. Further, Mututua has contributed between 30 and 100% of the point samples in every year of data collection and has been active in training all the other observers, ensuring great consistency over the entire 16-year period.

Data were collected as ten-minute focal samples (Altmann 1974) on all adult females in the study groups. Adult females within each group were sampled in random order during all active daylight hours, 07.00–18.00 (or 08.00–16.00 between 1988 and 1991). The result was approximately 35 000 focal samples of ten minutes' duration on 124 adult females in two subpopulations (six social groups). The subjects were born between 1962 (estimated) and February 11, 1995. Birth dates were estimated for 23 of the 124 subjects (those born before July 1971 in Alto's and before October 1977 in Hook's); for all other subjects, birth dates were known exactly. Sampling began in January 1984 on females that were already adult at that time; maturing females were added to the sampling schedule when they reached menarche. Focal subjects were between 4.5 and 27 years of age at the time of sampling.

Focal sampling

Focal samples involved the collection of "point" data every minute during the sample (signaled by a timer), as well as "all occurrences" data on agonistic interactions, mounts and mate guarding, and grooming. At each "point," the observer recorded the focal's activity (Table 6.1), her position (standing or sitting), the position of her infant and whether it was suckling, whether any other animals (other than her dependent infant)

Table 6.1 *Mutually exclusive and exhaustive activity categories of focal subjects*

Activity	Definition
Feed	Focal handles or processes food item, or puts food item into the mouth
Move	Focal walks, runs, or climbs
Groom	Focal systematically picks through the fur of another individual with the hands and sometimes the mouth (self-grooming is not recorded as groom)
Be groomed	Another individual picks through focal's fur
Other social	Focal engages in a social interaction other than grooming (agonistic interaction, greeting, play, etc.)
Rest	Focal is sedentary, not feeding, and not interacting

were within 5 m of her, and details about the identities of any such "neighbors." Three to four ten-minute samples per hour were collected in this manner for the duration of the study.

At each point for which "feeding" was recorded, the observer also recorded the type of food being eaten (Table 6.2); species was recorded only for the more common and well-known species. Data on foods consumed were available only for 1991–99 for Hook's subpopulation and for 1996–99 for Alto's subpopulation.

Data analysis

Time budgets

We estimated the proportion of time spent in each activity from the proportion of sample points for which that activity was recorded. Analysis of time budgets has limitations; time spent feeding, for instance, is not always a good proxy for calories consumed (e.g. see discussions in Alberts *et al.* [1996: 1270] and Altmann [1998: 106]). However, time budgets will broadly reflect the choices that animals make about the relative importance of various activities.

For the purposes of analysis, we combined the categories "moving" and "feeding" into a single category, "foraging." Baboons make a living by moving through their home range, selecting food items as they move. Thus, our activity category "moving" occurs, for the most part, as the baboons move to and between nutrient or water sources, while "feeding" occurs as the baboons harvest and ingest these nutrients. We view time spent feeding and moving combined as reflecting the amount of time that the baboons invest in obtaining food.

To measure the proportion of time spent in social activities, we combined "grooming," "being groomed," and "other social interactions." In practice, the large majority of social time was grooming time; other social interactions were typically brief events and together constituted approximately 2.5% of social time (approximately 0.25% of total time).

We also measured the proportion of time spent with no neighbor within 5 m. We view time spent with no neighbor as an indirect and inverse gauge of the time devoted to socializing. The presence of neighbors facilitates interactions and probably increases interaction rates, so that as time spent with no neighbor increases, socializing will generally decrease. We also view time spent with no neighbor as a gauge of the intensity of competition. Animals should tolerate greater proximity when foraging competition is less intense. Thus, an increase in time spent with no neighbors will

Table 6.2 *Types of food eaten by female baboons in Amboseli (see also Fig. 6.7). The 5% of feeding time not accounted for by these foods is accounted for by a variety of opportunistically consumed foods (Post 1982; Altmann 1998) and by unidentified foods. Water and vertebrate foods are included here for interest but occupy a tiny fraction of feeding time. Food types are ranked according to the relative proportion of feeding time devoted to them*

Type	Rank order	Description	Mean proportion of feeding time devoted to it (± s.e.)
Grass corms	1	Small perennial underground storage organs of grasses; the major representative is *Sporobolus rangei* (formerly *kentrophyllus*)	0.31 ± 0.03
Fruits	2	Fruits of shrubs and forbs; the major representative is *Trianthema ceratosepala*, but other species contribute importantly as well (see Table 6.3)	0.22 ± 0.03
Grass blades	3	Green leaves of grasses; baboons eat a wide variety of species	0.079 ± 0.013
Tree gum	4	The vast majority of gum consumed is from the fever tree, *Acacia xanthophloea*; occasionally gum of *Acacia tortilis* is consumed as well	0.065 ± 0.007
Seeds and material gleaned from dung	5	Seeds on ground and items gleaned from the dung of ungulates and elephants; baboons search primarily for seeds, but adult and larval invertebrates are sometimes abundant in the dung as well	0.057 ± 0.010
Flowers	6	Blossoms and buds from forbs, shrubs, and trees; three species represent the large majority of flowers eaten: *Acacia tortilis*, *Acacia xanthophloea*, and *Ramphicarpa montana* (an annual, highly rain-dependent forb)	0.053 ± 0.011
Grass blade bases	7	The meristematic tissue at the base of grass blades; to consume this, baboons will pull a blade of grass out of its sheath, bite off the lower portion of the blade, and discard the remainder; *Sporobolus consimilis* (elephant grass) is the main grass consumed in this way	0.040 ± 0.004

Table 6.2 (*cont.*)

Acacia seeds	8	Green and sometimes brown (dried) seeds removed from pods of *Acacia tortilis* or *Acacia xanthophloea*; green seeds of *tortilis* trees are available primarily in July through September, while green *xanthophloea* seeds are available in December–January.	0.036 ± 0.007
Grass seedheads	9	The baboons eat seedheads from a wide range of grass species	0.035 ± 0.010
Leaves of shrubs and forbs	10	Major contributors include *Lyceum* "*europeaum*" and *Salvadora persica*, but other species contribute as well	0.032 ± 0.004
Invertebrates	11	Grasshoppers, beetle larvae, and lepidopteran larvae are the major contributors	0.024 ± 0.007
Vertebrates	NA	Small vertebrates, including reptiles, birds, and mammals	Less than 0.001
Water	NA	During the dry season, permanent water holes and wells dug by the local Maasai people are the sources of water; during the wetter months, the baboons take advantage of seasonal rain pools	Less than 0.005

s.e., standard error.

generally reflect an increase in overall levels of foraging competition within the group.

Seasons

As noted above, the only reliable aspect of the yearly rainfall patterns in the study habitat was the long dry season, June through October. The remaining seven months of the year were highly variable from year to year. Rather than define seasons differently for each year according to the rainfall pattern in that year, we simply grouped the five months of the reliable long dry season, June through October, into a single period and the seven more variable months of November through May into a second period (the "wetter months"). Grouping November through May into a single season will undoubtedly obscure some behavioral variation that is dependent on rainfall; in particular, because those months were often dry, it will tend to reduce our ability to detect seasonal differences. Hence, this is a relatively conservative approach to examining patterns of seasonality.

Bivariate analysis of seasonal effects on activities

In our first analysis, we categorized each sample according to season, year, and subpopulation. We then calculated the total proportion of time spent in each activity for each subpopulation each year. This resulted in 27 "subpopulation-years" of data for the wetter months and 31 "subpopulation-years" of data for the long dry season (because we were missing complete "wetter months" data for four subpopulation-years). We next took the mean proportion of time spent in each activity for dry versus wetter months across all years, and used *t*-tests to examine seasonal differences in foraging, resting, socializing, and time spent with no neighbor within 5 m. Because successive years within a subpopulation may not be independent of each other, this procedure may result in some pseudo-replication. However, we also know that temporal changes in activity patterns have occurred over the years (Bronikowski & Altmann 1996), so that pooling across years will obscure important temporal differences. We dealt with the potential pseudo-replication by employing a P value for significance of $P = 0.01$ rather than the traditional $P = 0.05$.

Analysis of variance in time budgets

We also used a general linear model (using JMP™ software) to analyze variance in time spent in each activity. The predictor variables in our model were season (long dry versus wetter months), subpopulation, mean number of adult females in each group (as a measure of group size), total annual rainfall (total in millimeters fallen during the hydrological year), and year of study. We included subpopulation as a predictor variable because of the possibility that differences in home range between the two subpopulations contributed to differences in time budgets. We included mean social group size because a number of models predict that this should affect time budgets (e.g. Altmann [1980]; Dunbar [1992]). We included year of study because we knew from previous work that activity patterns change over the years, due to habitat changes, yearly variation in rainfall, and shifts in home range by the study groups (Bronikowski & Altmann 1996).

Analysis of variance in food eaten

We categorized baboon foods into the food component or plant part that the animals ate. We then analyzed variance in the proportion of feeding time devoted to each food type as a function of season, year of study, yearly rainfall, and subpopulation.

Phenology

Baboon foods in Amboseli fall into three categories of seasonality. Some show year-round availability. Others are highly seasonal, available only after the rains. The third category exhibits what we call "damped seasonality." That is, individual species can be highly seasonal in their productivity, but baboons can eat foods in this category in most months of the year (Table 6.3). This is because the "seasons" vary greatly between species and so collectively do not correspond to the "dry season–wetter months" dichotomy that we have presented. The consequence is that, by careful searching and by exploitation of many different species, baboons can eat foods in this category in most months of the year (Table 6.3).

The phenology of baboons' foods in Amboseli is well described (Post 1982: 12–14; Altmann 1998: 74–78). We supplemented these published descriptions with a new analysis of presence–absence data collected between 1985 and 1994 on fruits, *Acacia* pods, and the leaves of shrubs and forbs, three of the more seasonal foods in the diet. Phenological information from these various sources is summarized in Table 6.3.

Results

Time spent making a living changed with season and changed over time

Female baboons spent more time foraging and less time resting in the long dry season than during the wetter months (Table 6.4; Fig. 6.7 and 6.8). The

Table 8.3 Phenology of baboon foods in Amboseli (see also Post [1982: 12–14] and Altmann [1998: 74–88]). Foods are listed in rank order (according to proportion of feeding time devoted to them)

Type	Rank order	Basic phenology	Phenological description
Grass corms	1	Year round	Year-round availability for the predominant species, although corms may vary somewhat in nutrient content across seasons
Fruits	2	Damped seasonality	Some species highly seasonal, such as *Salvadora persica* (fruits from July or August until November or December) and *Commicarpus* spp. (fruits only after rains); other species available for many months, such as *Azima tetracantha* (fruits from May or June through January or February); two are available year round, at least in small quantities: *Withania somnifera* and *Trianthema ceratosepala*; this latter contributes the largest amount of fruit to the diet as baboons forage for it intensively during the dry season
Grass blades	3	Seasonal	New growth occurs immediately after rains begin; peak in February–May
Tree gum	4	Year round	Young trees are especially good sources, but trees exude gum all year round
Seeds and material gleaned from dung	5	Year round	Ungulate and elephant density in the baboons' home range varies seasonally, but dung and seeds are always abundant on the ground
Flowers	6	Damped seasonality	Three species are important: *A. tortilis* and *A. xanthophloea*, and *Ramphicarpa montana*; each is highly seasonal, but their peaks are distributed over the year: *tortilis* in January–April, *xanthophloea* in September–November, and *R. montana* after any rainy period
Grass blade bases	7	Year round	Blade bases of *Sporobolus consimilis*, a large perennial grass that is the largest contributor to this food type, are available year round
Acacia seeds	8	Damped seasonality	New pods/seeds are produced by *A. xanthophloea* in December–March, by *A. tortilis* in July–September; dry brown pods with dried seeds available all or most of the year
Grass seedheads	9	Seasonal	Development occurs approximately one month after rains begin
Leaves of shrubs and forbs	10	Seasonal	The preferred leaf is of *Lyceum* "*europaeum*," which is deciduous during dry periods. Many species of shrubs and forbs are evergreen, but the baboons prefer new leaves, which are rain-dependent in most species
Invertebrates	11	Damped seasonality	Peak availability of larvae is seasonal but varies across species

Table 6.4 *Analyses of variance in time spent foraging, resting, socializing, and alone*

Foraging			**Resting**	
Whole model: adjusted $R^2 = 0.35$, $P < 0.0001$			Whole model: adjusted $R^2 = 0.33$, $P < 0.0001$	
Effect	***F***	***P***	***F***	***P***
Season	**16.8**	**0.0001**	**17.97**	**< 0.0001**
Year	**11.54**	**0.0013**	**9.04**	**0.0041**
Group size	0.14	0.71	0.11	0.74
Rainfall	1.49	0.23	1.21	0.28
Subpopulation	0.0004	0.98	0.49	0.49
Alone (no neighbor)			**Socializing**	
Whole model: adjusted $R^2 = 0.088$, $P = 0.084$			Whole model: adjusted $R^2 = 0.16$, $P = 0.015$	
Effect	F	P	F	P
Season	0.04	0.84	3.08	0.085
Year	**9.36**	**0.0035**	**6.28**	**0.015**
Group size	**5.29**	**0.026**	0.04	0.84
Rainfall	1.02	0.32	0.7	0.41
Subpopulation	0.8	0.38	2.36	0.13

Bold type indicates terms with significant effects ($P < 0.01$) or strong trends.

clear but unsurprising inference is that quality of life is higher during the wetter months and that the dry season represents an ecological challenge for the animals.

Female activity profiles also changed substantially over time (Table 6.4; Fig. 6.8). From the mid 1980s to the mid 1990s, baboons in both subpopulations decreased foraging time and increased resting time (Fig. 6.8). This change coincided with the move that each subpopulation made to the western part of the Amboseli basin, where the fever tree woodland was large and healthy. From the middle to the end of the 1990s, however, foraging time increased for both subpopulations in each season, and resting time decreased. This may signal the end of a phase of relatively rapid density-independent growth that occurred immediately after each group moved. Alternatively, or in addition, the decrease in quality of life may be a consequence of the gradual decline of the fever tree woodlands that we observed in the western basin as the 1990s progressed.

The two subpopulations show remarkable similarity in their activity profiles; no effect of subpopulation on time spent in any activity is evident

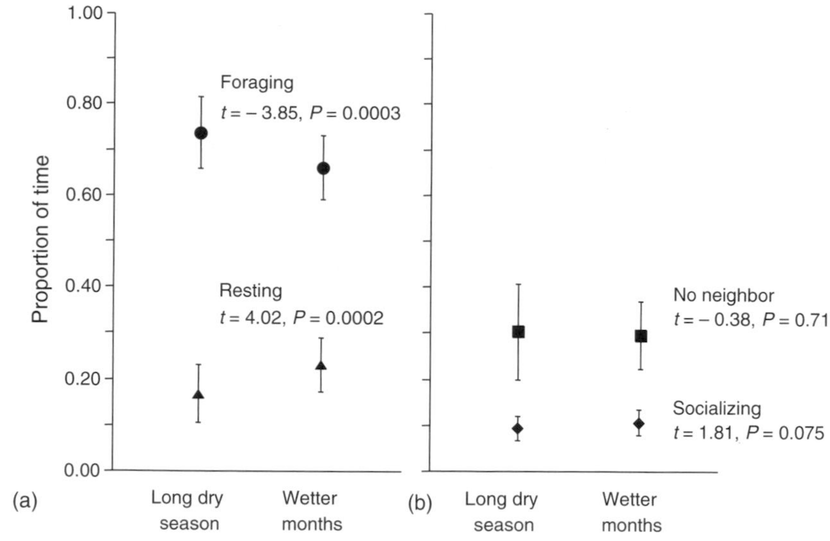

Figure 6.7 Seasonal differences in (a) time spent foraging and resting and (b) time spent socializing and with no neighbor. Each point represents the mean value (± standard deviation) of 31 group-years for long dry season and 27 group-years for wetter months (see text). Foraging time is calculated as the sum of moving and resting time. *t*-Tests are two-tailed.

in the analysis. While activity patterns varied significantly across years, this variability was not predicted by yearly differences in rainfall. Similarly, female group size, which varied from 6.5 to 18 adult females, did not predict time spent foraging or resting (Table 6.4).

Social time was unaffected by season but changed over time

Both the bivariate analysis and the analysis of variance indicated that season had no effect on time spent in social activities or on time spent with no neighbor (Figs. 6.7 and 6.9). However, both of these measures of socializing changed over time (Table 6.4). Specifically, as quality of life increased after the move west, time spent alone decreased and time spent socializing tended to increase in both seasons (Figure 6.9). Neither sub-population nor yearly rainfall affected social life. However, when groups were smaller, females showed a trend towards spending more time with no neighbor ($P = 0.026$) (Table 6.4).

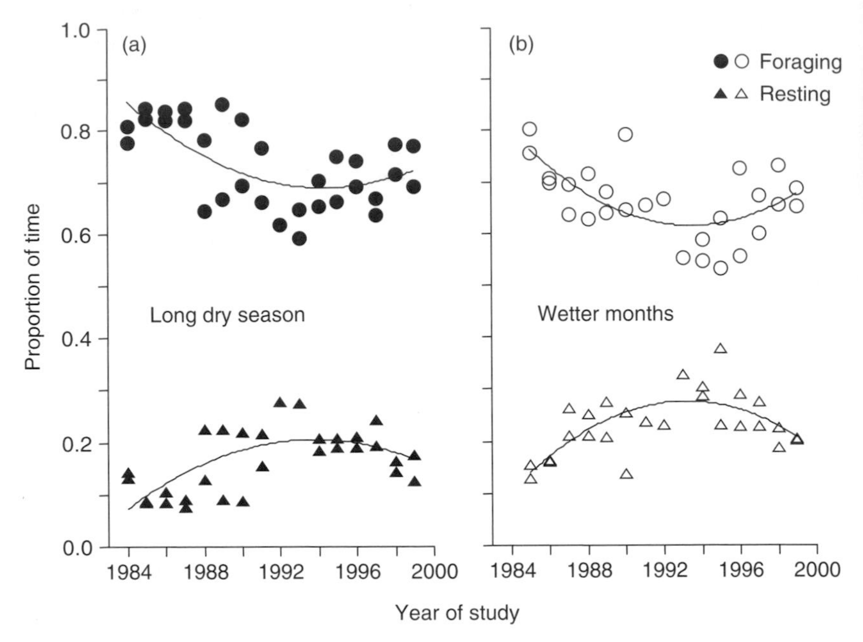

Figure 6.8 Amboseli baboons spend more time foraging (circles) and less time resting (triangles) in the dry season (a) than in the wetter months (b). Each point represents the yearly value for one subpopulation, and subpopulations are pooled in the graph because subpopulation explained no variance in activity budgets (Table 6.4). From 1984 to the mid 1990s, a gradual improvement in living conditions is evidenced by the decrease in foraging time and increase in resting time, but this reverses after the mid 1990s. These temporal effects are evident in both dry and wetter months (curves are fitted second-order polynomials; see Table 6.4 for analysis of variances).

Baboon diets had seasonal components but showed relative stability across seasons

Only four food types showed highly seasonal patterns of consumption in Amboseli: grass corms, grass blades, grass seedheads, and shrub/forb leaves (Fig. 6.10; Table 6.5) (grass seedheads are not significantly seasonal in our model because they show a pattern of consumption that is slightly shifted – by one month – relative to our definition of season). Shrub and forb leaves constitute a very small fraction of feeding time (Table 6.2; Fig. 6.10), indicating that grasses – their corms, blades, and seedheads – are the only foods in the Amboseli baboon diet that exhibit important seasonal patterns of consumption (Table 6.5; Fig. 6.10).

By far the largest set of food types in the baboon diet comprised those that showed damped seasonality. Each species shows a different temporal

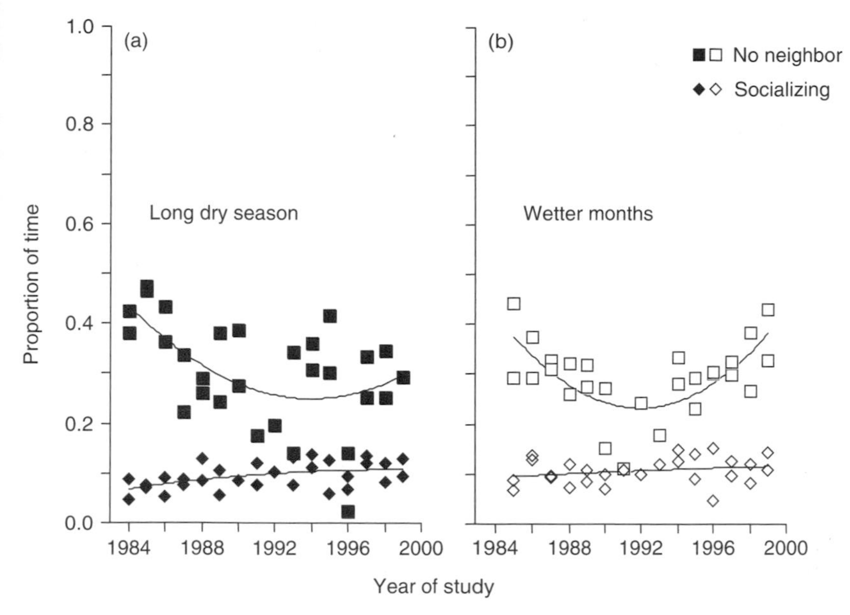

Figure 6.9 There is no significant effect of season on time spent alone (squares) or time spent socializing (diamonds), but a marked effect of year is evident. Proportion of time with no neighbors within 5 m reached a low point in the mid 1990s, when foraging time was at its lowest and resting time at its highest (see Fig. 6.5). Each point represents the yearly value for one subpopulation; subpopulations are pooled in the graph because subpopulation explained no variance in activity budgets (Table 6.4).

pattern of availability of, for instance, fruit, but the baboons mitigate this variability by exploiting a succession of species that are productive at different times of the year (Tables 6.3 and 6.5). Thus, the baboons achieve stability in their diet by carefully searching for and exploiting a large number of different species and plant parts across the year. They cannot achieve complete stability of food intake in this way; foods with damped seasonality still show considerable heterogeneity across months in their contribution to the diet (Fig. 6.11). However, this is not linked strictly to patterns of rainfall and the dry season because, as described above, different species have different phenologies. The result is that although the baboons eat different species of *Acacia* seeds in different months, they eat *Acacia* seeds in some quantity in almost every month. This is also true of flowers. Fruits show a complex pattern of availability across species, but fruits of some sort are available and consumed in every month (Table 6.3; Fig. 6.11).

Table 6.5 *Analyses of variance in proportion of feeding time devoted to each food type. Food types are shown in* ***bold face*** *if the model explained variance in their consumption; similarly, effects are shown in* ***bold face*** *if they explained a significant amount of the variance in feeding on that food type. Foods are ranked according to proportion of feeding time devoted to them. See Table 6.3 for detailed phenology*

Rank order	Food type	Phenology	Whole model		Effect of season		Effect of year		Effect of rainfall		Effect of subpopulation	
			Adjusted R^2	P	F	P	F	P	F	P	F	P
1	**Grass corms**	Year round	**0.78**	**< 0.0001**	**64.95**	**< 0.0001**	1.4	0.25	**7.61**	**0.0125**	2.06	0.17
2	Fruits	Damped seasonality	0.24	0.0567								
3	**Grass blades**	Seasonal	**0.53**	**0.0008**	**29.76**	**< 0.0001**	0.01	0.94	0.22	0.64	0.05	0.82
4	**Gum**	Year round	**0.43**	**0.005**	0.42	0.52	1.72	0.21	**20.09**	**0.0003**	0.05	0.82
5	Material gleaned from dung	Year round	0	0.95								
6	Flowers	Damped seasonality	0	0.5								
7	Grass blade bases	Year round	0.21	0.0736								
8	*Acacia* seeds	Damped seasonality	0	0.72								
9	**Grass seedheads**	Seasonal	**0.37**	**0.0105**	2.45	0.13	6.68	0.0182	1.44	0.24	0.24	0.63
10	**Shrub/forb leaves**	Seasonal	**0.72**	**< 0.0001**	**26.21**	**< 0.0001**	**9.07**	**0.0072**	1.32	0.26	**5.4**	**0.031**
11	Invertebrates	Damped seasonality	0.13	0.16								

Bold type indicates terms with significant effects or strong trends.

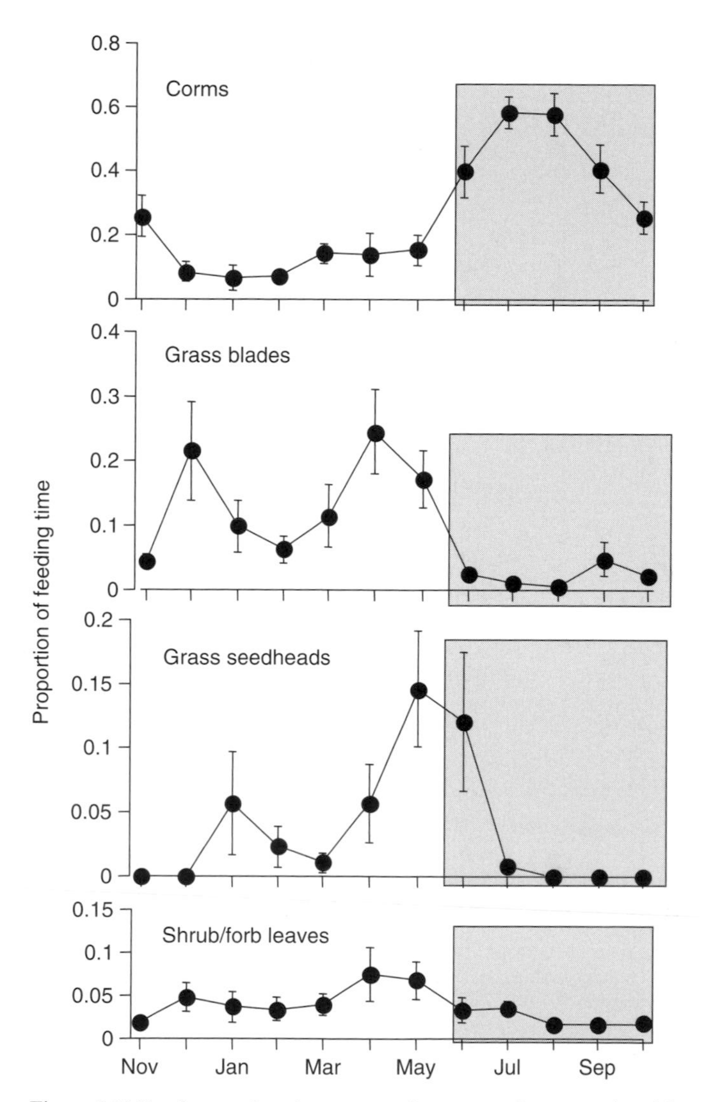

Figure 6.10 Food types that show seasonal patterns of consumption. Mean proportion of feeding time (± standard error [s.e.]) devoted to each food type is plotted against month of the year, beginning with November (the first wet month after the long dry season in Amboseli). Shaded boxes enclose the long dry season (June through October). See also Tables 6.2, 6.3, and 6.5.

Vertebrate animals accounted for a tiny fraction of the feeding time of female baboons in Amboseli (less than 0.1% of feeding time) (Table 6.2). Post (1982) recorded that 1% of feeding time in Amboseli was devoted to vertebrates and invertebrates combined, suggesting that in his study, as in

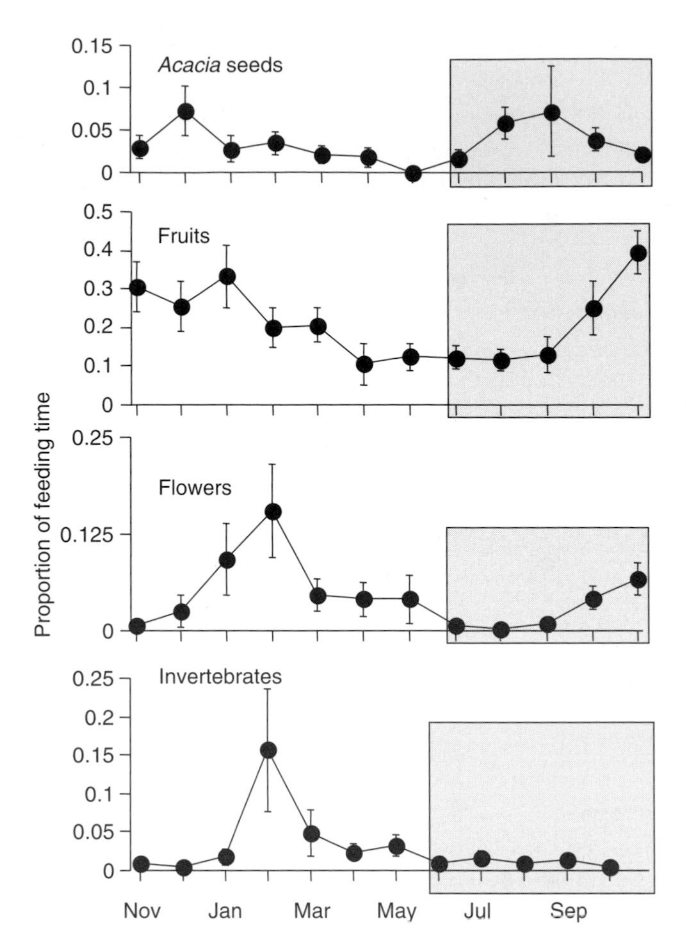

Figure 6.11 Patterns of consumption for food types that show damped seasonality in consumption patterns. In all cases, heterogeneity across months is evident, but consumption of the fruits, seeds, or flowers of at least some species occurs in both the long dry season and the wetter months. Conventions as in Fig. 6.7. See also Tables 6.2, 6.3, and 6.5.

ours, little feeding time was devoted to vertebrate prey, even for the adult males. Post (1982) notes that animal prey may be nutritionally important even if little time is devoted to it. This point may be relevant for invertebrates, which show measurable consumption in most months of the year. However, vertebrate prey in particular occupied so little feeding time for females in this study that it cannot be of great general importance.

Almost no variable in our model, other than season, affected the proportion of time spent on different food types (Table 6.5). Gum is consumed more heavily in drier years, and the leaves of shrubs and forbs have experienced a decrease in consumption over the course of the study. However, as noted, shrub/forb leaves occupy a very small proportion of feeding time overall, suggesting that this decrease may not be biologically significant for the animals. The two subpopulations showed no significant differentiation in diet.

Baboon diets changed over time

We did not observe any changes in time spent on different food types over the 1990s (our feeding data were available only for this decade; see Methods). However, we do have evidence to suggest that baboon diets have experienced substantial shifts over the three decades of our long-term study. Both Post (1982) and Altmann (1998) documented Amboseli baboon diets in the mid 1970s. In that period, *A. xanthophloea* products occupied substantially more feeding time than they did during the 1990s (Figure 6.12). This change in the importance of fever tree products is related to the change in the density of these trees in Amboseli. As the trees became less abundant, the baboons accommodated by altering their diets.

Discussion

The "fallback foods" strategy versus the "high-return foods" strategy

Female baboons in Amboseli certainly employed fallback foods: grass corms, available all year round, were the focus of intensive foraging activity only during the dry season, when key preferred foods (notably green grass blades and fruit) were scarce (Figs. 6.10 and 6.11) (see also Post [1982], Byrne *et al.* [1993], and Altmann [1998]). As a consequence, foraging time increased dramatically during the dry season. Grass corms require considerable processing time, so although they are reasonably rich in both protein and energy (Altmann *et al.* 1987; Byrne *et al.* 1993; Altmann 1998), their profitability is low. High-return foods played little role in the dry season diet of Amboseli females; vertebrate prey occupied only a tiny fraction of feeding time, and invertebrate prey were consumed in low quantities throughout the year, with a marked peak in one of the wetter months (Tables 6.2 and 6.5; Fig. 6.11).

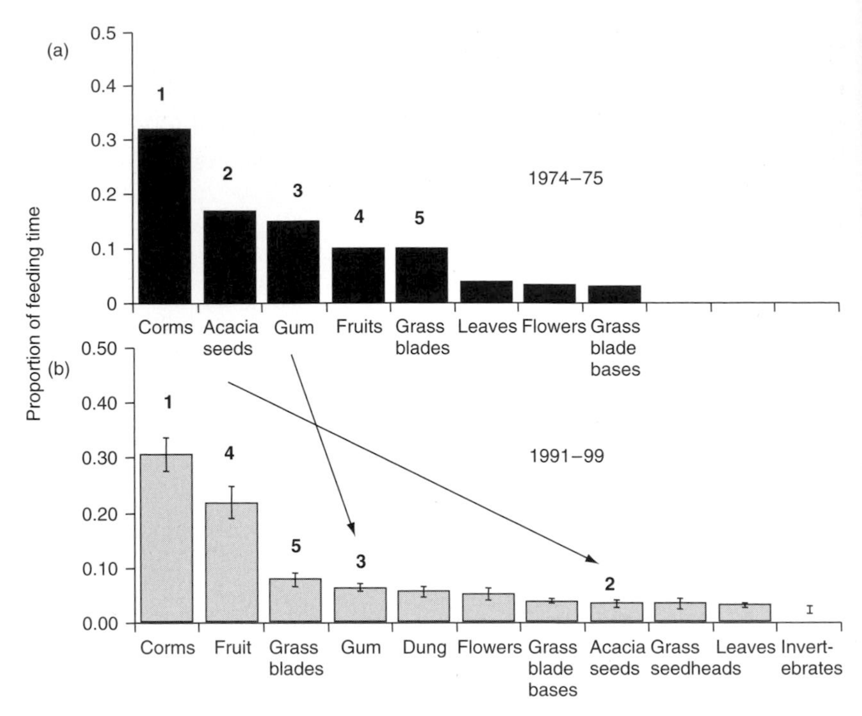

Figure 6.12 Proportion of feeding time spent on the ten major food types of Amboseli baboons. (a) Data reported by Post (1982) for the year 1974–75, with rank order of the food given above the bars for the top five foods (totaling 90% of feeding time). (b) Data from this study, pooling the two subpopulations, Hook's (1991–99) and Alto's (1996–99). Subpopulations are pooled in the graph because subpopulation explained no variance in feeding on various foods (Table 6.5). Numbers given above the bars indicate the rank order for the food type in Post's study. Note that in Post's study, *Acacia* gum and seeds were rank-ordered 2 and 3, respectively. By the 1990s both foods had dropped in importance, reflecting the fact that *Acacia* trees were less abundant in the 1990s than in the 1970s. Note also that in the 1990s, the top five foods constituted only 73% of feeding time.

Baboon foraging as a "handoff" strategy that mitigates seasonality

Only a few food types – notably the grasses – showed marked seasonality in their consumption by baboons, while many other important food types, including fruits, tree gum, material gleaned from dung, flowers, and invertebrates, were consumed in almost every month of the year. Indeed, even the grasses can be viewed as contributing to the diet in a constant manner across the year – the baboons simply switched from above-ground parts to below-ground parts during the dry season.

This suggests that the baboon foraging strategy might be thought of as a "handoff" strategy rather than a "fallback" strategy. Baboons achieve relative dietary stability partly by tracking carefully a large number of species and plant parts over the year and selectively exploiting foods as they become available. Even acknowledging that the succession of foods exploited – for instance, the fruits and flowers of different species – are not equivalent in their nutrient content (Altmann *et al.* 1987; Byrne *et al.* 1993; Altmann 1998), the result will be a constantly changing diet that probably allows the animals to maintain a relatively steady nutrient intake throughout the year. Detailed descriptions of the diets of other baboon populations support this notion (chacma baboons in Okavango, Botswana, and Kuiseb, Namibia [Hamilton *et al.* 1978]; yellow baboons in Mikumi, Tanzania [Norton *et al.* 1987]; chacma baboons in the Drakensberg Mountains, South Africa [Byrne *et al.* 1993]; see also Whiten *et al.* [1991]). This handoff strategy occurs even in populations with no fallback reliance on grass corms; in two of these populations, grass corms play a very small role in the baboon diet (Hamilton *et al.* 1978; Norton *et al.* 1987).

Baboons contrasted with vervet monkeys

The fact that baboons have almost entirely escaped reproductive seasonality suggests that the handoff strategy represents a very successful mode of adaptation to the savanna environment. In this regard, baboons represent an interesting and important contrast to vervet monkeys (*Cercopithecus aethiops*), the other widespread savanna monkey with which baboons often share habitat. Unlike baboons, vervet monkeys show relatively strong seasonality in birth and mating, with the birth peak in Amboseli occurring from October through January (Cheney *et al.* 1988). Further, Amboseli vervets are much more specialized feeders than baboons, focusing more intensively on *Acacia* products and relying on them as fallback foods particularly heavily in the dry season (Struhsaker 1967; Wrangham & Waterman 1981). Not surprisingly, the die-off of *A. xanthophloea* woodlands in the central basin of Amboseli resulted in the decline and eventually the local extinction of the vervet population in that area, although the vervet population persisted in parts of Amboseli in which fever trees persisted (Struhsaker 1973, 1976; Hauser *et al.* 1986; Isbell *et al.* 1990, 1991).

Thus, baboons and vervets have rather different modes of adaptation to the savanna environment in spite of extensive overlap in habitat and diet

(Struhsaker 1967; Wrangham & Waterman 1981; Altmann *et al.* 1987; Altmann 1998: Table 9.6). Vervets forage on fewer food types, restrict their foraging to a well-defended and relatively small territory, and employ a fallback strategy during the dry season. Baboons forage widely in large, undefended home ranges and employ a handoff strategy to utilize a broad and constantly changing set of foods. These differences may contribute importantly to the fact that vervets experience strong seasonal constraints on their reproduction while baboons cycle and conceive throughout the year.

Costs of fallback foods

During periods of food scarcity, most primate species probably lie somewhere along a continuum between a pure fallback strategy (i.e. relying extremely heavily on just one or a few food types) and a pure handoff strategy (i.e. moving from one food type to the next with no extra reliance on any particular type). Amboseli baboons clearly had elements of both strategies; along with their handoff approach, they also fell back on grass corms during the dry season. The baboon–vervet contrast suggests that a species' position on that continuum may have important consequences. In Amboseli, the purer fallback strategy (greater dietary specialization) of vervets was associated with greater vulnerability to local extinction as well as with a more seasonal physiology.

Reliance on fallback foods has multiple costs. One is an increase in foraging time for fallback foods that are time-consuming to process. Another was noted by Altmann (1998: 26–30): every food presents a "packaging problem" to the consumer, in that nutrients and toxins are packaged together and quantities of each vary from species to species and from season to season. Dietary diversity decreases during the dry season, at least for baboons (Post 1982; Norton *et al.* 1987), so the packaging problem is compounded by the fact that at this time of year, the animals have fewer alternative foods. *A. tortilis* seeds constitute important potential fallback foods for both vervets (Wrangham & Waterman 1981) and baboons (Fig. 6.11; Table 6.3). However, these seeds contain phenolics (including hydrolysable tannins) and trypsin inhibitor, both of which are toxic (Wrangham & Waterman 1981; Altmann *et al.* 1987; Altmann 1998). Thus, this easy-to-harvest and nutrient-rich food source is consumed in much lower quantities than one might expect, by both baboons and vervet monkeys, even though it reaches peak abundance during times when other foods are scarce (Wrangham & Waterman 1981; Altmann 1998). A third and critical cost of a fallback strategy, exemplified by the Amboseli vervets, is local extinction in the face of habitat change if the major fallback food declines.

The baboons' handoff strategy exhibits three characteristics that are probably critical to their relative success, particularly in the face of habitat change. First is their ability to range widely and even shift home ranges as the environment changes; vervets defend small territories and probably are unable to do this. Second is the fact that they are less reliant on any one food type than vervets are; they pursue a more full-blown handoff strategy. Third is their ability to utilize grasses to a greater extent than vervets. The first two characteristics will contribute to success in any habitat, while the third may be critical in determining how successfully a species adapts when woodland mosaic transitions to more open savanna.

The "dispensable social time" hypothesis versus the "social glue" hypothesis

Does social time represent "social glue," so that baboons conserve social time and sacrifice resting time when resources are scarce? This certainly appears to be true with respect to seasonal changes. Amboseli baboons conserved their social time, and reduced their resting time to accommodate the increased foraging demands of the season. However, on the larger scale of habitat change over time, the baboons sacrificed social time, and reduced their time with neighbors, in lower-quality habitats that demanded more foraging time (see also Bronikowski & Altmann [1996]).

Why might the baboons respond differently to seasonal changes in food availability than they do to habitat changes? They successfully absorbed seasonal demands on their foraging time budget without sacrificing social time, so why did they sacrifice social time when habitat quality deteriorated? One possible explanation is that seasonal changes were of smaller magnitude than the habitat changes that have occurred in Amboseli and that seasonal changes therefore taxed baboon energy reserves less than longer-term changes. This notion is supported by a comparison of foraging time differences within years (wet versus dry season foraging) with foraging time differences across years. On average, baboons spent 7.7% (range −4.5% to 22.2%) more time foraging in each dry season than they did in the corresponding "wetter months" (remember that these months are quite variable in rainfall, and that rains sometimes fail entirely). To contrast this with differences across years, we took the maximum yearly value for time spent foraging (81.3% for Alto's in 1985) and compared all other years with it (i.e. we subtracted the value for each year from this maximum). We found that the mean difference from the maximum time spent foraging was 12.9% (range 1.4% to 23.2%), much greater than mean seasonal differences. In other words, yearly changes in time spent foraging tended to be larger than changes within any one year.

A key prediction of the social glue hypothesis is that when groups are not able to maintain adequate time spent socializing, group cohesion will be lost (Dunbar 1992) and groups will presumably experience permanent fission. This prediction is not supported by our data. In fact, our social groups did fission, but not during periods when social time was limited. Instead, in each case they did so after they shifted to new habitats and experienced *decreases* in foraging time and *increases* in socializing time (Figs. 6.6, 6.8, 6.9). Indeed, in Hook's group, the fission occurred during 1995, the end of a three-year period when socializing time was at its peak and approached levels reported for Lodge group (Bronikowski & Altmann 1996), a food-augmented group with minimal nutritional or time constraints (cf. Bronikowski & Altmann [1996: Table 2] with this study, Figure 6.9). Further, in the years before they shifted their home ranges, the study groups consistently fell below Dunbar's estimated minimum social time necessary for group cohesion (Dunbar 1992: Equation 8), and they did so without experiencing fission. Even given the difficulties of estimating accurately a minimum social time requirement, this suggests that social time per se is not a major predictor of group cohesion.

However, Dunbar's prediction that animals will conserve social time in the face of food scarcity remains salient for interpreting seasonal changes in behavior. The difference we observed between seasonal and longer-term responses to habitat change may simply reflect the fact that baboons are fairly good at coping with seasonal changes but less successful at coping with more extreme changes in habitat quality. That is, baboons may indeed *attempt* to conserve social time as food availability fluctuates, but they are able to do this only within a fairly narrow range of habitat change – namely, the change experienced over the course of a year as the rains come and go.

Seasonal change versus habitat change as a selective force in primate evolution

Traditional hypotheses for human evolution attribute the emergence of unique human traits to movement into the savanna habitat, and to the particular challenges of that habitat, including marked seasonality (Foley

1987: Chapter 8, 1993; Potts 1998a, 1998b; Klein 1999: Chapters 4 and 5) (see also Chapter 17). In contrast, a recently articulated "variability hypothesis" suggests that unique human traits were selected for under a regime of constant habitat change (Potts 1996, 1998a, 1998b, 2002). The behavioral flexibility exhibited by humans is proposed to be a direct consequence of a long-term selection regime for traits that promoted survival in a fluctuating environment. Under this scenario, genes that promoted contingent behavioral responses that allowed adaptation to a range of habitats experienced strong positive selection because human ancestors experienced relatively dramatic habitat change over the course of the late Pliocene and early Pleistocene epochs.

In other words, Potts (1996) argues that it was long-term habitat change rather than the features of a particular habitat that ultimately selected for unique human traits. Under this hypothesis, seasonality would represent a relatively minor challenge to early hominids, while long-term habitat change would impose strong selection. Our data suggest that for Amboseli baboons, too, seasonal changes were relatively minor in their impact on behavior, while long-term habitat changes posed a greater challenge for the animals.

The success of the Amboseli baboon population in coping with challenging habitat change is notable. Their coping strategies have included decreasing the time devoted to socializing during hard times, modifying their diet as the habitat changed, and adaptively shifting their home range in the face of habitat deterioration (Bronikowski & Altmann 1996; Altmann & Alberts 2003). These strategies have been so successful that in recent decades, the baboon population in Amboseli has increased (Alberts & Altmann 2003), and several social groups have fissioned after growing in size (Altmann & Alberts 2003). The success of the baboons is in striking contrast to the local extinction experienced by vervets in areas of Amboseli that lost fever tree woodlands (Struhsaker 1973, 1976; Hauser *et al.* 1986; Isbell *et al.* 1991).

As noted earlier, baboons and humans share a number of key traits – a wide geographic distribution, success in but not restriction to savanna environments, and non-seasonal reproduction. We have also shown that baboons can be successful in the face of fairly dramatic environmental change, another key human trait under the variability hypothesis for human evolution (Potts 1996). Finally, they share three traits that Potts (1998a) proposes evolved in response to variability selection: (i) they are moderately highly encephalized, with a high neocortex-to-cortex ratio (Dunbar 1998); (ii) they exhibit a flexible locomotor system and readily utilize both arboreal and terrestrial habitats (Estes 1991; Fleagle 1999);

and (iii) they exhibit a highly flexible social system (Altmann & Altmann 1970; Barton *et al.* 1996; Dunbar & Dunbar 1977; Henzi & Barrett 2003).

Perhaps, then, baboons represent a model for understanding the behavioral plasticity of early hominids. If so, what traits might early hominids have exhibited? One would be handoff foraging, in which temporal variability in food abundance was mitigated by careful tracking and exploitation of food resources as they became available. Concomitant with this skill would be an ability to find alternatives when important foods became scarce as the habitat changed. A third trait would involve a well-buffered social structure in which individual relationships were serviced carefully. This might mean substantial investment in relationships, but the baboon model suggests that if forced to limit time investment during food scarcity, then alternative, equivalent modes of interacting might be pursued in order to maintain relationships with less cost. Finally, the baboon model suggests that a fourth very important trait would be the flexibility to actually alter one's own environment by finding and moving to more suitable habitats. As yet, the components of the variability hypothesis for hominid evolution have not been explored in a non-human primate system. Our analysis suggests that such an exploration could shed considerable light on the manner and consequences of the response of primates to environmental change.

Acknowledgments

We gratefully acknowledge financial support from the National Science Foundation (IBN-9985910 and its predecessors) and the Chicago Zoological Society to JA. We thank the Office of the President of the Republic of Kenya and the Kenya Wildlife Service for permission to work in Amboseli over the years. We thank the Institute of Primate Research for local sponsorship in Kenya, the wardens and staff of Amboseli National Park, and the pastoralist communities of Amboseli and Longido for continuous cooperation and assistance. Particular thanks go to the Amboseli fieldworkers who contributed to the data over the years, especially S. A. Altmann and A. Samuels. We also thank those who have contributed to the design and maintenance of the long-term Amboseli database, especially Karl O. Pinc, Stephanie Combes, Dominique Shimizu, and Jessica Zayas. We thank Carel van Schaik and Diane Brockman for helpful discussions and Charlie Janson for calculating vector lengths and P values for testing seasonality of reproduction.

References

Alberts, S. C. & Altmann, J. (2001). Immigration and hybridization patterns of yellow and anubis baboons in and around Amboseli, Kenya. *American Journal of Primatology*, **53**, 139–54.

— (2003). Matrix models for primate life history analysis. In *Primate Life Histories and Socioecology*, ed. P. M. Kappeler & M. E. Pereira. Chicago: University of Chicago Press, pp. 66–102.

Alberts, S. C., Altmann, J., & Wilson, M. L. (1996). Mate guarding constrains foraging activity of male baboons. *Animal Behaviour*, **51**, 1269–77.

Altmann, J. (1974). Observational study of behavior: sampling methods. *Behaviour*, **49**, 227–67.

— (1980). *Baboon Mothers and Infants*. Cambridge, MA: Harvard University Press.

Altmann, J. & Alberts, S. C. (2003). Intraspecific variability in fertility and offspring survival in a non-human primate: behavioral control of ecological and social sources. In *Biodemography of Fertility and Family Behavior*, ed. K. Wachter. Washington, DC: National Academy Press, pp. 140–69.

Altmann, J. & Muruthi, P. (1988). Differences in daily life between semi-provisioned and wild-feeding baboons. *American Journal of Primatology*, **15**, 213–21.

Altmann, J., Hausfater, G., & Altmann, S. (1985). Demography of Amboseli baboons, 1963–1983. *American Journal of Primatology*, **8**, 113–25.

Altmann, J., Alberts, S. C., Altmann, S. A., & Roy, S. B. (2002). Dramatic change in local climate patterns in the Amboseli basin, Kenya. *African Journal of Ecology*, **40**, 248–51.

Altmann, S. A. (1998). *Foraging for Survival*. Chicago: University of Chicago Press.

Altmann, S. A., & Altmann, J. (1970). *Baboon Ecology*. Chicago: University of Chicago Press.

Altmann, S. A., Post, D. G., & Klein, D. F. (1987). Nutrients and toxins of plants in Amboseli, Kenya. *African Journal of Ecology*, **25**, 279–93.

Barton, R. A., Byrne, R. W., & Whiten, A. (1996). Ecology, feeding competition and social structure in baboons. *Behavioral Ecology and Sociobiology*, **38**, 321–9.

Behrensmeyer, A. K. (1993). The bones of Amboseli: the taphonomic record of ecological change in Amboseli Park, Kenya. *National Geographic Research and Exploration*, **9**, 402–21.

Behrensmeyer, A. K. & Boaz, D. D. (1981). Late Pleistocene geology and paleontology of Amboseli National Park, Kenya. In *Palaeoecology of Africa and the Surrounding Islands*, Vol. 13 ed. J. A. Coetzee & E. M. van Zinderen Bakker, Sr. Rotterdam: A. A. Balkema, pp. 175–88.

Bentley-Condit, V. K., & Smith, E. O. (1997). Female reproductive parameters of Tana River yellow baboons. *International Journal of Primatology*, **18**, 581–96.

Bercovitch, F. B. & Harding, R. S. O. (1993). Annual birth patterns of savanna baboons (*Papio cynocephalus anubis*) over a ten-year period at Gilgil, Kenya. *Folia Primatologica*, **61**, 115–22.

Blumenschine, R. (1987). Characteristics of an early hominid scavenging niche. *Current Anthropology*, **28**, 383–407.

Bourliere, F. & Hadley, M. (1983). Present-day savannas: an overview. In *Ecosystems of the World 13: Tropical Savannas*, ed. F. Bourliere. Amsterdam: Elsevier Scientific, pp. 1–17.

Bronikowski, A. & Altmann, J. (1996). Foraging in a variable environment: weather patterns and the behavioral ecology of baboons. *Behavioral Ecology and Sociobiology*, **39**, 11–25.

Bunn, H. T., Bartram, L. T., & Kroll, E. M. (1988). Variability in bone assemblage formation from Hadza hunting, savenging, and carcass processing. *Journal of Anthropological Archaelogy*, **7**, 412–57.

Byrne, R. W., Whiten, A., Henzi, S. P., & McCulloch, F. M. (1993). Nutritional constraints on mountain baboons (*Papio ursinus*): implications for baboon socioecology. *Behavioral Ecology and Sociobiology*, **33**, 233–46.

Cheney, D. L., Seyfarth, R. M., Andelman, S. J., & Lee, P. C. (1988). Reproductive success in vervet monkeys. In *Reproductive Success*, ed. T. H. Clutton-Brock. Chicago: University of Chicago Press, pp. 384–402.

Cutler, A. H., Behrensmeyer, A. K., & Chapman, R. E. (1999). Environmental information in a recent bone assemblage: roles of taphonomic processes and ecological change. *Palaeogeography, Palaeoclimatology, Palaeoecology*, **149**, 359–72.

Dunbar, R. I. M. (1983). Theropithecines and hominids: contrasting solutions to the same ecological problem. *Journal of Human Evolution*, **12**, 647–58.

— (1991). Functional significance of social grooming in primates. *Folia Primatologica*, **57**, 121–31.

— (1992). Time: a hidden constraint on the behavioral ecology of baboons. *Behavioral Ecology and Sociobiology*, **31**, 35–49.

— (1998). The social brain hypothesis. *Evolutionary Anthropology*, **6**, 178–90.

Dunbar, R. I. M. & Dunbar, P. (1977). Dominance and reproductive success among female gelada baboons. *Nature*, **266**, 351–2.

— (1988). Maternal time budgets of gelada baboons. *Animal Behaviour*, **36**, 970–80.

Esikuri, E. S. (1998). Spatio-temporal effects of land use changes in a savanna wildlife area of Kenya. Ph. D. thesis, Virginia Polytechnic Institute and State University.

Estes, R. D. (1991). *The Behavior Guide to African Mammals*. Berkeley: University of California Press.

Fleagle, J. G. (1999). *Primate Adaptation and Evolution*, 2nd edn. San Diego: Academic Press.

Foley, R. (1987). *Another Unique Species: Patterns in Human Evolutionary Ecology*. New York: John Wiley & Sons.

Foley, R. A. (1993). The influence of seasonality on hominid evolution. In *Seasonality and Human Ecology: 35th Symposium Volume of the Society for the Study of Human Biology*, ed. S. J. Ulijaszek & S. S. Strickland. Cambridge: Cambridge University Press, pp. 17–37.

Gursky, S. (2000). Effect of seasonality on the behavior of an insectivorous primate, *Tarsius spectrum*. *International Journal of Primatology*, **21**, 477–95.

Hamilton, W. J. I., Buskirk, R. E., & Buskirk, W. H. (1978). Omnivory and utilization of food resources by chacma baboons, *Papio ursinus*. *American Naturalist*, **112**, 911–24.

Hastenrath, S. & Greischar, L. (1997). Glacier recession on Kilimanjaro, East Africa, 1912–89. *Journal of Glaciology*, **43**, 455–9.

Hauser, M. D., Cheney, D. L., & Seyfarth, R. M. (1986). Group extinction and fusion in free-ranging vervet monkeys. *American Journal of Primatology*, **11**, 63–77.

Hawkes, K., O'Connell, J. F., & Blurton-Jones, N. G. (1991). Hunting income patterns among the Hadza: big game, common goods, foraging goals and the evolution of the human diet. *Philosophical Transactions of the Royal Society, London, Series B*, **334**, 243–51.

Hay, R. L., Hughes, R. E., Kyser, T. K., Glass, H. D., & Liu, J. (1995). Magnesium-rich clays of the meerschaum mines in the Amboseli basin, Tanzania and Kenya. *Clays and Clay Minerals*, **43**, 455–66.

Henzi, P. & Barrett, L. (2003). Evolutionary ecology, sexual conflict, and behavioral differentiation among baboon populations. *Evolutionary Anthropology*, **12**, 217–30.

Isbell, L. A., Cheney, D. L., & Seyfarth, R. M. (1990). Costs and benefits of home range shifts among vervet monkeys (*Cercopithecus aethiops*) in Amboseli National Park, Kenya. *Behavioral Ecology and Sociobiology*, **27**, 351–8.

— (1991). Group fusions and minimum group sizes in vervet monkeys (*Cercopithecus aethiops*). *American Journal of Primatology*, **25**, 57–65.

Jolly, C. J. (1993). Species, subspecies, and baboon systematics. In *Species, Species Concepts, and Primate Evolution*, ed. W. H. Kimbel & L. B. Martin. New York: Plenum Press, pp. 67–107.

Kingdon, J. (1997). *The Kingdon Field Guide to African Mammals*. San Diego: Academic Press.

Klein, R. G. (1999). *The Human Career: Human Biological and Cultural Origins*, 2nd edn. Chicago: University of Chicago Press.

Koch, P. L., Heisinger, J., Moss, C., *et al.* (1995). Isotopic tracking of change in diet and habitat use in African elephants. *Science*, **267**, 1340–43.

Kummer, H. (1968). *Social Organization of Hamadryas Baboons*. Chicago: University of Chicago Press.

Lincoln, R., Boxshall, G., & Clark, P. (1998). *A Dictionary of Ecology, Evolution and Systematics*, 2nd edn. Cambridge: Cambridge University Press.

Malenky, R. K. & Wrangham, R. W. (1994). A quantitative comparison of terrestrial herbaceous food consumption by *Pan paniscus* in the Lomako Forest, Zaire, and *Pan troglodytes* in the Kibale Forest, Uganda. *American Journal of Primatology*, **32**, 1–12.

Melnick, D. J. & Pearl, M. C. (1987). Cercopithecines in multimale groups: genetic diversity and population structure. In *Primate Societies*, ed. B. B. Smuts, D. L. Cheney, R. Seyfarth, R. W. Wrangham, & T. T. Struhsaker. Chicago: University of Chicago Press, pp. 121–34.

Mitani, J. C. & Watts, D. P. (2001). Why do chimpanzees hunt and share meat? *Animal Behaviour*, **61**, 915–24.

Moss, C. J. (2001). The demography of an African elephant (*Loxodonta africana*) population in Amboseli, Kenya. *Journal of Zoology, London*, **255**, 145–56.

Muruthi, P., Altmann, J., & Altmann, S. (1991). Resource base, parity, and reproductive condition affect females' feeding time and nutrient intake within and between groups of a baboon population. *Oecologia*, **87**, 467–72.

Norton, G. W., Rhine, R. J., Wynn, G. W., & Wynn, R. D. (1987). Baboon diet: a five-year study of stability and variability in the plant feeding and habitat of the yellow baboons (*Papio cynocephalus*) of Mikumi National Park, Tanzania. *Folia Primatologica*, **48**, 78–120.

Overdorff, D. J. (1996). Ecological correlates to activity and habitat use of two prosimian primates: *Eulemur rubriventer* and *Eulemur fulvus rufus* in Madagascar. *American Journal of Primatology*, **40**, 327–42.

Post, D. G. (1981). Activity patterns of yellow baboons (*Papio cynocephalus*) in the Amboseli National Park, Kenya. *Animal Behaviour*, **29**, 357–74.

— (1982). Feeding behavior of yellow baboons (*Papio cynocephalus*) in the Amboseli National Park, Kenya. *International Journal of Primatology*, **3**, 403–30.

Potts, R. (1996). Evolution and climate variability. *Science*, **273**, 922–3.

— (1998a). Variability selection in hominid evolution. *Evolutionary Anthropology*, **7**, 81–96.

— (1998b). Environmental hypotheses of hominin evolution. *Yearbook of Physical Anthropology*, **41**, 93–136.

— (2002). Complexity and adaptability in human evolution. In *Probing Human Origins*, ed. M. Goodman & A. S. Moffat. Cambridge, MA: American Academy of Arts and Sciences, pp. 11–32.

Seyfarth, R. M. (1977). A model of social grooming among adult female monkeys. *Journal of Theoretical Biology*, **65**, 671–98.

Solbrig, O. T. (1996). The diversity of the savanna ecosystem. In *Biodiversity and Savanna Ecosystem Processes: A Global Perspective. Ecological Studies*, Vol. 21, ed. O. T. Solbrig, E. Medina, & J. F. Silva. Berlin: Springer-Verlag, pp. 1–27.

Stammbach, E. (1987). Desert, forest and montane baboons: multi-level societies. In *Primate Societies*, ed. B. B. Smuts, D. L. Cheney, R. Seyfarth, R. W. Wrangham, & T. T. Struhsaker. Chicago: University of Chicago Press, pp. 112–20.

Stanford, C. B. (1996). The hunting ecology of wild chimpanzees: implications for the evolutionary ecology of Pliocene hominids. *American Anthropologist*, **98**, 96–113.

Stanford, C. B., Wallis, J., Mpongo, E., & Goodall, J. (1994). Hunting decisions in wild chimpanzees. *Behaviour*, **131**, 1–18.

Strier, K. B. (1991). Diet in one group of woolly spider monkeys, or muriqis (*Brachyteles arachnoides*). *American Journal of Primatology*, **23**, 113–26.

Struhsaker, T. T. (1967). Ecology of vervet monkeys (*Cercopithecus aethiops*) in the Masai-Amboseli Game Reserve, Kenya. *Ecology*, **48**, 891–904.

— (1973). A recensus of vervet monkeys in the Masai-Amboseli Game Reserve, Kenya. *Ecology*, **54**, 930–32.

— (1976). A further decline in numbers of Amboseli vervet monkeys. *Biotropica*, **8**, 211–14.

Thompson, L. G., Mosley-Thompson, E., Davis, M. E., *et al.* (2002). Kilimanjaro ice core records: evidence of holocene climate change in tropical Africa. *Science*, **298**, 589–93.

Western, D. & Maitumo, D. (2004). Woodland loss and restoration in a Savanna park: a 20-year experiment. *African Journal of Ecology*, **42**, 111–21.

Western, D. & van Praet, C. (1973). Cyclical changes in the habitat and climate of an East African ecosystem. *Nature*, **241**, 104–6.

Whiten, A., Byrne, R. W., Barton, R. A., Waterman, P. G., & Henzi, S. P. (1991). Dietary and foraging strategies of baboons. *Philosophical Transactions of the Royal Society of London, Series B*, **334**, 187–97.

Williams, L. A .J. (1972). *Geology of the Amboseli Area. Degree Sheet 59, S. W. Quarter*. Nairobi: Ministry of Natural Resources, Geological Survey of Kenya.

Wrangham, R. W., & Waterman, P. G. (1981). Feeding behaviour of vervet monkeys on *Acacia tortilis* and *Acacia xanthophloea*: with special reference to reproductive strategies and tannin production. *Journal of Animal Ecology*, **50**, 715–31.

Wrangham, R. W., Conklin, N. L., Chapman, C. A., & Hunt, K. D. (1991). The significance of fibrous foods for Kibale Forest chimpanzees. *Philosophical Transactions of the Royal Society of London, Series B*, **334**, 171–8.

Wrangham, R. W., Conklin-Brittain, N. L., & Hunt, K. D. (1998). Dietary response of chimpanzees and cercopithecines to seasonal variation in fruit abundance. I. Antifeedants. *International Journal of Primatology*, **19**, 949–70.

Young, T. P. & Lindsay, W. K. (1988). Role of even-age population structure in the disappearance of *Acacia xanthophloea* woodlands. *African Journal of Ecology*, **26**, 69–72.

7 *Day length seasonality and the thermal environment*

RUSSELL HILL
Evolutionary Anthropology Research Group, Department of Anthropology, University of Durham, 43 Old Elvet, Durham, DH1 3HN, UK

Introduction

The importance of high ambient temperatures and intense solar radiation for the evolution of hominids in open savanna habitats has been the subject of considerable interest. A series of studies has considered the thermoregulatory advantages related to bipedalism (Wheeler 1991), loss of functional body hair (Wheeler 1992a), body size (Wheeler 1992b), physique (Wheeler 1993), and shade-seeking behavior (Wheeler 1994a). Furthermore, these papers have generated considerable debate (Porter 1993; Chaplin *et al.* 1994; Wheeler 1994b; do Amaral 1996; Wheeler 1996). It is surprising, therefore, that the importance of the thermoregulation in primate behavioral ecology has received comparatively little attention, with the body of former work focusing on other ecological factors such as food availability (Stelzner 1988). Nevertheless, a number of studies have reported primates to alter their activity schedules in response to thermoregulatory needs (baboons, *Papio* spp. [Stolz & Saayman 1970]; gelada, *Theropithecus gelada* [Iwamoto & Dunbar 1983]; pigtail macaques, *Macaca nemestrina* [Bernstein 1972]; sooty mangabeys, *Cercocebus atys* [Bernstein 1976]; chimpanzees, *Pan troglodytes* [Wrangham 1977]; gorillas, *Gorilla gorilla* [Fossey & Harcourt 1977]). However, in most cases, these studies have invoked post-hoc thermoregulatory interpretations, and few have examined explicitly the importance of the thermal environment under natural conditions.

The most detailed studies of thermoregulation in wild primates have been conducted on baboons (e.g. Stelzner & Hausfater [1986], Stelzner [1988], Brain & Mitchell [1999], Pochron [2000], and Hill [2005]). Like humans, baboons lack known mechanisms for effective brain cooling (such as carotid rete [Brain & Mitchell 1999]) and, as a consequence,

Seasonality in Primates: Studies of Living and Extinct Human and Non-Human Primates, ed. Diane K. Brockman and Carel P. van Schaik. Published by Cambridge University Press.

represent a useful analogy for examining thermoregulation in the context of hominid evolution. Furthermore, most mammals of similar body mass and activity patterns, and inhabiting the same arid environments, possess effective brain cooling mechanisms (Mitchell *et al.* 1987) and thus do not require the ready access to the water that is essential for evaporative cooling in primates (Mitchell & Laburn 1985). Terrestrial primates, therefore, are likely to be highly susceptible to the costs of thermoregulation in open habitats, and baboons are thus an ideal species for examining the importance of the thermal constraints on primate activity patterns.

In deep shade, baboons should maintain a steady net heat balance at temperatures of up to 40 °C (Funkhouser *et al.* 1967; Stelzner 1988). Nevertheless, baboons exposed to environmental conditions of clear sky, full sun, and low wind speeds will start to experience net heat gain at air temperatures above 15–20 °C (Stelzner 1988). The body mass of adult baboons (12–30 kg [Dunbar 1990]) inhibits rapid passive heat dissipation, such as that seen in small mammals (e.g. antelope ground squirrels, *Ammospermophilus leucurus* [Chappel & Bartholomew 1981]), or the extensive heat storage seen in some large mammals with brain cooling mechanisms (e.g. gemsbok, *Oryx gazella* [Taylor 1969]). Instead, baboons rely heavily on cutaneous and respiratory water loss to maintain a stable body temperature. Laboratory studies of heat stress indicate that approximately 80% of total evaporative cooling in baboons is due to cutaneous (rather than the more efficient respiratory) water loss (Hiley 1976; Funkhouser *et al.* 1967), indicating that access to water is an important constraint on thermoregulation. At Kuiseb, Namibia, the most arid environment in which baboons have been studied, and where there is little access to water, core body temperatures may fluctuate by as much as 5.3 °C and regularly exceed 41 °C (Brain & Mitchell 1999). This suggests that active body cooling through evaporation is employed only when water is available and that heat storage coupled with non-evaporative heat loss as air temperatures cool at night are possible under conditions of extreme water shortage. Nevertheless, prolonged water shortage can lead to significant loss of body mass and reduced plasma volume (Zurovsky & Shkolnik 1983). However, given the range of temperatures and water availability experienced by baboons in their natural environments, behavioral thermoregulatory mechanisms are likely to be employed well within the limits of physiological tolerance in most populations (Stelzner 1988). Indeed, the baboons at Kuiseb also "sandbathed" as a means of behavioral thermoregulation, with the frequency of sandbathing increasing on non-drinking days (Brain & Mitchell 1999).

Drinking is clearly an important behavioral thermoregulatory strategy, and Brain and Mitchell (1999) reported that body temperatures dropped by about 1.7 °C upon drinking at Kuiseb. It is not surprising, therefore, that a number of studies report baboons to center their ranging patterns around water sources in the face of thermal stress. Kummer (1968) describes hamadryas baboons moving to an isolated waterhole to drink and rest over the midday period, and at Honnet Nature Reserve, South Africa, chacma baboons drank during the hottest period of the day (Stoltz & Saayman 1970). Shade is another important behavioral response to thermoregulatory stress. Stelzner (1988) found that the baboons at Amboseli National Park, Kenya, responded to high heat stresses at midday by resting in shade whenever they encountered it. The baboons in this population did not actively seek shade but rather exploited it opportunistically, such that thermal constraints on habitat choice were apparent only at the microhabitat level. Similar behavioral patterns were observed at De Hoop Nature Reserve, South Africa (Hill 2005) where levels of both resting and grooming increased in response to thermal stress, and these activities were preferentially conducted in shade. Furthermore, air temperatures significantly influenced patterns of habitat choice at De Hoop, with the long summer days allowing the baboons to rest for long periods on cliff refuges over the midday period (Hill 2005). Variation in day length appears important in understanding the differences between the populations, with lower shade temperatures at De Hoop probably accounting for the elevated incidence of grooming under high temperatures in this population. This suggests that latitude and seasonal variation in day length may be important factors underlying variation in baboon thermoregulatory responses.

Seasonality also may be important in the thermoregulatory responses of baboons at Equatorial latitudes. Pochron (2000) found evidence of sun avoidance by the baboons of Ruaha National Park, Tanzania. Contrary to predictions, however, baboons avoided the sun across all activities during the cool, lush season but avoided the sun only while resting in the hot, dry season. Although temperature appeared to be driving sun avoidance in the dry season, humidity was shown to be responsible during the lush season. However, temperature, humidity, and wind speed are all important elements of the perceived environmental temperature experienced by baboons (Hill *et al.* 2004a). Nevertheless, constraints imposed by differential food availability between the two seasons at Ruaha would appear to drive the deviations from predictions based on thermoregulatory considerations alone, suggesting that seasonality might once again be important in constraining the thermoregulatory response.

Seasonality thus appears to be important in determining the extent to which animals are able to respond to the thermal environment. Although the importance of seasonality in food availability in constraining primate behavior has been relatively well studied (e.g. baboons [Post 1981]; samango monkeys, *Cerceopithecus mitis* [Lawes & Piper 1992]; red colobus, *Procolobus badius* [Clutton-Brock 1977]), seasonality in day length has often been overlooked, despite the fact that it represents an important constraint on behavior (Hill *et al.* 2003). The purpose of this chapter, therefore, is to examine the importance of temperature and day length in determining patterns of behavior in primates inhabiting savanna environments. Initially, the importance of the thermal environment in constraining daily behavioral patterns is assessed. Having demonstrated that temperature places significant constraints on behavior, the importance of climatic variables in explaining seasonal variation in activity is examined. Since time spent resting does not correlate with mean monthly temperatures as might be anticipated, with day length being the key constraint underlying seasonal variation in activity levels, the implications of day length seasonality in constraining behavioral flexibility are examined in relation to behavioral variation across sub-Saharan Africa.

Methods

This study focuses on a troop of chacma baboons (*Papio hamadryas ursinus*) at De Hoop Nature Reserve (20°24′E, 34°27′S), Western Cape Province, South Africa. Data were collected by means of instantaneous scan samples (Altmann 1974) at 30-minute intervals for a single study troop (VT) that ranged in size from 40 to 44 individuals over the course of the ten-month study period (March–December 1997). De Hoop is a coastal reserve and the vegetation is dominated by fynbos, a unique and diverse vegetation type comprising 80% of the Cape Floral kingdom, one of the six floristic regions of the world (Cowling & Richardson 1995). The reserve has a Mediterranean climate: annual rainfall during the study year was 395 mm, with a mean shade temperature of 17.4 °C. Due to its southerly latitude, De Hoop is highly seasonal, with substantial variation in both rainfall and temperature, as well as day length variation that is unparalleled at any other sub-Saharan African site (Fig. 7.1). Detailed climatic information was recorded at 30-minute intervals by an automatic weather station (see Hill *et al.* 2004a). More detailed information on the ecology of the reserve and data collection methods is given in Hill (1999).

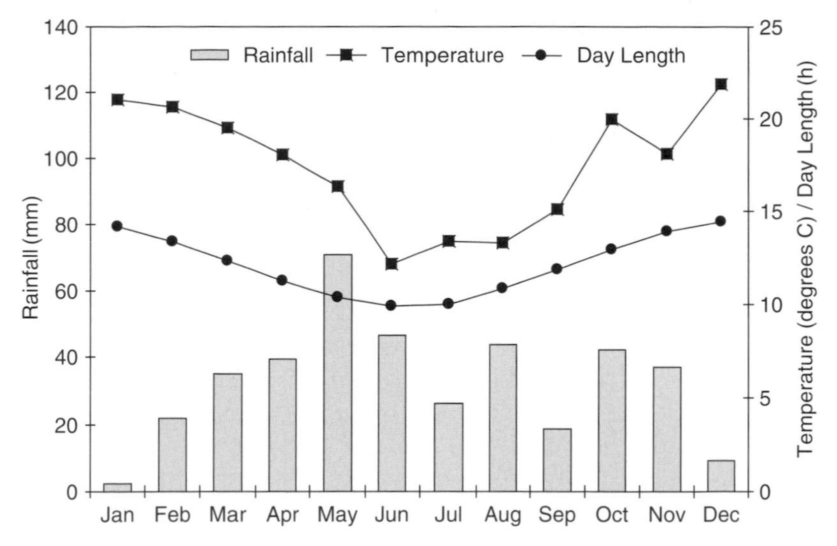

Figure 7.1 Seasonal variation in rainfall, temperature, and day length at De Hoop Nature Reserve, South Africa.

Results

Thermal constraints on behaviour

In order to assess the importance of the thermal environment, air temperatures were classified in terms of the baboons' "perceived environmental temperature" (PET), an index that accounts for the combined effects of solar radiation, wind speed, and humidity on the perceived air temperature (Hill *et al.* 2004a). These temperatures were then recoded into three categories: below ($T_{PET} < 25\,°C$), within ($25\,°C \leq T_{PET} \leq 30\,°C$), and above ($T_{PET} > 30\,°C$) the approximate thermal neutral zone (TNZ) for baboons (Elizondo 1977; Stelzner 1988). Such a classification system has been applied successfully to understanding patterns of baboon microhabitat choice and travel rates at Amboseli (Stelzner 1988), and behavioral patterns and habitat choice at De Hoop (Hill 2005).

Figure 7.2 displays the proportion of time spent resting in the three TNZ categories, with the analysis restricted to between 12.00 and 16.00, since thermal constraints are likely to be most significant during this period (Wheeler 1994a). This restriction also controls for possible time-of-day effects, since the high levels of resting on sleeping sites at dawn and dusk may result in an apparent association of these activities with low temperatures (Hill 1999). Figure 7.2 illustrates clearly that high levels of resting

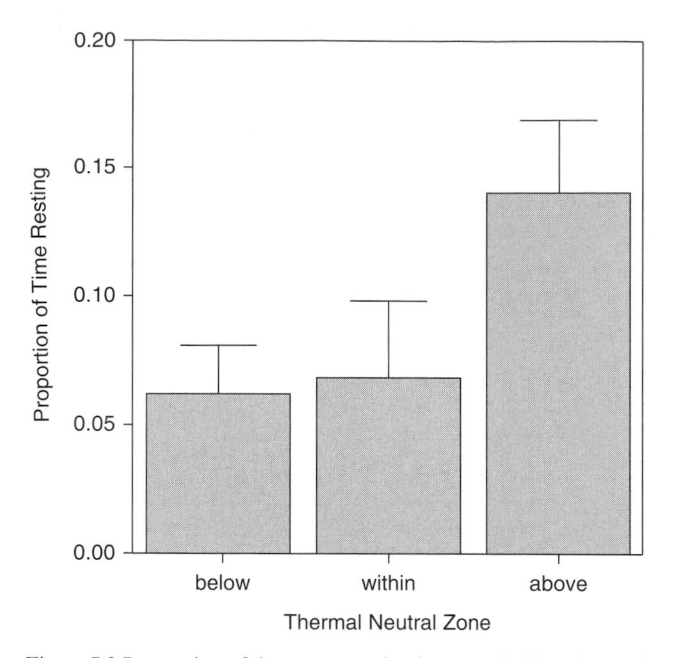

Figure 7.2 Proportion of time spent resting between 12.00 and 16.00 for perceived environmental temperature values below, within, and above the approximate thermal neutral zone for baboons.

occur under conditions of thermal stress, and the difference in activity levels between the TNZ categories is significant (analysis of variance [ANOVA]: $F_{(2,48)} = 12.58$, $P < 0.0001$). Post-hoc comparisons reveal that the main effect is for the above-TNZ category to differ from the other two (Schéffe: $P < 0.001$, in both cases). These results provide compelling support for high thermal loads constraining baboon behavioral options, such that the animals are forced into more sedentary activities as environmental temperatures increase.

Figure 7.3 displays the proportion of time spent in shade while resting relative to the TNZ categories. Again, the analysis is confined to the period from 12.00 to 16.00, and the below and within categories are combined in order to maintain sufficient sample sizes. The proportion of time spent in the shade differs significantly between the two categories (*t*-test: $t = -6.944$, df $= 21.46$, $P < 0.0001$), with proportionally more time spent in shade while resting under conditions of thermal stress. High "perceived environmental temperatures" thus place significant thermoregulatory constraints upon baboons, which force them to rest and seek shade in order to maintain a thermal balance with their environment.

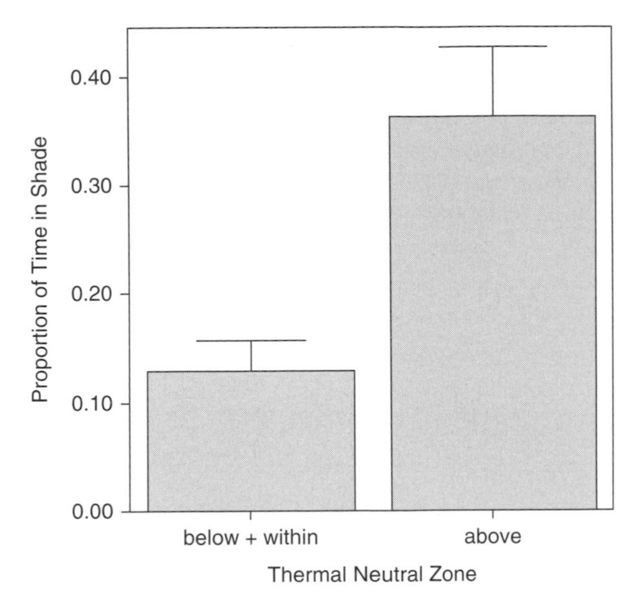

Figure 7.3 Proportion of time spent in shade while resting between 12.00 and 16.00 for perceived environmental temperatures below plus within and above the approximate thermal neutral zone for baboons.

Access to drinking water is an important constraint on baboon thermoregulatory behavior. Recorded observations of drinking during scan samples at De Hoop are few, but nevertheless it is possible to make a preliminary examination of the diurnal patterns of drinking behavior by baboons (Fig. 7.4). Although the data are limited, Figure 7.4 suggests a peak in drinking over the midday period, and the pattern of observations differs significantly from a uniform distribution across the day (Kolmogorov–Smirnov: $z = 1.657$, $n = 14$, $P < 0.01$). High temperatures over the midday period thus appear to elevate levels of drinking to facilitate evaporative cooling in response to thermoregulatory stress.

Seasonal analyses of behavior

Table 7.1 displays the least square multiple regression equations of ecological and behavioral variables on hours per day spent in feeding, moving, grooming, and resting behavior. A consistent feature of all four equations is that a positive function of day length is the main variable accounting for time spent in each activity. Feeding time is also a negative function of minimum monthly temperature, with a negative function of feeding time

Table 7.1 *Backwards regression equations of environmental variables on hours per day spent in activity, where MinT is the mean minimum monthly temperature (°C) and D is day length (hours). Independent variables not incorporated into the final models are T, the mean monthly temperature (°C), MaxT, the mean maximum monthly temperature (°C), RN_0, study month rainfall (mm), and RN_1, the rainfall in the month prior to the study month (mm)*

Activity	Equation	r^2	F	*P*
Feeding	F = 2.18 + 0.28 D − 0.17 MinT	0.803	$_{(2,7)}$ 14.30	0.003
Moving	M = 5.72 + 0.17 D − 0.90 F	0.718	$_{(2,7)}$ 8.90	< 0.02
Grooming	G = −0.90 + 0.21 D	0.808	$_{(1,8)}$ 33.74	< 0.0001
Resting	R = −5.08 + 0.61 D	0.923	$_{(1,8)}$ 96.24	< 0.0001

entering the equation for moving time. Nevertheless, the length of the day appears to be the primary constraint on the number of hours per day spent in each activity. For resting time in particular, this suggests that the primary thermal constraints on behavior relate to the scheduling of activity rather than the overall levels of that activity. Interestingly, if day length is excluded as an independent variable, then this results in an equation for resting time with maximum monthly temperature ($R = -2.91 + 0.20$ MaxT; $r^2 = 0.803$, $F_{(1,8)} = 32.51$, $P < 0.0001$). Although, in the absence of the previous analyses, this would apparently support the importance of the thermal environment in determining resting time, it in fact stresses the importance of including day length as an independent variable in these forms of analysis (Hill *et al.* 2004b).

The significance of the relationship with day length, however, does not preclude a possible thermoregulatory constraint imposed by daily temperatures. If high temperatures during the long summer days serve merely to schedule resting at times of greatest thermoregulatory stress (as opposed to constraining the animals to higher overall levels of inactivity), then we would predict that during the shorter days of winter, temperatures exceeding the TNZ should not result in elevated levels of resting, since the animals cannot afford to compromise on foraging. This appears to be the case. Figure 7.5 displays the proportion of time spent resting in the three TNZ categories between 12.00 and 16.00 for three winter months (June–August) and the three warmest months for which data are available (October–December). No significant differences in proportion of time

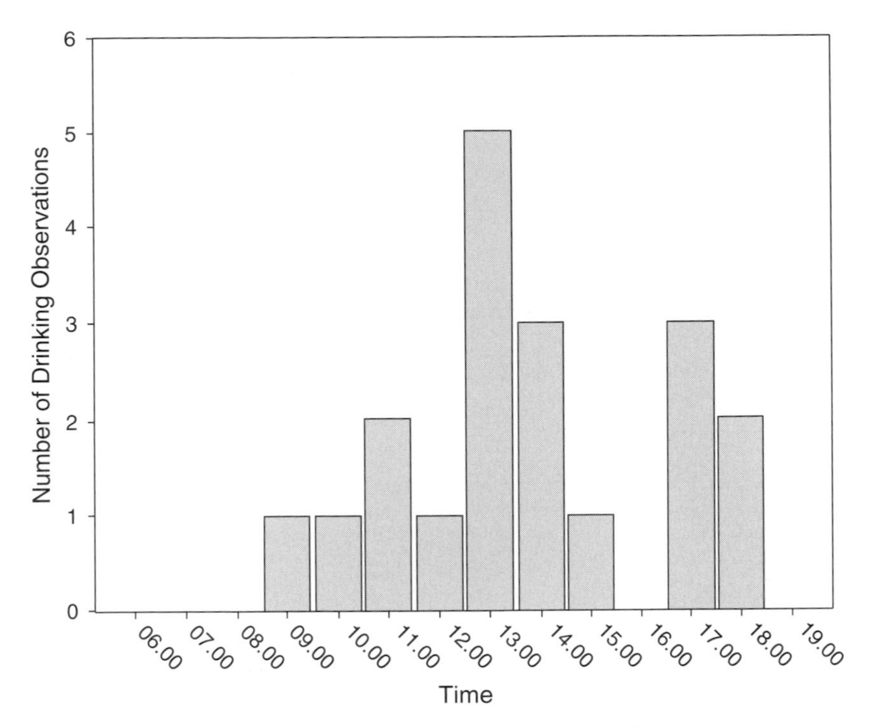

Figure 7.4 Distribution of drinking behavior across the day.

spent resting are observed between the TNZ categories during winter (ANOVA: $F_{(2,47)} = 0.44$, $P > 0.60$). However, the relationship for summer is similar to that for the data set as a whole (ANOVA: $F_{(2,48)} = 3.58$, $P < 0.05$), with the above category differing significantly from the below TNZ category (Schéffe: $P = 0.04$). This suggests, then, that the primary thermal constraints on behavior at temperate latitudes do indeed relate to the scheduling of activity rather than the overall levels of that activity, despite the importance of the thermoregulatory constraint.

Cross-populational determinants of resting time

If day length is the key ecological variable determining time allocation to specific activities at temperate latitudes, then one logical extension is that the degree of day length seasonality (and thus latitude) at a site could explain some of the differences in time budget allocation between populations. Time spent resting may be particularly susceptible to latitude.

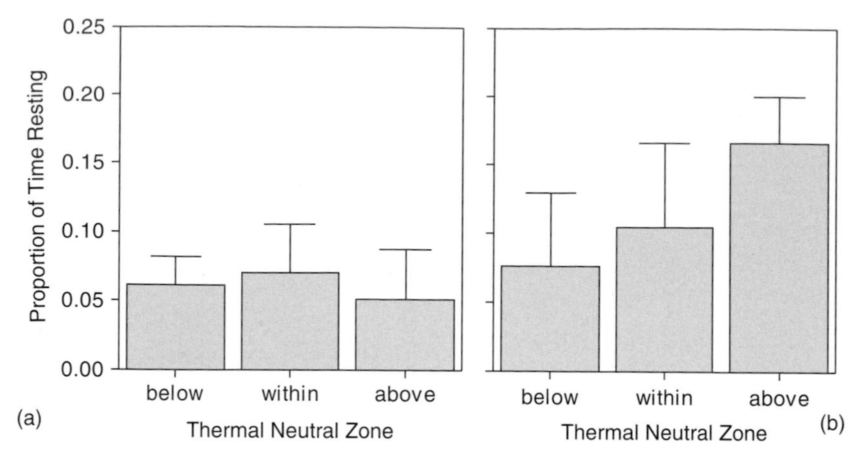

Figure 7.5 Proportion of time spent resting between 12.00 and 16.00 during (a) winter and (b) summer for perceived environmental temperature values below, within, and above the approximate thermal neutral zone for baboons.

Although longer days will initially allow for additional or more flexible foraging or social activities, as distance from the Equator increases, summer day lengths may become so long that they generate an excess of time that cannot be used profitably for essential activities. As a consequence, we would anticipate that the proportion of time spent resting (or in inactivity) should increase with latitude, all else held equal.

Time budget data are available from 16 baboon populations in order to examine this prediction. These data are primarily those from Dunbar (1992), supplemented with additional information from two other sites: Mkuzi, South Africa (Gaynor 1994), and De Hoop (Hill *et al.* 2003). Backwards regression analysis is used to determine the environmental variables that best explain variation in resting time between populations. While we might expect some of the environmental variables to be correlated with one other, the regression analysis reports their *independent* effects. The best-fit equation is:

$$\begin{aligned}\ln(\text{resting}) = & -9.63 + 1.11\ln(\text{rainfall}) + 2.10\ln(\text{PPI}) \\ & +1.20\ln(\text{temperature}) - 0.61\ln(\text{group size}) \\ & +0.19\ln(\text{latitude})\end{aligned}$$

($r^2 = 0.738$, $F_{(5,10)} = 5.62$, $P = 0.01$), where PPI is the plant productivity index (effectively the number of months of plant growth per year: see Hill & Dunbar 2002). It is clear that environmental parameters underlie a considerable degree of the variation in resting time between populations, suggesting that differences in habitat quality and, thus, food availability

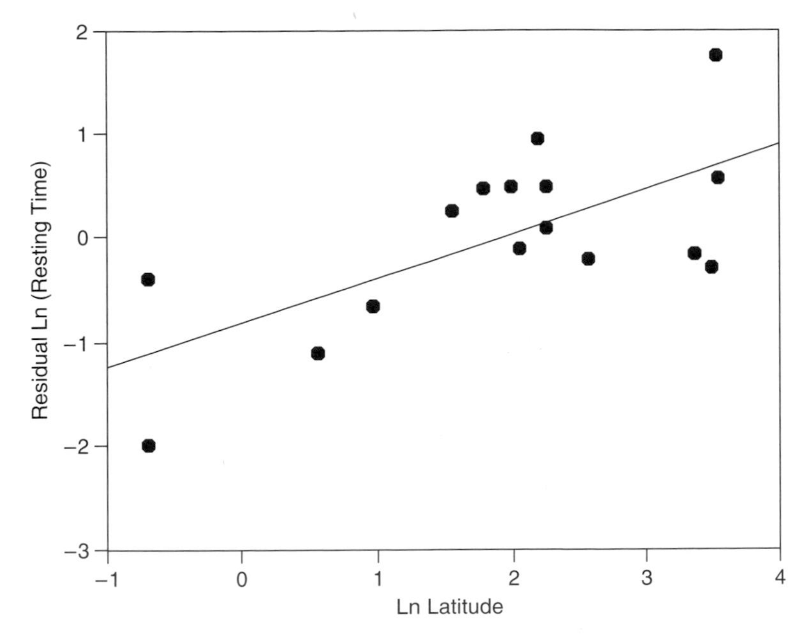

Figure 7.6 The relationship between residual percentage time spent resting (accounting for rainfall, temperature, primary productivity, and group size) and latitude across 16 baboon populations.

are a key factor underlying variation in time budget allocation between sites. However, a positive function of latitude is a significant component of the equation, although a positive function of temperature also enters the model. The relationship between residual resting time (controlling for the other factors in the equation) and latitude is displayed in Figure 7.6 ($r^2 = 0.448$, $F_{(1,14)} = 11.38$, $P = 0.005$). Thus, it is clear that with increasing distance from the Equator, populations spend more time resting over the course of the year, once differences in demography, climate, and environmental quality are held constant.

Discussion

The thermal environment acts as a significant constraint on baboon behavioral decision-making processes. Under conditions of thermal stress, baboons engage in more sedentary behaviors and seek shade in order to reduce the impact of solar radiation. The incidence of drinking also appears to be elevated at these times. These results confirm that the time period identified by Wheeler (1994a) as the period of greatest thermal stress (12.00–16.00) does indeed limit an animal's behavioral options when temperatures are high. Perhaps more interesting, though, is the fact that the strong relationships found at De Hoop are from such a southerly population, where mean annual temperatures are not that high relative to other African sites (see Dunbar [1992] for comparable data). The explanation for this lies in the interaction between the thermal environment and day length seasonality, and this also accounts for the observed differences in the thermoregulatory responses of the baboons at Amboseli and De Hoop.

At Amboseli, the baboons sought shade opportunistically as they encountered it and did not rest in shade for long periods (Stelzner 1988). In contrast, the baboons at De Hoop rested over the hottest parts of the day and sought shade on cliff refuges to conduct this activity (Hill 2005). Such a strategy should greatly reduce the water requirements of the animals in this population (see Wheeler [1994a] and Brain & Mitchell [1999]) and thus would appear to be the "optimal" solution to the problem of high daytime temperatures. So why do the baboons at Amboseli not exhibit similar behavioral patterns? It seems likely that, for this population, the demands of foraging often outweigh the costs of thermal loading, particularly in such a marginal habitat where time budgets are constrained. It is only the long summer day lengths at De Hoop that reduce the intensity of the foraging constraint over the midday period, allowing the animals to pursue a more optimal thermoregulatory strategy. As a consequence, the tradeoffs between temperature and activity are observed only during the long summer months. The primary thermal constraints thus appear to relate to the scheduling of specific activities in more temperate environments rather than delimiting the absolute levels of these activities.

Relaxed day length constraints at De Hoop probably also account for the pronounced deviations in patterns of habitat choice between the two populations. Stelzner (1988) found evidence for selection only at the microhabitat level at Amboseli, although the prediction that the baboons should have preferred more enclosed habitats in this population is likely to be confounded by predation risk. At De Hoop, however, the baboons were more likely to utilize cliffs over the midday period, when temperatures exceeded their TNZ (Hill 2005). The fact that cliffs are located in a few specific areas of the baboons' home range means that this has significant implications for their day journey routes and overall patterns of habitat choice. However, the baboons are able to spend enough time foraging due, again, to the long summer days: sufficient time exists before and after the periods of greatest thermal stress in which to obtain access to sufficient food resources.

The interaction between seasonal and ecological constraints also may account for the patterns of thermoregulatory behavior reported for the baboons at Ruaha (Pochron 2000). The baboons in this population avoided the sun during the cool, lush season but avoided the sun only while resting in the hot, dry season. However, the ecological constraints imposed by reduced food availability in the dry season may prevent the animals from following an optimal thermoregulatory response, such that sun avoidance occurs only when the animals are not foraging. In the cool, lush season, though, the increased food availability frees up more time to allow greater flexibility in the baboons' responses to temperature, such that they are able to avoid the sun while foraging, even though environmental temperatures are not as high.

Long summer day lengths are thus important in permitting a more flexible thermoregulatory strategy, and this is highlighted in the seasonal analysis, where day length is the key parameter underlying variation in the primary baboon activities. Latitude, and thus day length seasonality, is also important in determining the levels of non-foraging activity across populations. Although temperature also forms a significant component of this model, this does not necessarily imply a thermoregulatory effect (although it could be part of the explanation), since temperature is important in determining the proportion of fruit in the diet (Hill and Dunbar 2002) and, thus, also reflects habitat quality. In terms of day length, however, the primary importance of seasonal variation may lie in the fact that it produces a bottleneck in the short winter months, during which time the animals must still balance their time budgets, despite having potentially elevated energetic requirements. Although baboons may be able to compromise on certain activities (e.g. grooming) for a couple of months in order to free up sufficient time for foraging (Bronikowski and Altmann 1996; see also Chapter 6), short winter day lengths ultimately will constrain populations at temperate latitudes to smaller group sizes than those experiencing an identical set of ecological parameters at the equator (Hill *et al.* 2003). Since time spent in activities such as feeding and moving scales proportionally with group size, temperate populations must live in social contexts that allow them to conduct their essential activities during the winter bottleneck, such that group sizes will be smaller, all else equal. The thermoregulatory benefits of long days in the warm summer months therefore are offset by the potentially substantial costs of maintaining a warm body temperature during the cold winter months, when the number of hours available for foraging is limited.

The thermal environment clearly is an important ecological variable, but the ability of a primate population to respond to these thermoregulatory pressures is constrained by foraging requirements and the degree of seasonality experienced by that population. The greatest thermoregulatory constraints thus are experienced by populations in marginal Equatorial habitats, where there is little time left over from foraging in which to escape to shade and minimize exposure to solar radiation. However, in more productive environments, periods of relative food abundance may allow greater flexibility in the thermoregulatory response, although longer day lengths at more temperate latitudes provide the greatest opportunity to rest in deep shade for long periods at times of thermal stress. In the context of human evolution, the thermoregulatory advantages proposed in relation to bipedalism (Wheeler 1991), loss of functional body hair (Wheeler 1992a), body size (Wheeler 1992b), and physique (Wheeler 1993) would have been most significant for populations in Equatorial habitats, where food availability was at least seasonally restricted. For example, Wheeler (1991) argued that the thermoregulatory advantages of bipedalism would increase the potential time available for foraging in open environments. However, increased foraging times could be achieved as easily without the need for morphological change through migration of hominid populations away from Equatorial regions to latitudes where the intensity of the thermoregulatory constraint is relaxed. It may be that thermoregulatory considerations relating to the colder winter nights are of greater ecological precedence at these latitudes, such that opportunities to migrate were restricted. It is clear that the debate over the importance of thermoregulation in hominid evolution still has many issues that need to be addressed. However, it is also evident that the importance of thermoregulation as a constraint on primates in their natural environments is also a topic that has been greatly overlooked. Future studies on species such as the patas monkey (*Erythrocebus patas*), which have a thermoregulatory system comparable to that of humans (Gisolfi *et al.* 1982; Kolka & Elizondo 1983), would help to address both of these issues. Nevertheless, it is clear that temperature and day length variation are important ecological constraints that should be given greater precedence in future studies of primate behavioral ecology.

Acknowledgments

I am grateful to Diane Brockman and Carel van Schaik for inviting me to contribute to the symposium and this book, and to Diane, Carel, and Guy Cowlishaw for constructive comments on the manuscript. I thank Louise Barrett, Peter Henzi, and Cape Nature Conservation for their help and permission to work at De Hoop Nature Reserve, and Robin Dunbar for his support over the course of this study. Attendance at the XVIII Congress of the International Primatological Society in Adelaide was aided by a British Academy Conference Grant.

References

Altmann, J. (1974). Observational study of behaviour: sampling methods. *Behaviour*, **49**, 227–67.

Bernstein, I. S. (1972). Daily activity cycles and weather influences on a pigtail monkey group. *Folia Primatologica*, **18**, 390–415.

— (1976). Activity patterns in a sooty mangabey group. *Folia Primatologica*, **26**, 185–200.

Brain, C. & Mitchell, D. (1999). Body temperature changes in free-ranging baboons (*Papio hamadryas ursinus*) in the Namib Desert, Namibia. *International Journal of Primatology*, **20**, 585–98.

Bronikowski, A. M. & Altmann, J. (1996). Foraging in a variable environment: weather patterns and the behavioural ecology of baboons. *Behavioral Ecology and Sociobiology*, **39**, 11–25.

Chaplin, G., Jablonski, N. G., & Cable, N. T. (1994). Physiology, thermoregulation and bipedalism. *Journal of Human Evolution*, **27**, 497–510.

Chappell, M. A. & Bartholomew, G. A. (1981). Standard operative temperatures and thermal energetics of the antelope ground squirrel *Ammospermophilus leucurus*. *Physiology and Zoology*, **54**, 81–93.

Clutton-Brock, T. H. (1977). Some aspects of intraspecific variation in feeding and ranging behaviour in primates. In *Primate Ecology*, ed. T. H. Clutton-Brock. London: Academic Press, pp. 557–79.

Cowling, R. & Richardson, D. (1995). *Fynbos: South Africa's Unique Floral Kingdom*. Vlaeberg: Fernwood Press.

Do Amaral, L. Q. (1996). Loss of body hair, bipedality and thermoregulation. Comments on recent papers in the *Journal of Human Evolution*, **30**, 357–66.

Dunbar, R. I. M. (1990). Environmental determinants of intraspecific variation in body weight in baboons (*Papio* spp.). *Journal of Zoology, London*, **220**, 157–69.

— (1992). Time: a hidden constraint on the behavioural ecology of baboons. *Behavioral Ecology and Sociobiology*, **31**, 35–49.

Elizondo, R. (1977). Temperature regulation in primates. In *International Review of Physiology: Environmental Physiology II*, Vol. 15, ed. D. Robertshaw. Baltimore: University Park Press, pp. 71–118.

Fossey, D. & Harcourt, A. (1977). Feeding ecology of free-ranging mountain gorillas (*Gorilla gorilla beringei*). In *Primate Ecology*, ed. T. H. Clutton-Brock. London: Academic Press, pp. 415–49.

Funkhouser, G. E., Higgins, E. A., Adams, T., & Snow, C. C. (1967). The response of the savannah baboon (*Papio cynocephalus*) to thermal stress. *Life Science, Oxford*, **6**, 1615–20.

Gaynor, D. (1994). Foraging and feeding behaviour of chacma baboons in a woodland habitat. Ph. D. thesis, University of Natal.

Gisolfi, C. V., Sato, K., Wall, P. T., & Sato, F. (1982). In vivo and in vitro characteristics of eccrine sweating in patas and rhesus monkeys. *Journal of Applied Physiology*, **53**, 425–31.

Hiley, P. H. (1976). The thermoregulatory repsonses of the galago (*Galago crassicaudatus*), the baboon (*Papio cynocephalus*) and the chimpanzee (*Pan satyrus*) to heat stress. *Journal of Physiology, London*, **254**, 657–71.

Hill, R. A. (1999). Ecological and demographic determinants of time budgets in baboons: implications for cross-populational models of baboon sociobiology. Ph. D. thesis, University of Liverpool.

— (2005). Thermal constraints on activity scheduling and habitat choice in baboons. *American Journal of Physical Anthropology*, in press.

Hill, R. A. & Dunbar, R. I. M. (2002) Climatic determinants of diet and foraging behaviour in baboons. *Evolutionary Ecology*, **16**, 579–93.

Hill, R. A., Barrett, L., Gaynor, D., *et al.* (2003). Day length, latitude and behavioural (in)flexibility in baboons. *Behavioral Ecology and Sociobiology*, **53**, 278–86.

Hill, R. A., Weingrill, T., Barrett, L., & Henzi, S. P. (2004a). Indices of environmental temperatures for primates in open habitats. *Primates*, **45**, 7–13.

Hill, R. A., Barrett, L., Gaynor, D., *et al.* (2004b). Day length variation and seasonal analyses of behaviour. *South African Journal of Wildlife Research*, **34**, 39–44.

Iwamoto, T. & Dunbar, R. I. M. (1983) Thermoregulation, habitat quality and the behavioural ecology in gelada baboons. *Journal of Animal Ecology*, **53**, 357–66.

Kolka, M. A. & Elizondo, R. S. (1983). Thermoregulation in *Erythrocebus patas*: a thermal balance study. *Journal of Applied Physiology*, **55**, 1603–8.

Kummer, H. (1968). *Social Organisation of Hamdryas Baboons*. Chicago: University of Chicago Press.

Lawes, M. J. & Piper, S. E. (1992). Activity patterns in free-ranging samango monkeys (*Cercopithecus mitis erythrarchus* Peters, 1852) at the southern range limit. *Folia Primatologica*, **59**, 186–202.

Mitchell, D. & Laburn, H. P. (1985). The pathophysiology of temperature regulation. *Physiologist*, **28**, 507–17.

Mitchell, D., Laburn, H. P., Nijland, M. J. M., Zurovsky, Y., & Mitchell, G. (1987). Selective brain cooling and survival. *South African Journal of Science*, **83**, 598–604.

Pochron, S. T. (2000). Sun avoidance in the yellow baboons (*Papio cynocephalus cynocephalus*) of Ruaha National Park, Tanzania: variations with season, behaviour and weather. *International Journal of Biometeorology*, **44**, 141–7.

Porter, A. M. W. (1993). Sweat and thermoregulation in hominids. Comments prompted by the publications of P. E. Wheeler 1984–1993. *Journal of Human Evolution*, **25**, 417–23.

Post, D. G. (1981). Activity patterns of yellow baboons (*Papio cynocephalus*) in the Amboseli National Park, Kenya. *Animal Behaviour*, **29**, 357–74.

Stelzner, J. K. (1988). Thermal effects on movement patterns of yellow baboons. *Primates*, **29**, 91–105.

Stelzner, J. K. & Hausfater, G. (1986). Posture, microclimate, and thermoregulation in yellow baboons. *Primates*, **27**, 449–63.

Stoltz, L. & Saayman, G. S. (1970). Ecology and behaviour of baboons in the Transvaal. *Annuals of the Transvaal Museum*, **26**, 99–143.

Taylor, C. R. (1969). The eland and the oryx. *Scientific American*, **220**, 88–95.

Wheeler, P. E. (1991). The thermoregulatory advantages of hominid bipedalism in open equatorial environments: the contribution of increased convective heat loss and cutaneous evaporative cooling. *Journal of Human Evolution*, **21**, 107–15.

— (1992a). The influence of the loss of functional body hair on the energy and water budgets of the early hominids. *Journal of Human Evolution*, **223**, 379–88.

— (1992b). The thermoregulatory advantages of large body size for hominids foraging in savannah environments. *Journal of Human Evolution*, **223**, 351–62.

— (1993). The influence of stature and body form on hominid energy and water budgets: a comparison of *Australopithecus* and early *Homo* physiques. *Journal of Human Evolution*, **24**, 13–28.

— (1994a). The thermoregulatory advantages of heat storage and shade-seeking behaviour to hominids foraging in equatorial savannah environments. *Journal of Human Evolution*, **26**, 339–50.

— (1994b). The foraging times of bipedal and quadrupedal hominids in open equatorial environments (a reply to Chaplin, Jablonski & Cable, 1994). *Journal of Human Evolution*, **27**, 511–17.

— (1996). The environmental context of functional body hair loss in hominids (a reply to Amaral, 1996). *Journal of Human Evolution*, **30**, 367–71.

Wrangham, R. W. (1977). Feeding behaviour of chimpanzees in Gombe National Park, Tanzania. In *Primate Ecology*, ed. T. H. Clutton-Brock. London: Academic Press, pp. 504–37.

Zurovsky, Y. & Shkolnik, A. (1983). Water economy and body fluid distribution in the hamadray baboon (*Papio hamadryas*). *Journal of Thermoregulatory Biology*, **18**, 153–7.

8 *Seasonality in hunting by non-human primates*

JOHN C. MITANI
Department of Anthropology, University of Michigan, Ann Arbor
MI 48109–1092, USA

DAVID P. WATTS
Department of Anthropology, Yale University, New Haven
CT 06520–8277, USA

Introduction

Primates obtain most of their food from plants (Oates 1987), but some species are well known for their predatory behavior. Chimpanzees (*Pan troglodytes*) were the first non-human primates observed to hunt and eat meat in the wild (Goodall 1963). Subsequent field observations of baboons (*Papio* spp.) and capuchin monkeys (*Cebus capucinus*) have shown them to be proficient hunters (Harding 1973, 1975; Strum 1975, 1981; Hausfater 1976; Fedigan 1990; Perry & Rose 1994; Rose 1997, 2001). Given seasonal variations in primate feeding patterns (see Chapter 3), it is not surprising that primate predators display temporal variation in their tendencies to hunt. Studies of primate hunting seasonality generate considerable ecological and ethological interest and take on additional significance because of their potential to shed light on the evolution of meat-eating by early hominids (see Chapters 17 and 19). Systematic attempts to describe seasonal variation in hunting activity by non-human primates and efforts to identify its causal factors, however, have not been made.

In this chapter, we provide an overview of seasonal variation in primate predatory behavior. We focus on chimpanzees, baboons, and capuchin monkeys, three species for which sufficient observations exist to make comparisons. We begin by reviewing data on temporal variation in hunting frequency and success by each species. We proceed to discuss the factors that appear to affect this variation. Here, we consider several ecological factors that have been hypothesized to affect temporal variation in hunting (Table 8.1). We pay special attention to the hypothesis that

Seasonality in Primates: Studies of Living and Extinct Human and Non-Human Primates, ed. Diane K. Brockman and Carel P. van Schaik. Published by Cambridge University Press.

Table 8.1 *Factors hypothesized to affect primate hunting seasonality*

Factors	References
Ecological	
Seasonal shortages of plant foods create the need for fallback foods in the form of meat	Teleki (1973); Stanford (1996, 1998); Boesch & Boesch-Achermann (2000); Rose (2001)
Seasonal shortages in specific nutrients found in vertebrate prey lead to increased levels of hunting	Teleki (1973); Hausfater (1976); Takahata *et al.* (1984)
Temporal variation in fruiting activity increases the encounter frequency between primate predators and their prey by attracting them to common resources	Stanford (1998)
Dry seasons create open habitats and make it easier to capture prey	Fedigan (1990)
Wet conditions during the rainy season make it easier to capture prey	Boesch & Boesch-Achermann (2000)
Characteristics of prey	
Prey become more readily available and easier to capture during birth seasons	Takahata *et al.* (1984); Fedigan (1990); Rose (1997, 2001); Boesch & Boesch-Achermann (2000)
Characteristics of predators	
Differences in the seasonal availability of fertile females motivate males to hunt to procure meat, which they can use to swap for matings	Teleki (1973); Stanford *et al.* (1994b)
Large hunting parties make it easier for primate predators to capture prey, through either a "beater" effect or mobbing	Takahata *et al.* (1984); Mitani & Watts (2001); Mitani *et al.* (2002b)

increased hunting and meat-eating compensates for nutritional shortfalls caused by plant food scarcity, but we also entertain additional hypotheses that suggest variation in ecological conditions may make it easy for primate predators to capture prey in certain situations. In addition, we evaluate characteristics of prey and their primate predators that have been hypothesized to affect temporal variation in hunting activity (Table 8.1). These include birth seasonality of prey, which affects their differential availability, and social characteristics of predators such as variation in hunting group size and composition that influence the tendency or motivation to hunt. We conclude by identifying areas in need of further research and discussing the implications of primate hunting seasonality for the study of human evolution.

Throughout our discussion, we use the term "seasonality" broadly but focus primarily on variation in hunting behavior as a function of rainfall. Rainfall is an important source of variation affecting the abundance of plant foods, and different versions of the nutritional shortfall hypothesis invoke dry-season deficits in energy, protein, fat, and other nutrients to account for the evolution of hunting behavior (e.g. Stanford [1996, 1998]). Differences between wet and dry periods have also been suggested to play a role in facilitating prey capture in some circumstances (Fedigan 1990; Boesch & Boesch-Achermann 2000), and for this reason variation in rainfall figures prominently in some other explanations of seasonal hunting by primates (Table 8.1).

Chimpanzees

Chimpanzees (*Pan troglodytes*) have been well-studied at several sites throughout Africa. Long-term data on their hunting behavior exist for Gombe National Park, Tanzania, Mahale Mountains National Park, Tanzania, Taï National Park, Ivory Coast, and Ngogo in the Kibale National Park, Uganda. Red colobus monkeys (*Procolobus badius*) are the primary prey of chimpanzees at each of these sites, and most of our information regarding temporal variation in chimpanzee hunting involves predation on red colobus.

Gombe

Field observations indicate that the Gombe chimpanzees hunt during seasonal bursts of activity. The frequency of red colobus hunts showed considerable monthly variation during 11 years spanning 1982–92 (Stanford *et al.* 1994b) (Fig. 8.1a). In an early study, Teleki (1973) suggested that hunting success also varied temporally; a small sample ($n = 21$) of baboon, red colobus, and ungulate kills recorded during 30 months between 1968 and 1970 showed some monthly variation (see Fig. 1 in Teleki [1973]). A much larger sample of red colobus hunts derived over ten years between 1982 and 1991 revealed that successful predatory episodes tended to clump primarily between July and September (see Fig. 3 in Stanford *et al.* [1994a]).

A distinct dry season at Gombe occurs typically between the months of June and September (Goodall 1986). Using observations collected over 40 months between 1972 and 1974, Wrangham (1975) (cf. Wrangham & Bergmann Riss [1990]) reported that success rates did not differ between

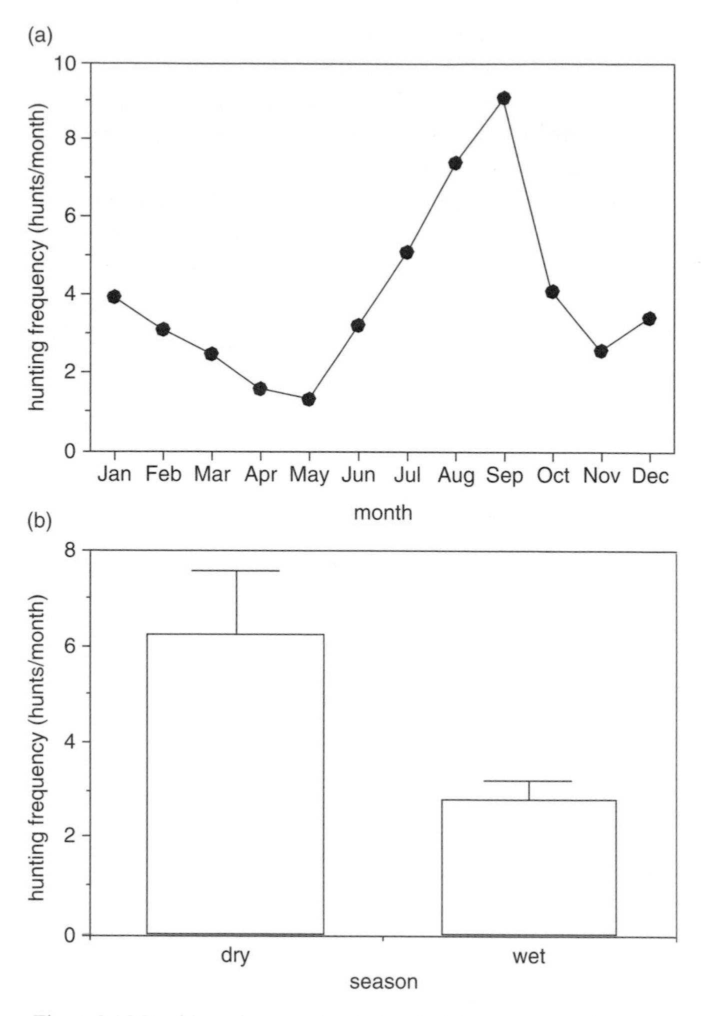

Figure 8.1 Monthly and seasonal variation in chimpanzee hunting frequency in the Gombe National Park, Tanzania. (a) Monthly variation. The mean number of hunts per month is shown ($n = 529$ hunts) (from Table 2 in Stanford *et al.* [1994b]). (b) Seasonal variation. The mean number of hunts observed per month (+1 Standard error[se]) in each season is shown ($n = 529$ hunts) (from Table 2 in Stanford *et al.* [1994b]).

dry and wet seasons. More recent data, derived from much larger samples of hunting attempts and predation events, paint a different picture. Both hunting frequency (Mann–Whitney U test: $Z = -2.21$, $n_1 = 4$, $n_2 = 8$, $P < 0.03$) (Fig. 8.1b) (Stanford *et al.* 1994b) and success (see Fig. 3 in Stanford *et al.* [1994a]) increased during the dry season.

Ecological factors and social characteristics of chimpanzee predators have been proposed to explain seasonal increases in hunting activity at Gombe (cf. Table 8.1). Seasonal shortages of food and specific nutrients may motivate chimpanzees to hunt more frequently during the dry season than in the wet season (Stanford 1996, 1998). In accord with this hypothesis, the body weight of chimpanzees at Gombe tends to decrease in the dry season (Wrangham 1975; Pusey *et al.* 2005). Alternatively, seasonal increases in hunting have been attributed to temporal variation in fruiting activity (Stanford 1998). Chimpanzees and red colobus monkeys eat fruit from some of the same tree species, and hunting frequency may increase when specific trees attract both predators and prey to the same areas. Finally, the availability of estrous females appears to increase during the dry season at Gombe (Wallis 1995), and the corresponding increase in hunting effort may be a part of a male mating strategy. Male chimpanzees at Gombe have been proposed to swap the meat they procure with estrous females from whom they obtain matings (Stanford *et al.* 1994b).

Mahale

Takahata and colleagues (1984: Table IV) reported temporal variation in the hunting behavior of the chimpanzees at Mahale. Observations conducted over 29 months between 1979 and 1982 revealed that the Mahale chimpanzees hunted primarily during six months. Fifty of 54 hunting attempts were made in May or between August and December (Fig. 8.2a). Hunting activity showed considerable heterogeneity among months, with a clear peak in September and October (Fig. 8.2a). Ungulates were pursued in the majority of these hunts (30 of 50 [60%] attempts) Takahata *et al.* 1984: Table IV), but the Mahale chimpanzees have eaten more red colobus monkeys during subsequent observations (Hosaka *et al.* 2001; Uehara *et al.* 1992). Hosaka and colleagues (2001) analyzed a larger, more recent data set based on observations collected over 120 months between 1979 and 1995. Their results revealed that successful hunting episodes occurred more uniformly across months than implied by the earlier Takahata *et al.* study (Fig. 8.2b). Nonetheless, capture frequency still showed considerable heterogeneity among months, with a clear August–September peak in hunting success (Fig. 8.2b).

Takahata and colleagues (1984) suggested that hunting activity correlated strongly with rainfall, with the majority of hunts taking place during the dry season. Closer scrutiny of their data, however, does not support this claim. Rainfall is sharply seasonal at Mahale; most rain falls during

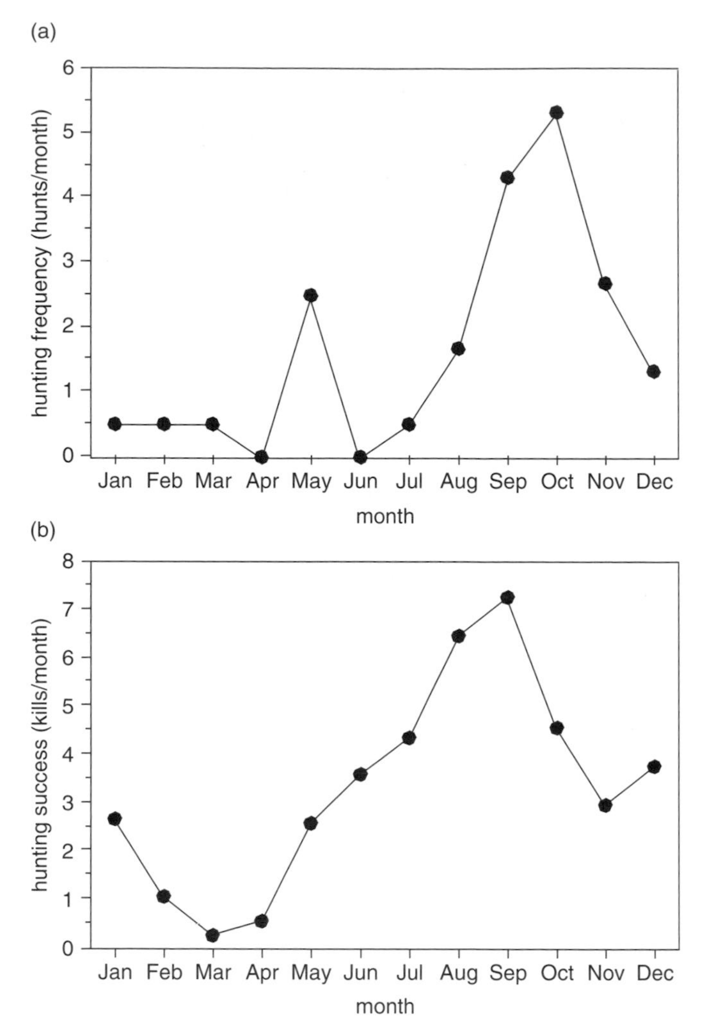

Figure 8.2 Monthly variation in chimpanzee hunting frequency and success at the Mahale Mountains National Park, Tanzania. (a) Hunting frequency. The mean number of hunts observed per month is shown ($n = 54$ hunts) (from Table 1 in Takahata *et al.* [1984]). (b) Hunting success. The mean number of kills observed per month is shown ($n = 431$ kills) (from Fig. 2 in Hosaka *et al.* [2001] and Hosaka, personal communication, 2001).

eight months typically between October and May (Takasaki *et al.* 1990). If one uses these months to demarcate the wet season, then Takahata *et al.*'s (1984) sample shows a peak in hunting activity during the transition from the dry to the rainy season between September and October (Fig. 8.2a). As

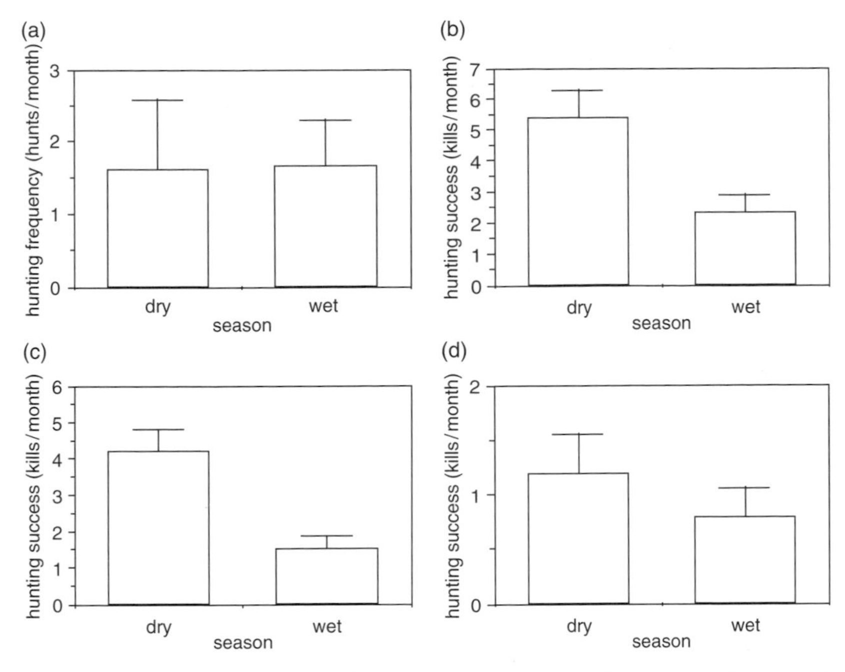

Figure 8.3 Seasonal variation in chimpanzee hunting frequency and success at the Mahale Mountains National Park, Tanzania. (a) Hunting frequency. The mean number of kills observed per month (+1 Standard error[se]) assigned to dry and wet seasons is shown ($n = 54$ hunts) (from Table I in Takahata *et al.* [1984]). (b) Hunting success. The mean number of kills observed per month assigned to dry and wet seasons is shown ($n = 431$ kills) (from Fig. 2 in Hosaka *et al.* [2001] and Hosaka, personal communication, 2001). (c) Hunting success for red colobus prey. Figure and data derived from Fig. 8.3b (d) Hunting success for mammalian prey other than red colobus. Figure and data derived from Fig. 8.3b.

a result, no significant difference in hunting frequency exists between wet and dry periods (Mann–Whitney U test: $Z = -0.17$, $n_1 = 4$, $n_2 = 8$, $P > 0.85$) (Fig. 8.3a). In contrast, long-term data collected over 17 years between 1979 and 1995 (Hosaka *et al.* 2001) indicate that chimpanzee hunting success increases significantly during the dry season at Mahale (Mann–Whitney U test: $Z = -2.21$, $n_1 = 4$, $n_2 = 8$, $P < 0.03$) (Fig. 8.3b). This effect holds specifically for predation on red colobus (Mann–Whitney U test: $Z = -2.71$, $n_1 = 4$, $n_2 = 8$, $P < 0.01$) (Fig. 8.3c) but not for predation on other mammalian taxa (Mann–Whitney U test: $Z = -0.77$, $n_1 = 4$, $n_2 = 8$, $P > 0.40$) (Fig. 8.3d).

Three factors have been implicated to explain seasonal variation in hunting frequency and success at Mahale (cf. Table 8.1). First, seasonal differences in prey availability have been proposed to account for some of the variation in chimpanzee hunting activity (Takahata *et al.* 1984). Bush pigs show a seasonal pattern of reproduction between October and January in east Africa (Kingdon 1979), a period that overlaps the September–October peak in hunting activity in Takahata *et al.*'s (1984) sample (Fig. 8.2a). Two other explanations invoke ecological factors that affect chimpanzee social and feeding behavior. First, hunting activity appears to increase during periods of seasonal fruit abundance, which in turn correlates with the formation of large chimpanzee parties (Takahata *et al.* 1984). Increased hunting activity for certain kinds of prey may result from a "beater" effect created by these large groups of chimpanzees. Large parties may flush prey, such as bush pigs and duiker, more frequently than the smaller parties that typically form during periods of fruit scarcity. Second, Takahata and colleagues (1984), citing the purported correlation between hunting activity and rainfall, speculated that the Mahale chimpanzees experience protein deficits during the dry season and that meat-eating increases partly to compensate for this and other, unspecified nutritional shortfalls.

Taï

Field observations at Taï conducted over 12 years between 1984 and 1995 reveal clear peaks in hunting frequency and success between September and October (see Fig. 8.1 in Boesch & Boesch-Achermann [2000]). These two months coincide with a seasonal peak in rainfall (see Fig. 1.3 in Boesch & Boesch-Achermann [2000]). Rainfall is in fact correlated strongly with both hunting frequency (Spearman $r = 0.75$, $P < 0.01$) (Fig. 8.4a) and success (Spearman $r = 0.71$, $P < 0.02$) (Fig. 8.4b) at Taï.

Three variables have been invoked to account for the seasonal pattern of hunting activity at Taï (Boesch & Boesch-Achermann 2000) (cf. Table 8.1). First, chimpanzees may be able to capture their primary prey, red colobus, more easily in the wet season than in the dry season because rain makes branches wet and slippery and monkeys frequently fall as chimpanzees pursue them. An increase in the availability of red colobus prey has been suggested as a second causal factor that leads to high levels of hunting activity at Taï. Red colobus at Taï have been reported to display a birth season during the peak months of hunting activity of September and October (Bshary 1995). Finally, the seasonal peak in hunting activity at Taï occurs at the end of the dry season, a period that has been argued to be a time of

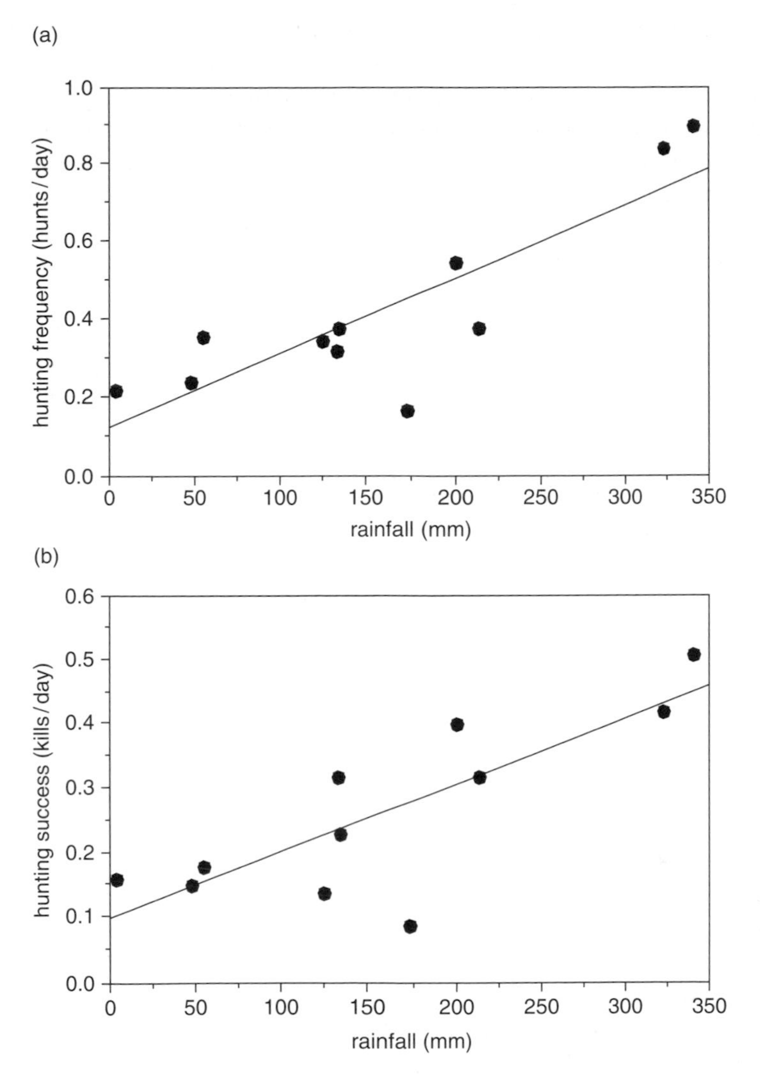

Figure 8.4 Relationships between rainfall and chimpanzee hunting frequency and success at the Taï National Park, Ivory Coast. (a) Rainfall and hunting frequency. (b) Rainfall and hunting success. Data from Fig. 8.1 in Boesch Boesch-Achermann (2000) and Boesch (personal communication, 2001).

low food availability (Doran 1997; Boesch & Boesch-Achermann 2000). Thus, hunger at the end of the food-poor dry season has been hypothesized to account for the increased level of hunting activity during the start of the rains in September and October (Boesch & Boesch-Achermann 2000).

Ngogo

Chimpanzees living in an unusually large community at Ngogo, Kibale National Park, Uganda, prey on several vertebrates. As is the case at other study sites, the Ngogo chimpanzees specialize in hunting red colobus monkeys (Mitani & Watts 1999, 2001; Watts & Mitani 2002a, 2002b). The frequency with which they hunted red colobus (Fig. 8.5a) and their success in capturing them (Fig. 8.5b) varied considerably across 16 months of observation between 1998 and 1999.

Our studies at Ngogo permit direct investigation of the effects of several factors on chimpanzee hunting frequency and success. In addition to recording observations of chimpanzee hunting behavior, we also collected data on rainfall, fruit availability, and chimpanzee party size and composition (Mitani & Watts, 2001; Mitani *et al.* 2002a; Mitani & Watts, unpublished data, 1998–99). Observations made by Struhsaker and Leland (1987) provide information on the timing of red colobus births. Fruit availability, the availability of estrous females, adult male party size and overall party size, displayed strong positive correlations with hunting frequency ($P \leq 0.05$ for all four comparisons) (Fig. 8.6) and success ($P < 0.01$ for all four comparisons) (Fig. 8.7). Neither rainfall nor the number of red colobus births per month showed any relationship with hunting frequency ($P \geq 0.25$ for both comparisons) (Fig. 8.6) or success ($P \geq 0.30$ for both comparisons) (Fig. 8.7).

We conducted a stepwise regression to examine further the influences of fruit availability, the availability of estrous females, overall party size, and adult male party size on hunting frequency and success. When all four independent variables are considered together, only overall party size and adult male party size help to explain variation in hunting frequency and success (frequency: $F_{2,13} = 9.78$, $P < 0.01$; success: $F_{2,13} = 26.39$, $P < 0.001$). Party size and the number of adult males show high correlations with each other and with the two other independent variables, as indicated by low tolerance scores (< 0.02); as a result, it is difficult to evaluate their independent effects. To circumvent the problem of collinearity, we conducted two additional analyses in which we removed male party size as an independent variable. Party size explained more of the variation in both of the dependent variables than did male party size, and collinearity among the three remaining independent variables was relatively low (tolerance values > 0.10). Results of these analyses confirm party size as the only significant predictor of hunting frequency and success (frequency: $F_{1,14} = 9.32$, $P < 0.01$; success: $F_{1,14} = 27.16$, $P < 0.001$).

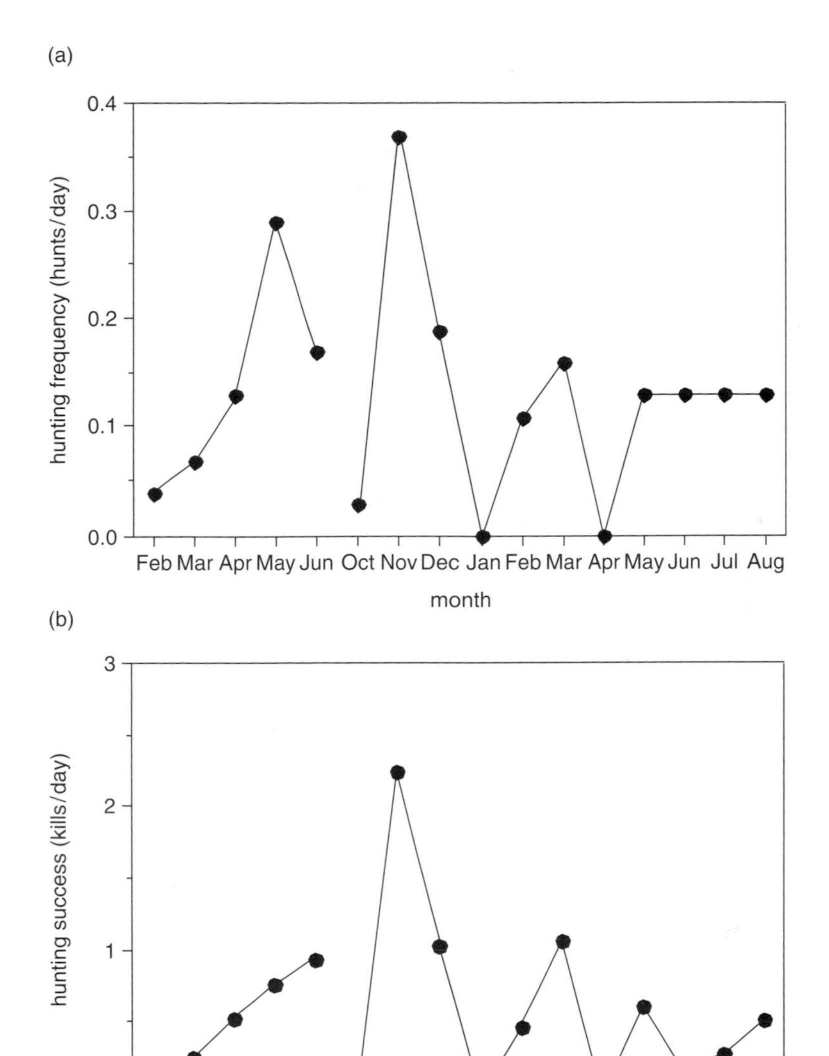

Figure 8.5 Monthly variation in chimpanzee hunting frequency and success at Ngogo, Kibale National Park, Uganda. (a) Hunting frequency. The mean number of hunts observed per day is shown ($n = 61$ red colobus hunts) (from Table 2 in Mitani & Watts [2001]). (b) Hunting success. The mean number of kills observed per month is shown ($n = 213$ red colobus kills) (Mitani & Watts, unpublished data, 1998–99).

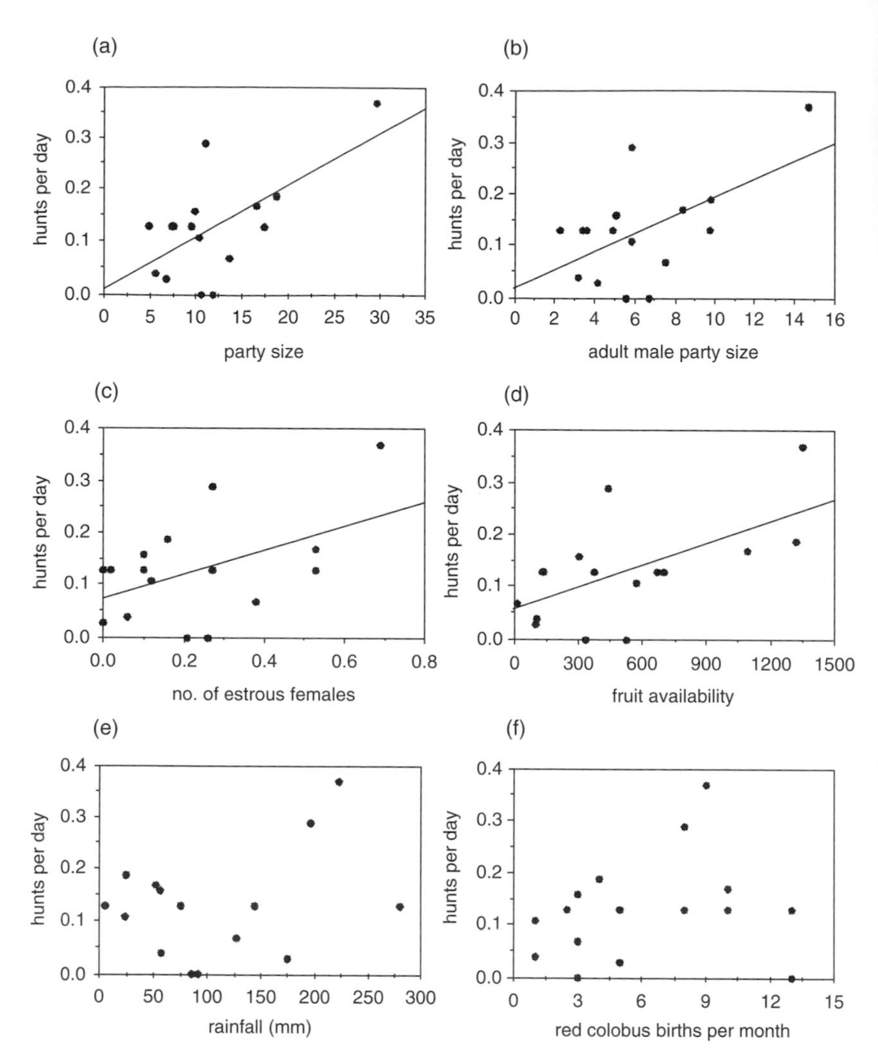

Figure 8.6 Correlates of chimpanzee hunting frequency at Ngogo, Kibale National Park, Uganda. (a) Party size (from Mitani & Watts [2001]; Mitani *et al.* [2002a]). (b) Adult male party size (from Mitani & Watts [2001]; Mitani *et al.* [2002a]). (c) Number of estrous females (from Mitani & [Watts 2001]; Mitani *et al.* [2002a]). (d) Fruit availability (adapted from Fig. 1a in Mitani & Watts [2001]). (e) Rainfall (adapted from Fig. 1b in Mitani & Watts [2001]). (f) Number of red colobus births (from Struhsaker & Leland [1987]; Mitani & Watts [2001]). All data points represent monthly averages ($n = 16$ months).

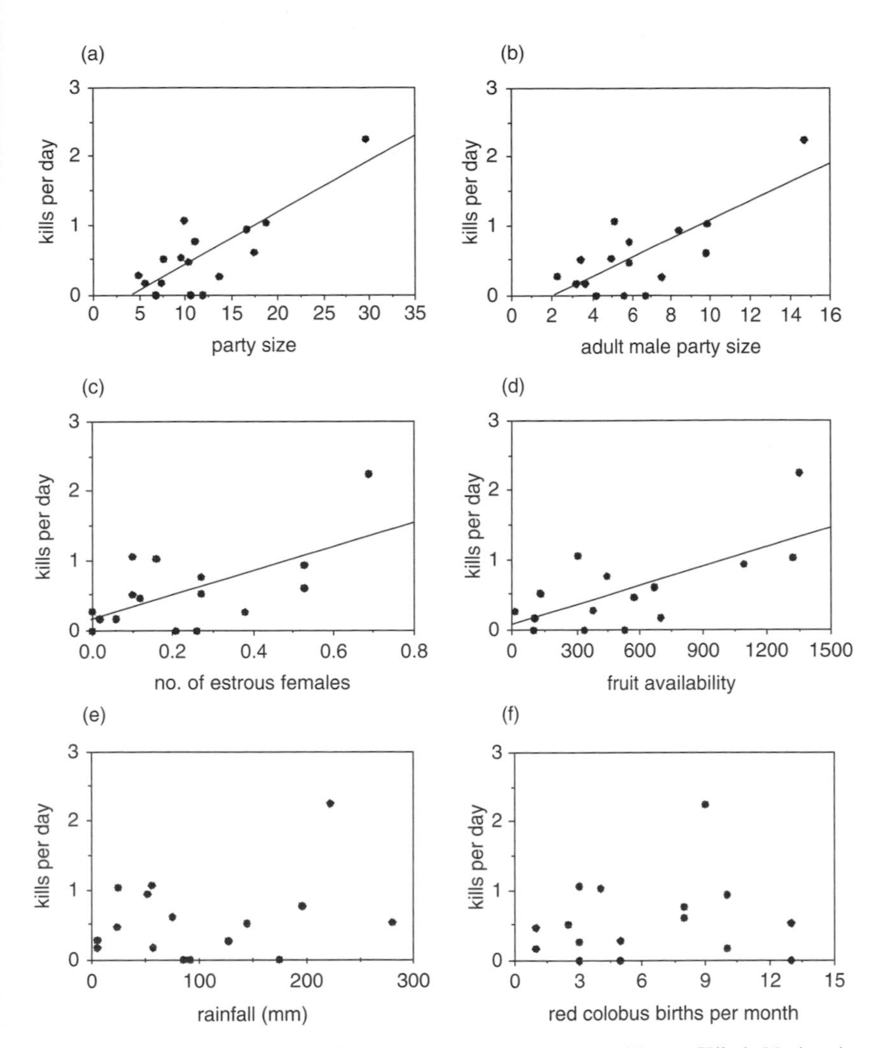

Figure 8.7 Correlates of chimpanzee hunting success at Ngogo, Kibale National Park, Uganda. (a) Party size (from Mitani & Watts [2001]; Mitani *et al.* [2002a]). (b) Adult male party size (from Mitani & Watts [2001]; Mitani *et al.* [2002a]). (c) Number of estrous females (from Mitani & Watts [2001]; Mitani *et al.* [2002a]). (d) Fruit availability (from Fig. 1a in Mitani & Watts [2001] and Mitani & Watts, unpublished data, 1998–99). (e) Rainfall (from Fig. 1b in Mitani & Watts [2001] and Mitani & Watts, unpublished data, 1998–99). (f) Number of red colobus births (from Struhsaker & Leland [1987]; Mitani & Watts [2001]). All data points represent monthly averages ($n = 16$ months).

Elsewhere, we have suggested that seasonal variation in hunting activity can be attributed to ecological factors, which in turn affect social characteristics of chimpanzee predators (Mitani & Watts 2001; Mitani *et al.* 2002b). This "ecological constraints" hypothesis begins by assuming that meat is a scarce and valuable resource that all chimpanzees clearly value. Meat is difficult to acquire, however, and chimpanzees base their decisions to hunt on the likelihood of success. Studies of chimpanzee predatory behavior show consistently that hunting success is correlated positively with hunting party size and the number of male hunters (Boesch & Boesch 1989; Stanford *et al.* 1994b; Mitani & Watts 1999; Hosaka *et al.* 2001; Watts & Mitani 2002a, 2002b). Male chimpanzee hunters appear to swamp red colobus prey defenses with strength in numbers; the hunting success of large parties with many male hunters exceeds that of small parties with few male hunters. Ecological factors constrain the formation of large parties, however. Low fruit availability heightens the ecological costs of feeding competition, leading to the creation of small foraging parties (Chapman *et al.* 1995; Wrangham 2000; Mitani *et al.* 2002a). Large chimpanzee parties form regularly only during more permissive periods of high fruit availability (Chapman *et al.* 1995; Wrangham 2000; Mitani *et al.* 2002a). These considerations provide a transparent explanation for why chimpanzees at Ngogo increase their hunting efforts during periods of high fruit abundance (Fig. 8.6d) and also provide a rationale for understanding the positive correlations between hunting frequency and success with overall and adult male party sizes (Fig. 8.6a, 8.6b, 8.7a, and 8.7b). In sum, temporal variation in chimpanzee hunting activity at Ngogo is associated with seasonal variations in fruit availability, which in turn appears to affect hunting decisions, frequency, and success.

Baboons

Hausfater's (1976) study of predatory behavior of baboons (*Papio cynocephalus*) conducted over 14 months between 1971 and 1972 in the Amboseli National Park, Kenya, provides the only quantitative data on hunting seasonality in this species. Hunting success displayed a consistent decline over the 14-month study period, and monthly success rates showed considerable heterogeneity over time (Fig. 8.8a). A distinct dry season occurs at Amboseli typically between the months of May and October (Altmann & Altmann 1970). Hausfater (1976) originally reported that baboon hunting success was significantly higher in the dry season than in the wet season,

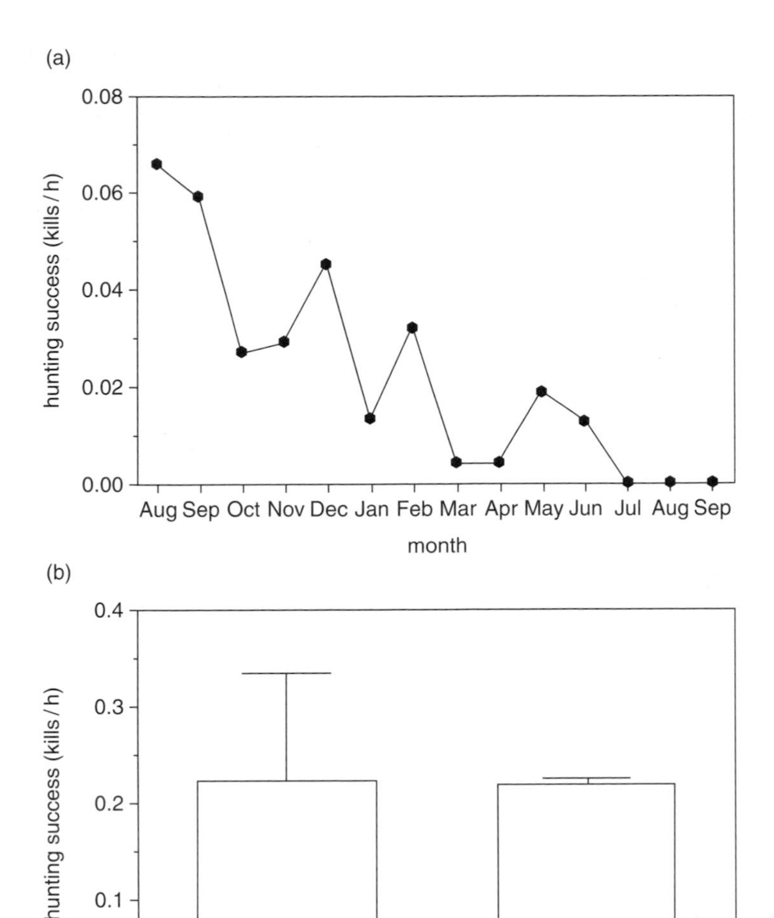

Figure 8.8 Temporal variation in baboon hunting success at the Amboseli National Park, Kenya. (a) Monthly variation in hunting success. The number of kills observed per hour of observation is shown ($n = 45$ kills) (from Table 2 in Hausfater [1976]). (b) Seasonal variation in hunting success. The mean number of kills observed per hour of observation (+1 Standard error[se]) in each season is shown. Data are from Fig. 8.8a.

but reanalysis of his data reveals no consistent seasonal difference (Mann–Whitney U test: $Z = -0.64$, $n_1 = 7$, $n_2 = 5$, $P > 0.50$) (Figure 8.8b). Hausfater (1976) speculated that temporal fluctuations in baboon predatory behavior could be attributed to specific deficits in vitamin B12. He was quick to caution, however, that this hypothesis was impossible to test given the lack of data on the nutritional content of baboon plant foods and prey. Altmann (1998) documented extensive seasonal variation in the array of foods and nutrients available to Amboseli baboons. He noted that arthropods probably serve as the main source of vitamin B12 and that arthropod abundance can vary greatly temporally. Whether any relationship exists between this variation and meat-eating remains unknown.

Capuchin monkeys

White-faced capuchin monkeys (*Cebus capucinus*) prey on a variety of vertebrates, including coatis, squirrels, mice, bats, birds, and lizards (Fedigan 1990; Perry & Rose 1994; Rose 2001). During 3703 hours of observation conducted over 24 months between 1991 and 1993, Perry (unpublished data) witnessed capuchins prey on other animals 61 times. Hunting success showed considerable monthly variation (Fig. 8.9a). Perry's Lomas Barbudal study site displays distinct wet and dry seasons between May–November and December–April, respectively, and capuchin hunting success does not differ between these periods (Fig. 8.9b). This result contrasts with those reported by Rose (1997), who observed capuchin predatory behavior at the Santa Rosa National Park, Costa Rica. Capuchins at Santa Rosa hunted more often (Fig. 8.10a) and with greater success (Fig. 8.10b) in the dry season than in the wet season.

Changes in prey availability, capuchin food abundance, and hunting conditions have been suggested to affect seasonal variations in hunting frequency and success at Santa Rosa (Fedigan 1990; Rose 1997, 2001). Capuchin prey, especially the young of coatis, squirrels, and parrots, become increasingly available during the dry season. All three species typically give birth towards the end of the dry season, and capuchins repeatedly visit, monitor, and search areas of high prey density during this time (Fedigan 1990; Rose 1997). This deliberate pattern of hunting is most evident when capuchins raid the nests of coatis, although it is also apparent when hunting squirrels (Rose 2001). Fruits and insects, the principal foods of capuchins, are not available widely during the dry season (Fedigan 1990), and capuchins may switch to hunting vertebrate prey to compensate for this seasonal shortfall (Rose 2001). Finally, pursuing mobile prey such as squirrels is easier in the open habitats created by the dry season, which make hunts more profitable at this time (Fedigan 1990).

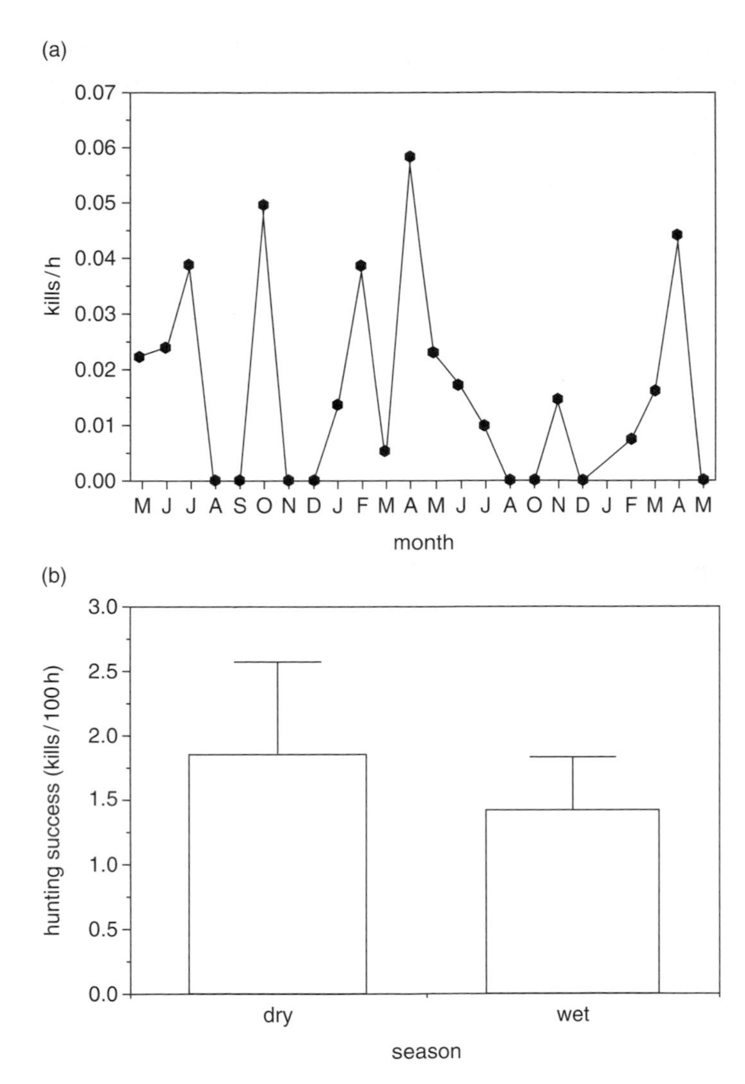

Figure 8.9 Temporal variation in capuchin hunting success at Lomas Barbudal Reserve, Costa Rica. (a) Monthly variation in hunting success. The number of kills made per hour observation is shown ($n = 61$ kills) (from S. Perry, unpublished data, 2001). (b) Seasonal variation in hunting success. The mean number of kills observed per 100 hours of observation in each Season (+1 Standard error[se]) is shown. Data taken from Fig. 8.9a.

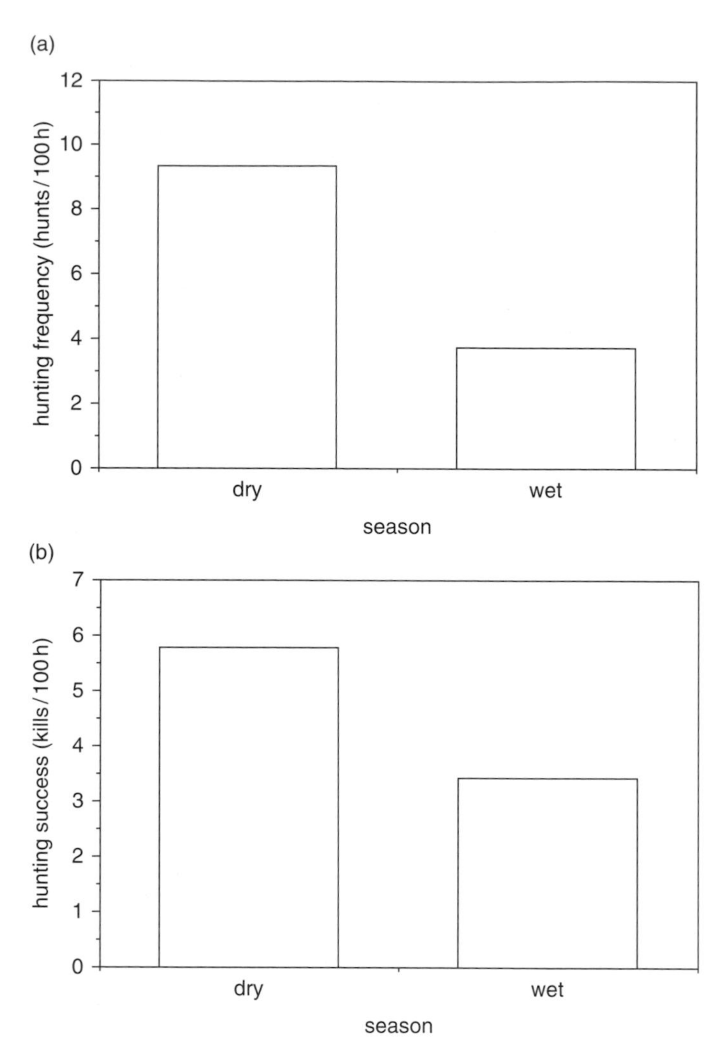

Figure 8.10 Seasonal variation in capuchin hunting frequency and success at Santa Rosa National Park, Costa Rica. (a) Hunting frequency. The number of hunts observed per 100 hours of observation in each season is shown (from Rose [1997]). (b) Hunting success. The mean number of kills observed per 100 hours of observation in each season is shown. Data taken from Fig. 8.9a.

Discussion

The preceding review reveals that despite considerable temporal variation in hunting frequency and success by primate predators, hunting seasonality depends strongly on the type of prey exploited. Seasonal effects can be shown consistently only for chimpanzees that hunt red colobus monkeys. Studies of chimpanzee predation on red colobus and observations of predatory behavior by baboons and capuchin monkeys suggest that several factors may affect hunting seasonality. Some are related to extrinsic ecological conditions, while others are associated with the characteristics of the prey and predators (Table 8.1).

Ecological factors frequently are invoked to account for seasonal variation in hunting activity. The low availability of preferred foods creates a need for fallback energy sources, and meat has been hypothesized as one such alternative food for chimpanzees and capuchins (Teleki 1973; Stanford 1996; Boesch & Boesch-Achermann 2000; Rose 2001). Another suggestion is that seasonal shortages of specific nutrients in the diets of chimpanzees and baboons lead to observed increases in hunting effort (Teleki 1973; Hausfater 1976; Takahata *et al.* 1984). These hypotheses, although simple and appealing, have been difficult to test given the paucity of data on food availability and the nutritional content of foods. Our own observations are not consistent with the version of the nutritional shortfall hypothesis that invokes energy as its currency. Instead of increasing their hunting efforts during periods of low fruit availability, chimpanzees at Ngogo show a decrement in hunting activity (Fig. 8.6d) (see also Mitani & Watts [2001]; Watts & Mitani [2002a, 2002b]). Recent data from Taï also reveal that the seasonal peak in hunting activity there coincides with a general onset in fruiting activity (Anderson 2001). Temporal changes in nutrient availability and intake in relation to hunting activity have not been investigated for any primate predator. Without these data, it is currently impossible to evaluate whether primate predators are driven to hunt by their need for specific nutrients.

Ecological conditions have also been hypothesized to facilitate prey capture and thereby affect the tendency of primate predators to hunt at some times rather than others (Table 8.1). While some speculate that dry seasons create open habitats, making it easier for primate predators to seize prey (Fedigan 1990), others suggest that conditions during the rainy season make it difficult for prey to escape and thus make them easier to capture (Boesch & Boesch-Achermann 2000). As with hypotheses that invoke low food availability, these "ease of capture" hypotheses are difficult to assess given the lack of quantitative data regarding how variation in ecological conditions, e.g. open versus closed habitats and wet versus dry conditions, affects hunting success. A similar lack of data on food choice and diet overlap preclude evaluating whether ecological factors facilitate hunting by bringing chimpanzee predators and their red colobus prey together at common food sources (Stanford 1998).

Characteristics of prey that result in their differential availability undoubtedly affect patterns of hunting seasonality (Takahata *et al.* 1984; Fedigan 1990; Rose 1997, 2001; Boesch & Boesch-Achermann 2000) (Table 8.1). Primate predators frequently prey selectively on members of the youngest age classes, and seasonal patterns of predation can be linked to the birth seasons of prey. This helps to explain seasonal patterns of predation on the young of coatis by capuchin monkeys (Rose 1997, 2001). At some sites, chimpanzees prey selectively on infant and juvenile red colobus monkeys (Uehara *et al.* 1992; Mitani & Watts 1999; Watts & Mitani 2002a). Differences in the seasonal availability of prey, however, cannot be invoked to explain the observed pattern of predation on these monkeys at Ngogo or Taï. Red colobus birth peaks do not appear to coincide with increased hunting activity at Ngogo (Fig. 8.6f), and the Taï chimpanzees do not prey selectively on immature red colobus (Boesch & Boesch 1989).

Social characteristics of chimpanzee predators provide a third set of hypotheses to explain hunting seasonality (Table 8.1). Observations at Gombe led Stanford and colleagues (1994b) to conclude that the single best predictor of the tendency for male chimpanzees to hunt is the presence of reproductively active females. They used this result along with the fact that male chimpanzees are the primary hunters (Stanford *et al.* 1994a; Mitani & Watts 1999; Boesch & Boesch-Achermann 2000; Watts & Mitani 2002a) and occasional observations of meat-sharing followed by mating (Stanford 1998) to argue for a provocative "meat-for-sex" hypothesis. According to this hypothesis, male chimpanzees are motivated to hunt in order to obtain meat that they can use to swap for matings (Stanford *et al.* 1994b). Female chimpanzees at Gombe show some reproductive seasonality, with conceptions occurring most frequently during the late dry season (Wallis 1995). This raises the possibility that males at Gombe increase their hunting efforts as a part of a strategy to increase their access to fertile females. Despite significant interest and publicity garnered by this idea, the hypothesized causal link between the availability of estrous females and increased male hunting activity has not yet been validated with repeated observations of meat-sharing between the sexes and subsequent mating. Our own analyses of meat-sharing and mating among the Ngogo chimpanzees showed that sharing meat with estrous females had no consistent effect on male copulatory frequency (Mitani & Watts 2001). Until relevant data become available from other sites, it seems prudent to withhold judgment on whether the presence of estrous females affects male chimpanzee hunting effort.

Increases in hunting frequency have also been attributed to a "beater effect" created by large foraging parties of chimpanzees (Takahata *et al.* 1984).

Large parties inadvertently flush prey and produce hunting opportunities typically unavailable to members of smaller parties. This hypothesis applies to predation on small, immobile prey, such as the young of bush pigs and duiker. Seasonality in predation on these prey, however, has not been demonstrated conclusively.

A final hypothesis proposes that ecological factors affect social characteristics of chimpanzee predators, which in turn influences seasonal variations in hunting activity. This ecological constraints hypothesis is based on our own observations of the hunting behavior of the Ngogo chimpanzees (Mitani & Watts 2001). Although our current data are largely consistent with this hypothesis, it is unclear whether it explains temporal variations in the hunting behavior of chimpanzees at other sites. Tests at these sites await systematic comparisons between food availability and chimpanzee hunting frequency and party size.

Future research and implications for human evolution

The preceding review points to large gaps in our understanding and identifies several areas in need of further study. First, why is robust empirical support for seasonal hunting limited only to chimpanzees that hunt red colobus prey? This appears all the more curious given that bush pig and duiker prey breed seasonally and that chimpanzees selectively hunt young, vulnerable individuals of these species. Second, studies that provide independent measures of food availability and assay the nutritional content of foods are needed to evaluate hypotheses that propose these as key factors that determine seasonal hunting by primate predators. Third, detailed observations on prey behavior and demography are required to assess whether differences in their availability affect temporal variation in primate hunting activity.

Pending resolution of these issues, can we use current information regarding hunting seasonality by primate predators to make inferences regarding meat-eating in our human ancestors? The presumed effects of seasonality on food availability have sometimes been used to explain why early humans made a transition to a meat-eating diet. For example, Foley (1987) suggested that some early hominids might have adapted to dry-season shortages in plant foods by hunting and eating meat. To support this hypothesis, Foley (1987) used comparative data, originally compiled by Dunbar (1983), to show that baboons at different study sites respond to dry-season food shortages by increasing their hunting effort. We have been unable to replicate the figures cited by Dunbar (1983: Table 7) and remain skeptical about claims that purport to show hunting seasonality by baboons. In addition, there are theoretical and empirical reasons to doubt each of the three crucial aspects of Foley's (1987) hypothesis. Does meat represent a viable fallback food? Is the dry season a period of food scarcity? Does the predicted relationship between hunting frequency and food scarcity hold?

The first point has been challenged by several investigators (Hawkes 1991; Stanford 1998; Wrangham *et al.* 1999), who argue that hunting is less energetically efficient than foraging for plants during food-poor times. In addition, studies at Ngogo (Mitani & Watts 2001) and Taï (Anderson 2001) indicate that fruit abundance varies independently of rainfall. This cautions us against using rainfall as a proxy variable for food availability, although it may be valid in drier habitats such as Gombe, where data on body weights suggest that the dry season may be a time of energetic stress (Wrangham 1975; Pusey *et al.* 2005). Finally, our own observations indicate that chimpanzees at Ngogo decrease their hunting activity during food-poor times (Fig. 8.6d) (Mitani & Watts 2001). Taken together, these observations indicate that scenarios that invoke seasonal food scarcity as a driving force leading to the evolution of meat-eating by humans will have to be re-evaluated.

Closer scrutiny of the way in which primate predators hunt provides additional insights into the utility of employing these data to reconstruct meat-eating by early humans. Hunting by contemporary primates is qualitatively different from the manner in which early humans acquired meat. For example, obvious differences exist between chimpanzees, who hunt arboreal monkeys, and the earliest hominids, who hunted or scavenged a wide variety of small to large prey on the ground (Plummer *et al.* 1999, 2001; Semaw 2000). While hunting by humans frequently involves planning and cooperation (e.g. Bailey [1991]), predation by primates typically is opportunistic, involving little collaboration among hunters (see reviews in Rose [1997], Uehara [1997], and Stanford [1998], but see also Boesch & Boesch [1989]). Primates differ from humans not only in the way in which they hunt but also in why they hunt. While seasonal periods of food scarcity might not have formed the initial impetus for the evolution of hunting (see above), meat-eating probably was obligatory during early human history. Current evidence, however, does not support any of the hypotheses that propose meat as a critical resource for primate predators. Our own ecological constraints hypothesis recognizes that meat is clearly valued by all chimpanzees but suggests that it does not a represent an absolutely *essential* resource necessary for successful survival and reproduction. This contention accords with the observation that not all chimpanzee populations regularly prey on other vertebrates (e.g. in Bossou) (Yamakoshi 1998). Additional hypotheses propose the same and indicate that hunting occurs under permissive ecological conditions when it is relatively easy for primates to capture prey. In sum, it is problematic to assume that non-human primate and early human hunting are homologous, and, as a result, it is questionable to use data on non-human primate hunting to make inferences about human evolution.

Hypotheses that propose scavenging as the major subsistence strategy employed by early humans argue that its profitability varied seasonally. During the dry season, food scarcity combined with high rates of prey mortality might have made scavenging a viable foraging strategy for small, unarmed hominids (Blumenschine 1987; Blumenschine & Cavallo 1992). The dearth of information available on primate scavenging (Hasegawa *et al.* 1983, Muller *et al.* 1995; Rose 2001), especially chimpanzees living in dry, open habitats, makes it difficult to use observations of primate hunting seasonality to assess the potential role of scavenging in human evolution. Here, paleoecological reconstructions of scavenging opportunities in the past may yield greater insights (e.g. Tappan [2001]; Van Valkenburgh [2001]). Equally difficult to evaluate is the hypothesis that seasonal deficits in specific nutrients, such as fat, affected early human hunting (Speth 1987, 1989). Testing this hypothesis will require additional study of the nutritional composition of foods consumed by primate predators (see above). In sum, current evidence regarding primate hunting seasonality is far from complete and cannot be used with certainty to address questions about hunting by our early human ancestors. Future research is likely to be rewarded richly.

Acknowledgments

Our fieldwork at Ngogo has been sponsored by the Ugandan National Parks, Uganda National Council for Science and Technology, and the Makerere University. We are grateful to G. I. Basuta, J. Kasenene, and the staff of the Makerere University Biological Field Station for logistic support. We thank C. Businge, J. Lwanga, A. Magoba, G. Mbabazi, G. Mutabazi, L. Ndagizi, H. Sherrow, J. Tibisimwa, A. Tumusiime, and T. Windfelder for assistance in the field and the editors and W. Sanders for comments on the manuscript. Our research at Ngogo has been generously supported by grants from the Detroit Zoological Institute, the Louis B. Leakey Foundation, the National Geographic Society, the National Science Foundation (NSF) (SBR-9253590 and BCS-0215622), University of Michigan, and Wenner-Gren Foundation to JCM and LSB. Leakey Foundation and National Geographic Society to DPW. We are grateful to D. Anderson, C. Boesch, K. Hosaka, K. Hunley, N. Newton-Fisher, S. Perry, and J. Speth for providing advice and access to their unpublished material.

References

Altmann, S. (1998). *Foraging for Survival*. Chicago: University of Chicago Press.

Altmann, S. & Altmann, J. (1970). *Baboon Ecology*. Chicago: University of Chicago Press.

Anderson, D. 2001. Tree phenology and distribution, and their relation to chimpanzee social ecology in the Taï National Park, Côte d'Ivoire. Ph. D. thesis, University of Wisconsin, Madison.

Bailey, R. (1991). *The Behavioral Ecology of Efe Pygmy Men in the Ituri Forest, Zaire*. Ann Arbor: Museum of Anthropology, University of Michigan.

Blumenschine, R. (1987). Characteristics of an early hominid scavenging niche. *Current Anthropology*, **28**, 383–407.

Blumenschine, R. & Cavallo, J. (1992). Scavenging and human evolution. *Scientific American*, **267**, 90–6.

Boesch, C. & Boesch, H. (1989). Hunting behavior of wild chimpanzees in the Taï National Park. *American Journal of Physical Anthropology*, **78**, 547–73.

Boesch, C. & Boesch-Achermann, H. (2000). *The Chimpanzees of the Taï Forest: Behavioural Ecology and Evolution*. Oxford: Oxford University Press.

Bshary, R. (1995). Rote Stummelaffen, *Colobus badius*, and Dianameerkatzen, *Cercopithecus diana*, in Taï Nationalpark, Elfenbeinküste: Wozu assoziieren sie? Ph. D. thesis, Ludwig Maximilian University, Munich.

Chapman, C., Wrangham, R., & Chapman, L. (1995). Ecological constraints on group size: an analysis of spider monkey and chimpanzee subgroups. *Behavioral Ecology and Sociobiology*, **36**, 59–70.

Doran, D. (1997). Influence of seasonality on activity patterns, feeding behavior, ranging, and grouping patterns in Taï chimpanzees. *International Journal of Primatology*, **18**, 183–206.

Dunbar, R. (1983). Theropithecines and hominids: contrasting solutions to the same ecological problem. *Journal of Human Evolution*, **12**, 647–58.

Fedigan, L. (1990). Vertebrate predation in *Cebus capucinus*: meat-eating in a neotropical monkey. *Folia Primatologica*, **54**, 196–205.

Foley, R. (1987). *Another Unique Species*. New York: John Wiley & Sons.

Goodall, J. (1963). Feeding behaviour of wild chimpanzees: a preliminary report. *Symposium of the Zoological Society of London*, **10**, 39–48.

— (1986). *The Chimpanzees of Gombe*. Cambridge, MA: Belknap Press.

Harding, R. (1973). Predation by a troop of olive baboons (*Papio anubis*). *American Journal of Physical Anthropology*, **38**, 587–91.

— (1975). Meat-eating and hunting in baboons. In *Socioecology and Psychology of Primates*, ed. R. Tuttle. The Hague: Mouton Publishers, pp. 245–57.

Hasegawa, T., Hiraiwa-Hasegawa, M., Nishida, T., & Takasaki, H. (1983). New evidence on scavenging behavior in wild chimpanzees. *Current Anthropology*, **24**, 231–2.

Hausfater, G. (1976). Predatory behavior of yellow baboons. *Behaviour*, **56**, 44–68.

Hawkes, K. (1991). Showing off: tests of an hypothesis about men's foraging goals. *Ethology and Sociobiology*, **12**, 29–54.

Hosaka, K., Nishida, T., Hamai, M., Matsumoto-Oda, A., & Uehara, S. (2001). Predation of mammals by the chimpanzees of the Mahale Mountains, Tanzania. In *All Apes Great and Small*, Vol. 1, ed. B. Galdikas, N. Briggs, L. Sheeran, G. Shapiro, & J. Goodall. New York: Kluwer Academic, pp. 107–30.

Kingdon, J. (1979). *East African Mammals*. Chicago: University of Chicago Press.

Mitani, J. & Watts, D. (1999). Demographic influences on the hunting behavior of chimpanzees. *American Journal of Physical Anthropology*, **109**, 439–54.

— (2001). Why do chimpanzees hunt and share meat? *Animal Behaviour*, **61**, 915–24.

Mitani, J., Watts, D., & Lwanga, J. (2002a). Ecological and social correlates of chimpanzee party size and composition. In *Behavioral Diversity in Chimpanzees and Bonobos*, ed. C. Boesch, G. Hohmann, & L. Marchant. Cambridge: Cambridge University Press, pp. 102–11.

Mitani, J., Watts, D., & Muller, M. (2002b). Recent developments in the study of wild chimpanzee behavior. *Evolutionary Anthropology*, **11**, 9–25.

Muller, M., Mpongo, E., Stanford, C., & Boehm, C. (1995) A note on scavenging by wild chimpanzees. *Folia Primatologica*, **65**, 43–7.

Oates, J. (1987). Food distribution and foraging behavior. In *Primate Societies*, ed. B. Smuts, D. Cheney, R. Seyfarth, R. Wrangham, & T. Struhsaker. Chicago: University of Chicago Press, pp. 197–209.

Perry, S. & Rose, L. (1994). Begging and transfer of coati meat by white-faced capuchin monkeys. *Primates*, **35**, 409–15.

Plummer, T., Bishop, L., Ditchfield, P., & Hicks, J. (1999). Research on Late Pliocene Oldowan sites at Kanjera South, Kenya. *Journal of Human Evolution*, **36**, 151–70.

Plummer, T., Ferraro, J., & Ditchfield, P. (2001). Late Pliocene Oldowan excavations at Kanjera South, Kenya. *Antiquity*, **75**, 809–10.

Pusey, A., Oehlert, G., Williams, J., & Goodall, J. (2005). The influence of ecological and social factors on body mass of wild chimpanzees. *International Journal of Primatology*, **26**, 3–32.

Rose, L. (1997). Vertebrate predation and food sharing in *Cebus* and *Pan*. *International Journal of Primatology*, **18**, 727–65.

— (2001). Meat and the early human diet: insights from Neotropical primate studies. In *Meat-Eating and Human Evolution*, ed. C. Stanford & H. Bunn. Oxford: Oxford University Press, pp. 141–59.

Semaw, S. (2000). The world's oldest stone artefacts from Gona, Ethiopia: their implications for understanding stone technology and patterns of human

evolution between 2.6–1.5 million years ago. *Journal of Archaeological Science*, **27**, 1197–214.
Speth, J. (1987). Early hominid subsistence strategies in seasonal habitats. *Journal of Archaeological Science*, **14**, 13–29.
— (1989). Early hominid hunting and scavenging: the role of meat as an energy source. *Journal of Human Evolution*, **18**, 329–43.
Stanford, C. (1996). The hunting ecology of wild chimpanzees: implications for the evolutionary ecology of Pliocene hominids. *American Anthropologist*, **98**, 96–113.
— (1998). *Chimpanzee and Red Colobus*. Cambridge: Harvard University Press.
Stanford, C., Wallis, J., Matama, H., & Goodall, J. (1994a). Patterns of predation by chimpanzees on red colobus monkeys in Gombe National Park, 1982–1991. *American Journal of Physical Anthropology*, **94**, 213–28.
— (1994b). Hunting decisions in wild chimpanzees. *Behaviour*, **131**, 1–20.
Struhsaker, T. & Leland, L. (1987). Colobines: infanticide by adult males. In *Primate Societies*, ed. B. Smuts, D. Cheney, R. Seyfarth, R. Wrangham, & T. Struhsaker. Chicago: University of Chicago Press, pp. 83–97.
Strum, S. (1975). Primate predation: interim report on the development of a tradition in a troop of olive baboons. *Science*, **187**, 755–7.
— (1981). Processes and products of change: baboon predatory behavior at Gilgil, Kenya. In *Omnivorous Primates*, ed. R. Harding & G. Teleki. New York: Columbia University Press, pp. 255–302.
Takahata, Y., Hasegawa, T., & Nishida, T. (1984). Chimpanzee predation in the Mahale Mountains from August 1979 to May 1982. *International Journal of Primatology*, **5**, 213–33.
Takasaki, H., Nishida, T., Uehara, S., *et al.* (1990). Summary of meterological data at Mahale Research Camps, 1973–1988. In *The Chimpanzees of the Mahale Mountains*, ed. T. Nishida. Tokyo: University of Tokyo Press, pp. 291–300.
Tappan, M. (2001). Deconstructing the Serengeti. In *Meat-Eating and Human Evolution*, ed. C. Stanford & H. Bunn. Oxford: Oxford University Press, pp. 13–32.
Teleki, G. (1973). *The Predatory Behavior of Wild Chimpanzees*. Lewisburg, PA: Bucknell University Press.
Uehara, S. (1997). Predation on mammals by the chimpanzee (*Pan troglodytes*). *Primates*, **38**, 193–214.
Uehara, S., Nishida, T., Hamai, M., *et al.* (1992). Characteristics of predation by chimpanzees in the Mahale Mountains National Park, Tanzania. In *The Chimpanzees of the Mahale Mountains*, ed. T. Nishida. Tokyo: University of Tokyo Press, pp. 143–58.
Van Valkenburgh, B. (2001). The dog-eat-dog world of carnivores: a review of past and present carnivore dynamics. In *Meat-Eating and Human Evolution*, ed. C. Stanford & H. Bunn. Oxford: Oxford University Press, pp. 101–21.
Wallis, J. (1995). Seasonal influence on reproduction in chimpanzees of Gombe National Park. *International Journal of Primatology*, **16**, 435–51.
Watts, D. & Mitani, J. (2002a). Hunting behavior of chimpanzees at Ngogo, Kibale National Park, Uganda. *International Journal of Primatology*, **23**, 1–28.

— (2002b). Hunting and meat sharing by chimpanzees at Ngogo, Kibale National Park, Uganda. In *Behavioral Diversity in Chimpanzees and Bonobos*, ed. C. Boesch, G. Hohmann, & L. Marchant. Cambridge: Cambridge University Press, pp. 244–54.

Wrangham, R. (1975). The behavioral ecology of chimpanzees in Gombe National Park, Tanzania. Ph. D. thesis, Cambridge University.

— (1977). Feeding behaviour of chimpanzees in Gombe National Park, Tanzania. In *Primate Ecology*, ed. T. Clutton-Brock. London: Academic Press, pp. 503–38.

— (2000). Why are male chimpanzees more gregarious than mothers? A scramble competition hypothesis. In *Primate Males: Causes and Consequences of Variation in Group Composition*, ed. P. Kappeler. Cambridge: Cambridge University Press, pp. 248–58.

Wrangham, R. & Bergmann Riss, E. (1990). Rates of predation on mammals by Gombe chimpanzees, 1972–1975. *Primates*, **31**, 157–70.

Wrangham, R., Jones, J., Laden, G., Pilbeam, D., & Conklin-Brittain, N. L. (1999). The raw and the stolen: cooking and the ecology of human origins. *Current Anthropology*, **40**, 567–94.

Yamakoshi, G. (1998). Dietary responses to fruit scarcity of wild chimpanzees at Bossou, Guinea: possible implications for ecological importance of tool use. *American Journal of Physical Anthropology*, **106**, 283–95.

9 *Human hunting seasonality*

REBECCA BLIEGE BIRD
Department of Anthropological Sciences, Stanford University, Stanford CA 94305–2117, USA

DOUGLAS W. BIRD
Department of Anthropological Sciences, Stanford University, Stanford CA 94305–2117, USA

Introduction

Human hunting strategies, like those of many non-human primates, vary seasonally with fluctuations in prey abundance, encounter rates, and profitability (Winterhalder 1981; Smith 1991). Temporality in resource supply has profound social effects as well, and some of the earliest studies of hunter–gatherers emphasized the impact of seasonality on settlement size, mobility, general economic organization, and even property rights, religion, family structure, and the sexual division of labor (Mauss & Beuchat 1906; Thomson 1936). For Mauss and Beuchat (1906), seasonality meant temperature: they suggested that Inuit families were organized very differently in the summer than in the winter as a result of the nature of changes in foraging opportunities. For Thomson (1936), seasonality meant rainfall, commenting that the effect of distinct wet and dry seasons in northern Australia might lead one to think that they were observing two different "tribes" of people. Anthropological interest in seasonality and its effects on human social organization has waned since then, frustrated by an inability to find correlations between seasonality and human behavior. Our goal in this chapter is to explore the utility of two approaches to understanding the relationship between seasonality and social behavior. One attempts to use comparative ecological data across groups to explain differences in aspects of social and economic behavior such as mobility and land tenure decisions; the other examines how different individuals within a group may respond differently to resource seasonality. Toward this end, we focus on a single case study: explaining why we see seasonal sex differences in hunting game among Western Desert Australian aborigines.

Seasonality in Primates: Studies of Living and Extinct Human and Non-Human Primates, ed. Diane K. Brockman and Carel P. van Schaik. Published by Cambridge University Press.

Hunter–gatherer responses to seasonality

One of the simplest models predicting how foragers respond to changing spatial and temporal patterns of resource availability is the contingent prey choice model (Stephens & Krebs 1986). This model predicts that if prey types are distributed in a fine-grained manner in space, then temporal reductions in encounter rates with higher-ranked prey cause foragers to become less selective and to broaden the number of prey types that they include in their diet. If resources are not distributed randomly, then patch choice models predict that a reduction in the overall foraging efficiency within a habitat (such as when a forager can no longer gain energy by foraging as it travels between food patches) would cause foragers to either remain in a patch longer or travel farther in order to reach more profitable patches (Charnov 1976). Humans often respond to predictable seasonal variability in food supply by broadening or narrowing the diet and by changing mobility patterns (Binford 1980; Smith 1991). However, they also rely on sociocultural flexibility to mediate their responses to seasonality: investing in technology to reduce handling costs, changing social structure (Bahuchet 1988), using alternative hunting methods (Bailey & Aunger 1989), and maintaining long-distance exchange relationships with other groups (Cashdan 1985).

However, mobility remains one of the most widely recognized effects of seasonality on human subsistence strategies. The relationship between spatial and temporal resource patchiness and mobility patterns in anthropology is commonly analyzed with the "forager–collector model" (Binford 1980). The forager–collector model specifies that mobility patterns vary along a continuum from "forager" (high residential mobility) to "collector" (high logistical mobility), according to the spatial and temporal patchiness of staple resources. The basic assumptions of the model are those long recognized in foraging theory: the degree to which resources are distributed randomly should have large effects on mobility, settlement, and subsistence patterns (Cashdan 1992). While not drawn explicitly from foraging theory, Binford's model links a spatially and temporally homogeneous distribution of resources with high residential mobility, constant low levels of production, low logistical mobility, little storage, and a generalized and expedient tool technology. At this end of the continuum, "foragers" map themselves on to resources, moving people to food in response to temporal and spatial fluctuations in abundance. As resources become more patchy in space or time, people begin to pin their camps more permanently to particular locations and become "collectors." Collectors adjust to a heterogeneous environment where resources are acquired asynchronously by transporting goods to people (increasing logistic mobility). Ultimately, this influences the degree to which foragers rely on food storage and other complex, specialized technology to even out spatial and temporal fluctuations in supply.

The forager–collector model is commonly evaluated using broad environmental correlates of resource patchiness (such as latitude, average annual rainfall, and temperature), along with measures of residential mobility (such as average number of camp moves per year) across different study groups (Thomas 1983; Binford 2001). The availability of water, in particular, is often promoted as a key influence on seasonal patterns of foraging and mobility, especially in more arid areas. Daily water requirements for humans should make seasonal fluctuations in the effective environment of savannas and deserts, especially salient. As Taylor (1964) notes, foragers in these environments are especially "tethered" to key resources.

While we expect that seasonality in rainfall will have a profound effect on behavior in more arid areas, we should not expect a simple correlation between the two. Binford's (2001) data show that even under highly controlled circumstances, people respond to seasonality in manifold and elaborate ways. For example, if we measure the degree of seasonal variation in rainfall as the ratio of rainfall in the driest month to that of the wettest, limit our analysis to savanna and desert Australian foragers, and attempt to predict estimates of the average number of residential moves per year, then we still find no predictable correlation: $r = 0.143$, $t = -0.578$, $P = 0.5711$ (Fig. 9.1) (data from Binford [2001]: Tables 4.01, 5.01, and 8.04).

In the case of seasonal mobility, Kelly (1995:126–30) suggests that our inability to predict mobility based on very broad environmental variables lies in the fact that foragers tethered to water respond in maddeningly complex ways to very local conditions. The factors that determine the availability of water (local substrate, subsurface geology, topography, local evapotranspiration, etc.) are extremely variable and localized in space (Mather 1962). Since all other subsistence resources that people rely on are also tethered in intricate ways to the same variability in access to water, then the layers of environmental complexity make descriptive generalizations highly problematic as heuristic models. However, if we attempt to solve the complexity issue by adding additional variables to an inductive model, then we lose explanatory power: eventually, the model becomes a specific description of local conditions. Thomas (1983) makes this point clear for different groups of western Great Basin Shoshone. While each group experienced similar broad climatic regimes, their social

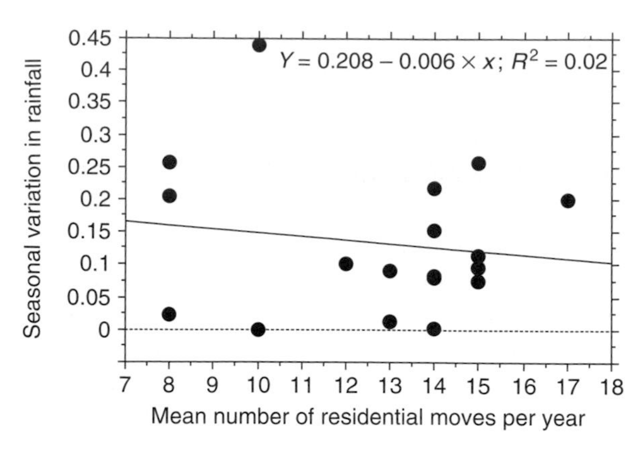

Figure 9.1 The relationship between the amount of seasonal variation in rainfall (calculated as the ratio of the amount of rainfall in the driest month to the amount of rainfall in the wettest month) (see Binford [2001]: Table 4.01) and residential mobility for 17 Australian Aboriginal societies living in regions of savannas and deserts that experience less than 1000 mm of rainfall per year (groups listed in Binford [2001]: Tables 8.01 and 5.01).

organization and mobility patterns differed considerably. All three groups responded to seasonal variability in their staple food source, piñon pine nuts, by storing them through the winter. However, those living in regions without water catchment areas, and with smaller, unpredictably producing piñon groves, were more mobile and lacked individual ownership for specific piñon groves. In each case, if we wanted to account for each group's specific land tenure and mobility pattern using the forager–collector model, then we would need to incorporate such specific climatic, geologic, and social variables that the model would fit only that single case.

Given these empirical and methodological problems, we are suspicious of inductive generalizations about hunter–gatherer seasonality that attempt to explain variability between groups using broad environmental generalizations. However, we do not argue that what Binford (2001: 11) refers to as "autocorrelated dynamics of social organization" relative to seasonality are simply arbitrary or incomprehensible relative to ecological variables. Rather, we suggest that the evidence of seasonal fluctuations in local conditions is a critical source of data to evaluate the utility of theoretically deduced hypotheses about the goals and constraints that structure variability in human behavior within groups. We may gain more insight into the complexity of human responses to seasonality by incorporating the idea that foraging strategies are constrained and facilitated by individual social and reproductive tradeoffs. Below, we provide a more detailed analysis of sex differences in hunting strategies to illustrate this approach, beginning with some theoretical approaches to sex differences in non-human primates.

Seasonality and sex differences: within-group analyses

The primate context

As many chapters in this book have shown, primate behavior flexes responsively to accommodate dynamic environmental variability in resource supply. However, the circumstances that shape foraging flexibility are complex and not predictable, given broad generalizations about rainfall, latitude, and temperature (see Chapter 3). While primates react to changing costs and benefits of foraging for particular food items, their responses are not designed solely to maximize foraging efficiency. Primates must always trade off food intake with other fitness-related pursuits, such as finding mates, caring for offspring, and competing with others for social position. Differences in the nature of such tradeoffs are likely to lie at the root of some of the complexity we see in primate responses to seasonality. Individuals within groups may respond to temporal fluctuations in food supply differently if they face different social and reproductive tradeoffs with acquiring certain types of foods; this is often particularly apparent between males and females (Boinski 1988) and between old and young individuals (Altmann 1998).

Responses to seasonal changes in the abundance of higher-ranked resources are usually mediated by one or more of these tradeoffs. Physical tradeoffs can make some alternative seasonal resources quite costly to exploit. For example, among *Cebus* living sympatrically in the seasonal rainforests of Cocha Cashu (Terborgh 1983), species with more robust jaw and tooth morphology (*C. apella*) broaden the diet to include palm nuts, while those with more gracile masticatory apparatus (*C. albifrons*) range further to exploit increasingly rare patches of fruit. Responses to seasonality are also influenced by social tradeoffs: primates that aggregate in response to predation may find it less costly to respond to seasonal reductions in fruit productivity by traveling further, while primates living in smaller groups that respond to predation through more cryptic strategies might respond by broadening their diet. Finally, primate responses to seasonality are also mediated by reproductive tradeoffs. Where baboons are forced to guard estrous female mates, males reduce the time they spend foraging (Alberts 1996); squirrel monkey males that must invest in competitive interactions with other males forage as time-minimizers, attempting to minimize the time it takes to reach a certain nutritional threshold, while females forage as energy-maximizers, especially when pregnant and lactating (Boinski 1988). Females often face tradeoffs with energy-maximizing when such strategies conflict with putting on fat, nursing infants, or conserving energy.

Among chimpanzees, where male reproductive strategies involve alliance formation and grouping, males find solitary foraging strategies (ant dipping and nut cracking) more costly (Boesch & Boesch 1984). Female chimps generally spend more time foraging for ants and termites, nut cracking, and hunting terrestrial prey, while males spend more time hunting arboreal prey (McGrew 1992). The difference between the male and female foraging patterns is most profound during seasons when males are hunting arboreal monkeys, fruit is abundant, and groups are larger (see Chapter 8). Since larger groups nearly always contain more females, and more estrous females, male hunting has sometimes been interpreted as a reproductive strategy involving the exchange of meat for sex (Stanford 1996). Although estrous females may be more likely than non-estrous females to get meat (Tutin 1979), sharing is directed more often toward other males (Nishida *et al.* 1992); when it is directed toward estrous females, it does not measurably impact male copulatory frequency (Mitani & Watts 2001). Stanford *et al.* (1994) further note that individual hunters actually receive less meat by joining a monkey hunt than they could by pursuing prey alone. Why, then, should male hunters prefer to acquire less meat from arboreal monkeys to use as a tool to acquire sex or allies, when they might more easily acquire more meat from solitary pursuits of terrestrial game? The data so far seem to be consistent with an explanation that males seasonally hunt arboreal monkeys as part of a mating investment strategy involving the competitive display of quality to other males via the "honest" nature of hunting as a signal of dominance. Honest signals are those that are tied intrinsically to the quality being displayed. Signals may be kept honest through a number of mechanisms, one of which is the differential cost or benefit of the display (costly signaling) (Zahavi 1975; Grafen 1990; Johnstone 1997).

The honest signaling hypothesis proposes that males hunt monkeys in order to create alliances and gain the respect of rivals (Nishida *et al.* 1992), but the primary benefits of the hunt lie not in the subsequent sharing and consumption of meat (else why not simply take terrestrial mammals?) but in the hunting display itself. Hunting arboreal monkeys by chasing them through the treetops is energetically costly, difficult to accomplish, and inherently dangerous, and observers often find that some males are better hunters than others (Stanford 1996). If monkey-hunting success allows males to acquire or maintain dominance rank, or intimidates others into letting them gain access to females during times when they are most likely to be fertile, then it might function as a sexually selected signal of competitive ability, more akin to branch-shaking displays than dipping for termites. This hypothesis squarely places only arboreal monkey hunting as a sexually selected male display, because it satisfies the conditions of an honest signal of quality and it differs significantly from what females do. Other types of hunting that both males and females engage in, such as the capture of terrestrial prey, may be more explicable as an energy- or protein-maximizing foraging strategy. This does not mean that females might never engage in monkey hunting (they occasionally do) or that monkey hunting never provides a net nutritional benefit. The nutritional benefits gained through sharing and distribution of the prey are erratic with respect to the hunter: many times, the hunter does not eat the prey or control distributions. With the honest signaling hypothesis, the problematic focus on sharing as the avenue of benefit for hunters is replaced by a focus on the information value of the hunting display itself. The costly signaling model predicts that males may hunt monkeys only seasonally because (i) the benefits of display are higher when groups are larger; (ii) the costs of acquisition are higher because more individuals other than the hunter consume meat; and (iii) reproductive competition among males nets higher payoffs when more females are in estrous, as might occur if estrous cycles are even moderately seasonal.

Individual variation among human foragers

The possibility that reproductive strategies might influence foraging decisions among humans has not gone unexplored, but we still know very little about how foraging strategies vary within groups and less still about how different individuals respond to seasonal fluctuations in prey abundance and distribution. The data that we do have tend to suggest that sex differences in foraging (often referred to as the "sexual division of labor") often vary seasonally and that the foraging decisions of both sexes are affected by reproductive tradeoffs. Hadza women respond to seasonal availability of roots and tubers depending upon their current reproductive and provisioning tradeoffs: nursing and pregnant women and women with children focus on berries, while postmenopausal women exploit roots intensively (Hawkes *et al.* 1989, 1995, 2001a, 2001b). In South America, Hiwi women spend more time foraging when roots are in season, except when they are pregnant or nursing, while men devote more effort to foraging when large amounts of fruit are available (Hurtado & Hill 1990). In Australia, women often hunt small animals seasonally: in Arnhem Land, women are active hunters of birds and fish in the dry season and provide nearly 30% of all game acquired (Altman 1987).

Similar patterns are suggested for other Aboriginal groups, particularly those in the Western Desert, where women are very active hunters (Cane 1987; Tonkinson 1991). Here, seasonal fluctuations in the availability and profitability of game animals seem to correspond to changes in men's and women's hunting strategies. Men seem particularly drawn to hunting for larger animals, while women prefer smaller game. We hypothesize that these differences are a function of sex differences in social and reproductive tradeoffs, particularly in the benefits of skill display. One way to test this hypothesis is to use seasonal variability in key prey species as an independent variable predicting men's and women's foraging choices. If men's choices are biased toward acquiring costly prey or prey that honestly signal some quality of the acquirer, while women's choices are biased toward those that provide better for maximizing foraging efficiency given the tradeoffs of caring for a family, then seasonal variability in encounter rates or profitability of "signaling" prey versus "provisioning" prey may predict seasonal variability in men's and women's hunting strategies. Among Western Desert Martu, key prey species vary seasonally in abundance and the method of capture: goanna lizard and bustard are the most important of these. While women pursue goannas year round using different methods, but never pursue bustards, men focus on bustards and pursue goannas only seasonally, when they are tracked on the surface. Is this difference a result of the influence of sex-specific reproductive strategies on prey choice? To answer this question, we turn toward a more detailed investigation of hunting seasonality among the Martu.

Case Study: hunting seasonality in the arid grasslands of Australia's Western Desert

Methods

The term "Martu" conventionally refers to foraging groups whose traditional estates are located in the northwest section of Australia's Western Desert (Walsh 1990; Tonkinson 1974, 1991). Our study focused on Martu currently living in Parnngurr (23.1° S, 123.5° E), comprising a core population of about 100 individuals. All Martu participants, including the children, spent most of their lives in the desert, and the formative years of those aged 35 years and older were spent as full-time foragers. Foraging is defined as time spent searching, pursuing (including tracking and extracting an individual prey), collecting, and processing wild foods. Martu participants traveled by vehicle from Parnngurr (on trips that ranged from 1 to 22 days) to field camps of their choosing, and then from those camps walked or drove to foraging locations. Data are available from 101 camp-days over three field seasons between 2000 and 2002, with observational data from all months but October and November. During foraging trips, we conducted detailed focal individual follows: each researcher accompanied a single individual and noted all time allocated to travel, search, pursuit, collecting, and processing, along with the weight of each item (if game) or parcel (if fruit, vegetable, or insects) captured at the end of foraging. A total of 763 focal individual follows (481 female, 282 male) are used in the analysis of hunting presented below (children's foraging activities are currently being analyzed; see Bird & Bliege Bird [2005]. In addition to the focal follows, we recorded the duration of all foraging episodes and the weights (by item or type) of all food captured by all camp or trip participants. A total of 2086 foraging hours were recorded by 28 different women and 15 different men. Energy values were taken from published sources analyzing the composition of aboriginal foods (Brand Miller *et al.* 1993). Edible weights for animals were calculated in the field by weighing uncooked individuals and asking foragers to discard the waste material from those same individual animals into a receptacle. On foraging trips, individuals averaged 1702 ± 210 (standard error [SE]) kcal per forager per day, not including those in camp who did no foraging (only the smallest children and the researchers). We supplied an average of 300 kcal per day per participant, primarily in the form of flour and sugar.

Temporality in resources and foraging behavior

In the northwestern region of the Western Desert, tropical moisture flows in from the north, producing widespread rainfall in the hottest part of the year (December–February). In between these occasional storms, isolated thunderstorms can produce localized heavy rainfall. Outside of the hot season, very little rain falls, except during the height of the Austral winter (June–August), when southerly storm systems occasionally stretch far enough north to produce light rainfall over wide regions of the desert (Fig. 9.2).

As detailed by Walsh (1990), before European establishment of permanent wells in the heart of their estates, Martu mobility and group size were tethered to the availability and distribution of rainfall. Generally,

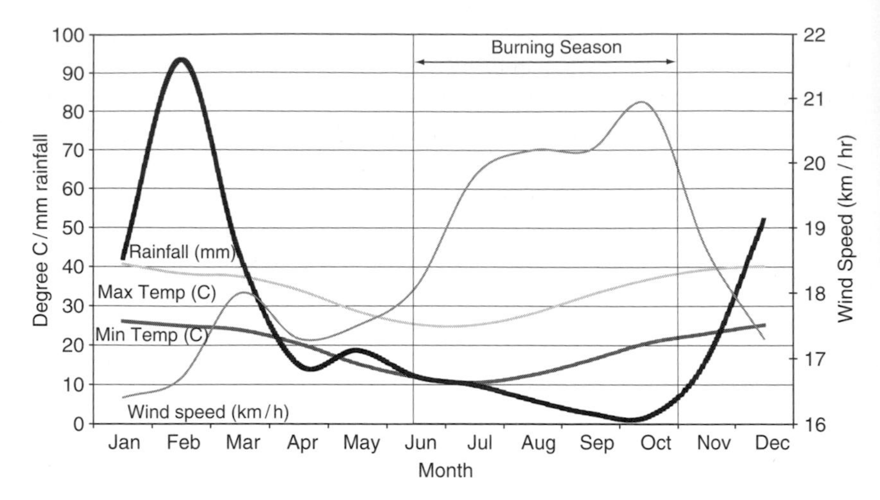

Figure 9.2 Climatic averages from 1970 to 1990 for the northern section of the Western Desert (the Great Sandy Desert), collected from the recording station at Telfer Gold Mine, located approximately 100 km north of Parnngurr.

residential mobility was highest during *Yalijarra* (January–April, the hot/wet season), when widespread rains filled dispersed ephemeral claypans and watercourse soaks. When these emptied as *Wantajarra* (May–August, the cool/dry season) progressed, people would resort to more permanent water sources (rockholes, the Rudall River system, and "native wells"). During *Wantajarra*, residential mobility decreased and group size generally increased. As temperatures began to increase through *Tulparra* (September–December, the hot/dry season), and if there was no rain through December, then residential mobility declined dramatically and groups would retreat to core reserve areas of their estates, where they had access to large, shady rockholes in deep gorges or permanent springs.

Table 9.1 provides a general description of seasonal variability in Mardu resource use. Today, Martu depend on animal resources for the majority of their foraged foods. While the relative importance of collecting plants, especially grass and tree seeds, has declined with an increasing reliance on flour, the Martu insist that hunting has always occupied most of men's, women's, and children's foraging time (see Bird & Bliege Bird 2005). During our study period, large and small goanna lizards (perenti *Veranus gigantius*, sand goanna *V. gouldii*, and ridge-tail goanna *V. acanthurus*), bustards (*Ardeotis australis*), feral cats, snakes (especially *Aspidites* spp.), and grubs (*Cossidae*) provided 68% of the daily calories, while bush fruits, roots, and seeds provided 13%. The remainder (19%) was supplied by store carbohydrates (mostly flour and sugar). Nearly all meat came from game under

Table 9.1 *Martu seasonal calendar of resources*

	Yalijarra				*Wantajarra*				*Tulparra*			
Martu resource	Jan	Feb	Mar	Apr	May	Jun	Jul	Aug	Sep	Oct	Nov	Dec
Primary animals												
Marlu and Karlaya – kangaroo and emu (spear)	I	t	t	t	t	t	t	t	t	I	I	I
Marlu and karlaya – kangaroo and emu (gun)	t	t	t	t	t	t	t	t	t	t	t	t
Kipara – bustard	T	T	t	t	T	T	T	T	T	T	T	T
Yaliparra – perenti goanna	T	T	T	T					t	t	t	t
Parnapunti – sand goanna	T	T	T	T/B	B	B	B	B	T/B	T	T	T
Lunkuta – skink					T	T	T	T				
Feral cat	T	T	T	T	T	T	T	T	T	T	T	T
Lunki – *Cossid larvae* spp.	X	X	X	X	X	X	X	X	X	X	X	X
Primary plants												
Tree seeds – *Acacia spp.*	x	x							X	X	X	X
Grass seeds – woolybutt				X	X	x	x	x				
Kanjamarra – vigna yam					X	X	X	X				
Jinjiwirri – *Solanum centrale*				X	X	X						
Wamala – *Solanum* spp.	V			X	X	X	X	X			V	V

Martu seasons as they correspond roughly with the Gregorian calendar in an average year. B, burrowed and hunted with fire; I, intercepted at waterholes; t, tracked but rarely encountered; T, tracked; T/B, transitional month, goanna active on surface only at midday; V, variably available, depending upon rainfall; x, reduced availability; X, usually available. Timing of availability of most plant resources varies according to local onset of rainfall: two growth periods may be possible for *Solanum* if winter rains fall. See also Walsh (1990).

10 kg in body size. Larger game, such as kangaroo (*Macropus rufa* and *M. robustus*) and emu (*Dromaius novaehollandiae*) was acquired only occasionally. Feral camels were encountered widely but were shot rarely. While Martu consider approximately 150 plant and animal species edible (Walsh 1990), during our study, which took place during three very wet years, four species supplied 81% of all calories acquired: sand goanna (*V. gouldii*), bustard (*Eupodotis australis*), bush tomato (*Solanum diversiflorum*), and bush raisin (*S. centrale*).

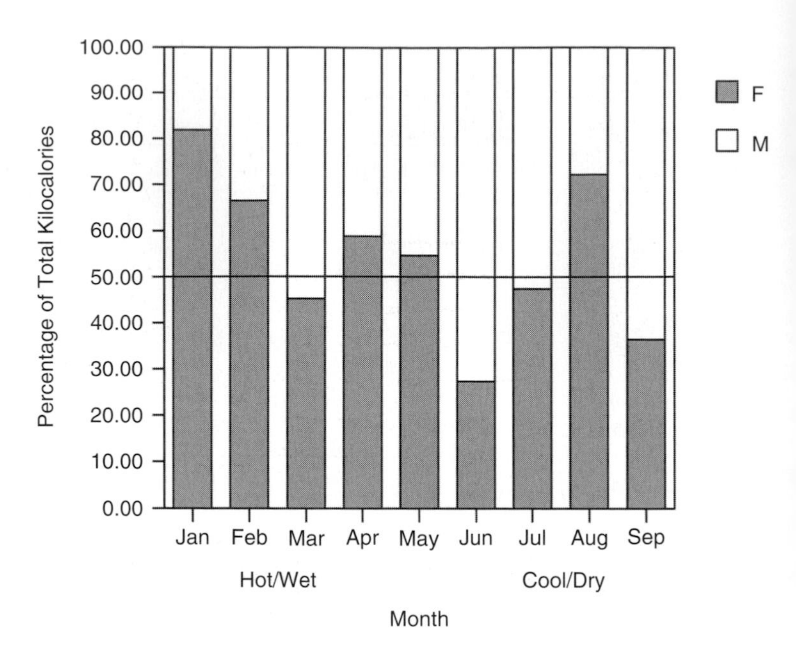

Figure 9.3 Variability in amount of foraged foods (kcal) acquired by men and women in each month sampled during 2000–02. Data are derived from a total of 763 adult foraging follows: 481 women and 282 men.

While goanna and bustards can be acquired throughout the year, acquisition patterns vary seasonally. In both seasons, there are two primary "hunt types:" *wana* (digging stick) hunting and gun hunting. *Wana* hunting involves searching primarily for goanna and other smaller prey using a digging stick. During months when the average high temperature is above 36 ° C (generally October through March), goannas are tracked on the surface during the day. When the days become too cool, the lizards begin to burrow. Martu often describe *Wantajarra* as the burning season: hunters burn off tracts of old-growth spinifex grass in order to search for recent burrows, and then probe with a *wana* for the lie. Gun hunting uses small-gauge rifles or throwing sticks (spears are used today only for traditional punishment) and focuses on larger, mobile game, particularly bustards. Bustards are a more variable resource year round but tend to be more scarce at the end of *Yalijarra*, when they disperse to mate.

Over the sampled period (January through September), women produced 45% of the meat calories and 52% of all calories through foraging. The percentage of total foraging production attributed to women varied by month, from a high of 82% of all calories at the beginning of *Yalijarra* to 27% of all calories at the beginning of *Wantajarra* (Fig. 9.3). While

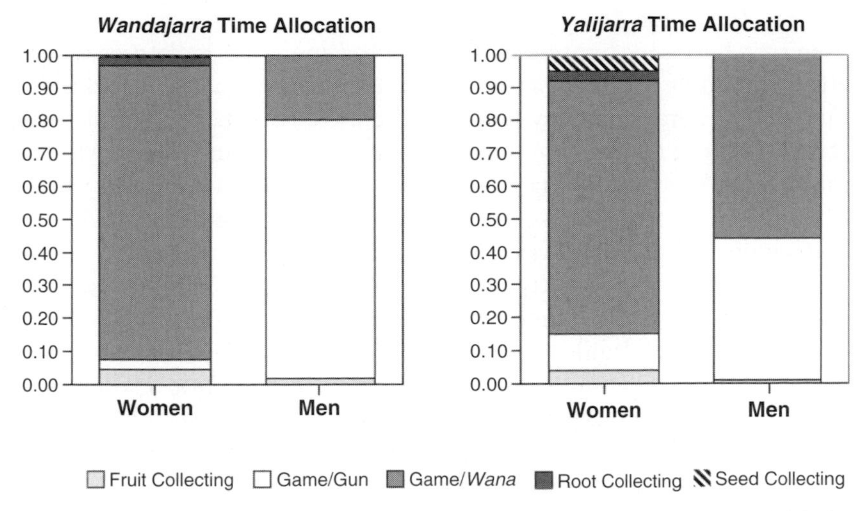

Figure 9.4 Seasonal variability in time allocation to different foraging activities by Martu men and women in 2000–02 during *Yalijarra* (the hot/wet) and *Wantajarra* (the cool/dry). Data cover 2086 forager-hours – 1305 during *Wantajarra* and 781 during *Yalijarra* – with similar samples of men and women in both seasons.

women are culturally proscribed from using spears and spear-throwers, they are not prevented from hunting larger game using other methods, and women and children often help men to spot game on gun hunts. However, women prefer to spend their time *wana* hunting: the proportion of time that women allocate to *wana* hunting does not change seasonally (Fig. 9.4). Men spend a significantly greater proportion of foraging time *wana* hunting in *Yalijarra* and more time gun hunting for larger game in *Wantajarra* (Fig. 9.4). *Wana* hunting carries lower average return rates than gun hunting, but with less variability: non-parametric statistics that ignore the effects of outliers reveal that *wana* hunting is ranked higher than gun hunting ($U = 33429$, $P = 0.0001$). This is because 58% of gun hunts fail to acquire any calories, and 74% fail of gun hunts to acquire the targeted large prey, while only 11% of *wana* hunts fail.

Explaining contemporary variability

The question we will consider here is why men today choose to spend more time *wana* hunting for smaller game in *Yalijarra* and more time gun hunting for larger game in *Wantajarra*.

Hypothesis 1

Men switch to *wana* hunting in *Yalijarra* because it is a more predictably efficient hunting option. This hypothesis considers gun hunting and *wana* hunting as characterized by mutually exclusive search. This means that foragers do not search simultaneously for *wana* prey and gun prey; they must choose to search for one or the other. If this is the case, then we would expect gun hunting to be less efficient than *wana* hunting during *Yalijarra*, and a more efficient option in *Wantajarra*.

In *Yalijarra*, the mean return rate for gun hunting (1073 ± 347 kcal/h) is higher than for *wana* hunting (643 ± 52 kcal/h). The difference is significant if we include outliers (degrees of freedom [DF] = 184, $t = 2.215$, $P = 0.0280$) but non-significant if we do not ($U = 1964$, $P = 0.0701$). In *Wantajarra*, when men spend the least time *wana* hunting and the most time gun hunting, the mean return rate for gun hunting (2752 ± 1027 kcal/h) is again higher than for *wana* hunting (544 ± 29 kcal/h; DF = 283, $t = 2.949$, $P = 0.0035$), but the extreme variability of gun hunting returns causes the rank to reverse when outliers are ignored: *wana* hunting becomes significantly more efficient than gun hunting ($U = 6215$, $P = 0.0001$).

These results suggest that, on average, gun hunting is more efficient than *wana* hunting in both seasons, but that *wana* hunting is more frequently a better option than gun hunting in *Wantajarra*. These results do not support the hypothesis that men switch to *wana* hunting in *Yalijarra* because it is a more efficient hunting option overall.

Hypothesis 2

Here we consider the possibility that men and women search simultaneously for all possible prey all the time, and gun hunting and *wana* hunting are not characterized by mutually exclusive sets of prey. In this case, the contingent prey choice model predicts that reductions in encounter rates with higher-ranked prey items are likely to cause foragers to broaden the diet to include lower-ranked items. If Martu men are switching to goanna following reductions in encounter rates with bustard during *Yalijarra*, then we should expect search failures for bustard to be higher in *Yalijarra* than in *Wantajarra*.

Prey rankings in *Yalijarra* and *Wantajarra* are presented in Table 9.2. The highest-ranked prey item in both seasons is bustard, with skink ranked second. Goanna ranks fourth in both seasons. Search failures for bustard were significantly higher in *Yalijarra* than in *Wantajarra*: 61% of *Yalijarra* bustard searches and 20% of *Wantajarra* searches failed because no game was encountered (chi squared [χ^2] = 13.48, $P = 0.0002$). There were no significant seasonal differences in bustard pursuit failure: 33% of

Table 9.2 *Prey ranking*

Prey	Season	No. of pursuits	On-encounter return rate (kcal/h)	Rank	Overall returns (kcal/h)
Bustard	*Wantajarra* (cool dry)	88	22 392	1	Women, 733; men, 2125
Skink		26	20 131	2	
Cat		14	6295	3	
Goanna		475	5318	4	
Bush tomato		12	4217	5	
Small lizard		6	3248	6	
Bush raisin		19	3150	7	
Gum		1	1500	8	
Flowers		30	1437	9	
Snake		8	1074	10	
Grub		87	630	11	
Perenti		5	501	12	
Kangaroo		14	416	13	
Yam		8	382	14	
Bandicoot		2	0	15	
Bilby		1	0	16	
Echidna		1	0	17	
Rabbit		6	0	18	
Bustard	*Yalijarra* (hot wet)	32	24 796	1	Women, 944; men, 859
Skink		1	7098	2	
Bush tomato		9	4835	3	
Goanna		87	4647	4	
Perenti		18	3687	5	
Snake		8	1736	6	
Cat		4	1203	7	
Grub		11	745	8	
Corm		5	285	9	
Yam		1	276	10	
Kangaroo		1	0	11	

Yalijarra pursuits and 46% of *Wantajarra* pursuits failed ($\chi^2 = 0.464$, $P = 0.4956$). The chance of failure from both causes was also not significantly different by season: 77% of *Yalijarra* and 57% of *Wantajarra* hunts experienced either search or pursuit failure ($\chi^2 = 3.092$, $P = 0.0787$).

This suggests that men may be sensitive to encounter rates with bustard and widen their diet breadth as the prey choice model predicts. However, this would explain men's seasonal prey choice only if search failure in both seasons caused men to increase the time they spend *wana* hunting.

Hypothesis 3

If men are switching to *wana* hunting only as a result of reductions in encounter rates with bustard, then they should be just as likely to *wana* hunt following a search failure in *Wantajarra* as in *Yalijarra*.

To test this hypothesis, we examined men's time allocation to gun hunting on days that they also switched to *wana* hunting, excluding those hunts where men tried to combine both activities. There were significant seasonal differences in time that men spent *wana* hunting following failure to encounter bustard (Table 9.3). In both seasons, when men were successful in capturing bustard, they spent just over 40 minutes per day also hunting goanna. However, failure to encounter prey had very different effects on men's seasonal time allocation to goanna hunting. In *Yalijarra*, men spent significantly more time *wana* hunting after failing to encounter bustard (154 minutes/day) than when they failed to encounter bustard in *Wantajarra* (50 minutes/day: $DF = 26$, $t = 3.078$, $P = 0.0049$). This was also the case for pursuit failures, although the sample size (three failed pursuits in the wet season) is too small to draw meaningful statistical inferences. Only failure in *Yalijarra* caused men to spend more time *wana* hunting. This suggests that while men may be paying attention to seasonal changes in the encounter rate with higher-ranked animals, they choose to respond to this only in *Yalijarra*, when goanna are among many animals tracked on the surface. This may mean either that the benefits to men of hunting goanna in *Yalijarra* are higher, possibly due to a preference for tracking, or that there are fewer opportunity costs to *wana* hunting, since there is no burning of large areas of grassland that frighten larger game away.

Hypothesis 4

If *wana* hunting is preferred in *Yalijarra* because prey are tracked, why would this bias men toward preferring to hunt goanna? One possibility is that men are "variance-prone" foragers, preferring prey that offer greater variance. Previous work examining variance and hunting suggests that high-variance activities can hold the potential to discriminate skill among foragers if more skilled individuals are able to demonstrate lower variance around their mean returns or a lower failure rate than other individuals (Bliege Bird *et al.* 2001). Are goanna harvests associated with higher variance and higher failure rates in *Yalijarra*?

Table 9.3 *Minutes that men spent searching and pursuing goanna on days when bustard was:*

	Captured successfully				*Encountered, but pursuit failed*				*Not encountered, search failed*			
	Count	Mean	SD	SE	Count	Mean	SD	SE	Count	Mean	SD	SE
Yalijarra	6	44	34	14	3	153	151	87	16	154	102	26
Wantajarra	25	41	66	13	20	53	56	13	12	50	64	18
		Mean differential	t	P		Mean differential	t	P		Mean differential	t	P
		3.48	0.124	0.902		99.65	2.262	0.034		103.708	3.078	0.005

SD, standard deviation; SE, standard error.

Table 9.4 *Variance and hunting strategies*

Season	Mean harvest (kcal)	*N*	Variance	CV	SE
***Wana* hunts (mostly goanna)**					
Wantajarra	834	212	537 145	0.879	50.3
Yalijarra	1265	124	1 602 293	1.001	113.7
Gun hunts (mostly bustard)					
Wantajarra	2864	118	58 635 525	2.674	705
Yalijarra	2353	28	44 216 150	2.826	1257

CV, coefficient of variation; SE, standard error.

To test this hypothesis, we examine variability in the total harvest (in kilocalories) of goanna acquired through *wana* hunting in each season (Table 9.4). Both the average size of the harvest and the variance are higher by all measures in *Yalijarra*. For comparison, we also include the mean harvest size and variance associated with hunts solely for larger game.

However, *wana* hunt variance is still an order of magnitude lower than gun hunt variance in *Yalijarra*. Further complicating the issue is that pursuit success rates for goanna prey are higher in *Yalijarra*: both men and women have poorer success at capturing individual prey in *Wantajarra*. A multiple logisitic regression model (Table 9.5) of sex and season shows a strong effect of season: hot/wet season goanna + perenti pursuits are more than three times more likely to be successful than cool/dry season hunts. There is little effect of sex: both men and women have poor goanna pursuit success in *Wantajarra* and do much better in *Yalijarra* (Table 9.6). While *Yalijarra wana* hunt harvests are more variable than *Wantajarra* hunts, foragers arc actually more likely to have a successful pursuit. This does not suggest that men switch to *wana* hunting during *Yalijarra* because it offers higher pursuit failures than during *Wantajarra*.

Discussion and conclusions

Martu men do not respond to seasonal resource variation in the same way that women do. Men often pass over opportunities to obtain very high return rates from seasonal fruits and roots. They spend very little time hunting goanna unless they are unsuccessful in the search for larger game, but only during *Yalijarra*, when goanna are tracked. The inability to find bustard does not influence men to hunt like women during *Wantajarra*, when goanna are hunted with fire and dug from burrows.

Table 9.5 *Logistic model coefficients table for goanna pursuit outcome by sex and season*

	Coefficient	SE	Coefficient/SE	χ^2	P	R	Coefficient	95% lower CL	95% upper CL
Success: constant	–0.624	0.104	–5.977	35.721	<0.0001	–0.213	0.536	0.437	0.658
Season: *Yalijarra*	1.139	0.241	4.72	22.283	<0.0001	0.165	3.123	1.946	5.011
Sex: male	–0.318	0.237	–1.344	1.807	0.1789	0	0.728	0.458	1.157

CL, confidence limit; SE, standard error.

Table 9.6 *Goanna pursuit outcome by season*

	Wantajarra		*Yalijarra*				
	No. of pursuits	%	No. of pursuits	%	Total	Chi-squared test	
Women							
Fail	251	64	29	43	280	DF	1
Success	140	36	39	57	179	χ^2	11.305
Total	391		68		459	*P*	0.0008
Men							
Fail	64	76	5	26	69	DF	1
Success	20	24	14	74	34	χ^2	17.43
Total	84		19		103	*P*	<0.0001

DF, degrees of freedom.

We suggest two possible explanations for this. First, *wana* hunting during *Yalijarra* may not be entirely exclusive with continuing to track larger game, while *Wantajarra wana* hunts may actually preclude hunting larger game. *Wantajarra* season hunts involve burning large tracts of grassland, which drives away larger game, leaving only the smaller burrowed animals within a radius of about five 5 km. This facilitates tracking small game but increases the time that men must spend tracking larger game and likely increases the chance of losing the animal. During *Yalijarra*, men who *wana* hunt often encounter the fresh tracks of other animals as they track goanna and perenti. When they do, they often pursue the larger game (cat, kangaroo, bustard, or emu).

Second, men may prefer mobile game in all seasons because only tracking has the potential to discriminate skill levels among hunters. While Martu ascribe no political power to skilled trackers, they do make overt distinctions in skill for male and female hunters. A *miltilya* is a good hunter and can refer to either sex and any acquirer of any type of meat, large or small. As one Martu woman put it, "*Miltilya* women are those who always come back with much meat. A *miltilya* man can track and hunt meat better than other men." In 1927, Spencer and Gillen noted that the Aranda men and women in the Central desert

> know the track of every beast and bird ... In [tracking] the men vary greatly ... Whilst they can all follow tracks which would be indistinguishable to the average white man, there is a great difference in their ability to do so when they become obscure ... a really good one will unerringly follow them up on horse- or camel-back.

If women gain status through hunting, as men do, then are they gaining status the same way? According to Martu, they are not. Men gain *miltilya* notoriety through their ability to more consistently find animals tracked over long distances where the track often disappears; women gain status through their productive or provisioning capacity. This difference is visible in the way that foraging returns vary across individual men and women. Martu men's hunting is unpredictable according to time spent, but individual hunters have significantly different rates of failure; women's hunting is significantly more predictable with time, and there is a much weaker individual effect. Much of this may be due to the fact that women often hunt cooperatively, pooling their prey as they do so, while men rarely hunt cooperatively and share instead with non-hunting group members.

The general implications of our data are relevant to understanding the nature of human sex differences in foraging strategies. Seasonal variation in the behavior of prey and availability of alternative resources affect the basic nature of the sexual division of labor in ways that suggest some intriguing similarities between human and chimpanzee male and female hunting decisions. We have suggested previously that many foraging sex differences could be the result of sex differences in the signals of quality sent as a form of social competition (Bliege Bird *et al.* 2001; Smith *et al.* 2003). Women may compete to gain notoriety as consistent provisioners, while men may compete to demonstrate intrinsic hidden qualities related to gaining social benefits. Women may not compete as men do because the costs of doing so are high relative to their provisioning goals; men do not compete as women do because provisioning competes with their goal of demonstrating hidden qualities. Men may avoid hunting like women during *Wantajarra* because seasonal variation in hunting methods creates high costs for a strategy designed to gain maximum benefits from display of tracking skill. Our results offer only preliminary tests of this hypothesis; further work will focus directly on testing the proposition that Mardu men and women forage differently and gain status differently as a result of different reproductive goals.

Our data may also serve as a cautionary tale about the generality of primate responses to seasonality and the importance of treating groups as composed of individuals with differing foraging goals and reproductive tradeoffs. Generalizations that reduce behavior to some group average, or that fail to capture the details of environmental variability, trivialize the variability that begs explaining and obscure the complex relationship between seasonality and behavior. Inductive approaches that look merely for correlations between ecological and social variables demonstrate only how variable are both human and non-human primate responses to seasonality. An alternative way to approach seasonality is to take the variability as a source of data for which we can evaluate hypotheses derived from well-established theoretical propositions, as we have done here (see reviews in Winterhalder & Smith [2000]). Paying attention to such intragroup differences may better inform our understanding of how seasonality affects primate groups and allow us to design better models of the evolution of hominid behavior in seasonal environments.

References

Alberts, S. (1996). Mate guarding constrains foraging activity of male baboons. *Animal behavior*, **51**, 1269–77.

Altman, J. (1987). *Hunter Gatherers Today: An Aboriginal Economy in North Australia*. Canberra: AIATSIS.

Altmann, S. (1998). *Foraging for Survival*. Chicago: University of Chicago Press.

Bahuchet, S. (1988). Food supply uncertainty among the Aka Pygmies (Lbaye, Central African Republic). In *Coping with Uncertainty in Food Supply*, ed. I. de Gariene and G. A. Harrison. Oxford: Clarendon Press, pp. 118–49.

Bailey, R. & Aunger, R. J. (1989). Net hunters vs. archers: variation in women's subsistence strategies in the Ituri forest. *Human Ecology*, **17**, 273–97.

Binford, L. R. (1980). Willow smoke and dog's tails: hunter–gatherer settlement systems and archaeological site formation. *American Antiquity*, **45**, 1–17.

— (2001). *Constructing Frames of Reference: An Analytical Method for Archaeological Theory Building Using Hunter–Gatherer and Environmental Data Sets*. Berkeley: University of California Press.

Bird, D. W. & Bliege Bird, R. (2005). Mardu children's hunting strategies in the Western Desert, Australia: implications for the evolution of human life histories. In *Hunter–Gatherer Childhoods*, ed. B. Hewlett and M. Lamb. New York: Aldine de Gruyter, in press.

Bliege Bird, R., Bird, D. W., & Smith, E. A. (2001). The hunting handicap: costly signaling in human male foraging strategies. *Behavioral Ecology and Sociobiology*, **50**, 9–19.

Boesch, C. & Boesch H. (1984). Possible causes of sex differences in the use of natural hammers by wild chimpanzees. *Journal of Human Evolution*, **13**, 415–40.

Boinski, S. (1988). Sex differences in the foraging behavior of squirrel monkeys in a seasonal habitat. *Behavioral Ecology and Sociobiology*, **23**, 177–86.

Brand Miller, J., James, K., & Maggiore, P. (1993). *Tables of Composition of Australian Aboriginal Foods*. Canberra: Aboriginal Studies Press.

Cane, S. (1987). Australian aboriginal subsistence in the Western Desert. *Human Ecology*, **15**, 391–433.

Cashdan, E. (1985). Coping with risk: reciprocity among the Basarwa of Northern Botswana. *Man*, **20**, 454–74.

— (1992). Spatial organization and habitat use. In *Evolutionary Ecology and Human Behavior*, ed. E. A. Smith and B. Winterhalder. New York: Aldine de Gruyter, pp. 237–66.

Charnov, E. (1976). Optimal foraging: the marginal value theorem. *Theoretical Population Biology*, **9**, 129–36.

Grafen, A. (1990). Biological signals as handicaps. *Journal of Theoretical Biology*, **144**, 517–46.

Hawkes, K., O'Connell, J. F., & Blurton Jones, N. (1989). Hardworking Hadza grandmothers. In *Comparative Socioecology: The Behavioral Ecology of Humans and Other Mammals*, ed. V. Standen & R. A. Foley. London: Blackwell, pp. 341–66.

— (1995). Hadza children's foraging: juvenile dependency, social arrangements and mobility among hunter-gatherers. *Current Anthropology*, **36**, 688–700.

— (2001a). Hunting and nuclear families: some lessons from the Hadza about men's work. *Current Anthropology*, **42**, 681–709.

— (2001b). Hadza meat sharing. *Evolution and Human Behavior*, **22**, 113–42.

Hurtado, A. M. & Hill, K. (1990). Seasonality in a foraging society: variation in diet, work effort, fertility, and sexual division of labor among the Hiwi of Venezuela. *Journal of Anthropological Research*, **46**, 293–346.

Johnstone, R. A. (1997). The evolution of animal signals. In *Behavioural Ecology: An Evolutionary Approach*, ed. J. R. Krebs & N. B. Davies. Oxford: Blackwell, pp. 155–78.

Kelly, R. L. (1995). *The Foraging Spectrum: Diversity in Hunter–Gatherer Lifeways*. Washington, DC: Smithsonian Books.

Mather, J. R. (1962). *Average Climatic Water Balance Data of the Continents. Part II: Asia (Excluding USSR)*. Centerton, NJ: Thornthwaite Associates Laboratory of Climatology.

Mauss, M. & Beuchat. H, (1906). *Seasonal Variations of the Eskimo: A Study in Social Morphology*. London: Routledge.

McGrew, W. C. (1992). *Chimpanzee Material Culture: Implications for Human Evolution*. Cambridge: Cambridge University Press.

Mitani, J. & Watts, D. (2001). Why do chimpanzees hunt and share meat? *Animal Behaviour*, **61**, 915–24.

Nishida, T., Hasegawa, T., Hayaki, H., Takahata, Y., & Uehara, S. (1992). Meat sharing as a coalition strategy by an alpha male chimpanzee. In *Topics in Primatology*, Vol. 1, ed. T. Nishida, W. C. McGrew, P. Marler, M. Pickford & F. B. M. de Waal. Tokyo: University of Tokyo Press, pp. 159–74.

Smith, E. A. (1991). *Inujjuamiut Foraging Strategies: Evolutionary Ecology of an Arctic Hunting Economy*. New York: Aldine.

Smith, E. A., Bliege Bird, R., & Bird, D. W. (2003). The benefits of costly signaling: Meriam turtle hunters and spearfishers. *Behavioral Ecology*, **14**, 116–26.

Spencer, B. & Gillen, F. J. (1927). *The Arunta: A Study of a Stone Age People*. London: Macmillan.

Stanford, C. (1996). The hunting ecology of wild chimpanzees: implications for the evolutionary ecology of Pliocene hominids. *American Anthropologist*, **98**, 96–113.

Stanford, C., Wallis, J., Mpongo, E., & Goodall, J. (1994). Hunting decisions in wild chimpanzees. *Behavior*, **131**, 1–20.

Stephens, D. & Krebs, J. R. (1986). *Foraging Theory*. Princeton: Princeton University Press.

Taylor, W. (1964). Tethered nomadism and water territoriality: an hypothesis. *Proceedings of the 35th International Congress of Americanists*, **2**, 197–203.

Terborgh, J. (1983). *Five New World Primates: A Study in Comparative Ecology*. Princeton: Princeton University Press.

Thomas, D. H. (1983). On Steward's models of Shoshonean sociopolitical organization: a great bias in the Great Basin? In *The Development of Political Organization in North America*, ed. E. Tooker. Philadelphia: American Ethnological Society, pp. 59–68.

Thomson, D. (1936). *Donald Thomson in Arnhem Land*. Sydney: Gordon and Gotch.

Tonkinson, R. (1974). *The Jigalong Mob: Aboriginal Victors of the Desert Crusade*. Menlo Park, CA: Cummings.

— (1991). *The Mardu Aborigines: Living the Dream in Australia's Desert*. New York: Holt, Rhinehart and Winston.

Tutin, C. (1979). Mating patterns and reproductive strategies in a community of wild chimpanzees (*Pan troglodytes schweinfurthii*). *Behavioral Ecology and Sociobiology*, **6**, 29–38.

Walsh, F. (1990). An ecological study of traditional aboriginal use of "country": Martu in the Great and Little Sandy Deserts, Western Australia. *Proceedings of the Ecological Society of Australia*, **16**, 23–37.

Winterhalder, B. (1981). Foraging strategies in the boreal forest: an analysis of Cree hunting and gathering. In *Hunter Gatherer Foraging Strategies*, ed. B. Winterhalder & E. A. Smith. Chicago: University of Chicago Press, pp. 66–98.

Winterhalder, B. & Smith, E. A. (2000). Analyzing adaptive strategies: human behavioral ecology at twenty-five. *Evolutionary Anthropology*, **9**, 51–72.

Zahavi, A. (1975). Mate selection: a selection for handicap. *Journal of Theoretical Biology*, **53**, 205–14.

Part IV *Seasonality, reproduction, and social organization*

10 *Seasonality and reproductive function*

DIANE K. BROCKMAN
Department of Sociology and Anthropology, University of North Carolina at Charlotte, Charlotte NC 28223, USA

CAREL P. VAN SCHAIK
Anthropologisches Institut, University of Zurich, Winterhurerstrasse 190, CH-8057, Zurich, Switzerland

Introduction: types of seasonality

Primates, like other mammals, exhibit varying patterns of reproductive seasonality, spanning the continuum from sharply delineated seasonal periods of mating and births to absolute non-seasonality, where mating and births are distributed broadly throughout the year. Both more qualitative reviews (Lancaster & Lee 1965; Lindberg 1987; Whitten & Brockman 2001) and recent quantitative reviews (Di Bitetti & Janson 2000) (see also Chapter 11) show that seasonal birth distributions are the norm rather than the exception for primates, particularly for species residing at higher latitudes, where food resources undergo pronounced annual seasonal fluctuations (see Chapter 11). Seasonal variation in the frequency of births is also a fairly common phenomenon in human populations (see Chapter 13), although its adaptive significance may have been more pronounced in earlier hominins than in contemporary humans. However, while documentation of the temporal patterning of reproduction and its regulation in primates have advanced steadily over the past few decades, we are a long way from having the detailed interspecific comparisons needed to answer the ultimate question, "Why be seasonal, and if so, how?"

Answers to this question invariably center on how resources are used and are allocated in support of reproductive effort (Drent & Daan 1980; Stearns 1989, 1992) under differing environmental regimes. In seasonal environments, we expect that it is in a female's best interest to align the costliest portion of her reproductive cycle with seasonal food peaks, so that

Seasonality in Primates: Studies of Living and Extinct Human and Non-Human Primates, ed. Diane K. Brockman and Carel P. van Schaik. Published by Cambridge University Press.

she can acquire the essential resources to compensate for peaks in energy expenditure, thereby enhancing her overall fitness (Sadleir 1969). Because lactation is the energetically most costly portion of the reproductive cycle in mammals, births are timed to coincide with seasonal increases in food abundance (Bronson 1989). Previous studies of vertebrates (reviewed in Jönsson [1997]), including primates (van Schaik & van Noordwijk 1985; Koenig *et al.* 1997; Richard *et al.* 2000; Gould *et al.* 2003; Lewis & Kappeler 2005) show, however, that this pattern of reproductive timing is by no means universal, and that females employ widely divergent resource-acquisition and -allocation tactics in support of reproductive effort.

These divergent tactics have been modeled using two different conceptual dichotomies: capital versus income breeding (Drent & Daan 1980; Stearns 1989, 1992) or classic versus alternative reproductive timing (van Schaik & van Noordwijk 1985). The terms "capital breeding" and "income breeding" were initially developed to distinguish between the tactical rules that birds employ when deciding to increase clutch size during the breeding season (Drent & Daan 1980). A capital breeder increases clutch size contingent upon absolute levels of available capital (e.g. energy stores), an additional egg being laid only if the female has acquired a critical threshold level of condition, which declines as the breeding season progresses. Decisions to increase clutch size in income breeders, in contrast, rest upon the rate of accumulation of capital, such that a female adds another egg to her clutch if the rate of accumulation exceeds a critical threshold level, irrespective of laying date (Jönsson 1997).

Co-opted by life-history theorists (Stearns 1989, 1992), the terms took on a new meaning that focused on decisions concerning the whole reproductive cycle rather than the production of additional offspring. Thus, for Stearns (1992: 221, 222), a capital breeder is one that "uses stored energy for reproduction" whereas an income breeder "uses energy acquired during the reproductive period rather than stored energy for reproduction." Phrased in this way, the capital versus income breeding dichotomy has proved to be a useful paradigm for explaining *seasonal* breeding strategies in vertebrates, e.g. birds (Jönsson 1997) and seals (Boyd 2000), particularly the role played by seasonal fluctuations in resources in scheduling reproductive events. Although it is now well established, this terminology has less predictive power when it comes to explaining reproductive tactics among mammals whose reproductive cycle exceeds, sometimes by far, the annual season of increased food abundance. Most primates, of course, fall in this category: females can be pregnant for six months or more; then lactate for six months or more; and in the extreme case of the orangutan, the reproductive cycle spans seven to eight years (see Chapter 12). The ability

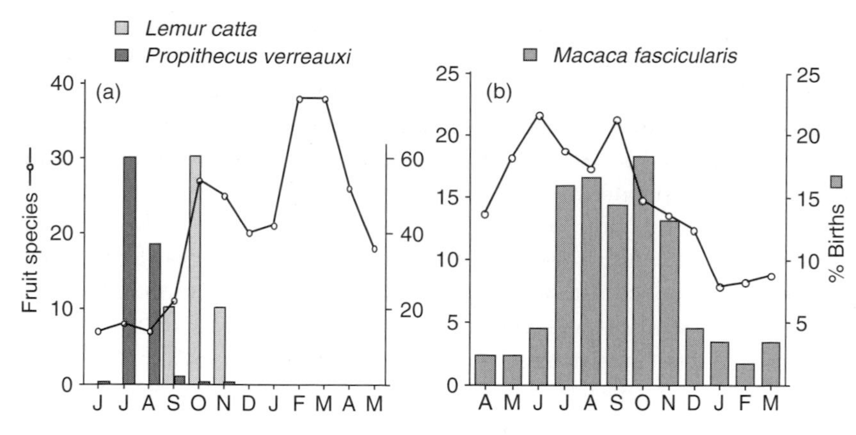

Figure 10.1 Scheduling of reproductive timing relative to food abundance wherein birth peaks precede or follow the annual food peak in *Lemur catta* and *Propithecus verreauxi*: (a) characteristic of income breeders and *Macaca fascicularis*; (b) characteristic of capital breeders.

of the income versus capital breeding model to accommodate such multi-seasonal reproductive patterns has, until now, been limited because it was not evident whether the model's predictions concerned conception, pregnancy, or lactation.

Another, independently derived approach, the classic versus alternative timing model (van Schaik & van Noordwijk 1985), has been particularly useful in explaining the interaction between the timing of reproduction and the environment in species with multiseasonal reproductive cycles. This model suggests that primates employ divergent reproductive tactics depending upon the degree of environmental seasonality and predictability of seasonal food peaks (Fig. 1.2). The classic pattern of reproductive timing occurs in seasonal environments when food peaks are totally predictable, allowing females to time the most expensive portion of their reproductive effort (midlactation) to coincide with, or immediately follow, seasonal peaks in food abundance. Hence, as illustrated in Fig. 10.1a, birth peaks precede or coincide with the annual food peak (e.g. strepsirrhines, including *Lemur catta* [Sauther 1998] and *Propithecus* [Richard *et al.* 2000, 2002], New World and Old World monkeys [van Schaik & van Noordwijk 1985]). However, (i) when the occurrence of annual food peaks is unpredictable, as in regions with mast-fruiting, or (ii) when infant dependency is longer than one year, females resort to an alternative reproductive strategy, wherein pregnancy is timed to peak food abundance and the birth peak follows annual food peaks (e.g. *Macaca fascicularis* [van Schaik & van Noordwijk 1985]) (Fig. 10.1b). We can further expect this type of response in environments with very limited seasonality, because in such cases the costs of deferring conception when out of synchrony with environmental seasonality are likely to exceed any benefits.

If we consider only the energetically most expensive part of the reproductive cycle (mid to late lactation; cf. Altmann [1980]), then the classic response corresponds to income breeding and the alternative response corresponds to capital breeding. We use the capital versus income terminology as defined above and further develop their predictive value for other aspects of reproductive function in seasonal and aseasonally breeding primates (see below).

In this chapter, we address the question, "How do capital and income breeders differ in their reproductive responses to environmental seasonality?" Answers to this question depend upon our ability to recognize the two tactics. For the benefit of readers with limited background in reproductive physiology, we provide a basic review of reproductive function in primates in Box 10.1. We then present and test, to the extent possible, predictions concerning the impact of the income–capital continuum on (i) environmental effects on reproductive physiology, (ii) maternal condition, and (iii) aspects of infant performance (Table 10.1). Results of our review show that the capital versus income model requires modification to accommodate the gradations of reproductive responses found in primates, the income– capital continuum model reflecting more accurately the complexity of responses found in this group. We also show the existence of numerous other differences between these various response types (Table 10.1).

Box 10.1 Reproductive function

Physiologically, reproductive function in mammals, including primates, is characterized by hormonal regulation of the various phases of the reproductive cycle, including ovulation, pregnancy, and lactation. During the follicular phase of the ovarian cycle, estrogens (E) produced by the ovaries increase gradually until they peak at mid cycle, initiating a preovulatory surge in luteinizing hormone (LH). Once ovulation has occurred, E and LH decrease while the transformation of the ruptured follicle into a corpus luteum stimulates the secretion of progesterone (P_4) during the second, or luteal, phase of the cycle (Dixson 1998). Periovulatory elevations in E at mid cycle are linked behaviorally to the female's initiation of sexual behavior (proceptivity) and her willingness to allow males to copulate (receptivity) (Dixson 1998). Male reproductive function also is regulated hormonally, principally through the actions of testosterone (T) and its metabolites, the latter being central to the activation and maintenance of male sexual behavior and the development of secondary sex characteristics (Dixson 1998). Seasonal variation in testes volume is androgen-dependent (Sade 1964; Dixson 1976; Bogart *et al.* 1977), breeding season increases in testes volume and T concentrations being indicative of seasonal variations in Leydig cell activity (Neaves 1973) and, thus, spermatogenesis (Graham 1981).

Table 10.1 *Predictions concerning (i) environmental effects on reproductive physiology/endocrine system, (ii) maternal condition and reproductive state relative to peak food availability, and (iii) aspects of infant performance in income and capital breeders*

Measure	Income breeders: predictions (external cuing)	Capital breeders: predictions (internal cuing)
(i) *Photoperiod*: responsiveness of females and males to photoperiod	Females: yes; entrain to natural light environment and abundant food; still seasonal in captivity	Females: no; entrain to food abundance; non-seasonal in captivity
	Males: yes or entrain to female reproductive state, or both (seasonal variation in T)	Males: no; entrain to female reproductive state (less seasonal variation in T)
(i) *Endocrine responses* to food abundance	Females/males: some endocrine response to food abundance (E & T elevations)	Females/males: strong endocrine response to food abundance (E & T elevations)
(ii) *Conception* and food abundance	No clear relationship; may coincide with food scarcity	Coincides with, or occurs soon after, increase in food abundance (or after internal condition threshold is passed)
(ii) *Fat accumulation during pregnancy* (estrus vs. postpartum states)	Little (although opportunistic increase in energy intake during pregnancy expected)	Substantial: before and during pregnancy whenever possible
(ii) *Abortion rates*: sensitivity to food scarcity during pregnancy	Lower: on average, low rates of resorption or abortion	Higher: on average, high rates of resorption or abortion
(iii) *Interannual variation in birth rate*	Lower: on average, low interannual variation in birthrates	Higher: on average, high interannual variation in birthrates
(iii) *Infant mortality rates*: sensitivity to food scarcity during lactation	Higher: depending on food abundance during lactation	Lower: even if food relatively scarce during lactation

E, estrogen; T, testosterone.

Recognizing income versus capital breeding

The different reproductive responses to seasonality are most clearly recognized in two ways: (i) physiologically, by determining whether in the absence of fertilization ovarian cyclicity is seasonal (income) or continuous (capital), and (ii) whether the timing of conceptions remains seasonal in captivity in natural light conditions (income) or not (capital). This requires morphological and hormonal information for males and females, including for females data on vaginal morphology/cytology and periovulatory elevations in E and P_4, and for males data on seasonal changes in testes mass and T variation. Hormonal data typically come from studies of captive populations, but the development of field endocrinology techniques using excreted steroid assays (reviewed in Whitten *et al.* [1998]) has improved our understanding of hormone–behavior relationships in wild populations.

The frequency of ovarian cyclicity is expected to be less in seasonal income breeders than in capital breeders. In the absence of conception, lactation, or senescence, income breeders will experience at most a few ovulatory cycles per year, while capital breeders in similar reproductive condition will exhibit monthly ovulatory cycles throughout the year. This difference derives from the fact that external cues (e.g. photoperiod) activate ovarian activity in income breeders, whereas endogenous factors (e.g. condition thresholds) govern the onset and acceleration of ovarian activity in capital breeders, day-to-day variation in energy intake, and expenditure, resulting in a more variable "stuttering" pattern of endocrine response in females.

Phrased in this way, differentiating between capital and income breeding appears to be fairly straightforward, but in practice we find that some income breeding species actually may be able to respond more like capital breeders, depending upon their specific life-history traits relative to condition thresholds (Fig. 10.2). Thus, strict income breeders, for whom deviating from the optimal timing of birth is very costly regardless of current condition, have a very narrow time-to-conception window, their probability of conceiving being dependent upon achieving a minimal condition threshold (Fig. 10.3). In relaxed income breeders on the other hand, females in good condition are able to afford deviations from the optimum birth timing; as a result, relaxed income breeders exhibit more variability in conception rate and timing. Relaxed income breeding probably evolved because the fitness costs (in terms of infant mortality) of deviating from the optimum of conception (and hence of birth and

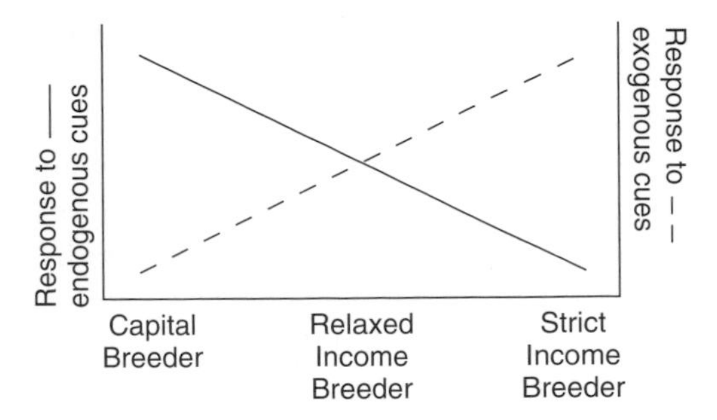

Figure 10.2 Ovarian responsiveness of primates to environmental cues. Exogenous (e.g. photoperiod) and endogenous (e.g. condition thresholds) cues activate ovarian activity in strict income and capital breeders, respectively, while a combination of cues activates ovarian responses in relaxed income breeders.

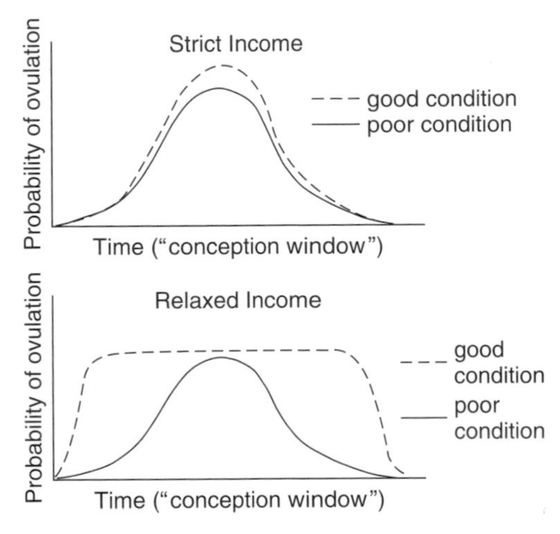

Figure 10.3 Comparisons of conception windows in strict and relaxed income breeders relative to condition thresholds. Note that conception windows broaden in relaxed income breeders, but not in strict income breeders, as they are in better condition.

lactation) could be compensated up to a point by being in superior physical condition.

Males of capital breeders are expected to respond directly to female reproductive state. However, especially in strict income breeders, in which mating seasons would be expected to be delineated sharply, males may respond to the same external cues that females use, so as to be ready to breed at the first sign of female ovarian activity. Hence, for males, we need data on seasonal changes in testes mass (e.g. spermatogenesis) and T variation.

As shown in Table 10.2 (see also, Chapter 11), we see two main classes of capital breeders, i.e. species that tend to cycle year round in the absence of fertilization and show a complete absence of seasonality, especially in captivity. The first group consists of large-bodied species that typically have long periods of infant dependency spanning multiple seasons. The second group contains species from Southeast Asia that are faced with extreme resource unpredictability, such that neither can afford to rely on exogenous cues alone to signal the onset of reproductive function.

Income breeders (Table 10.2), i.e. the species exhibiting fewer ovarian cycles per year, range from strictly seasonal periods of cyclicity both in the wild and in captivity (strict income breeders) to a more variable pattern when cyclicity becomes more flexible (relaxed income breeders) as females are able to maintain positive energy balance under captive conditions when food is non-limiting.

Information on seasonal variation in testes mass in capital breeders is conspicuous by its absence for reasons that are not entirely clear. One might speculate, however, that in light of the year-round cyclicity of capital breeding females (Table 10.2), males would be expected to maintain spermatogenesis throughout the year, as is the case for humans (Griffin 1988), in order to respond to female reproductive state. In contrast, male income breeders experience seasonal variation in testicular function similar to the seasonal ovarian activity observed in females. Our results show that, with few exceptions (e.g. *Brachyteles archnoides*, *Cercopithecus aethiops*), males experience testicular recrudescence just before (e.g. *Eulemur mongoz*, *Propithecus verreauxi*, *Macaca mulatta*) or coincident with (*Eulemur coronatus*, *E. fulvus*) ovarian cyclicity, indicating that male income breeders are responding to the same exogenous signals as females use to cue the onset of reproduction. *B. archnoides*, however, appears to show attributes of both income (females) and capital breeding (males). The fact that females exhibit a fairly tight time-to-conception window (e.g. 65% conceptions occur November–February) while males exhibit no seasonal variation in fecal testosterone (fT) concentration suggests that female reproduction is strongly condition-dependent, the onset of ovarian activity associated with wet season increases in food availability (Strier 1996). The lack of reproductive seasonality in males may be a bet-hedging strategy wherein individual males are continuously fertile, being able to impregnate any female achieving a minimal condition threshold for conception.

Table 10.2 *Recognizing income (I) and capital (C) breeding physiologically*

Species	Measure	Cycles/year (natural light); conceptive season in months	Variation in testes mass/T (relative to onset of ovarian cycles)?	Lessening of seasonality in captivity?[a]	Reference
Strict income breeders					
Lemur catta	Vaginal cytology; sT; mating	3; May (25 °S)	Yes; 3 months prior	Slight; 1 to 2[b] (32 °N)	Sauther (1991, 1998); Evans & Goy (1968); Bogart *et al.* (1977)
Varecia variegata	Vaginal morphology and uE/uP; sT; mating	2–3; May–July (15 °S)	?	No: 3 to 2 (32 °N)	Bogart *et al.* (1977); Shideler *et al.* (1983); Morland (1991)
Eulemur coronatus	Vaginal cytology; testes mass; conception–birth 125 days; mating	3; November–March (February–March conception 36 °N)	Yes; coincident (December)	No; 2 to 2 (36 °N) (conceptions)	Kappeler (1987); Mittermeier *et al.* (1994)
Saimiri sciureus	sE_2/sP_4; conception: births – 170 days gestation; testes biopsy	6.29 monthly cycles (March–June, 38 °N; July–September, 3.5 °S)	Yes; entrained to females	No; 3–3 to 4 (January–March, 25 °N)	Schiml *et al.* (1996, 1999); Baldwin (1968); DuMond & Hutchinson (1967)
Macaca fuscata	Menstruation and swelling; conception: births – 173 days gestation	2.6 (0–6); October–December (35 °N)	?	No; 2 to 2 (October–December, 28 °N)	Takahata (1980); Fedigan & Griffin (1996)
Relaxed income breeders					
E. fulvus myottensis (=fulvus)	Vaginal cytology and morphology; mating	4–5; April–June (18 °S)	Yes; coincident	Yes; 3 to 6 (36 °N)	Harrington (1975); Cranz *et al.* (1986)

Table 10.2 (*cont.*)

Species	Measure	Cycles/year (natural light); conceptive season in months	Variation in testes mass/T (relative to onset of ovarian cycles)?	Lessening of seasonality in captivity?[a]	Reference
E. f. rufus	fEt and f5-P-3OH; fT	?; May–June (20 °S)	Yes; coincident	Yes: 2 to 3 (November–January, 53 °N)	Ostner & Heistermann (2002); Ostner *et al.* (2002)
E. macaco	sE/sP_4; vaginal cytology sT	3: November–January (32 °N)	Yes; 1 month prior	?	Bogart *et al.* (1977)
E. mongoz	Vaginal cytology and sE_2/sP_4; testes mass	2–4: April–June (12 °S)	Yes; 1 month. prior	Yes: 3 to 4 (late December–March, 36 °N)	Tattersall (1976); Perry *et al.* (1992)
Propithecus verreauxi	FE_2/fP_4 (wild); conception: births – 164 days gestation (36 °N); fT	2; January–March (25 °S)	Yes; 1 month prior	Yes: 2 to 8 (36 °N)	Brockman & Whitten (1996, 2003); Brockman *et al.* (1998); Brockman (unpublished data)
Brachyteles arachnoides	fE/fP_4, conception: births – 7 months gestation; fT	3–6; November–February (20 °S)	No	?4 months: September, November, February, March (22.5 °S)	Strier & Ziegler (1997); Strier (1996); Strier *et al.* (1999); Pissinatti *et al.* (1994)
Macaca mulatta	fE_2/fP_4, menses, sexual swellings, sLH, FSH, E2; sT, testes mass	2–6; fall/winter	Yes; 1 month prior (sT) and coincident (testes mass)	Yes; year round; entrained to females	Walker *et al.* (1984); Gordon *et al.* (1976); Vandenbergh (1969)
M. sylvana	Conception: births – 165 days gestation	?; November-January (36 °N)	?	Yes; 3 to 5 (September–April, 39 °N)	Roberts (1978)
Cercopithecus aethiops	Perineum coloration; conception; PdG; testes biopsy	?; May–August (0 °N)	Year round (1 ° S)	Yes; 5 to year round (36 °N)	Whitten (1982); Eley *et al.* (1986)

entellus (Ramnagar, Napal)		(27 °N)		(32 ° N; 38 °N)	(1988)
Capital breeders					
Daubentonia madagascariensis	Vaginal morphology; conceptions: births – 157 days gestation	3; October–February (15 °S)	?	Yes; 5 to year round (36 °N)	Sterling (1994); Feistner & Taylor (1998); Brockman (unpublished data); Beattie *et al.* (1992)
M. nemestrina	Sexual swelling; mating	Year round (2 °–6°S)	?	–	Oi (1996)
M. fascicularis	Pregnancy; conception: births – 167 days gestation	Year round (32 °N); January–May (3 °N)	?	–	Kavanagh & Laursen (1984); van Schaik & van Noordwijk (1985)
M. arctoides	Conception: births – 162 days gestation	Year round (July–April, 1.5 °S)	?	–	Smith (1984)
Pan troglodytes	Anogenital swellings	Year round (4 °S)	–	–	Wallis (1995); Nishida *et al.* (1990)
Pan paniscus	Conception: births – 240 days gestation	Year round (0 °S)	–	–	Furuichi *et al.* (1998)
Gorilla g. beringei	Mating; conception: births – 8.5 months	10 + ; Year round (1 °S)	–	–	Watts (1991, 1998)
Pongo	uE1C and uT	Year round, but elevations during December–February Mast fruiting	Year round	–	Knott (1999)
Homo	sLH, FSH, E_2, P, T	Year round	Year round	–	Ojeda (1988); Griffin (1988)

[a] Numbers of months in which 80 + % of conceptions occurred.
[b] months.
E_2, estradiol; f5-p-3OH, fecal 5α-pregnane-3α-o1-20-one; fE, fecal estrogen; fE_2, fecal estradiol; fEt, fecal total estrogen; fP_4, fecal progesterone; FSH, follicle-stimulating hormone; fT, fecal testosterone; P, progesterone; PdG, pregnanediol glucuronide; sE, serum estrogen; sE_2, serum estradiol; sLH, serum luteinizing hormone; sP_4, serum progesterone; sT, serum testosterone; T, testosterone; uE, urinary estrogen; uE1C, urinary estrone conjugate; uE_t, urinary total estrogen; uP, urinary progesterone; uT, urinary testosterone.

Seasonality of reproductive function: predictions and evaluations

Environmental cues and endocrine physiology

Predictions

The environmental cues (e.g. photoperiod, rainfall, food abundance) driving these divergent reproductive tactics are operationally distinct. Income breeders are expected to rely on external cues such as photoperiod to time reproductive events (Colwell 1974). On the other hand, capital breeders would need to rely on internal cues (e.g. fat stores exceeding a threshold value or positive energy balance exceeding a given period; see Chapter 13) to time reproductive events. Capital breeders are therefore less dependent, if at all, upon variation in photoperiod.

We derive two predictions from the effects of environmental cues on endocrine aspects of physiology in income and capital breeders (Table 10.1). First, endocrine responses (E, P_4, T, testes mass) to photoperiod are expected to be stronger in income breeders than in capital breeders. Endocrine responses of female income breeders will be strongly entrained to seasonal variation in photoperiod, while those of males (T, testes mass) will be influenced either by similar photoperiodic cues or by the reproductive state of females (e.g. elevated E levels). Endocrine responses of female capital breeders will respond especially to changes in energy balance of the female, while those of males will be cued by the reproductive state of females if they are variable to begin with. Second, endocrine responses of income breeders are expected to be weakly entrained to seasonal variation in food abundance while those of capital breeders will be influenced strongly by seasonal peaks in food abundance.

Evaluation

The tightly delimited breeding and birth seasons seen in several strepsirrhine primates (income breeders; see Table 10.2) shows that, as predicted, exogenous cues (e.g. photoperiod) strongly regulate reproduction in members of this group, even in captivity (e.g. *Varecia variegata*, *Lemur catta*, *Eulemur coronatus*) (Fig. 10.4), indicative of strict income breeding. While this may indeed be the case for many species, there appear to be some that become substantially less seasonal in captivity when food is abundant (e.g. *Hapalemur griseus*, *Propithecus verreauxi*, *Daubenontia madagascariensis*) (Fig. 10.5), suggesting that regulation of reproduction involves the interaction of both exogenous and endogenous cues in relaxed income breeders. In the case of *Daubentonia*, the taxon with the slowest life history among

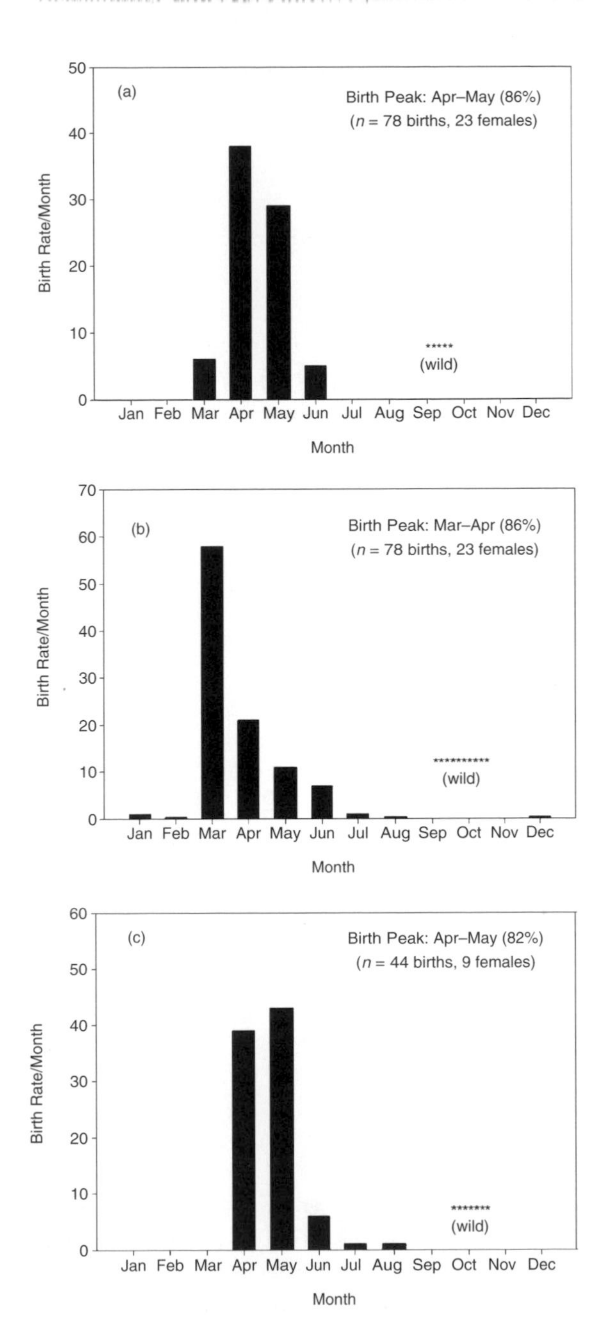

Figure 10.4 Timing of birth peaks in strict income breeders housed at the Duke University Primate Center: (a) *Varecia variegata*, (b) *Lemur catta*, and (c) *Eulemur coronatus*.

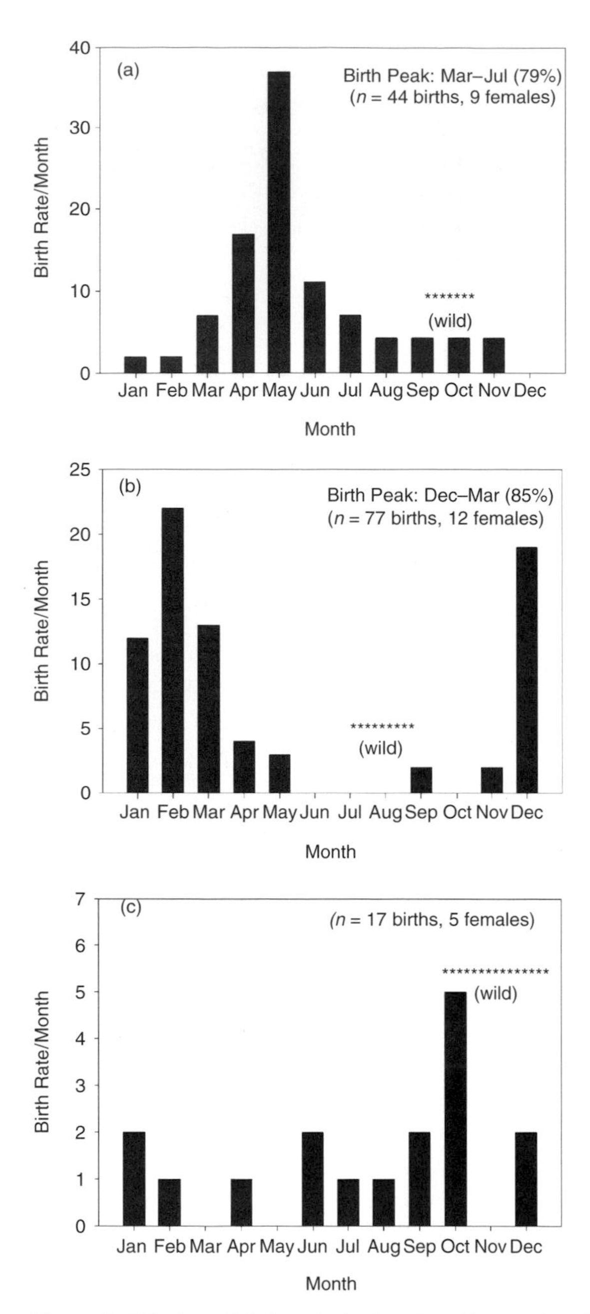

Figure 10.5 Timing of birth peaks in the relaxed income breeding (a) *Hapelemur griseus* and (b) *Propithecus v. coquereli*, and (c) the capital breeding *Daubentonia madagarcariensis* housed at the Duke University Primate Center.

extant lemurs, exogenous cues may not be involved at all in timing reproduction, indicative of capital breeding.

Some of the best evidence for a role of both external and internal cues to time reproductive events in income breeders derives from captive studies linking shifts in photoperiod to seasonal changes in the onset and termination of estrus and testes mass in strepsirrhine primates. Decreasing day length has been shown to be the prime trigger of reproduction in a number of lemurs (e.g. see Rasmussen [1985]), but endogenous and social factors are also known to impact the degree of endocrine synchrony within groups (reviewed in Izard [1990]), as expected for relaxed income breeders. The synergistic interaction of exogenous and endogenous cues regulating reproductive physiology is best illustrated in van Horn's (1975) study of captive ring-tailed lemurs (*Lemur catta*). Results showed that decreasing day length, which begins two months before the breeding season, reactivates ovarian function in quiescent females (and testicular function in males, as measured by testes volume [Vick & Periera 1989]), but the termination of ovarian activity depends upon both exogenous and endogenous factors, lengthening photoperiods and pregnancy both acting to inhibit the continuation of cycles (van Horn 1975).

In contrast, the onset of reproductive function in sifaka (*Propithecus verreauxi*) appears to be regulated by both exogenous (photoperiod) and endogenous cues (fat stores). Exogenous entrainment of reproduction in wild sifaka at Beza Mahafaly Special Reserve (BMSR) is most evident in the species' extremely sharp two-month birth peak (Fig. 10.1a) and significant breeding versus birth season variation in fecal estradial (fE_2) and T concentrations (Fig. 10.6a and b) (Brockman & Whitten 1996; Brockman *et al.* 1998). Fecal T concentrations peak in December, coincident with the austral summer solstice of decreasing day length and the wet season flush of young leaves, but one month prior to the early January–late February period of estrus-related estradiol (E_2) elevations (Fig. 10.6c). The onset of estrus-related E_2/progesterone (P_4) peaks, on the other hand, occurs three weeks following the summer solstice (Brockman & Whitten 1996), coincident with abundant young leaves and increasing fruiting cycles. This suggests that female responsiveness to exogenous cues may be dependent upon their ability to reach a minimum condition threshold to initiate ovulation, indicative of relaxed income breeding tactics. Support for this idea comes from studies of captive sifaka, showing that 85% of conceptions (extrapolated from birth dates and gestation of $164 + / - 10$ days [Brockman & Whitten 2003]) occur over a four-month period from mid June to mid September (36 ° N natural light), with the onset of initial cycles coincident with the summer solstice period of decreasing day length. The

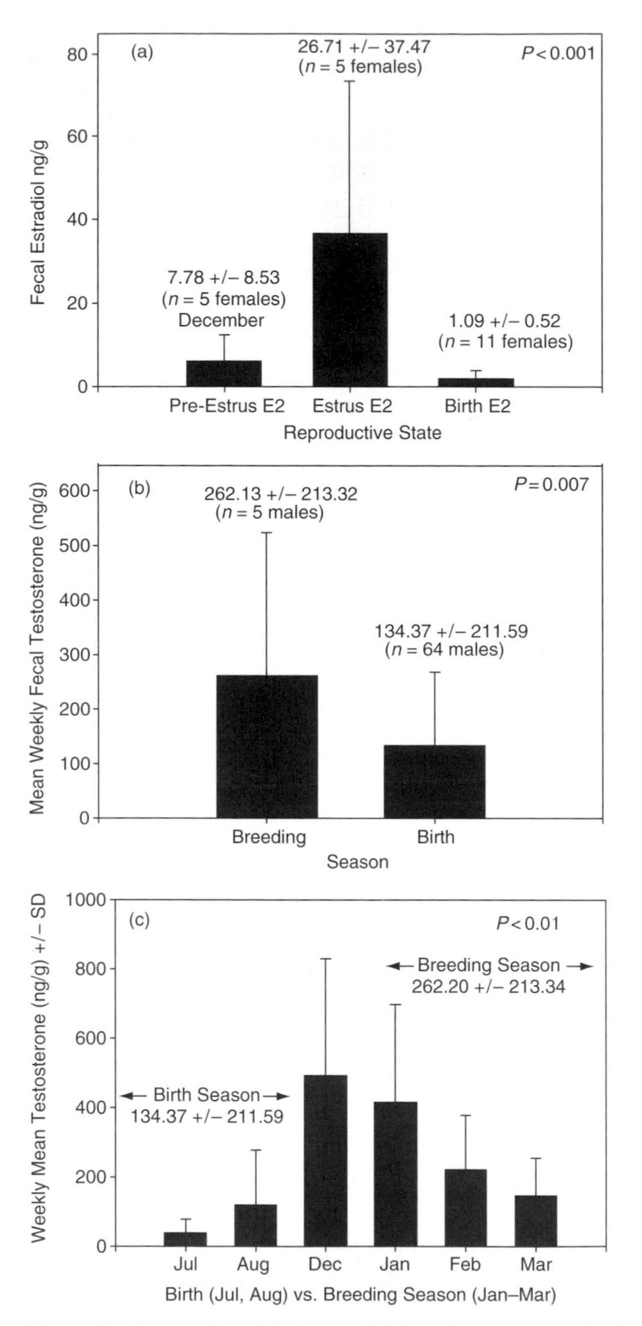

Figure 10.6 Seasonality of fecal estradiol (a) and testosterone (b, c) concentrations in *Propithecus verreauxi* at Beza Mahafaly, Madagascar.

remaining 15% conceptions occur in mid October, November, March, and May, most likely associated with pairings with new males (Brockman & Whitten 2003).

The impact of endogenous cues on female reproductive function is clearly evident in large-bodied capital breeders such as the orangutan. Orangutans live in seasonal but unpredictable environments of Southeast Asia, a region noted for its occasional periods of mast fruiting. Births in large-bodied primates are typically less seasonal than those of their small-bodied cousins (see Chapter11). The large body size of orangutans places them in a scheduling bind vis-à-vis the optimum time to give birth, a consequence of both a prolonged reproductive cycle (e.g. seven to eight years) and a long period of infant dependence. Knott's (2001) (see also Chapter 12) studies of the energetics of reproduction in wild orangutans show that ovarian function in non-pregnant orangutans is linked strongly with temporal variation in fruit availability: urinary estrone (E_1) concentrations in females with large infants increase when there is high fruit availability and decrease during periods of low fruit availability, with conceptions occurring during fruiting peaks (see also Fox [1998]).

Seasonal influences on conception, pregnancy, lactation, and infant mortality

Predictions

Species employing capital versus income breeding tactics are predicted to vary in their decision rules and sensitivity to food scarcity during different parts of the reproductive cycle, regardless of whether the cycle spans a single or multiple seasons (Table 10.1). Although the predictions consider capital and income breeding as discrete responses to seasonality, they refer to the extremes. In particular, relaxed income breeders are expected to be intermediate.

Predictions of the model include the following:

1. Condition thresholds for ovulatory cycles and, hence, conception are expected to be more critical among capital breeders than income breeders. Conception should have little or no relationship to food abundance in strict income breeders, whereas in capital breeders conception can occur only after an internal condition threshold is passed, often coincident with increased food abundance. As noted above, however,

relaxed income breeders should be intermediate: conception may be triggered by both exogenous (e.g. photoperiod) and endogenous (e.g. fat stores) cues, resulting in a more flexible set-point-to-conception window (Fig. 10.2). Tests of this prediction require information on either the temporal relationship between conception and food peaks and/or female condition (e.g. body weight) during conception.

2. Female income breeders, whose pregnancy may occur during periods of food scarcity, often will be unable to store major fat reserves in anticipation of lactation, being limited to utilizing those energy resources acquired during the mid-lactational period of food abundance. Capital breeders, on the other hand, are expected to use the period of food abundance that leads to pregnancy in the first place to store as many reserves as possible during pregnancy. This prediction calls for information on female condition, particularly fat accumulation or increased weight during pregnancy or, in the absence of these data, estrus versus postpartum changes in body weight.
3. Female sensitivity to food scarcity during pregnancy and, hence, likelihood of resorption or abortion of the fetus is expected to be less among income breeders than capital breeders. For income breeders, food scarcity during gestation is not a predictor of the ability to sustain the infant during lactation, whereas for a capital breeder, an inability to store sufficient reserves during gestation predicts difficulties during lactation, especially if food supply will be predictably low, as in mast fruiting areas (see Chapter 12). Hence, capital breeders will show a higher propensity to resorb or abort, whereas income breeders will experience lower rates of resorption or abortion. Tests of this prediction require data on rates of pregnancy loss (e.g. abortion, stillbirths), as evidenced by changes in paracallosal skin during pregnancy, an aborted conceptus, and/or hormonal evidence of spontaneous abortion.
4. Interannual variation in birth rates is expected to be higher in capital breeders than strict income breeders (at least, among those species that are able to complete their reproductive cycle within a year) because pregnancy in capital breeding females is much more contingent upon food availability than among strict income breeders. The latter would have similar birth rates across years and therefore will vary more in infant survival rates. Tests of this prediction ideally involve comparisons of species with different breeding responses coinhabiting the same geographic area.
5. Female sensitivity to food scarcity during lactation and its impact on infant mortality are expected to be higher in income breeders than capital breeders. Leaner income breeding females are expected to have

higher rates of infant mortality during periods of food scarcity because their infants will grow rapidly in order to be independent by the onset of the next lean season, thereby subjecting mothers to unacceptably high energetic costs during lactation. Female capital breeders typically give birth after the peak in food abundance and, relying less on food peaks, will have lower infant mortality rates even if food is relatively scarce during lactation. Tests of this prediction include information on infant mortality (or survivorship).

The data required for comprehensive tests of these predictions, especially while controlling for the possible confounding effects of life history and environmental seasonality, are not available for most species. Some are so rare that they will have to be collected specifically for such tests. In what follows, we have therefore compiled representative examples of the degree to which the relevant data illustrate income (e.g. strict and relaxed) versus capital breeding tactics.

Evaluation: conception and food abundance

As shown in Table 10.3, our predictions are largely upheld. Conception in the small-bodied strict income breeders (e.g. *Microcebus*, *Phaner*, *Leontopithecus*) appears to have little or no relationship to increased food abundance or body condition. In fact, female mouse lemurs (*Microcebus*) experience weight loss before the onset of ovarian activity (Wrogemann *et al.* 2001) as females are emerging from torpor. In contrast, the onset of ovarian activity in seasonal relaxed income (e.g. *Propithecus*, *Brachyteles*, *Macaca mulatta*) breeders and aseasonal capital breeders (e.g. *Papio anubis*, *M. fascicalaris*, *Pongo*) is linked strongly to condition thresholds, such that the decision to conceive is made based upon acquiring sufficient reserves to support pregnancy.

Field assessments of income versus capital breeding rely on the temporal relationship between estrus/conceptions and food peaks and on variation in female condition and/or body mass during the reproductive cycle, the latter being a more direct measure of the effect of condition on conception. Gould *et al.* (2003: 190) report that ring-tailed lemurs (*Lemur catta*) at BMSR are most likely income breeders rather than capital breeders because “ … females do not strongly rely upon resources such as fat stores during reproduction; rather they use maximum resources obtained from the environment when in the process of gestation and lactation.” *L. catta* at this site gestate during the dry season of fruit scarcity and lactate and wean infants during the wet season period of increasing fruit availability, thus “ … it is unlikely that they can easily store resources for

Table 10.3 *The effect of condition on ovarian activity and conception in income and capital breeders*

Species	Condition effect on conception?	Measure: body weight/condition, 1; conceptions at food peak, 2	Reference
Strict income breeders (SI)			
Microcebus murinus	No	1	Wrogemann *et al.* (2001); Perret & Aujard (2001)
M. rufus	No	1	Wrogemann *et al.* (2001)
Phaner fucifer	No	1	Schulke (2003)
Lemur catta	No	2	Gould *et al.* (2003)
Leontopithecus chrysomelas	No	2	De Vleeschouwer *et al.* (2003)
Cercopithecus aethiops	No	2	Lee (1987)
Relaxed income (RI) or capital (C) breeders			
Propithecus verreauxi	Yes (RI)	1	Richard *et al.* (2000)
Eulemur fulvus rufus	Yes (RI)	1	Overdorff *et al.* (1999)
Callithrix jacchus	Yes (RI)	1	Tardif & Jaquish (1997)
Saguinus oedipus	Yes (RI)	1	Kirkwood (1983)
Leontopithecus rosalia	Yes (RI)	1	Bales *et al.* (2001)
Aotus azarai	Yes (RI)	2	Fernandez-Duque *et al.* (2002)
Brachyteles archnoides	Yes (RI)	2	Strier (1996); Strier & Ziegler (1994)
Alouatta seniculus	Yes (RI)	2	Crockett & Rudran (1987)
Presbytis entellus (Ramnagar)	Yes (RI)	2	Koenig *et al.* (1997); Ziegler *et al.* (2000)
Erythrocebus patas	Yes (RI)	2	Chism *et al.* (1984)
Macaca f. fuscata	Yes (RI)	2	Takahashi (2002)
M. sinica	Yes (RI)	1	Dittus (1998)
M. mulatta (Cayo Santiago, Uttar Pradesh, India)	Yes (RI)	2	Rawlins & Kessler (1986); Southwick *et al.* (1965)
Papio anubis	Yes (C)	1	Bercovitch (1987)
Macaca fascicularis	Yes (C)	2	Van Schaik & van Noordwijk (1985)
Macaca silenus	Yes (C)	2	Kumar & Kurup (1985)
Papio cyocephalus	Yes (C)	2	Altmann (1980); Chapter 6 of this book
Theropithecus gelada	Yes (C)	2	Dunbar (1984)
Pongo	Yes (C)	2	Knott (2001); Chapter 12 of this book

Table 10.3 (*cont.*)

Species	Condition effect on conception?	Measure: body weight/ condition, 1; conceptions at food peak, 2	Reference
Gorilla gorilla	? (C)	2	Tutin (1994); Watts (1998)
Pan trolodytes	Yes (C)	2	Nishida *et al.* (1990); Wallis (1995)
P. paniscus	Yes (C)	2	Furuichi *et al.* (1998)
Homo	Yes (C)	1	Dufour & Sauther (2002); Chapter 13 of this book

reproduction" (Gould *et al.* 2003: 190). Data on ring-tailed lemur infant mortality in the first year of life in drought years (80%) versus non-drought years (52%) (Gould *et al.* 2003) further support the idea that ring-tailed lemurs are strict income breeders (see also Table 10.2).

Sifaka (*Propithecus verreauxi*) at the same site are reported to be capital breeders (Richard *et al.* 2000, 2002) based upon seasonal changes in body mass. The finding that heavier females of the previous mating season were more likely than lighter females to give birth the following birth season, coupled with increased wet-season female–female competition, leads Richard *et al.* (2000: 381) to conclude that " ... energy acquisition and storage are critically important in the life history strategies of female sifaka and that 'capital breeding' may be a feature of sifaka reproductive strategies." The authors acknowledge, however, that the form of capital breeding in sifaka is not as pronounced as that seen in other vertebrates (e.g. emperor penguins, some pinnipeds) (Richard *et al.* 2000); we would agree, arguing instead that sifaka are relaxed income breeders relying on both photoperiodic cues and condition thresholds to initiate ovarian activity. Evidence in support of this comes from studies of captive sifaka (Table 10.2.) showing that under non-food-limiting conditions, conceptions occur in eight months of the year, 85% of these in June–August (see also Fig. 10.5b) (Brockman, unpublished data).

Relaxed income breeding also characterizes some larger-bodied primates, including woolly spider monkeys (*Brachyteles archnoides*) (see Table 10.3). Female woolly spider monkeys experience three to six ovarian cycles before conception, the onset of copulatory activity coinciding with the early wet season of increasing food abundance (Strier 1996). Although conceptions occur in all but three months (July, September, May) of the year, 75% of them are concentrated in the November–February early and late rainy seasons, suggesting strong seasonality. Conceptions in captive populations are, as yet, too few to confirm the degree to which reproduction becomes even more relaxed when food is plentiful, but preliminary data from the Rio de Janeiro Primate Center (Table 10.2) (Pissinatti *et al.* 1994) suggest that captive *Brachyteles* are as seasonal as those in the wild.

In contrast, medium-bodied terrestrial species such as olive baboons (*Papio anubis*) exhibit the classic hallmarks of a capital breeder (Table 10.3); the reproductive cycle of olive baboons averages one year and nine months, including gestation and lactation periods of about 6 and 14 months, respectively (Rowe 1996). Bercovitch's (1987) studies of the relationship between body weight and reproductive condition (e.g cycling, pregnancy, lactation) in wild olive baboons at Gilgil, Kenya, shows that females experience energetic costs associated with weight loss during lactation and that the resumption of postlactational cycling is condition-dependent, females having to surpass a minimum weight threshold to resume ovarian activity.

Evaluation: fat accumulation during pregnancy

With few exceptions (*Propithecus*), data on pregnancy-related fat accumulation in income versus capital breeding primates are extremely rare. This paucity of data is due to field biologists often being reluctant to sedate and weigh/measure pregnant females or risk injuring newborns to obtain postpartum versus estrus body-weight estimates as a proxy for gestation-related fat accumulation. For the few data that exist, we find that, as predicted (Table 10.1), strict (e.g. *Cheirogaleus medius* [Müller 1999]; *Microcebus murinus* [Fietz 1998]; *M. rufus* [Atsalis 1999]) and relaxed income breeders (*Propithecus verreauxi* [Lewis, unpublished data]) experience little weight gain or fat accumulation during pregnancy. Results of Lewis's studies (unpublished data) of sifaka at Kirindy showed that while females had virtually identical body weights during the estrus and pregnancy periods (mean estrus: 3.54 kg +/ − 0.35 standard deviation [SD]; mean pregnancy: 3.62 kg +/ − 0.43; not significant [NS], $n = 5$ females), females actually lost 10% of their subcutaneous body fat during pregnancy (mean estrus: 4.0 mm +/− 0.79 SD; mean pregnancy: 3.0 mm +/− 1.0 SD; NS, $n = 5$ females), suggesting that pregnancy can be quite costly for sifaka. While there are virtually no data on fat accumulation across the reproductive cycle in capital breeders (reviewed in Dufour & Sauther [2002]), a few studies have reported on variation in maternal weight during the reproductive cycle. In this regard, we find that in contrast to income breeders, capital breeders exhibit significant weight gains during gestation, although a portion of this increase undoubtedly derives from the placenta and amniotic fluid. Captive aye-aye females experience a 6% increase in body weight during gestation (mean estrus-related weight: 2.61 kg +/−1.26 SD; mean pregnancy weight: 2.76 kg +/− 0.90 SD; $P = 0.005$, 10 pregnancies, 4 females) (Brockman & Krakauer, unpublished data), while human females experience pregnancy-related weight gains ranging from 14% (rural Gambians) to 20% (urban Scottish) above the pre-pregnant weight (Dufour & Sauther 2002). Although these results suggest that gestation may be less costly in capital breeders than in income breeders, it is not always the case in rural populations dependent upon subsistence agriculture. Gambian women suffer annual periods of nutrition stress as a consequence of seasonal food shortages, resulting in only a 0.4-kg gain in body fat during pregnancy (Dufour & Sauther 2002).

Evaluation: sensitivity to food scarcity during pregnancy and abortion

With the exception of *Papio cynocephalus* (Altmann & Alberts, unpublished data), direct evidence of pregnancy failure (e.g. hormonal evidence, aborted conceptus) is rarely if ever documented in wild populations because these events often occur early in gestation and are extremely difficult to observe. Studies of captive primates, however, yield substantially more information on prenatal mortality rates (Table 10.4), and show that our predictions are largely upheld, particularly in those species housed at the same facility (e.g. macaques, baboons). Rates of prenatal mortality are lower in income breeders than in capital breeders, and significantly so in captive relaxed income breeders versus capital breeders, regardless of geographic location (relaxed income mean: 19.6 +/− 4.5 SD; $n = 6$ taxa; capital mean: 40.4 +/− 21.9 SD; $n = 6$ taxa; $P = 0.04$) (Hendrickx & Nelson 1971). This difference is maintained in macaques housed at the same facility, including relaxed income (e.g. *Macaca mulatta*, *M. radiata*) and capital breeding species (e.g. *M. fascicularis*, *M. nemestrina*, *M. arctoides*) (Fig. 10.7a). An examination of Hendrie *et al.*'s (1996) larger macaque data set shows, however, that while *M. mulatta* (relaxed income breeders) and *M. fascicularis* (capital breeders) may have comparable prenatal mortality rates (17.0% versus 17.8%, respectively) and pregnancy trajectories (Fig. 10.7b), they differ substantially when one examines early pregnancy loss (e.g. gestational day [GD] 18–70): the capital breeder experiences higher rates of fetal loss (*c.* 13.5%) than the relaxed income breeder does (8.9%). Early and mid gestation are distinctly different time periods when disruptions to the establishment of pregnancy are most likely (GD 18–30) and when placental abruptions and abortion

Table 10.4 *Female sensitivity to food scarcity during pregnancy and the likelihood of abortion in income and capital breeders*

Species	Are females food-sensitive and likely to abort?	Measure: % prenatal mortality (captivity)	Reference
Strict income breeders			
Varecia variegata	No	7.0 (11/157)	Bernischke *et al.* (1981)
Relaxed income (RI) or capital breeders (C)[a]			
Saimiri sciureus	Yes (RI)	19.5 (36/149)	Cho *et al.* (1994)
Presbytis entellus	Yes (RI)	26.6 (4/15)	Hendrickx & Nelson (1971)
Cercopithecus aethiops	Yes (RI)	13.3 (6/45)	Hendrickx & Nelson (1971)
Cercocebus atys	Yes (RI)	19.6 (10/51)	Hendrickx & Nelson (1971)
Macaca mulatta	Yes (RI)	16.6 (840/5068)	Hendrie *et al.* (1996)
M. radiata	Yes (RI)	21.7 (44/203)	Hendrie *et al.* (1996)
M. fascicularis	Yes (C)	17.8 (81/455)	Hendrie *et al.* (1996)
M. nemestrina	Yes (C)	38.1 (61/160)	Hendrickx & Nelson (1971)
M. arctoides	Yes (C)	30.0 (20/70)	Hendrickx & Nelson (1971)
Papio spp.	Yes (C)	60.5 (26/43)	Kuehl *et al.* (1992)
Papio anubus	Yes (C)	23.0	Kuehl *et al.* (1992)
Papio cynocephalus	Yes (C)	10.0[b]	Altmann *et al.* (1988)
Homo	Yes (C)	73.0	Bolage (1990)

[a] $P = 0.04$ relaxed income versus capital breeders.
[b] Data from wild populations.

(GD 51–125) occur most often (Hendrie *et al.* 1996). The importance of these stages of gestation for capital breeding but not income breeding females resides in the fact that these females rely on energy stores acquired during pregnancy to support lactation. A female that finds herself without sufficient energy stores to continue the current reproductive effort is more likely to abandon the pregnancy during early to mid gestation (GD 71–100), thus avoiding risks to her own survival and her future reproductive effort.

Evaluation: interannual variation in birth rates

Evidence that interannual variation in birth rates (IBR) is higher in capital breeders than in income breeders derives from a comparative analysis of two income breeding species inhabiting Beza Mahafaly Special Reserve, Madagascar (*Lemur catta*, strict income; *Propithecus verreauxi*, relaxed

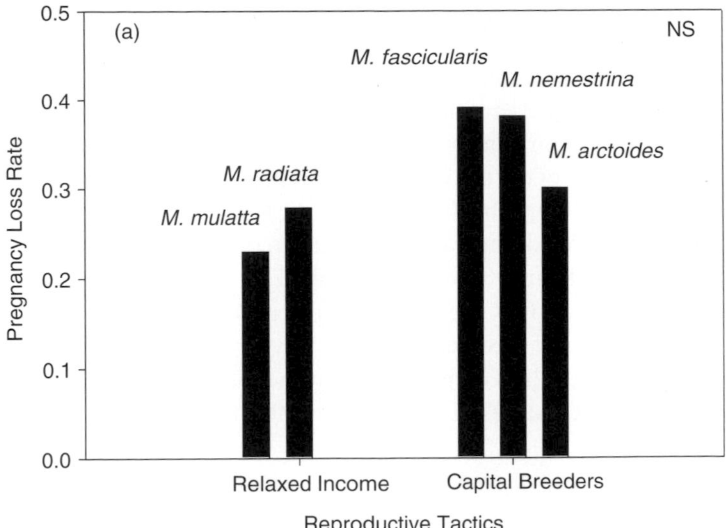

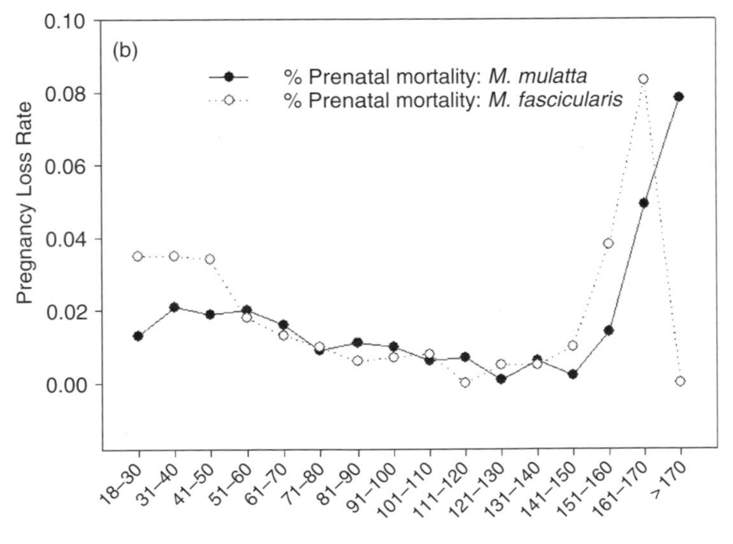

Figure 10.7 Comparisons of pregnancy loss rates in relaxed income breeding versus capital breeding macaques (Hendrickx & Nelson 1971) (a), and of the time course of fetal mortality (b) in a capital breeder (*Macaca fascicularis*) and a relaxed income breeder (*M. mulatta*) (Hendrie *et al.* 1996). Note that the capital breeders experience higher rates of early pregnancy loss relative to the relaxed income breeder.

income) versus a small capital breeder (*Macaca fascicularis*) from Ketambe Research Area, Gunung Leuser National Park, northern Sumatra. Strict income breeding females are expected to have more similar birth rates across years than relaxed income breeding or capital breeding females, whose birth rates vary with intra-annual variations in food abundance. Results of an analysis of difference in average IBR in good years (e.g. above-average birth rates) versus those in bad years (e.g. below-average birth rates) show that a gradation occurs in the direction predicted (as calculated from citations): ring-tailed lemurs (strict income) exhibit the least variation in IBR (i.e. 15%, $n = 6$ years) (Gould *et al.* 2003), followed by sifaka (relaxed income; 21%, $n = 14$ years) (Richard *et al.* 1991), while long-tailed macaques (capital) exhibit the highest IBR value (33%, $n = 12$ years) (van Noordwijk & van Schaik 1999) Fig. 10.8i).

Evaluation: sensitivity to food scarcity during lactation and infant mortality

Using the case studies discussed above, we examined variation in infant mortality rates (IMR) in strict income (*L. catta*), relaxed income (*P. verreauxi*), and capital breeding (*M. fascicularis*) females to assess the degree to which female sensitivity to food scarcity during lactation impacts interannual variation in infant mortality. We predicted that differences in average IMRs in good years (i.e. when food is abundant during gestation and lactation) versus bad years (i.e. when food is scarce during gestation and lactation) in a capital breeder such as *M. fascicularis* would be much less than those in strict and relaxed breeders such as *L. catta* and *P. verreauxi.* This difference exists because capital breeders are able to make adjustments earlier in the cycle and, in fact, do not even initiate cycling during periods of food scarcity, or they abort early when they lose condition. Results of an analysis of differences in average IMR in good years versus bad years show that a gradation occurs in the direction predicted (as calculated from citations): ring-tailed lemurs (strict income) exhibit the greatest variation in IMR (e.g. 47%; Gould *et al.* [1999]), followed by sifaka (relaxed income 29%; Richard *et al.* [1991, 2002]), while long-tailed macaques (capital) exhibit the least amount of variation in IMR (e.g. 3%; van Noordwijk & van Schaik [1999]) (Fig. 10.8ii).

Discussion

In this chapter, we have shown that a modified income–capital breeding model – the *income-capital continuum model* – has strong heuristic value for

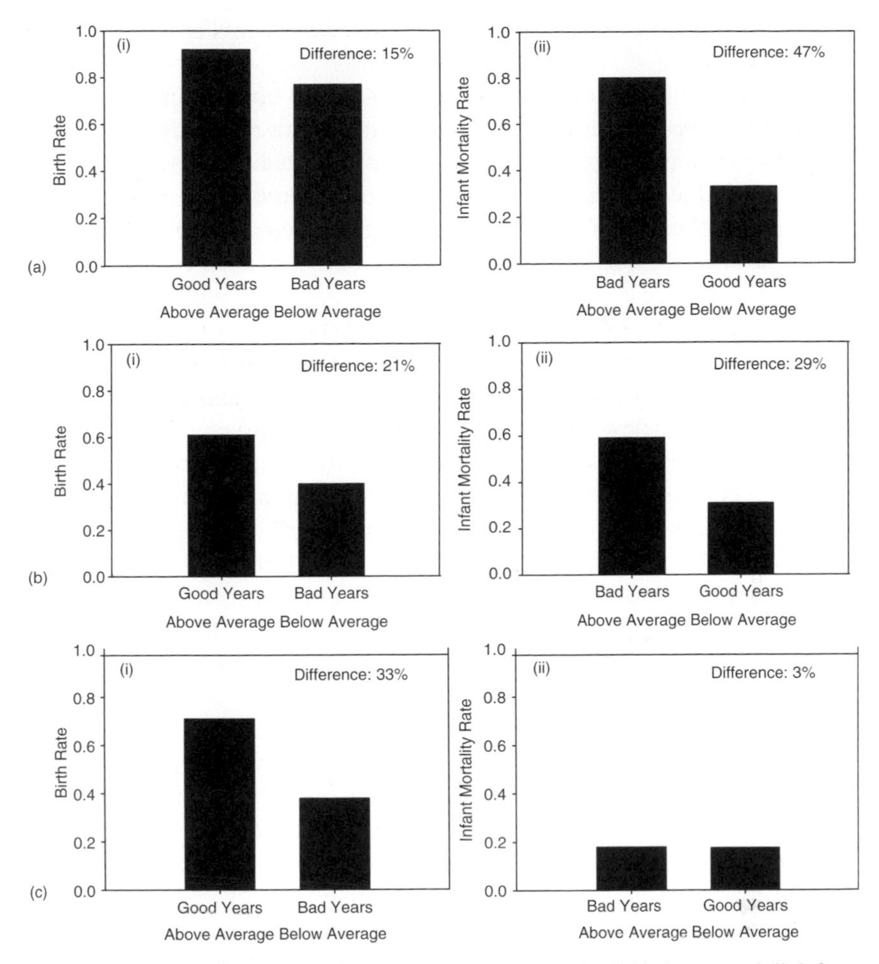

Figure 10.8 Comparisons of interannual variation in (i) birth rates and (ii) infant mortality rates in (a) a strict income breeder (*Lemur catta*), (b) a relaxed income breeder (*Propithecus verreauxi*), and (c) a capital breeder (*Macaca fascicularis*) breeder in good (e.g. high birth rates and food abundance) versus bad (e.g. low birth rates and food abundance) years.

predicting the nature of reproductive timing in primates, including the reproductive tactics of large-bodied aseasonal vertebrates (e.g. primates) whose reproductive cycle often far exceeds the annual season of increased food abundance. Its value is predicated upon the following two conditions: (i) that reproduction is defined clearly in terms of which component of the annual reproductive cycle is being emphasized (e.g. pregnancy, lactation), and (ii) that testable predictions are generated regarding the impact of the capital–income continuum on the respective components of the reproductive cycle.

This chapter indicates that the income–capital continuum has strong predictive power. It allows us to differentiate between the strict income breeding strepsirrhine primates and the capital breeding apes in all phases of the reproductive cycle; it also allows us to predict the reproductive responses of those intermediate species (e.g. relaxed income breeders) for whom both exogenous and endogenous factors appear to regulate reproductive function (e.g. sifaka).

As predicted by the income–capital continuum, the reproductive responses of strict income breeders reviewed here include (i) a narrow conception window that is cued exogenously (e.g. by photoperiod), even in captivity; (ii) little or no fat accumulation during pregnancy; (iii) lower prenatal mortality rates; and (iv) little or no interannual variation in birth rates; but (v) the greatest interannual variation in infant mortality rates. Capital breeders, on the other hand, (i) have a variable conception window that is cued endogenously through food abundance, condition thresholds, or energy balance; (ii) accumulate fat reserves during pregnancy; (iii) have higher prenatal mortality rates; (iv) have the highest interannual variation in birth rates; and (v) have the least variation in infant mortality rates.

Income breeders among primates vary substantially in the degree to which exogenous and endogenous cues regulate ovarian activity, with some species actually being able to respond more like capital breeders, depending upon their specific life-history traits relative to condition thresholds. The reproductive responses of relaxed income breeders (e.g. *Propithecus verreauxi*, *Brachyteles archnoides*) are differentiated from strict income breeders (e.g. *Lemur catta*, *Varecia variegata*) by: (i) an exogenously cued conception window that becomes broader in captivity when food is always abundant, a physiological consequence of ovarian activity being linked to both photoperiod and condition thresholds such that the decision to conceive is based ultimately upon acquiring sufficient reserves to support pregnancy; (ii) no fat accumulation, or the loss of subcutaneous fat, during pregnancy; (iii) low prenatal mortality rates; (iv) interannual variation in births rates that is intermediate between those of strict income and capital breeders; and (v) interannual variation in infant mortality rates that is intermediate between those of strict income breeders and capital breeders. Thus, relaxed income breeders appear to exhibit a mosaic of reproductive responses that are derived from both income breeding and capital breeding tactics. Responses of relaxed income breeders rely initially upon exogenous cues (probably mainly photoperiod) to reactivate reproductive function one to two months before the onset of the breeding season, similar to that seen in strict income breeders. However, in contrast to strict income breeders, it appears that relaxed income breeders may have an added "decision benefit" derived from information regarding condition thresholds. As a result, females are buffered against initiating ovarian activity when the energy stores are not sufficient to ensure a successful reproductive outcome; or, conversely, they are able to initiate reproduction when they are in excellent condition outside the regular mating season. Our only direct evidence in support of this idea comes from captive data on seasonal species in which the tight reproductive schedule seen in the wild becomes much more relaxed in captivity when food is abundant, albeit still seasonal (e.g. *Hapelemur*, *Propithecus*). These results suggest that the mechanisms regulating reproductive function in seasonal species interact in interesting and complex ways to yield the unique patterning observed here in primates.

We therefore propose that the current capital versus income model (Stearns 1992) be reconfigured to accommodate the gradation in reproductive responses observed in primates and that this modified version be called the *income–capital continuum model*.

In conclusion, we believe that this modified model offers a robust approach for answering the ultimate question posed earlier: "Why be seasonal, and if so, how?" We have provided a theoretical framework (and preliminary tests of that framework) for addressing the "how?" of reproductive seasonality in primates, ranging from strict income breeders (e.g. *Lemur catta*), to relaxed income breeders (e.g. *Propithecus verreauxi*), to capital breeders (e.g. apes and humans). We need many additional tests of our predictions for elucidating seasonality in reproductive function in other vertebrates, including a broader array of primates, before this new framework can be accepted fully, but the first round of testing has been very favorable to the new model.

Additional tests of our model will require information that is currently lacking for most primate species, including seasonality of births in the wild and in captivity, the relationship between condition thresholds and ovulation, estrus versus pregnancy fat accumulation, abortion rates, interannual variation in birth rates, and infant mortality. We need coordinated studies of both captive and wild populations to test this modified model more thoroughly. It is clear, nevertheless, that primate reproductive responses in the face of temporally varying food abundance are surprisingly diverse.

Acknowledgments

DKB's fieldwork has been made possible by support from a Grant-in-Aid of Research from Sigma Xi, grants from the Williams Fund of Yale University, the Conservation Fund of the American Society of Primatologists, Chicago Zoological Society, Jersey Wildlife Preservation Trust International, the Boise Fund of Oxford University, the Art and Sciences Research Council at Duke University, the Margot Marsh Biodiversity Foundation, and the National Science Foundation (SBR-3903531; BCS-9905985). CvS's fieldwork was made possible by long-term support from the Wildlife Conservation Society. We thank Rebecca Lewis, Brandie Littlefield, and Patricia L. Whitten for their helpful comments.

References

Altmann, J. (1980). *Baboon Mothers and Infants*. Cambridge: Harvard University Press.

Altmann, J., Hausfater, G, & Altmann, S. (1988). Determinants of reproductive success in savannah baboons (*Papio cynocephalus*). In *Reproductive Success*, ed. T. H. Clutton-Brock. Chicago: University of Chicago Press, pp. 403–18.

Atsalis, S. (1999). Seasonal fluctuations in body fat and activity levels in a rain-forest species of mouse lemur, *Microcebus rufus*. *International Journal of Primatology*, **20**, 883–910.

Baldwin, J. D. (1968). The social behavior of adult male squirrel monkeys (*Saimiri sciureus*). *Folia Primatologica*, **9**, 281–314.

Bales, K., O'Herron, M., Baker, A. J., & Dietz, J. M. (2001). Sources of variability in numbers of live births in wild golden lion tamarins (*Leontopithecus rosalia*). *American Journal of Primatology*, **54**, 211–21.

Beattie, J. C., Feistner, A. T. C., Adams, N. M. O., Barker, P., & Carroll, J. B. (1992). First captive breeding of the aye-aye, *Daubentonia madagascariensis*. *Dodo*, **28**, 23–30.

Bercovitch, F. B. (1987). Female weight and reproductive condition in a population of olive baboons (*Papio anubis*). *American Journal of Primatology*, **12**, 189–95.

Bernischke, K., Kumamoto, A. T., & Bogart, M. H. (1981). Congenital anomalies in *Lemur variegatus*. *Journal of Medical Primatology*, **10**, 38–45.

Bogart, M. H., Kumamoto, A. T., & Lasley, B. L. (1977). A comparison of the reproductive cycle of three species of lemur. *Folia Primatologica*, **28**, 134–43.

Bolage, C. E. (1990). Survival probability of human conceptions from fertilization to term. *International Journal of Fertility*, **35**, 75–94.

Boyd, I. L. (2000). State-dependent fertility in pinnipeds: contrasting capital and income breeders. *Functional Ecology*, **14**, 623–30.

Brockman, D. K. & Whitten, P. L. (1996). Reproduction in free-ranging *Propithecus verreauxi*: estrus and the relationship between multiple partner matings and fertilization. *American Journal of Physical Anthropology*, **100**, 57–69.
— (2003). Hormonal basis of reproductive competition in female *Propithecus v. coquereli*: mothers and daughters in conflict? *Lemur News*, **8**, 13–15.
Brockman, D.K., Whitten, P.W., Richard, A.F., & Schneider, A. (1998). Reproduction in free-ranging male *Propithecus verreauxi*: the hormonal correlates of mating and aggression. *American Journal of Physical Anthropology*, **105**, 137–51.
Bronson, F. H. (1989). *Mammalian Reproductive Biology*. Chicago: University of Chicago Press.
Chism, J., Rowell, T., & Olson, D. (1984). Life history patterns of female patas monkeys. In *Female Primates: Studies by Women in Primatology*, ed. M. Small. New York: Alan R. Liss, pp. 175–90.
Cho, F., Hamano, M., & Ohto, H. (1994). Breeding of squirrel monkeys (*Saimiri sciureus*) under indoor-caged conditions from 1981–1994. *Tsukuba Primate Center News*, **13**, 5–8.
Colwell, R. K. (1974). Predictability, constancy, and contingency of periodic phenomena. *Ecology*, **55**, 1148–53.
Cranz, C., Ishak, B., Brun, B., & Rumpler, Y. (1986). Study of the morphological and cytological parameters indicating oestrus in *Lemur fulvus myottensis*. *Zoo Biology*, **5**, 379–86.
Crockett, C. M., & Rudran, R. (1987). Red howler monkey birth data I. Seasonal variation. *American Journal of Primatology*, **13**, 347–68.
De Vleeschouwer, K., Leus, K., & van Elsacker, L. (2003). Characteristics of reproductive biology and proximate factors regulating seasonal breeding in captive golden-headed tamarins (*Leontopithecus chrysomelas*). *American Journal of Primatology*, **60**, 123–37.
Di Bitetti, M. S. & Janson, C. H. (2000). When will the stork arrive? Patterns of birth seasonality in neotropical primates. *American Journal of Primatology*, **50**, 109–30.
Dittus, W. P. J. (1998) Birth sex ratios in toque macaques and other mammals: integrating the effects of maternal condition and competition. *Behavioral Ecology and Sociobiology*, **44**, 149–60.
Dixson, A. F. (1976). Effects of testosterone on the sternal cutaneous glands and genitalia of the male greater galago (*Galago crassicaudatus crassicaudatus*). *Folia Primatologica*, **26**, 207–13.
— (1998) *Primate Sexuality: Comparative Studies of the Prosimians, Monkeys, Apes, and Human Beings*. New York: Oxford University Press.
Drent, R. H. & Daan, S. (1980). The prudent parent: energetic adjustments in avian breeding. In *The Integrated Study of Bird Populations*, ed. H. Klomp & J. W. Woldendrop. New York: North Holland, pp. 225–52.
Dufour, D. L. & Sauther, M. L. (2002). Comparative and evolutionary dimensions of the energetics of human pregnancy and lactation. *American Journal of Human Biology*, **14**, 584–602.
DuMond, F. V. & Hutchinson, T. C. (1967). Squirrel monkey reproduction: the fatted male phenomenon and seasonal spermatogenesis. *Science*, **158**, 1067–70.

Dunbar, R. I. M. (1984). *Reproductive Decisions: An Economic Analysis of Gelada Baboon Social Strategies*. Princeton: Princeton University Press.

Eley, R. M., Else, J. G., Gulamhusein, N., & Lequin, R. M. (1986). Reproduction in the vervet monkey (*Cercopithecus aethiops*): I. Testicular volume, testosterone, and seasonality. *American Journal of Primatology*, **10**, 229–35.

Evans, C. S. & Goy, R. W. (1968). Social behavior and reproductive cycles in captive ring-tailed lemurs (*Lemur catta*). *Journal of Zoology, London*, **156**, 187–97.

Fedigan, L. M. & Griffin, L. (1996). Determinants of reproductive seasonality in the Arashiyama west Japanese macaques. In *Evolution and Ecology of Macaque Societies*, ed. J. E. Fa & D. G. Lindburg. New York: Cambridge University Press, pp. 369–88.

Feistner, A. T. C. & Taylor, T. (1998). Sexual cycles and mating behaviour of captive aye-ayes, *Daubentonia madagascariensis*. *Folia Primatologica*, **69** (**Suppl. 1**), 409.

Fernandez-Duque, E., Rotundo, M., & Ramirez-Llorens, P. (2002). Environmental determinants of birth seasonality in night monkeys (*Aotus azarai*) of the Argentinean Chaco. *International Journal of Primatology*, **23**, 639–56.

Fietz, J. (1998). Body mass in wild *Microcebus murinus* over the dry season. *Folia Primatologica*, **69** (**Suppl**.), 183–90.

Fox, E. A. (1998). The function of female mate choice in the Sumatran orangutan, *Pongo pygameus abelii*. Ph. D. thesis, Duke University.

Furuichi, T., Idano, G., Ihobe, H., *et al.* (1998). Population dynamics of wild bonobos (*Pan paniscus*) at Wamba. *International Journal of Primatology*, **19**, 1029–43.

Gordon, T. P., Rose, R. M., & Bernstein, I. S. (1976). Seasonal rhythm of plasma testosterone levels in the rhesus monkey (*Macaca mulatta*): a three year study. *Hormones and Behavior*, **7**, 229–43.

Gould, L. G., Sussman, R. W., & Sauther, M. L. (1999). Natural disasters and primate populations: the effects of a 2-year drought on a naturally occurring population of ring-tailed lemurs (*Lemur catta*) in southwestern Madagascar. *International Journal of Primatology*, **20**, 69–84.

Gould, L. G., Sussman, R. W., & Sauther, M. L. (2003). Demographic and life-history patterns in a population of ring-tailed lemurs (*Lemur catta*) at Beza Mahafaly Reserve, Madagascar: a 15-year perspective. *American Journal of Physical Anthropology*, **120**, 182–94.

Graham, C. E. (1981). Endocrine control of spermatogenesis in primates. *American Journal of Primatology*, **1**, 157–65.

Griffin, J. E. (1988). Male reproductive function. In *Textbook of Endocrine Physiology*, ed. J. E. Griffin & S. R. Ojeda. New York: Oxford University Press, pp. 165–85.

Harley, D. (1988). Patterns of reproduction and mortality in two captive colonies of hanuman langur monkeys. *American Journal of Primatology*, **15**, 103–14.

Harrington, J. E. (1975). Field observations of social behavior of *Lemur fulvus fulvus* E. Geoffroy 1812. In *Lemur Biology*, ed. I. Tattersall & R. W. Sussman. New York: Plenum Press, pp. 259–79.

Hendrickx, A. G. & Nelson, V. G. (1971). Reproductive failure. In *Comparative Reproduction of Nonhuman Primates*, ed. E. S. E. Hafez. Springfield, Il: Charles C. Thomas, pp. 403–25.

Hendrie, T. A., Petterson, P. E., Short, J. J., *et al.* (1996). Frequency of prenatal loss in a macaque breeding colony. *American Journal of Primatology*, **40**, 41–53.

Izard, M. K. (1990). Social influences on the reproductive success and reproductive endocrinology of prosimian primates. In *Socioendocrinology of Primate Reproduction*, ed. T. E. Ziegler & F. B. Bercovitch. New York: Wiley-Liss, pp. 159–86.

Jönsson, K. I. (1997). Capital and income breeding as alternative tactics of resource use in reproduction. *Oikos*, **78**, 57–66.

Kappeler, P. M. (1987). Reproduction in the crowned lemur (*Lemur coronatus*) in captivity. *American Journal of Primatology*, **12**, 497–503.

Kavanagh, M. & Laursen, E. (1984). Breeding seasonality among long-tailed macaques, *Macaca fascicularis*, in Peninsular Malaysia. *International Journal of Primatology*, **5**, 17–29.

Kirkwood, J. K. (1983). Effects of diet on health, weight and litter size in captive cotton-top tamarins, *Saguinus o. oedipus*. *Primates*, **24**, 515–20.

Knott, C. D. (1999). Reproductive, physiological and behavioral responses of orangutans in Borneo to fluctuations in food availability. Ph. D. thesis, Harvard University.

— (2001). Female reproductive ecology of the apes. In *Reproductive Ecology and Human Evolution*, ed. P. T. Ellison. New York: Aldine de Gruyter, pp. 429–63.

Koenig, A., Borries, C., Chalise, M. K., & Winkler, P. (1997). Ecology, nutrition, and timing of reproductive events in an Asian primate, the hanuman langur (*Presbytis entellus*). *Journal of Zoology, London*, **243**, 215–35.

Kuehl, T. J., Kang, I. S., & Silver-Khodr, T. M. (1992). Pregnancy and early reproductive failure in the baboon. *American Journal of Primatology*, **28**, 41–8.

Kumar, A. & Kurup, G. U. (1985). Sexual behavior of the lion-tailed macaque, *Macaca sylenus*. In *The Lion-Tailed Macaques: Status and Conservation*, ed. P. G. Heltne. New York: Alan R. Liss, pp. 109–30.

Lancaster, J. B. & Lee, R. B. (1965). The annual reproductive cycle in monkeys and apes. In *Primate Behavior: Field Studies of Monkeys and Apes*, ed. I. DeVore. New York: Holt, Reinhart, and Winston, pp. 486–513.

Lee, P. C. (1987). Nutrition, fertility and maternal investment in primates. *Journal of Zoology, London*, **213**, 409–22.

Lewis R. J. & Kappeler, P. M. (2005). Seasonality, body condition and the timing of reproduction in *Propithecus verreauxi verreauxi* in the Kirindy Forest. *American Journal of Primatology*, 2005, in press.

Lindburg, D. G. (1987). Seasonality of reproduction in primates. *Comparative Primate Biology*, **2**, 167–218.

Mittermeier, R. A., Tattersall, I., Konstant, W. R., Meyers, D. M., & Mast, R. (1994). *Lemurs of Madagascar*. Washington, DC: Conservation International.

Morland, H. S. (1991). Social organization and ecology of black and white ruffed lemurs (*Varecia variegata*) in lowland rainforest, Nosy Mangabe, Madagascar. Ph. D. thesis, Yale University.

Müller, A. E. (1999). Aspects of social life in the fat-tailed dwarf lemur (*Cheirogaleus medius*): inferences from body weights and trapping data. *American Journal of Primatology*, **49**, 265–80.

Neaves, W. B. (1973). Changes in testicular Leydig cells and in plasma testosterone levels among seasonally breeding rock hyrax. *Biology of Reproduction*, **8**, 451–66.

Nishida, T., Takasaki, H., & Takahata, Y. (1990). Demography and reproductive profiles. In *The Chimpanzees of the Mahale Mountains: Sexual and Life History Strategies*, ed. T. Nishida. Tokyo: University of Tokyo Press, pp. 63–98.

Oi, T. (1996). Sexual behavior and mating system of the wild pig-tailed macaque in West Sumatra. In *Evolution and Ecology of Macaque Societies*, ed. J. E. Fa & D. G. Lindburg. New York: Cambridge University Press, pp. 342–68.

Ojeda, S. R. (1988). Female reproductive function. In *Textbook of Endocrine Physiology*, ed. J. E. Griffin & S. R. Ojeda. New York: Oxford University Press, pp. 129–64.

Ostner, J. & Heistermann, M. (2002). Endocrine characteristics of female reproductive status in wild redfronted lemurs (*Eulemur fulvus rufus*). *General and Comparative Endocrinology*, **131**, 274–83.

Ostner, J., Kappeler, P. M., and Heistermann, M. (2002). Seasonal variation and social correlates of androgen excretion in male redfronted lemurs (*Eulemur fulvus rufus*). *Behavioral Ecology and Sociobiology*, **52**, 485–95.

Overdorff, D. J., Merenlender, A. M., Talata, P., Telo, A., & Forward, Z. A. (1999). Life history of *Eulemur fulvus rufus* from 1988–1998 in southwestern Madagascar. *American Journal of Physical Anthropology*, **108**, 295–310.

Perret, M. & Aujard, F. (2001). Regulation by photoperiod of seasonal changes in body mass and reproductive function in gray mouse lemurs (*Microcebus murinus*): differential responses by sex. *International Journal of Primatology*, **22**, 5–24.

Perry, J. M., Izard, M. K., & Fail, P. A. (1992). Observation on reproduction, hormones, copulatory behavior and neonatal mortality in captive *Lemur mongoz* (Mongoose Lemur). *Zoo Biology*, **11**, 81–97.

Pissinatti, A., Coimbra-Filho, A. F., & dos Santos, J. L. (1994). Muriqui births at the Rio de Janeiro Primate Center. *Neotropical Primates*, **2**, 9–10.

Rasmussen, D. T. (1985). A comparative study of breeding seasonality and litter size in eleven taxa of captive lemurs (*Lemur* and *Varecia*). *International Journal of Primatology*, **6**, 501–17.

Rawlins, R. C. & Kessler, M. J. (1986). Demography of the free-ranging Cayo Santiago macaques. In *The Cayo Santiago Macaques: History, Behavior and Biology*, ed. R. G. Rawlins & M. J. Kessler. Albany: State University of New York Press, pp. 47–72.

Richard, A. F., Rakotomanga, P., & Schwartz, M. (1991). Demography of *Propithecus verreauxi* at Beza Mahafaly, Madagascar: sex ratio, survival and fertility, 1984–1988. *American Journal of Physical Anthropology*, **84**, 307–22.

Richard, A. F., Dewar, R. E., Schwartz, M., & Ratsirarson, J. (2000). Mass change, environmental variability and female fertility in wild *Propithecus verreauxi*. *Journal of Human Evolution*, **39**, 381–91.

— (2002). Life in the slow lane? Demography and life histories of male and female sifaka (*Propithecus verreauxi verreauxi*). *Journal of Zoology, London*, **256**, 421–36.

Roberts, M. S. (1978). The annual reproductive cycle of captive *Macaca sylvana*. *Folia Primatologica*, **29**, 229–35.

Rowe, N. (1996). *A Pictorial Guide to the Living Primates*. East Hampton, NY: Pogonias Press.

Sade, D. E. (1964). Seasonal cycle in size of testes of free-ranging *Macaca mulatta*. *Folia Primatologica*, **2**, 171–80.

Sadleir, R. M. F. S. (1969). *The Ecology of Reproduction in Wild and Domestic Mammals*. London: Methuen.

Sauther, M. L. (1991). Reproductive behavior of free-ranging *Lemur catta* at Beza Mahafaly Special Reserve, Madagascar. *American Journal of Physical Anthropology*, **84**, 463–77.

— (1998). Interplay of phenology and reproduction in ring-tailed lemurs: implications for ring-tailed lemur conservation. *Folia Primatologica*, **69 (Suppl. 1)**, 309–20.

Schiml, P. A., Mendoza, S. P., Saltzman, W., Lyons, D. M., & Mason, W. A. (1996). Seasonality in squirrel monkeys (*Saimiri sciureus*): social facilitation by females. *Physiology and Behavior*, **60**, 1105–13.

— (1999). Annual physiological changes in individually housed squirrel monkeys (*Saimiri sciureus*). *American Journal of Primatology*, **47**, 93–103.

Schulke, O. (2003). To breed or not to breed: food competition and other factors involved in female breeding decisions in the pair-living nocturnal fork-marked lemur (*Phaner furcifer*). *Behavioral Ecology and Sociobiology*, **55**, 11–21.

Shideler, S. E., Lindburg, D. G., & Lasley, B. L. (1983). Estrogen–behavior correlates in the reproductive physiology and behavior of the ruffed lemur (*Lemur variegatus*). *Hormones and Behavior*, **17**, 249–63.

Smith, (1984). Non-seasonal breeding patterns in stumptailed macaques (*Macaca arctoides*). *Primates*, **25**, 117–22.

Southwick, C. H., Beg, M. A., & Siddiqi M. R. (1965). Rhesus monkeys in North India. In *Primate Behavior: Field Studies of Monkeys and Apes*, ed. I. DeVore, New York: Holt, Rinehart and Winston, pp. 111–59.

Stearns, S. C. (1989). Trade-offs in life-history evolution. *Functional Ecology*, **3**, 259–68.

— (1992). *The Evolution of Life Histories*. New York: Oxford University Press.

Sterling, E. J. (1994). Evidence for nonseasonal reproduction in wild aye-ayes (*Daubentonia madagascariensis*). *Folia Primatologica*, **62**, 46–53.

Strier, K. B. (1996). Reproductive ecology of female muriquis (*Brachyteles arachnoides*). In *Adaptive Radiations of Neotropical Primates*, ed. M. A. Norconk, A. L. Rosenberger, & P. A. Garber. New York: Plenum Press, pp. 511–32.

Strier, K. B. & Ziegler, T. E. (1994). Insights into ovarian function in wild muriqui monkeys (*Brachyteles arachnoides*). *American Journal of Primatology*, **32**, 31–40.

— (1997). Behavioral and endocrine characteristics of the reproductive cycle in wild muriqui monkeys, *Brachyteles arachnoides*. *American Journal of Primatology*, **42**, 299–310.

Strier, K. B., Ziegeler, T. E., & Wittwer, D. J. (1999). Seasonal and social correlates of fecal testosterone and cortisol levels in wild male muriquis (*Brachyteles archnoides*). *Hormones and Behavior*, **35**, 125–34.

Takahashi, H. (2002). Female reproductive parameters and fruit availability: factors determining onset of estrus in Japanese macaques. *American Journal of Primatology*, **51**, 141–53.

Takahata, Y. (1980). The reproductive biology of a free-ranging troop of Japanese monkeys. *Primates*, **21**, 303–29.

Tardif, S. D. & Jaquish, C. E. (1997). Number of ovulations in the marmoset monkey (*Callithrix jacchus*): relation to body weight, age and repeatability. *American Journal of Primatology*, **42**, 323–9.

Tattersall, I. (1976). Group structure and activity rhythm in *Lemur mongoz* (Primates, Lemuriformes) at Ampijoroa, Madagascar. *Anthropological Papers of the American Museum of Natural History, NY*, **52**, 193–216.

Tutin, C. E. G. (1994). Reproductive success story: variability among chimpanzees and comparisons with gorillas. In *Chimpanzee Cultures*, ed. R. W. Wrangham, W. C. McGrew, F. B. M. de Waal, & P. G. Heltne. Cambridge: Harvard University Press, pp. 181–93.

Vandenberg, J. (1969). Endocrine coordination in monkeys: male sexual responses to the female. *Physiology and Behavior*, **4**, 261–4.

Van Horn, R. N. (1975). Primate breeding season: photoperiodic regulation in captive *Lemur catta*. *Folia Primatologica*, **24**, 203–20.

Van Noordwijk, M. A. & van Schaik, C. P. (1999). The effects of dominance rank and group size on female lifetime reproductive success in wild long-tailed macaques, *Macaca fascicularis*. *Primates*, **40**, 105–30.

Van Schaik, C. P. & van Noordwijk, M. A. (1985). Interannual variability in fruit abundance and the reproductive seasonality in Sumatran long-tailed macaques (*Macaca fascicularis*). *Journal of Zoology, London*, **206**, 533–49.

Vick, L. G. & Periera, M. E. (1989). Episodic targeting aggression and the history of lemur social groups. *Behavioral Ecology and Sociobiology*, **25**, 3–12.

Walker, M. L., Wilson, M. E., & Gordon, T. P. (1984). Endocrine control of the seasonal occurrence of ovulation in rhesus monkeys housed outdoors. *Endocrinology*, **114**, 1074–81.

Wallis, J. (1995). Seasonal influence on reproduction in chimpanzees of Gombe National Park. *International Journal of Primatology*, **16**, 435–51.

Watts, D. P. (1991). Mountain gorilla reproduction and mating behavior. *American Journal of Primatology*, **24**, 211–25.

— (1998). Seasonality in the ecology and life histories of mountain gorillas (*Gorilla gorilla beringei*). *International Journal of Primatology*, **19**, 929–48.

Whitten, P. W. (1982). Female reproductive strategies among vervet monkeys. Ph. D. thesis, Harvard University.

Whitten, P. W. & Brockman, D. K. (2001). Strepsirrhine reproductive ecology. In *Reproductive Ecology and Human Evolution*, ed. P. T. Ellison. New York: Aldine de Gruyter, pp. 321–50.

Whitten, P. W., Brockman, D. K., & Stavisky, R. C. (1998). Recent advances in non-invasive techniques to monitor hormone–behavior interactions. *Yearbook of Physical Anthropology*, **41**, 1–23.

Wrogemann, D., Radespiel, U., & Zimmerman, E. (2001). Comparison of reproductive characteristics and changes in body weight between captive populations of rufus and gray mouse lemurs. *International Journal of Primatology*, **22**, 91–108.

Ziegler, T., Hodges, K., Winkler, P., & Heistermann, M. (2000). Hormonal correlates of reproductive seasonality in wild female hanuman langurs (*Presbytis entellus*). *American Journal of Primatology*, **51**, 119–34.

11 *Seasonality of primate births in relation to climate*

CHARLES JANSON
Department of Ecology and Evolution, State University of New York, Stony Brook NY 11794–5245, USA

JENNIFER VERDOLIN
Department of Ecology and Evolution, State University of New York, Stony Brook NY 11794–5245, USA

Introduction

It is well established that most primate populations show at least some seasonality in the frequency of births (Lancaster & Lee 1965; Lindburg 1987; Di Bitetti & Janson 2000). There are two major questions concerning this seasonality: (i) what determines the narrowness of the peak (if there is one)?, and (ii) what determines when the peak occurs? Because these two questions deal with distinct kinds of hypotheses and data analyses, we will treat them separately. In this chapter, we attempt to cover the background, current hypotheses, and observed patterns for each topic by analyzing quantitative birth data on 70 wild populations of primates; there are many more data sets on captive primates, but we do not deal with those here. To streamline the presentation, the details of how we acquired the data set and performed statistical analyses are given in Appendix 11.1. We conclude with a brief summary of the results and implications for primate responses to seasonal variation in general.

Before discussing adaptive hypotheses for birth seasonality, it is important to distinguish seasonality from synchrony. The patterns and causes of these two phenomena are distinct. High seasonality in births necessarily will produce a high level of synchrony, but the reverse is not true. Births can be synchronized within a group but less synchronized between groups. Factors that favor or cause within-group synchrony of births, such as predation on infants (Boinski 1987) and infanticide (Butynski 1982), may show little seasonality and thus may not select for high population-wide seasonality of births (Di Bitetti & Janson 2000). Therefore, these selective factors will not be discussed further in this chapter. Because it is

Seasonality in Primates: Studies of Living and Extinct Human and Non-Human Primates, ed. Diane K. Brockman and Carel P. van Schaik. Published by Cambridge University Press.

uncommon for authors currently to report the contributions of each group to the total data set of birth dates, we have not tried to separate the effects of within-group synchrony from those of seasonality in this analysis. When sample sizes are small or limited to a single study group, high synchrony could be confounded with high seasonality.

Narrowness of the birth peak

Early discussions of birth seasonality were dominated by the questions of whether particular species demonstrated birth seasons or birth peaks (Lancaster & Lee 1965). Birth seasons were defined as strict periods within which births occurred and outside of which they did not. Birth peaks were defined as a tendency to cluster births during some months of the year, even if births occurred in all months. In part because of the erratic quality of the data available, most reviews had to be based on qualitative statements about the degree of seasonality. For instance, until 1999 no frequencies of observed births per month were published for Malagasy primates, so that conclusions about the apparent extreme seasonality of births in Malagasy primates had to rely on semi-quantitative statements such as "all births occur during June and July." The increasing availability of complete data on numbers of births per month in many primate species has allowed researchers to test whether their populations show significant departures from randomness in births across seasons.

Despite the rough scale of the data available, early reviews noted the clear pattern that birth seasons were more likely to occur in primates that live in the temperate zone than in tropical species (Lancaster & Lee 1965). This pattern suggested the important influence that seasonality of resource production has on primate birth seasonality. However, more refined analysis of this hypothesis has had to wait until recently (Di Bitetti & Janson 2000; Sinclair *et al.* 2000), perhaps for lack of suitable indices for measuring seasonality. Indeed, the apparently simple problem of testing for "significant" seasonality hides a subtle level of statistical complexity.

Measures of birth seasonality

Many studies argue for the presence of seasonality in births by testing for uneven frequencies of births across months. However, the non-parametric tests typically used on birth data by primatologists are not designed specifically for testing seasonality. Any factor that produces strongly synchronized births, such as bouts of infanticide, can produce marked peaks in birth activity, which may differ significantly from an even distribution of births per month, even if the births are not seasonal (e.g. Rudran [1973]: Horton Plains area). In addition, the statistics commonly used in previous studies of birth seasonality, such as the chi-squared or G statistic applied to monthly birth frequencies (e.g. Altmann [1980]), are very inefficient because they ignore the underlying continuity of the time variable. Thus, subtle seasonal peaks may exist but may fail to be detected in these categorical tests.

The problem of time continuity is solved with tests such as the Kolmogorov–Smirnov test, in which the categories are ordered or arrayed in an unambiguous way and thus yield a more powerful test (Siegel 1956). Several publications on birth seasonality have used this test to compare the cumulative distribution of births over a year with that expected from a uniform distribution (e.g. Sinclair *et al.* [2000]). However, there remains an added problem – although time is continuous, seasons do not graph well on to a simple linear axis, because they repeat themselves every year. Although data typically are graphed according to the calendar year, from January to December, it is quite possible that the frequencies for January are more similar to those for December than to those in March. A Kolmogorov–Smirnov test that analyzes data starting with the month of January and ending with December will generally yield different results than one that starts in July and goes through to June. There is no simple objective way to decide where to "break" the cycle of months to provide a proper test of the hypothesis of uniformity.

Luckily, both the problem of time continuity and that of recurrent seasonality can be solved with the use of circular statistics (Batschelet 1981). A brief introduction to circular statistics is given in Appendix 11.2. The use of circular statistics can produce different results from those of other tests. For instance, Bercovitch and Harding (1993) claim that births of baboons at Gilgil, Kenya, are not seasonal, based on a chi-squared statistic testing homogeneity across 12 months treated as categories. However, a reanalysis of their data using circular statistics shows modest seasonality that differs significantly from a uniform distribution ($P < 0.02$) (Table 11.1). The same data tested with a Kolmogorov–Smirnov test would have been marginally different from a uniform distribution ($0.04 < P < 0.05$) if the axis started with January, but would not have been ($P > 0.20$) if the year had be taken to start in April. We employ circular statistics throughout this paper when testing for uniformity of births across time.

One parameter used in circular statistics, the mean vector length (r), is well-suited to use as an index of seasonality. The parameter r is a measure of how unevenly observations are spread across a cycle. In the case of

Table 11.1 *Hypotheses for the width and relative timing of birth peaks in primates*

Selective factor	Justification/predictions	References
Hypotheses concerning the width of the birth peak		
Food production	Energy limits adult or infant survival	Altmann (1980); Lindburg (1987); Goldizen *et al.* (1988)
Diet	Fruit is more seasonal than leaves	Strier *et al.* (2001)
Body mass	Affects fat storage and life history	Bronson (1989)
Geography	Continental differences in climate	Van Schaik & van Noordwijk (1985); Wright (1999)
Hypotheses concerning the timing of the birth peak relative to peak food availability		
Weanling survival: income-I breeders	Weaning occurs at peak food	Altmann (1980); Crockett & Rudran (1987)
Maternal survival: income-II breeders	Birth occurs at peak food	Boinski (1987); Goldizen *et al.* (1988)
Maternal condition: capital breeders	Birth occurs at end of peak food	Van Schaik & van Noordwijk (1985); Bercovitch & Harding (1993); Richard *et al.* (2000); Strier *et al.* (2001)

births, r is 0 when the frequencies are spread perfectly evenly across months, whereas r is 1.0 when all births occur at exactly the same time. Although the mean vector length depends slightly on sample size and more so on the degree of precision of the recorded birth data (see Appendix 11.2), it is still relatively comparable among studies. In addition, with a few caveats (see Appendix 11.1), the same parameter can be calculated on climate data such as rainfall or temperature, or phenological data of food availability, to provide simple objective measures of resource seasonality as well (see Chapter 2). Thus, we can attempt here a broad comparative study of birth seasonality across all primates and test a variety of hypotheses with a single broad data set.

Hypotheses for birth seasonality

Restating the early hypotheses about resource seasonality, we should expect relatively lower r values (reduced birth seasonality) in species that live in less seasonal environments and rely on (or can reproduce while using) less seasonal resources (Bercovitch & Harding 1993; Watts 1998) (Table 11.1). A test of these hypotheses with New World primates was consistent with the expectation that birth seasonality depends on the seasonality of resource production. Birth seasonality is lowest for folivorous species (all in one genus, *Alouatta*), and among tropical populations at low latitudes, where the period of plant production is not constrained by cold temperatures (Di Bitetti & Janson 2000).

Another factor affecting birth seasonality is body mass, or, more precisely, life history parameters correlated with body mass (Table 11.1) (Bronson 1989). Very small species (< 0.3kg) generally have short reproductive cycles (gestation plus infant dependency) and may be able to fit two such cycles into a single year. Because these two cycles are usually separated by five to six months, they will lead to relatively low measured birth seasonality, even if both birth peaks occur during the period of relatively high food abundance. Species of moderate size (0.3–3 kg) can reproduce only once a year and should be expected to time their births to maximize fitness. If they use seasonal resources, then they should show highly seasonal birth patterns. Larger-bodied species (>3 kg) should, in general, not be able to fit a reproductive cycle into a single year. Because their energetic expenditure is spread out over a longer period and may necessarily include a period of relative food scarcity, the exact timing of births may be less important to fitness (Bronson 1989). Thus, we predict that birth seasonality will increase with body mass across small-sized species but will decrease with body mass in moderate-sized to large species (see also Chapter 10). Overall, we expect birth seasonality to be related negatively to body mass. If a significant quadratic effect of mass is observed, then its parameter should be negative (so that the curve is concave down). In addition, species with bimodal birth peaks should be restricted to small-bodied primates. These patterns were found to occur in New World primate species (Di Bitetti & Janson 2000).

In addition to the above hypotheses, it has been suggested that particular geographical areas may produce distinctive selection pressures on seasonal births (Table 11.1). For instance, Southeast Asian primates have to cope with greater between-year variation in food supply because of community-wide fruit masting. They should still be seasonal, but opportunistic, with conception following the fruiting peaks in particular years (van Schaik & van Noordwijk 1985). Thus, in long-term studies, seasonality may be reduced because of between-year variation in the timing of masting. In Madagascar, strong seasonality of climate and food production has been argued to favor extreme breeding seasonality

Table 11.2 *Species and major data used in this chapter. Additional data (latitude, body masses, diet category) are available from CHJ. Dates correspond to day no. throughout a year (January 1 = 1, December 31 = 365). Superscripts after the data on date of peak fruiting indicate the authors' scaling of the quality of the chronological data (see Appendix 11.2)*

Species	Locality	Mean birth date	Gestation(days)	Weaning age(days)	*r* birth	*r* rain	*r* temperature adjusted	Predicted date of maximum productivity	*r* fruit phenology	Date of peak fruiting	References
Alouatta caraya	Corrientes, Argentina	199.26	190	325	0.3239	0.283	0.2487	18.73	–	–	Di Bitetti & Janson (2000)
Alouatta fusca	Estação Biológica Caratinga, Brazil	191.56	190	300	0.0891	0.556	0.1148	362.88	0.211	8.1[3]	Strier *et al.* (2001)
Alouatta palliata	La Pacifica, Costa Rica	330.21	187	325	0.0738	0.508	0.0402	221.40	0.511	87.4[1]	Di Bitetti & Janson (2000)
Alouatta palliata	Santa Rosa, Costa Rica	62.11	187	325	0.3232	0.54	0.0402	222.43	0.287	95.4[1]	Di Bitetti & Janson (2000)
Alouatta seniculus	Hato Masaguaral, Venezuela	23.64	191	300	0.2462	0.567	0.0503	192.41	0.116	240.0[2]	Crockett & Rudran (1987)
Alouatta seniculus	La Macarena, Colombia	252.06	191	340	0.1515	0.397	0.0512	159.30	0.4894	120.2[3]	Di Bitetti & Janson (2000)
Aotus azarai	Guaycolec, Formosa, Argentina	305.47	133	179	0.9481	0.199	0.271	19.23	0.1251	21.2[1]	Fernandez-Duque *et al.* (2002)
Aotus trivirgatus	Manu National Park, Peru	324.55	133	179	0.3385	0.276	0.0337	35.09	0.2829	6.7[4]	Wright (1985)

Arctocebus calabarensis	Makokou, Gabon	15.21	134	115	0.1062	0.185	0.0597	351.94	0.3365	297.7[2]	Charles-Dominique (1977)
Ateles belzebuth	La Macarena, Colombia	329.41	226	650	0.4905	0.397	0.0512	159.30	0.4894	120.2[3]	Klein (1971)
Ateles geoffroyi	Barro Colorado Island, Panama	262.39	226	821	0.5385	0.34	0.0308	241.11	0.2248	113.1[4]	Di Bitetti & Janson (2000)
Ateles paniscus	Manu National Park, Peru	98.34	226	650	0.3737	0.276	0.0337	35.09	0.2829	6.7[4]	Di Bitetti & Janson (2000)
Brachyteles arachnoides	Estação Biológica Caratinga, Brazil	210.45	226	638.75	0.6377	0.556	0.1148	362.88	0.211	8.1[2]	Strier *et al.* (2001)
Callicebus moloch	Manu National Park, Peru	251.13	163.5	140	0.8157	0.276	0.0337	35.09	0.2829	6.7[4]	Di Bitetti & Janson (2000)
Callithrix flaviceps	Sao Paulo, Brazil	334.58	148	65	0.1047	0.556	0.1148	362.88	0.211	8.1[2]	Di Bitetti & Janson (2000)
Callithrix jacchus	Northeastern Brazil	25.24	148	84	0.1958	0.455	0.0516	123.31	–	–	Di Bitetti & Janson (2000)
Cebuella pygmaea	Pacaya-Samiria, Peru	347.46	136	90	0.1462	0.126	0.0212	43.68	–	–	Di Bitetti & Janson (2000)
Cebus apella	Iguazu, Argentina	347.52	160	416	0.825	0.069	0.2249	14.17	0.3739	331.8[4]	Di Bitetti & Janson (2000)
Cebus apella	La Macarena, Colombia	89.03	160	416	0.4939	0.397	0.0512	159.30	0.4894	120.2[3]	Di Bitetti & Janson (2000)

Table 11.2 (*cont.*)

Species	Locality	Mean birth date	Gestation(days)	Weaning age(days)	*r* birth	*r* rain	*r* temperature adjusted	Predicted date of maximum productivity	*r* fruit phenology	Date of peak fruiting	References
Cebus capucinus	Barro Colorado Island, Panama	88.51	160	415	0.6071	0.34	0.0308	241.11	0.2248	113.1[4]	Di Bitetti & Janson (2000)
Cebus capucinus	Santa Rosa, Costa Rica	263.61	160	415	0.3189	0.54	0.0402	222.43	0.287	95.4[1]	Di Bitetti & Janson (2000)
Cebus olivaceus	Hato Masaguaral, Venezuela	173.78	160	415	0.655	0.567	0.0503	192.42	0.116	240.0[2]	Di Bitetti & Janson (2000)
Cercopithecus aethiops	Amboseli, Kenya	334.97	163	365	0.8454	0.492	0.0512	35.98	0.1082	262.6[3]	Struhsaker (1967)
Cercopithecus ascanius	Kibale, Uganda	45.63	153	630	0.2503	0.122	0.0305	197.73	0.115	148.6[4]	Struhsaker (1977)
Cercopithecus campbelli lowei	Abidjan, Ivory Coast	360.68	180	549	0.93935	0.388	0.0588	153.96	–	–	Bourliere *et al.* (1970)
Cercopithecus mitis kolbi	Kenya	39.25	140	439	0.7929	0.277	0.0558	98.42	0.0939	352.8[3]	Omar & De Vos (1971)
Cercopithecus mitis stuhlmanni	Kibale, Uganda	55.04	140	439	0.6575	0.122	0.0305	197.73	0.115	148.6[4]	Rudran (1978); Butynski (1988)
Colobus badius	Kibale, Uganda	130.58	170	773.8	0.3707	0.122	0.0305	197.73	0.115	148.6[4]	Struhsaker (1975)
Erythrocebus patas	Laikipia district, Kenya	2.09	165	212	0.8512	0.099	0.0305	143.82	–	–	Chism *et al.* (1984)

Eulemur fulvus rufus	Ranomafana National Park, Madagascar	262.39	118	135	0.996	0.262	0.1643	31.18	0.1484	271.5[4]	D. Overdorff (2001, unpublished data)
Eulemur mongoz	Anjamena, Madagascar	293.9	118	135	0.9925	0.872	0.0908	32.48	0.164	336.6[2]	Curtis & Zaramody (1999)
Eulemur rubriventer	Ranomafana National Park, Madagascar	281.82	118	135	0.917	0.262	0.1643	31.18	0.1484	271.5[4]	D. Overdorff (2001, unpublished data)
Euoticus elegantulus	Makokou, Gabon	24.12	135	.	0.3813	0.185	0.0597	351.94	0.3365	297.7[2]	Charles-Dominique (1977)
Galago alleni	Makokou, Gabon	21.5	134	–	0.3425	0.185	0.0597	351.94	0.3365	297.7[2]	Charles-Dominique (1977)
Galago demodovii	Makokou, Gabon	48.694	112	49	0.4612	0.185	0.0597	351.94	0.3365	297.7[2]	Charles-Dominique (1977)
Galago moholi	Nylsvley Reserve, South Africa	330.19	124	78.5	0.4331	0.500	0.2751	361.01	–	–	Pullen *et al.* (2000)
Gorilla gorilla	Rwanda	203.92	256	1278	0.0507	0.115	0.0130	53.70	–	–	Watts (1998)
Hapalemur griseus	Beza Mahafaly, Madagascar	333.92	100	120	0.8750	0.720	0.1520	74.42	0.323	36.8[2]	C. Grassi (2001, unpublished data)
Lagothrix lagotricha	Macarena, Colombia	265.37	225	315	0.5888	0.397	0.0512	159.30	0.4894	120.2[3]	Di Bitetti & Janson (2000)

Table 11.2 (*cont.*)

Species	Locality	Mean birth date	Gestation(days)	Weaning age(days)	*r* birth	*r* rain	*r* temperature adjusted	Predicted date of maximum productivity	*r* fruit phenology	Date of peak fruiting	References
Lemur catta	Beza, Mahafaly Madagascar	250.63	135	179	0.9652	0.72	0.1520	74.42	0.323	36.8[2]	M. Sauther (2001, unpublished data)
Leontopithecus rosalia	Poço dos Antas, Brazil	309.22	129	90	0.6772	0.135	0.1034	377.20	–	–	Di Bitetti & Janson (2000)
Macaca cyclopis	Taiwan	129.56	165	186	0.9135	0.635	0.1553	193.13	–	–	Wu & Lin (1992); Hsu *et al.* (2001)
Macaca fascicularis	Ketambe, Indonesia	241.98	162	427	0.6094	0.088	0.0367	43.21	0.2386	230.7[4]	Van Schaik & van Noordwijk (1985)
Macaca fuscata	Japan	141.94	165	286	0.9227	0.264	0.4890	209.97	0.5994	23.3[3]	Kawai *et al.* (1967)
Macaca maurus	Karaenta Nature Reserve, Sulawesi	128.68	165	195	0.4975	0.56	0.0162	20.74	–	–	Okamoto *et al.* (2000)
Macaca nemestrina	West Sumatra, Indonesia	211.12	167	365	0.2479	0.359	0.0456	45.94	–	–	Oi (1996)

Macaca nemestrina	Lima Belas, Malaysia	262	167	365	0.2546	0.070	0.0201	368.95	0.2131	178.9[3]	Caldecott (1986)
Macaca radiata	Bangalore, India	69.31	162	365	0.8647	0.503	0.1077	214.68	–	–	Rahaman & Partha-sarathy (1969)
Macaca silenus	Anaimalai Station, Tamil Nadu, India	22.68	174	365	0.3286	0.246	0.0874	235.66	0.5218	143.8[4]	Kumar & Kurup (1985)
Macaca sinica	Polonnaruwa, Sri Lanka	10.72	165	365	0.6140	0.298	0.0977	397.34	0.239	227.2[1]	Dittus (1977)
Macaca sylvanus	Algeria	137.66	165	224	0.9404	0.266	0.4793	262.22	–	–	Menard & Vallet (1993)
Macaca sylvanus	Gibraltar	176.73	165	224	0.8943	0.513	0.2928	315.22	–	–	MacRoberts & MacRoberts (1966)
Macaca thibetana	Mount Emei, China	76.82	165	217	0.8427	0.588	0.7204	205.05	–	–	Zhao & Deng (1988)
Pan troglodytes	Gombe and Mahale, Tanzania	136.18	228	1825	0.0878	0.456	0.0183	28.86	–	–	Goodall (1983); Nishida *et al.* (1990)
Papio anubis	Kenya	54.24	180	584	0.1854	0.23	0.0666	102.52	–	–	Bercovitch & Harding (1993)
Papio cynoce-phalus	Tana River, Kenya	328.47	175	365	0.1424	0.548	0.1319	42.46	0.2278	70.6[3]	Bentley-Condit & Smith (1997)
Papio cynoce-phalus	Amboseli, Kenya	282.8	175	487	0.1372	0.462	0.0673	28.32	0.1082	262.6[3]	J. Altmann & S. Alberts (2002, unpublished data)
Papio ursinus	South Africa	327.5	187	420	0.4185	0.483	0.1965	367.73	–	–	Lycett *et al.* (1999)

Table 11.2 (*cont.*)

Species	Locality	Mean birth date	Gestation(days)	Weaning age(days)	*r* birth	*r* rain	*r* temperature adjusted	Predicted date of maximum productivity	*r* fruit phenology	Date of peak fruiting	References
Perodicticus potto	Makokou, Gabon	272.59	193	150	0.7698	0.185	0.0597	351.94	0.3365	297.7[3]	Charles-Dominique (1977)
Presbytis aygula	West Java	70.87	168	–	0.3468	0.275	0.0029	24.43	–	–	Ruhiyat (1983)
Presbytis thomasi	Ketambe, Indonesia	230.19	168	636	0.0406	0.088	0.0367	43.21	0.2386	230.7[4]	L. Sterck, R. Steenbeek, & S. Wich (2002, unpublished data)
Procolobus verus	Taï National Park, Ivory Coast	353.43	169	365	0.5372	0.253	0.0385	203.38	0.417	1.5[3]	Korstjens (2001)
Propithecus diadema	Ranomafana National Park, Madagascar	157.68	179	547	0.8869	0.262	0.1643	31.18	0.1484	271.5[4]	S. Pochron & P. Wright (2001, unpublished data)
Saguinus fuscicollis	Manu National Park, Peru	343.38	149	91	0.6257	0.276	0.0337	35.09	0.2829	6.7[4]	Di Bitetti & Janson (2000)

Saimiri oerstedi	Corcovado National Park, Costa Rica	71.8	170	150	0.9371	0.271	0.0402	225.46	0.2419	67.6[1]	Di Bitetti & Janson (2000)
Semnopithecus entellus	Nepal	83.42	168	416	0.8593	0.725	0.3096	200.04	0.1786	119.0[3]	Koenig *et al.* (1997)
Theropithecus gelada	Sankaber, Ethiopia	52.89	170	450	0.0901	0.718	0.0925	190.59	–	–	Dunbar & Dunbar (1975)
Trachypithecus pileata	Bangladesh	68.4	170	380	0.7707	0.576	0.1326	199.21	0.155	178.5[2]	Stanford (1991)
Trachypithecus senex monticola	Horton Plains, Sri Lanka	184.4	170	219	0.0876	0.059	0.1284	292.79	–	–	Rudran (1973)
Trachypithecus senex vetulus	Polonnaruwa, Sri Lanka	164.85	170	219	0.4976	0.298	0.0977	397.34	0.239	227.2[1]	Rudran (1973)

of lemurs (Wright 1999). To account for these possibilities, we include geographic area as an added variable in our analyses. In particular, we ask whether there is something special about birth seasonality in Malagasy primates. More specifically, we ask: "Is their seasonality what would be expected of relatively small-bodied primates at relatively high latitudes?" Including geography also allows us to test whether the patterns found previously in New World species hold for other major primate radiations.

Choice of variables

We shall use the vector length r as a measure of the width of the birth peak (see Appendix 11.2). To test the hypotheses mentioned above, we need data on four independent variables: food type consumed, seasonality of food production, body mass, and geographic area. Of these, all but the seasonality of food production can be indexed by readily available data published in reviews (e.g. Harvey *et al.* [1987]; Ross & Jones [1999]).

Assessing the seasonality of food production ideally requires repeated quantitative measurement of food availability as perceived by primates for the same localities for which we possess birth data. However, quantitative data on resource production or biomass are available for relatively few primate species, and even for these, the adequacy and comparability of different measures of food availability have been questioned (e.g. Chapman *et al.* [1994]). Thus, we use three approaches to test this general hypothesis. First, because latitude can serve as a simple proxy for seasonality, we examine for all 70 populations how birth seasonality varies with latitude. Second, we use measures of climatic seasonality that should correlate with the seasonality of food production. Of the climatic measures, temperature seasonality is correlated highly with latitude, but rainfall seasonality is not in this sample and could help to explain regional differences in birth seasonality. To account for the fact that biological activity increases as a power function of temperature, we calculated expected production based on both an unweighted measurement of temperature and a weighted measure that parallels known differences in biological activity (see Appendix 11.1 for details). Ideally, we would have included seasonal variation in the intensity of solar radiation, because previous work suggests that the phenology of tropical plants may be well correlated with this measure (van Schaik *et al.* 1993). Unfortunately, data on this aspect of climate are not easily available for most primate study sites or from weather stations close to them.

Finally, we were able to find phenological data for resources corresponding to the main diet type for the 49 populations; for these, we use mean vector length r of monthly food availability. Although this measure ought to approximate most closely the selection pressure of resource variation on birth timing, the underlying phenology data are not always comparable. All the measures focus on plants with a given edible phenophase (ripe fruit, open flowers, flush leaves), but these are weighted in a variety of ways (see Appendix 11.1; see also Chapter 2) and are of varying duration and quality in terms of total number of plants sampled. For this reason, we included indexes of phenological methods and study quality when evaluating the effects of phenological scores on birth seasonality.

Results

Nearly two-thirds of the variation in birth seasonality, as measured by the mean vector length r, was explained by combined effects of differences in diet, body mass, latitude, and continent (Table 11.3). To streamline the reporting of statistics, from here on all statistical results for the effects of particular independent variables are given holding all the remaining variables constant statistically using a general linear model. Birth seasonality increased with increasing latitude and latitude squared (Fig. 11.1). The combined effect of linear and quadratic terms for latitude was significant ($F[2,59] = 12.42$, $P < 0.0001$). However, latitude squared alone was a better predictor of birth seasonality than the linear effect of latitude alone

Table 11.3 *Results of multiple analysis of covariance (ANCOVA) of narrowness of birth peak, as measured by the vector r (see Appendix 11.2), as a function of diet, latitude, log_{10} of female body mass, and geographic area*

Source	DF	Sum of squares	F ratio	Probability > F
Model	10	3.917	9.1950	<0.0001
Diet	4	1.633	9.586	<0.0001
Latitude	1	0.00023	0.0054	0.942
Latitude^2	1	0.0472	1.108	0.297
Log_{10}(mass)	1	0.2975	6.985	0.011
Continent	3	0.4399	3.443	0.022
Error	59	2.5132		

DF, degrees of freedom.

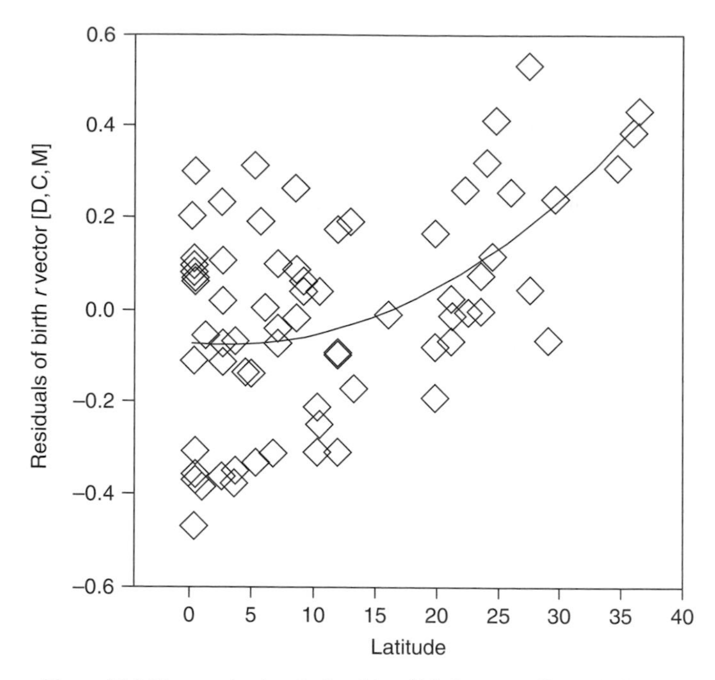

Figure 11.1 The quadratic relationship of birth seasonality to latitude, after adjusting for body mass [M], diet [D], and land mass [C], is highly significant statistically (see Table 11.4).

(Table 11.4). When both linear and quadratic terms for latitude were included, the linear term was not significantly different from 0 (Table 11.4). This pattern reflects the fact that *r* did not vary with latitude among tropical locations within 15° of the Equator ($F[1,47]=0.35$, $P>0.5$). We performed all subsequent regressions with latitude squared rather than latitude. It is likely that the relationship of birth seasonality to latitude reflects the underlying relationship between latitude and resource seasonality, although that hypothesis needs to be tested directly (see below).

Birth seasonality decreased with increasing body mass, as expected (Fig. 11.2, Table 11.4). If both linear and squared terms for body mass were included, then the squared term was not significantly different from 0 (Table 11.4). The squared term for body mass was not significantly different from 0 for any individual geographic area except Asia, but the term was positive, the opposite of the predicted negative (concave down) relationship. We did not include the square of body mass in any subsequent analysis.

Table 11.4 *Results of various statistical tests discussed in the text*

Dependent variable	Independent variable	Variables controlled	F statistic	DF	*P*
Birth seasonality	Latitude	C, D, M	23.699	1, 59	<0.001
Birth seasonality	$(\text{Latitude})^2$	C, D, M	25.27	1, 59	<0.001
Birth seasonality	Latitude	C, D, M, L2	0.005	1, 58	0.94
Birth seasonality	ln(Body mass)	C, D, L2	7.18	1, 60	0.010
Birth seasonality	(ln[Body mass])2	C, D, L2, M	1.42	1, 59	0.24
Birth seasonality	Diet	C, L2, M	9.78	4, 60	<0.001
Birth seasonality	Continent	D, M, L2	3.53	3, 60	0.020
Birth seasonality	Continent	D, M, L2, N	3.49	3, 57	0.021
Birth seasonality	Continent	D, M, R, R2, T, T2	2.01	3, 57	0.123
Birth seasonality	Continent	D, M, R, R2, Ta, Ta2	2.06	3, 57	0.123
Birth seasonality	$(\text{Latitude})^2$	D, M, C, R,R2, Ta, Ta2	2.98	1, 56	0.089
Fruiting seasonality	Continent	L, L2	0.73	3, 20	0.54
Fruiting seasonality	Continent	R, R2, T, T2	0.36	3, 18	0.79
Fruiting seasonality	Continent	R, R2, Ta, Ta2	0.34	3, 18	0.79
Leaf flush seasonality	Continent	L, L2	0.26	3, 15	0.86
Leaf flush seasonality	Continent	R, R2, Ta, Ta2	0.50	3, 13	0.69
Flowering seasonality	Continent	L, L2	1.39	3, 12	0.29
Flowering seasonality	Continent	R, R2, Ta, Ta2	1.65	3, 10	0.24
Birth seasonality	Continent	D, M, F	5.33	3, 39	0.004
Birth seasonality	Seasonality of main item in diet	D, M, C	5.05	1, 39	0.03

C, continent; D, diet; F, fruiting seasonality; L, latitude; L2, $(\text{latitude})^2$; M, ln(mass); N, sample size of births; R, rainfall seasonality; R2, R^2; T, temperature seasonality; T2, T^2; Ta, adjusted temperature seasonality; Ta2, Ta^2. The degrees of freedom (DF) for testing resource seasonality was equal to the number of study sites with corresponding data, whereas for testing correlates of birth seasonality, it was equal to the number of primate populations with available data.

Birth seasonality varied significantly among dietary types (Fig. 11.3, Table 11.4). Gum- and insect-eating species, whether in the New World (marmosets) or in the Old World (galagos and lorises), had much lower seasonality than species with other dietary types. In part, this difference was caused by a more pronounced bimodal birth pattern in the five gum-eating species, three of which show significant bimodality ($P < 0.05$). Although this bimodal birth pattern could be due to small body mass alone, the two small-bodied insectivorous species (a galago and a loris)

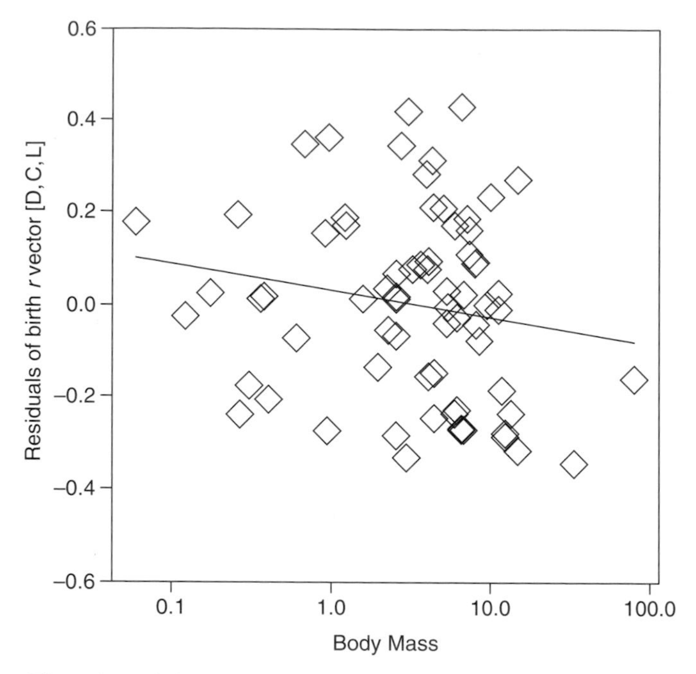

Figure 11.2 Birth seasonality (adjusted for latitude [L], diet [D], and land mass [C]) declines significantly with body mass in a linear relationship (see Table 11.4).

did not show bimodal birth peaks, and small-bodied frugivores showed relatively weak bimodality. Folivores had intermediate values of birth seasonality and differed significantly from gum feeders and frugivores but not from insectivores or omnivores. Frugivores (including frugivore–insectivores) had the highest values of birth seasonality and differed significantly from all other diet types, except omnivores.

Birth seasonality differed significantly among land masses (Fig. 11.4, Table 11.4). The only pair of areas that differed significantly in pair-wise a-posteriori comparison ($P < 0.05$, Tukey–Kramer honesty significant difference [HSD] test) was America and Madagascar. Madagascar had much higher levels of birth seasonality than other areas, even after controlling for the effects of body mass, diet, and latitude. There are at least three alternative reasons that primates on these distinct areas might differ in birth seasonality for a given body mass, diet, and latitude. First, there is a potential bias due to sample size. Because the mean vector is bounded between 0 and 1, small random data sets will have larger expected mean vectors than will large random data sets (Batschelet 1981). Many birth

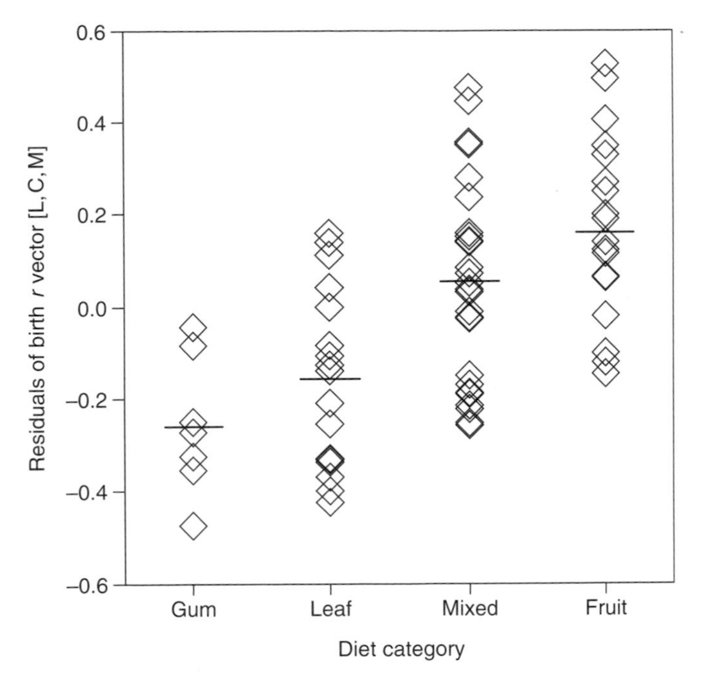

Figure 11.3 Birth seasonality (adjusted for latitude [L], body mass [M], and land mass [C]) differs significantly among major dietary types (see Table 11.4). Mixed diets typify "omnivorous" primates, which generally rely on fruits or seeds for energy and leaves for protein or as a fallback resource.

data on Malagasy species are from small, unpublished data sets (Table 11.2), and thus there could an upward bias in mean birth seasonality. However, when sample size is included in the multiple regression along with the other predictors, the effect of continental area is still significant (Table 11.4), and Madagascar still differs significantly from South America ($P < 0.05$, a-posteriori Tukey–Kramer HSD).

Second, the degree of resource seasonality at a given latitude might differ among land masses. In this case, primates in all areas might respond equally to resource compression, but the latter is more pronounced at a given latitude on some land masses than others.

Third, there might be intrinsic historical differences among the radiations of primates in the various regions, such that those of some lineages are more sensitive to resource seasonality. A possible cause could be differences in metabolism, with the lemurs capable of lowering their metabolic rates during periods of inactivity or low food abundance, but at a cost

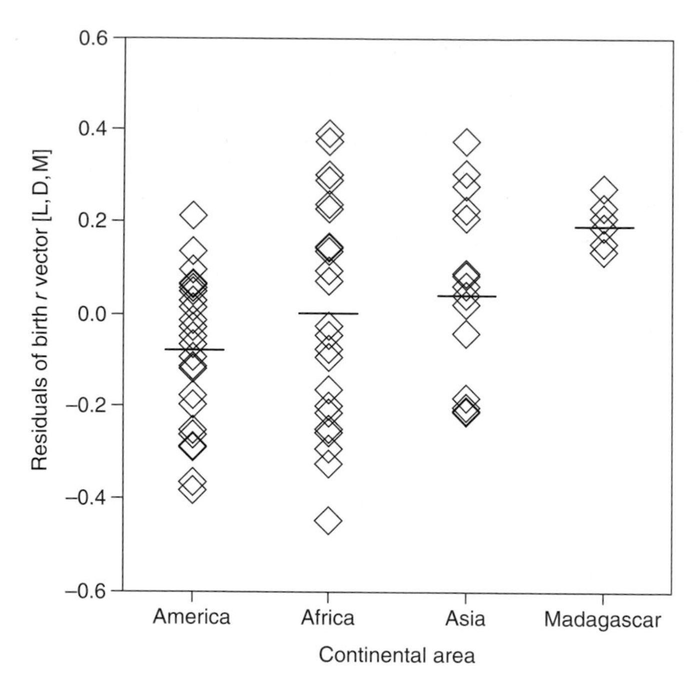

Figure 11.4 Birth seasonality (adjusted for latitude [L], body mass [M], and diet [D]) differs significantly among land masses (see Table 11.4). A-posteriori comparisons show that Madagascar differs significantly from other areas.

that they cannot reproduce at that time and may take several weeks to months to come into reproductive condition.

These hypotheses can be tested indirectly by using more immediate proxies than latitude for resource seasonality. Plant production in general increases with both rainfall and temperature, so indices of the seasonality of either or both of these climatic factors should bear a closer relationship to plant productivity than should latitude alone. We tested the effects of these climatic variables on birth seasonality by including for all the primate study sites the mean vector lengths for rainfall and temperature (either unadjusted or adjusted for biological activity), and their squared values, instead of latitude squared, in an analysis similar to that reported in Table 11.3. Using these climatic variables, the differences among land masses in birth seasonality are reduced and they are no longer significantly heterogeneous (Table 11.4), ($P = 0.12$, regardless of whether adjusted or unadjusted temperatures are used). However, while the differences among Africa, South America, and Asia are not significant, Madagascar remains different from South America ($P = 0.048$, Tukey–Kramer HSD). When climatic seasonality variables and latitude are used together in the multiple regression, the effects of latitude become insignificant (Table 11.4). These results suggest that most of the effect of latitude and land mass on birth seasonality is due to differences in climatic seasonality, which in turn ought to influence resource seasonality. However, primates on Madagascar may retain greater birth seasonality than expected on other land masses, even given the very seasonal environment on Madagascar and small mean body masses of extant lemurs.

There remains the possibility that primate food resources are more seasonally compressed in Madagascar than on other land masses for a given level of climatic seasonality. One test of this hypothesis is to compare directly the seasonality of food production among continents, holding either latitude or climatic variables constant. Controlling either for latitude or climatic variables, there were no significant differences among continents in seasonality of any resource category (Table 11.4) (r values for fruit phenology from Table 11.1; others are unpublished data). Madagascar was not consistently more seasonal than the other continents for any resource category. Thus, Madagascar does not appear to differ markedly from the remaining land masses in resource seasonality once latitude or climatic variables are accounted for.

To see whether continent-specific patterns of resource seasonality might still account for differences in birth seasonality among continents, we regressed a primate population's birth seasonality against both land mass and the phenological seasonality of the major food item in its diet (counting insects and gums as similar to flush leaves and omnivores as frugivores). In this case, the differences among continents are still significant (Table 11.4) ($P = 0.004$), and Madagascar differs from all the other continents ($P = 0.002$). Qualitatively similar results are obtained if (i) the analysis is restricted only to the 32 frugivorous and omnivorous species, using fruiting phenology to index resource seasonality; (ii) the analysis is limited to only the 41 species for which high-quality phenological data exist (scores of 2 or higher; see Appendix 11.1); and (iii) folivores are treated as responding to fruit phenology rather than flush leaf phenology. The effect of resource seasonality on birth seasonality (holding other variables constant) was significant (Table 11.4) ($P = 0.03$), but only if the major diet type for folivores was leaves rather than fruit.

These results suggest that differences in measured resource phenology do not account for the difference in birth seasonality between Madagascar and other continents. Although better data on resource phenology are desirable to test this hypothesis, the significant difference in birth seasonality between Madagascar and other continents held even when restricting the analysis only to high-quality phenological data. It is possible that differences in seasonality of some other resource are the driving force behind the tighter breeding synchrony of Malagasy primates, but data are currently not available to perform a comparative analysis of this possibility. Thus, we conclude tentatively that primates in Madagascar differ fundamentally from primates on other continents in their reproductive responses to resource seasonality.

Timing of births relative to expected plant productivity

In nearly all discussions of birth seasonality, it is assumed that the timing of births should be related to the timing of food production (van Schaik & van Noordwijk 1985; Bronson 1989; Wright 1999; Di Bitetti & Janson 2000). Precisely which nutrient limits reproduction may vary from species to species, but most commonly it is assumed to be energy; for lack of more precise predictions and data, we will not distinguish between possible dietary components here. We also will not discuss or test hypotheses that have been suggested to promote within-species birth synchrony, such as predation on infants (Boinski 1987), as these should not generally inform us about the timing of births (see above).

Even if reproduction is coordinated with the period of peak energy availability, the precise timing of births relative to this peak may vary among species for several reasons, including life history strategy and body size (Table 11.1). With respect to life histories, species can be divided into "income" breeders, which use current nutrient intake to fuel their reproduction, and "capital" breeders, which use stored nutrients to subsidize their reproductive effort (Richard *et al.* 2000) (see also Chapter 10). Income breeders should give birth so that the energetically most stressful part of reproduction will match the period of maximum nutrient availability. If natural selection acts to minimize mortality of the infants, then infant weaning should match the period of greatest availability of weaning foods (Altmann 1980; Stanford 1991; Wright 1999; Di Bitetti & Janson 2000). These species will be called income-I breeders. This category should be comprised of species that are small, with low adult survival that favors high investment in offspring production, and that depend on highly seasonal resources, such as insects, flush leaves, or fruits. If instead natural selection acts to minimize stress on the reproducing female, then such species will be referred to as income-II breeders (Table 11.1). Such species are likely to be larger than those in the first category but still rely on seasonally predictable food resources. For most primate species, the period of maximum maternal energy demand is thought to occur during mid to late lactation (Altmann 1980). The period of maximum plant flower and leaf production usually occurs shortly after the first solstice (see Chapter 2). Although fruit production is less predictable, the peak of flowering is followed by a peak of fruiting some time later. Thus, income breeders should give birth shortly before the first solstice, so that late lactation or weaning occurs one to several months after the first solstice. All other things being equal, income-I breeders should give birth earlier than income-II breeders, because weaning necessarily occurs later than does mid-lactation.

Capital breeders (Table 11.1) store up nutrients during the period of high food availability, and their ability to conceive depends on body condition (Richard *et al.* [2000]). Thus, it would be expected that their births would follow the first solstice (and its high food availability) by a period equal to the delay in conception plus the gestation period (Nishida *et al.* 1990; Strier *et al.* 2001). For most monkeys, gestation is about five months, so to a first approximation, one would expect birth peaks six or more months after the solstice. Relative to the following resource peak, the birth peak of capital breeders will in general appear poorly timed for weaning.

Before testing these predictions, we need to be able to categorize species as income or capital breeders. Van Schaik and van Noordwijk (1985) suggested that species that live in very unpredictable environments cannot use meteorological cues to predict the period of high food availability. In this case, they must behave as capital breeders. Alternatively, capital breeders may simply be species that have the capacity to store enough nutrients to help their reproductive efforts. In this case, capital breeding should be correlated with large body size, as the lower per-mass metabolic demand of larger species (Kleiber's law) dictates that large species will be able to store more nutrients relative to their needs than will small species. An exception might be small species that can go into torpor, but primates do not reproduce while in torpor. Thus, small-bodied species normally will be income breeders, regardless of environmental predictability.

A test of these predictions is to examine the timing of primate births relative to the solstice as a function of latitude. This relationship is shown in Fig. 11.5. This graph shows four major results. First, within the Equatorial tropics (10.5 °S–10.5 °N latitude), there is no apparent relationship between birth timing and the solstices. This lack of fit may be due to

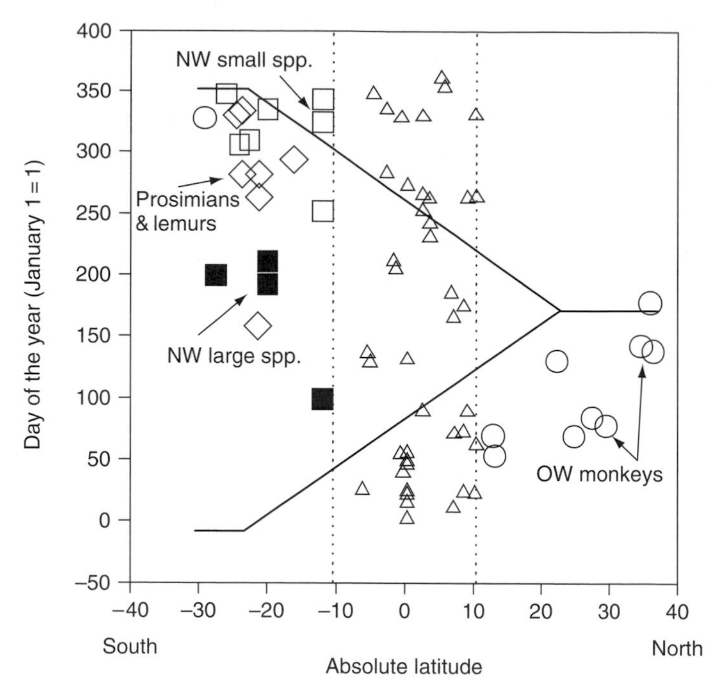

Figure 11.5 The pattern of timings of primate mean birth dates and the timing of the solstices as a function of latitude for the entire data set ($n = 70$ populations). For species within the strict tropics (<10.5° N or S), there was no obvious relationship between birth dates and the solstices. Outside the strict tropics, birth dates tended to cluster within 10–90 days before the first solstice for cercopithecines (Old World [OW] monkeys) and small-bodied platyrrhines (New World [NW] small spp.) and strepsirrhine species (Prosimians and lemurs). Large-bodied platyrrhines (NW large spp.) and strepsirrhines (single data point near NW large spp.) instead tended to give birth about 180–200 days after the first solstice, as expected if they are capital breeders (see text).

the relatively close spacing of the solstices, combined with measurement error in estimating mean birth date. Alternatively, the period of peak food production may not be well correlated with the solstices near the Equator, especially if the food in question is fruits (see Chapter 2). Second, non-Equatorial populations of four distinct radiations of primates all time their births close to the first solstice at their latitude. These include prosimians, lemurs, small-bodied New World monkeys, and large-bodied cercopithecoids (including colobines). Third, one group of primates times births to about six months after the first solstice at their latitude: the large-bodied New World primates, including both folivorous and frugivorous types, and one large-bodied folivorous lemur, a close relative of which is known to be a capital breeder (Richard *et al.* 2000) (but see Chapter 10).

For species of a given reproductive strategy, the precise timing of births relative to peak food availability should depend on body size, because larger species have longer gestation and weaning periods. For income breeders (species that give birth before the peak of food production), larger species should give birth earlier, relative to peak food availability, than should smaller species, but the reverse pattern is true of capital breeders. To test this pattern, we restricted ourselves initially to species that predominantly breed outside the strict tropics (10.5° or more from the Equator), as the timing of peak food availability within the strict tropics was difficult to ascertain from climatic data (see Chapter 2). By using the first solstice as a predictor of peak food availability, we could use the largest possible sample size to test for detailed patterns. For strepsirrhines, including both African prosimians and Malagasy lemur species, smaller species gave birth significantly closer to the first solstice (Fig. 11.6), and the same pattern held for small-bodied species in the New World (<2 kg) (Fig. 11.7). As expected, given the relatively close match between climatic

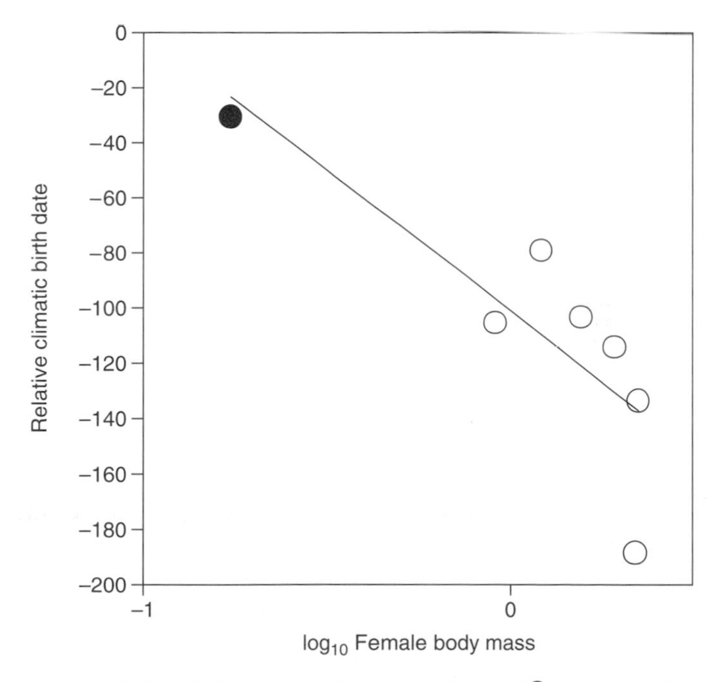

Figure 11.6 The relative climatic birth dates of lemurs (○) and prosimians (●) become earlier for larger species outside the strict tropics. The relative climatic birth date is the absolute mean birth date minus the mean date of peak rainfall and adjusted temperature (see Appendix 11.1).

and phenological data in the New World (see Chapter 2), plotting relative phenological birth dates against body mass gives much the same result as Fig. 11.7, with a marked negative relationship (not plotted).

Large-bodied species (>3 kg) in the New World were broadly consistent with capital breeding, as their birth dates were about one gestation length after the solstices, but there was no discernible trend across the small range of body sizes available (Fig. 11.7). Capuchin monkeys (Fig. 11.7) followed neither pattern, giving birth too late to agree with the trend for income breeders yet not delayed enough to appear as capital breeders. They are likely to be income-II breeders, as they possess a very long adult lifespan (Judge & Carey 2000), which should favor reducing reproductive stress on the mother. Extra-tropical species of cercopithecines did not show any trend of relative birth dates with increasing body size (Fig. 11.8); however, known gestation lengths vary by only about 20 days across the span of

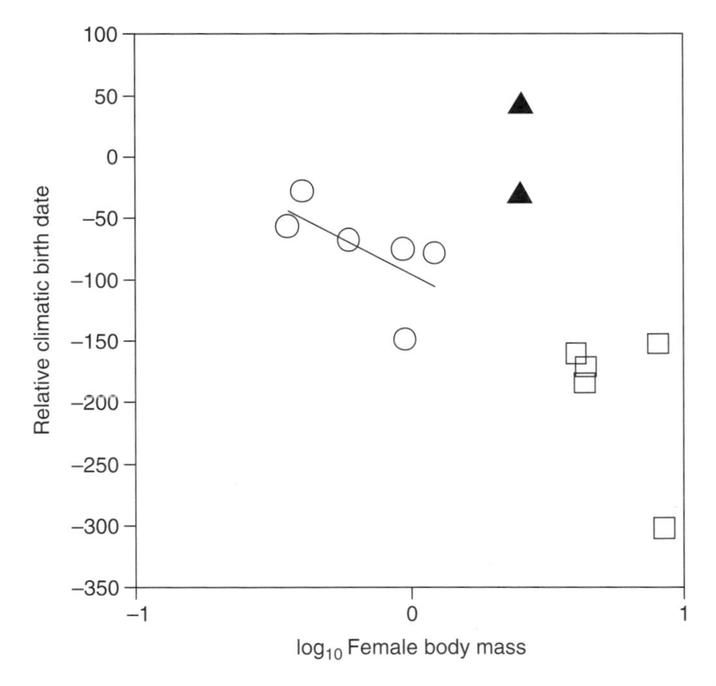

Figure 11.7 Relative climatic birth dates (see Fig. 11.6) become earlier for larger New World primates of mass less than 2 kg (○), outside the strict tropics. Capuchin monkeys (▲) give birth later than expected for income-I breeders, but too early for capital breeders. Large-bodied atelines (□) show no trend of birth dates with increasing body mass, but the values are generally consistent with capital breeders.

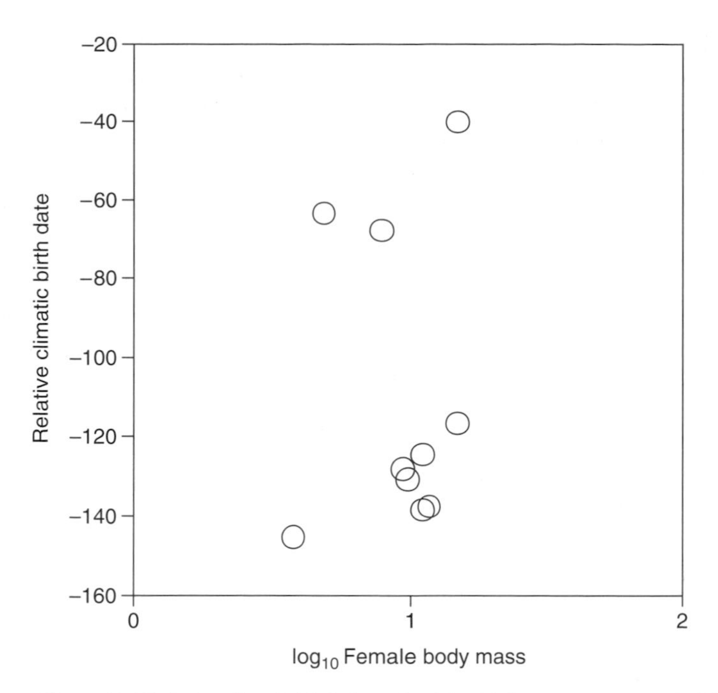

Figure 11.8 Relative climatic birth dates (see Fig. 11.6) are not correlated with body mass in Old World cercopithecines outside the strict tropics. Birth dates appear to cluster into two groups: one at about −50 days (consistent with income-I breeders) and another at about −130 days (consistent with capital breeders).

sizes available, so little trend is expected if these species are capital breeders.

To resolve further the absolute timing of births among primate species, it is necessary to know what resources each species is responding to and the phenology of those resources. Thus, it may be expected that species of distinct diets may differ in their timing of births relative to the solstices. Outside the tropics, phenological patterns of various resources are clustered relatively tightly around the first solstice (see Chapter 2), so that the patterns of births relative to the solstices should be (and, in general are) similar to those expected if the timing of resource availability were known. However, such resource predictability does not hold within the tropics, at least for the phenology of fruit production. Thus, to understand the absolute timing of births within the tropics requires using data on the actual phenology of the resources relevant to each primate species in each study site.

Data on phenology were gathered from literature sources, usually from the same study site as the birth data (Appendix 11.1). Only high-quality data were used (scores of 2, 3, or 4) (see Table 11.1 and Appendix 11.1). Using data from Old World primates, there is a significant circular correlation between mean birth date and mean date of food availability (using fruit availability for all species except insectivores) (Fig. 11.9) ($\chi^2 = 14.07$, degrees for freedom [DF] = 4, $P = 0.007$). The fit and statistics were not improved when data on leaf availability were substituted for fruit availability for the three species that are folivores ($\chi^2 = 4.48$, DF = 4, $P = 0.35$). The residuals of the linear correlation were not predicted well by body mass, and nor was there any overt evidence for a distinction between income and capital breeders (e.g. prosimians versus monkeys). For the tropical New World primates, there was a good fit between mean birth date and measured date of peak food availability (using fruit phenology for all species), although in this case it was important to account for income versus capital breeders (Fig. 11.10). If the birth dates of capital breeders are shifted by −208.5 days (the mean gestation period for

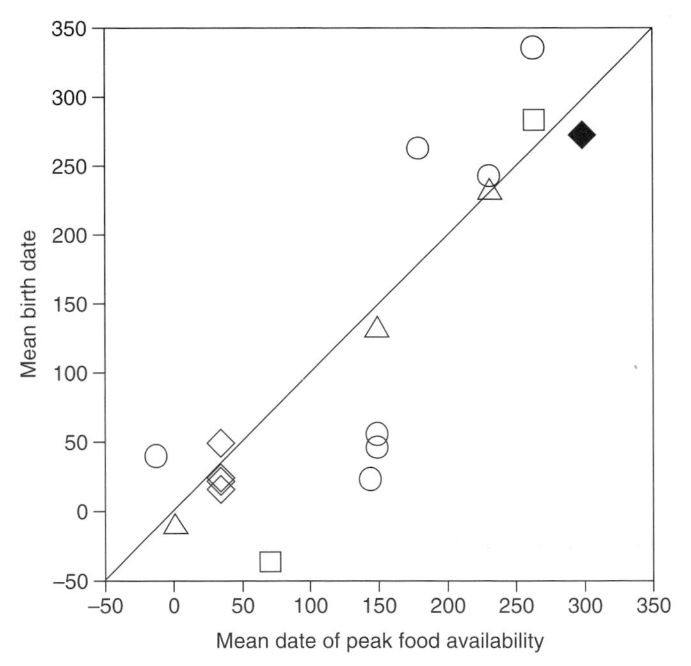

Figure 11.9 For tropical Old World primates, mean birth dates are correlated well with mean dates of peak food production, based on fruit phenology for all species, except for peak of new leaf production for insectivorous prosimians (◇). No taxonomic or dietary group shows systematic deviations from the expected trend for income-II breeders (solid line). ○, *Cercopithecus* and *Macaca*; □, baboons; △, colobines; ◆, *Periodicticus*.

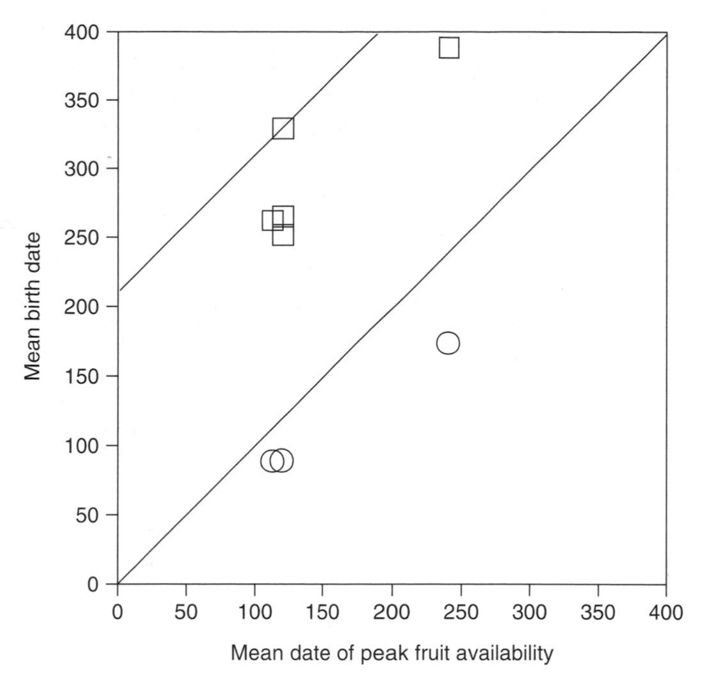

Figure 11.10 Mean birth dates of tropical New World primates are correlated with the mean dates of fruit production. Capuchin monkeys (○) give birth close to the mean date of fruit production, consistent with income-II breeders. Large-bodied atelines (□) give birth 150–210 days after the peak of fruit production, consistent with capital breeders.

atelines), to make them comparable to income breeders, the circular correlation between mean birth date and peak fruit availability is high ($r = 0.84$), but is not significant (Chi-square = 6.69, DF = 4, $P = 0.15$), due in large part to the small number of independent phenology data (three study sites).

Summary and conclusions

Utility of r *measure*

The use of circular statistics in our analysis demonstrates several advantages. The average vector length, r, allows not only an objective measure of seasonality of births but also sensitive statistical tests and a simple way to calculate the correct mean date of births. By doubling the values of the ordinate variable, r can be used to detect bimodal patterns (see Appendix 11.2).

In addition, the same method can be adapted to calculate the degree of seasonality and the mean date of climatic variables or resource availability (see Appendix 11.1).

There are some minor drawbacks to r as a measure of seasonality. First, if the data are in fact distributed randomly, then the expected value of r increases as the sample size of observations decreases. Thus, when using r itself as the dependent variable for comparative analyses, it is possible that some of the variation in apparent seasonality between species is only a reflection of differences in sample size. In our analysis, we controlled for this possibility by using sample size as a covariate in some of our analyses. In all cases, the effect of sample size was small and not statistically significant in the context of other independent variables.

Second, the calculation of r assumes that the dependent variable reflects counts of observations. In this case, an increase in this variable is equally important regardless of the absolute level of the variable – a change from 0 to 1 is weighted equally with a change from 45 to 46. However, when some other scale is used for the dependent variable (e.g. temperature, fruit biomass), then the underlying assumption of scale equality may not be correct. In addition, negative values of the dependent variable are not allowed. For these reasons, we used two calculations of seasonality for temperature. The first assumed equality of scaling; each unit was weighted the same regardless of the current level of the variable and negative values were set to zero. The second calculation allowed for increasing biological activity at higher levels of temperature (see Appendix 11.1 for the formula used). In our analyses, it made little difference which calculation we used for the temperature effect, but the second, biologically active version yielded slightly higher correlations with birth seasonality. It could be argued that rainfall should also be scaled so that any month with more than 200 mm of rainfall is set equal to 200 mm, although we did not employ such a transformation. Higher levels of rainfall are unlikely to promote increased productivity and may even reduce productivity due to increased cloudiness (cf. van Schaik *et al.* [1993]).

Effects of diet versus body size on birth seasonality

Because diet is related closely to body mass in primates (Kay 1984), teasing apart their independent effects on birth seasonality was not always easy. In a previous analysis of birth seasonality for only New World primates (Di Bitetti & Janson 2000), the low seasonality of births in very small-bodied species was attributed to their small size, although it could also have been caused by their use of gums as a major food source. Gums are induced to flow by wounding tree and vines and thus should be less seasonal than many other plant resources, which typically show strong annual phenological patterns even in the tropics. With the larger data set used here, gum-feeders are significantly less seasonal than frugivores or omnivores, even after controlling for body mass, latitude, and continental region. Among the five species of primates of body mass less than 0.5 kg, only one in our analysis does not rely on plant exudates. This species, *Saguinus fuscicollis*, is also the only highly seasonal breeder in this set, reinforcing the idea that diet is more important than body mass in determining bimodal birth patterns and consequent low overall birth seasonality.

Among larger species, the effects of diet and mass are easier to dissect because there is substantial overlap in body mass among folivores, frugivore–insectivores, and omnivores. In this case, it was clear that folivores had lower birth seasonality than omnivores or frugivore–insectivores of equivalent mass, latitude, and continental origin. Leaves are considered to be more uniformly available through time, although significant seasonality in production of new leaves is known from many tropical areas (see Chapter 2). Larger species are less seasonal breeders than small species in a given dietary category, probably because selection on birth timing is less stringent when infant dependency periods cannot be completed within a single season of high food abundance.

Geographic variation in the narrowness of the birth peak

As in a previous study on New World primates (Di Bitetti & Janson 2000), we found that births are more seasonal the further a primate population lives from the Equator, although there is little effect of latitude between 0 and 15°. Presumably, this effect of latitude reflects the increasingly seasonal production of food. When temperature and rainfall seasonality, which should correlate with food production, are included in the analysis, the effect of latitude becomes statistically insignificant. When we used the most direct measure of resource seasonality, the vector r of production of the main food item in the diet of a given primate population, it correlated significantly with birth seasonality even holding constant diet, mass, and continent (Table 11.4). When latitude or climatic variables are added to the analysis after this measure of resource seasonality, none of the added terms (either linear or quadratic) contributes significantly to explaining variation in birth seasonality (Janson & Verdolin, 2005, unpublished data). Thus, it appears that the effects of latitude and climate on birth seasonality are mediated via resource seasonality. Although other, more direct effects of climate on birth seasonality have been suggested (e.g. cold temperatures [Zhao & Deng 1988] and dry periods [Crockett & Rudran 1987]), our analysis does not support a consistent effect of climate on birth seasonality beyond that mediated through resource availability.

Continental differences

Variation in narrowness of the birth peak across primate populations was explained largely by variation in female body mass, diet, and latitude. Even after adjusting for these three variables, however, there remained significant differences among continental areas in the width of the birth peak. The major and only consistent difference among continents is that primates in Madagascar were significantly more seasonal in their births than species on other continents. This difference is not apparently due to continent-specific differences in the seasonality of climatic variables such as rainfall and temperature (controlling for latitude); nor was it apparently due to differences in resource phenology between Madagascar and other areas. Thus, it is possible that primates on Madagascar respond to a given level of resource seasonality with narrower birth peaks than do other primates of similar diet, body mass, and latitude. What could cause such a difference?

Malagasy primates appear to have relatively low resting metabolism compared with most other taxa discussed in this chapter (Ross 1992). This low metabolism may itself be an adaptive response to a history of extreme or very predictable seasonality (e.g. Pereira [1993]). Whatever the cause, the low basal metabolism may influence reproductive strategies. If lemurs enter a period of reduced metabolism during the period of food scarcity, then this may preclude mating and conception during that period, or perhaps doing so would raise the survival cost of reproduction to the point that it would not be favored. If mating and conception are not performed during the lean period, then there may be a time delay between the first indicators of increased food production and the start of reproductive activity. Such a delay may be shorter or non-existent in species that maintain high year-round metabolism.

Birth timing and energy availability

Analysis of absolute birth dates suggests that outside the Equatorial tropics, mean birth dates occur shortly before the date of the first solstice

(Fig. 11.5). As this time is likely to be the period of maximum insolation, this pattern is consistent with the prediction that plant productivity should peak during periods of peak insolation (see Chapter 2), assuming that birth peaks are indeed related to peak plant productivity. Within the Equatorial tropics ($<10.5°$), mean birth dates bore little or no relationship to the solstices (Fig. 11.5) and yet were predicted quite well by actual dates of peak food availability, as determined by quantitative phenological studies (Figs. 11.9 and 11.10). Thus, the general notion that birth timing is related to energy availability is well supported across all the primates of different radiations, sizes, and dietary types.

Birth timing and female reproductive strategies

Differences in relative birth timing did appear among species. The timing of births for small-bodied primates (<3 kg) was predicted well by the weaning hypothesis (Figs. 11.5–11.7). Because of longer infant dependency periods in larger species in this mass range, they had consistently earlier births relative to the peak of inferred food production than did smaller species within the set of small-bodied primates. No small-bodied species appeared to use the strategy of storing nutrients and giving birth following the major period of food production. These species therefore all appear to be income-I breeders, fueling their reproductive expenditures by current food intake so as to maximize infant survival (Richard *et al.* 2000) (see also Chapter 10).

In contrast, primates of body mass greater than 3 kg used a diversity of strategies. Some species appeared to be income breeders but timed births close to the peak of food production, perhaps to minimize nutritional stress on the mother rather than minimizing stress on the offspring. These income-II species appear to be selected to maximize maternal survival while reproducing using seasonally predictable food sources. This timing was characteristic of capuchin monkeys and perhaps most tropical cercopithecine monkeys and prosimians (Fig. 11.9), as well as a few non-tropical cercopithecines (Figs. 11.5 and 11.8).

The remaining species appeared to be capital breeders (Richard *et al.* 2000), storing up nutrients during times of food abundance and initiating reproduction about five to six months after the peak availability of food (Figs. 11.5–11.8). These species include all the large New World monkeys, the lemur genus *Propithecus* (but see Chapter 10), probably the great apes (not shown), and a diverse assortment of Old World cercopithecines. Oddly, the latter did not include the species from Southeast Asian forests that inspired the hypothesis of opportunistic breeding in response to fruit masting years (van Schaik & van Noordwijk 1985). The only factor uniting these diverse species of capital breeders is large body size (for their phylogenetic group) and correspondingly long interbirth intervals. Perhaps capital breeding is favored simply because selection on weaning time or maternal energetics is very weak, given the long period over which maternal investment is provided. However, a variety of cercopithecine populations apparently behaved as income breeders (see Figs. 11.8 and 11.9), but they shared no obvious characteristic and nor did they differ obviously from the apparent capital breeders.

Although not generally possible to test rigorously in this analysis, several additional factors have been suggested to affect the timing of primate births. First, where climatic or resource peaks are broad, females under the weaning hypothesis can delay birth until closer to the resource peak, thereby reducing the conflict between the mother's interests and those of her offspring over birth timing. Conversely, narrower peaks of food production, associated with higher latitudes and altitudes, should cause birth dates to occur earlier relative to the mean date of peak food production, so that weaning can still occur before food production declines. A couple of studies on individual species have found earlier birth dates in populations living in more seasonal habitats (Kawai *et al.* 1967; Menard & Vallet 1993). However, in our analysis, income breeders living at higher latitudes appeared, if anything, to time their births closer to the solstice, rather than anticipating it more (Fig. 11.5).

Second, diet and phylogeny (life history) will affect the length of juvenile dependency via juvenile growth rates (e.g. see Harvey *et al.* [1987]). Thus, distinct primate radiations might show distinct relationships between birth dates and peak food production. However, such differences were not obvious in our comparisons (Figs. 11.5–11.8).

Third, different dietary items (insects, young leaves, mature fruits) reach phenological peaks at different times. In areas with strongly seasonal rainfall, generally young leaves emerge first following the period of drought, followed closely by rapid increases in insect numbers and later a peak in fruit abundance closer to the peak of rainfall (see Chapter 2). During periods of leaflessness, sap flow should also be curtailed, thus reducing the opportunities for gum feeding. Thus, we may expect gum feeders and folivores to breed earlier relative to the peak in rainfall, and frugivore–insectivores to be intermediate, followed by more strict frugivores. In practice, in our analyses, the timing of peak fruit abundance was the best absolute predictor of mean birth date for all dietary types, except insectivores, which matched best to the peak of new leaf flush. This pattern suggests that variation in easily assimilated energy (fruit) governs the birth timing of folivores, just as it does that of frugivores.

Fourth, food may not be the only seasonal factor influencing birth timing. In temperate-zone primate species, the difficulty for an infant to thermoregulate means that births must occur within the warm or dry period (Zhao & Deng 1988; Henzi *et al.* 1992). This factor could explain the apparent mismatch between mean birth date (mid May) and peak fruit availability (mid January) in Japanese macaques (Table 11.1) and the significant negative correlation between mean birth date and elevation seen in *Macaca thibetana* (Zhao & Deng 1988).

Challenges for the future

To date, nearly all tests of adaptive timing and seasonality of births have been indirect, using the same basic kind of comparative analysis presented here. Direct tests of adaptation would examine whether survival or growth outcomes of infants depended on their birth timing within a population. A few such analyses have been performed, but not enough to draw any general conclusions. We appeal to researchers with long-term population data sets to examine the fitness consequences of birth dates using individual-level data.

Although many of the smaller-bodied species examined here appeared to fall neatly into one or another set of predicted behaviors of birth timing relative to expected or measured food availability, it was more difficult to understand the birth timing of larger species, especially among the Old World monkeys. Cercopithecine species with relatively high levels of birth seasonality ($r > 0.6$) appeared to divide into those with mean birth dates about 50 days versus those with mean birth dates about 150 days before the climatically predicted period of peak food availability (e.g. see Figs. 11.8 and 11.9). These would appear to correspond to income versus capital breeders, but the reasons for the differences between the species are presently obscure. If capital breeders do correspond to species adapted to irregular phenological cycles, then far better and longer-term data on phenology will be needed to test this idea.

Finally, future tests of birth seasonality and timing will, we hope, be able to rely on larger data sets from long-term studies of known animals. These should be reported in such a way that it is possible to tease apart the contributions of single groups and study years to the overall pattern or seasonality. Cases in which high levels of infanticide may be producing high levels of birth synchrony in single groups (e.g. Rudran [1978];

Butynski [1982]) would then be less likely to confound measures of birth seasonality. With better data on birth timing, quantitative measures of availability of different nutrients, and the length of infant dependency periods, we look forward to more refined and direct tests of the causes of birth timing.

Appendix 11.1

Data sources for birth data

We combed the primate literature for raw data on birth dates. For this analysis, we chose data sets that conformed to several criteria. First, data from captive animals were not used, but we did use data from provisioned populations as long as the provisioning did not constitute a major fraction of the diet. Second, at least one complete year of observation was required, during which all births were recorded and dates estimated to the nearest month. If sampling was not uniform across months for the entire sample or any subsample (for which we could extract the birth data), then we excluded the study in this analysis. Although it is still technically possible to test for seasonality, the mean vector length used in circular statistics would be biased and cannot be compared with other studies in which sampling is uniform across months. If more than one year of observation was available, but the total sample was not an even multiple of 12 months, then the earliest months of data were truncated to provide the largest integer number of years of complete data. Third, at least six total births were needed. Although it is dangerous to conclude much from samples of fewer than ten births, we included three studies with smaller samples because two were from wild species in Madagascar, on which virtually no quantitative information on birth months has been published, and the other was on the night monkey, for which the sample size is not likely to increase. The summary values used in this chapter are provided in Table 11.1. The raw data distributions are available upon request from the author (CJ), except for a few unpublished data sets.

For each data set of monthly birth frequencies, we calculated the mean vector length r, which is a measure of the narrowness of the distribution of the data within the year. A value of 1.0 means that all the data occurred in the same month, while a value of 0.0 means that the data were spaced perfectly evenly over the months. For data sets with small total frequencies, chance coincidence of observations in the same or nearby months can produce spuriously high r values, even if births are in fact non-seasonal.

Therefore, in some analyses, we included the sample size of births as a covariate to control for this possible effect. Probability values for the calculated r values were taken from Batschelet (1981). The intermediate steps in calculating r also permitted an estimation of the mean birth date, although the monthly resolution of the raw data limits the precision of this estimate.

Climatic data

For each primate population for which we could find quantitative data on birth months, we obtained data on rainfall and temperature seasonality. Whenever possible, we used data from the same study site as for the birth data. Otherwise, we used long-term average climate data from online sources. The most common of these was Worldclimate.com, which bases its averages on the data set Global Historical Climatology Network (GHCN) version 1 of the National Climatic Data Center (NCDC). These records were partially validated during compilation; a more accurate but much more limited data set (GHCN version 2, also from the NCDC) was consulted when possible. In those cases where we obtained data from these sources and later found papers with climate data from the actual study site, the values for mean date and seasonality of rainfall and temperature were in very close agreement. In all but one case, the location used for long-term climate averages was within one degree (about 110 km) of the location of the study site. The localities for the climate data reported in Table 11.2 are available from CHJ upon request.

The long-term monthly climate averages for rainfall and temperature were subjected to the same calculations as monthly birth frequencies to produce a mean vector length and mean date for rainfall and temperature. Because the values of rainfall and temperature are not frequencies, the resulting r values cannot be tested statistically. Also, their validity as measures of climatic seasonality for plant production depends on two assumptions: (i) that zero is a meaningful lower bound and (ii) that the relationship between plant productivity and rainfall or temperature is linear. The first assumption seems reasonable enough – plant production is essentially zero in the absence of rain and below 0 °C. The second assumption is less certain. It seems reasonable to argue that plant production is linearly proportional to rainfall over the ranges of values typically observed in primate habitats (i.e. excluding extreme deserts and very wet rain forests). However, biological activity usually increases exponentially (the Q_{10} effect) with temperature up to around 30 °C. To adjust for this effect, we calculated the mean vector length and mean date both for unadjusted temperature T and for adjusted temperature $T^* = 10 \times (2^{T/10} - 1)$. For $T = 0$, 10, 20, and 30 °C, the corresponding values of T^* are 0, 10, 30, and 70.

Phenology data

Data were found for 28 primate study sites covering a total of 49 primate populations for which birth data were also available. We used the data reported in Chapter 2 for eight of these sites, recalculated the values in Chapter 2 for an additional six sites, and added data for 14 sites not covered. Food availability was estimated in at least four different ways: (i) number of species, (ii) density of plants or the percentage of a fixed sample, (iii) summed crown volume or diameter at breast height (DBH) values for plants with a given phenophase, and (iv) biomass. Thus, in analyses of birth seasonality or timing that involved phenological data, we included as a covariate the method used to estimate food availability for each population, in case the method influenced the degree of measured seasonality in food production (as found in Chapter 2). Because of the relatively small total sample sites of study sites with phenological measures, it was not feasible to perform analyses using only one method of estimating food availability. Because studies also varied by total duration and total sample size of trees monitored, we also assigned a rank score to the quality of the different phenological studies, ranging from 0 to 4, based on the following criteria: 0, partial year only or months missing (data interpolated) or total sample less than 50 individual plants (when known); 1, single or multiple full years using number of species; 2, single full year using density-related index of availability (number of trees); 3, single full year using biomass-adjusted indexes (summed DBH or crown volume of food trees or fruit-fall traps); 4, as 2 or 3, but with multiple years of data. This quality score was also used as a covariate in the analyses involving phenological data; in some cases, analyses were restricted to populations with quality scores of 2–4.

Statistical considerations

When we had values of birth dates from more than one population within one degree of latitude, we averaged the values, unless the birth mean vector lengths differed by more than 0.25 (e.g. see Rudran [1973]). For this chapter, no attempt was made to adjust the tests for phylogeny, as many of the tests involve relatively complex general linear models including two categorical variables (continent, diet). Because of uneven taxonomic sampling (see Table 11.1) and the particular structure of primate phylogeny, some of our conclusions may be biased. Although the mean vector length value for squirrel monkeys (*Saimiri oerstedi*) was an outlier in a previous analysis of New World primates (Di Bitetti & Janson 2000), this was not so in the current analysis and it is included in all results here. Excluding it made no difference to any of the qualitative conclusions reported.

Statistical analyses were performed with JMP 3.2. Conventional parametric tests of general linear models were used because specific non-parametric versions of similar tests are not available. In most cases, visual inspection of the residuals did not reveal marked heteroscedasticity, so transformation of the variables was not needed. However, the distribution of body masses was very uneven on a linear scale; nearly all masses were between 0.1 and 10 kg, but two outliers occurred near 30 and 80 kg. To reduce the effect of these outliers, we used log-transformed body masses in this analysis. The type-I error rate was set to 0.05.

Appendix 11.2

Circular statistics make use of the fact that any repeating continuous axis (e.g. months or compass directions) can be graphed as a circle of angles. In this case, each observation is graphed as a vector (an arrow starting from the point $X = 0$, $Y = 0$) of length one with the angle given by 360° multiplied by the position of the observation along the axis, divided by the length of a complete repeat of the axis. In the case of seasons, the total length of the axis is one year, or 365 days. For instance, a birth observed on January 31 would be given the angle $(360 \times 31/365) = 30.6°$. When all the vectors are added together, they sum to a single vector of total length L and angle A. The length of the vector divided by the number of observations is denoted r and is a measure of how closely spaced the observations are along the axis – if all the N observations are identical, then the length L will equal N, so that $r = L/N = 1.0$. If the observations are spaced evenly along the axis, then arrows pointing in opposite directions will cancel each other out so that $L = 0$ as does r. Thus, r is a convenient measure of how concentrated the observations are along the axis. Although it might seem that L should depend on where one starts the axis, in fact it does not – regardless of what absolute angle is assigned to a given observation, similar angles among observations will increase L and dissimilar angles (in particular, those separated by exactly one-half the total repeat length of the axis) will cancel each other.

The value of r can be tested against the null hypothesis of no clustering of observations along the axis, using tables in Batschelet (1981) (the Raleigh test). In addition to the usefulness of r as a measure of seasonal clustering, the angle A of the final vector is a precise measure of the mean or average of the positions of the observations along the axis, whether they cluster at one end or the other or in the middle of the axis. The only potential drawback in the use of r is that it works only for distributions of data with a single mode along the axis – bimodal distributions will tend to give r values close to 0 even if the data are clearly not spread uniformly over the axis. However, this drawback is easily solved by doubling the angle calculated for each observation, and thus in essence examining the presence of clustering within each half-repeat of the axis (Batschelet 1981).

Additional complications arise when attempting to perform correlations or regressions using a circular variable. Because the axis for one variable repeats itself, the expected graph is not a straight line but a spiral graphed on to the cylinder made by connecting the ends of the axis to each other. There is no simple solution to this problem. A variety of techniques, several of which are reviewed in Batschelet (1981), have been suggested to measure and draw statistical inferences from circular correlations. The method we employ is that of Jupp and Mardia (1980), as described in Batschelet (1981) (in the original paper, the formula for calculating the correlation coefficient contains an error, which is corrected in Batschelet's presentation).

References

Altmann, J. (1980). *Baboon Mothers and Infants.* Cambridge, MA: Harvard University Press.

Batschelet, E. (1981). *Circular Statistics in Biology*. London: Academic Press.

Bentley-Condit, V. K. & Smith, E. O. (1997). Female reproductive parameters of Tana river yellow baboons. *International Journal of Primatology*, **18**, 581–96.

Bercovitch, F. B. & Harding, R. S. O. (1993). Annual birth patterns of savanna baboons (*Papio cynocephalus annubis*) over a ten-year period at Gigil, Kenya. *Folia Primatologica*, **61**, 115–22.

Boinski, S. (1987). Birth synchrony in squirrel monkeys (*Saimiri oerstedi*): a strategy to reduce neonatal predation. *Behavioral Ecology and Sociobiology*, **21**, 393–400.

Bourliere, F., Hunkeler, C., & Bertrand, M. (1970). Ecology and behavior of Lowe's Guenon (*Cercopithecus campbelli lowei*) in the Ivory Coast. In *Old World Monkeys*, ed. J. R. Napier and P. H. Napier. New York: Academic Press, pp. 297–350.
Bronson, F. H. (1989). *Mammalian Reproductive Biology*. Chicago: University of Chicago Press.
Butynski, T. M. (1982). Harem-male replacement and infanticide in the blue monkey (*Cercopithecus mitus stuhlmanni*) in the Kibale Forest, Uganda. *American Journal of Primatology*, **3**, 1–22.
— (1988). Guenon birth seasons and correlates with rainfall and food. In *A Primate Radiation: Evolutionary Biology of the African Guenons*, ed. A. Gautier-Hion, F. Bourliere, J.-P. Gautier, J. Kingdon. Cambridge: Cambridge University Press, pp. 284–322.
Caldecott, J. O. (1986). An ecological and behavioural study of the pig-tailed macaque. *Contributions to Primatology*, **21**, 1–259.
Chapman, C. A., Wrangham, R. W., & Chapman, L. J. (1994). Indices of habitat-wide fruit abundance in tropical forests. *Biotropica*, **26**, 160–71.
Charles-Dominique, P. (1977). *Ecology and Behavior of Nocturnal Primates: Prosimians of Equatorial West Africa*. New York: Columbia University Press.
Chism, J., Rowell, T., & Olson, D. (1984). Life history patterns of female patas monkeys. In *Female Primates: Studies by Women Primatologists*, ed. M. F. Small. New York: Alan R. Liss, pp. 175–90.
Crockett, C. M. & Rudran, R. (1987). Red howler monkey birth data I: seasonal variation. *American Journal of Primatology*, **13**, 347–68.
Curtis, D. J. & Zaramody, A. (1999). Social structure and seasonal variation in the behaviour of *Eulemur mongoz*. *Folia Primatologica*, **70**, 79–96.
Di Bitetti, M. S. & Janson, C. H. (2000). When will the stork arrive? Patterns of birth seasonality in Neotropical primates. *American Journal of Primatology*, **50**, 109–30.
Dittus, W. P. J. (1977). The social regulation of population density and age–sex distribution in the toque monkey. *Behaviour*, **63**, 281–322.
Dunbar, R. I. M. & Dunbar, P. (1975). Social dynamics of gelada baboons. *Contributions to Primatology*, **6**, i–viii, 1–157.
Fernandez-Duque, E., Rotundo, M., & Ramirez-Llorens, P. (2002). Environmental determinants of birth seasonality in night monkeys (*Aotus azarai*) of the Argentinean chaco. *International Journal of Primatology*, **23**, 639–56.
Goldizen, A. W., Terborgh, J. W., Cornejo, F., Porras, D. T., & Evans, R. (1988). Seasonal food shortage, weight loss, and the timing of births in saddle-back tamarins (*Saguinus fuscicollis*). *Journal of Animal Ecology*, **57**, 893–901.
Goodall, J. (1983). Population dynamics during a 15 year period in one community of free-living chimpanzees in the Gombe National Park, Tanzania. *Zeitschrift Tierpsychologie*, **61**, 1–60.
Harvey P. H., Martin, R. D., & Clutton-Brock, T. H. (1987). Life histories in comparative perspective. In *Primate Societies*, ed. B. B. Smuts, D. L. Cheney, R. M. Seyfarth, R. W. Wrangham, & T. T. Struhsaker. Chicago: University of Chicago Press, pp. 181–96.

Henzi, S. P., Byrne, R. W., & Whiten, A. (1992). Patterns of movement by baboons in the Drakensberg mountains: primary responses to the environment. *International Journal of Primatology*, **13**, 601–28.

Hsu, M. J., Agoramoorthy, G., & Lin, J. F. (2001). Birth seasonality and inter-birth intervals in free-ranging Formosan macaques, *Macaca cyclopis*, at Mt. Longevity, Taiwan. *Primates*, **42**, 15–25.

Judge, D. S. & Carey, J. R. (2000). Postreproductive life predicted by primate patterns. *Journals of Gerontology*, **A55**, B201–9.

Jupp, P. E. & Mardia, K. V. (1980). A general correlation coefficient for directional data and related regression problems. *Biometrika*, **67**, 163–73.

Kawai, M., Azuma, S., & Yoshiba, K. (1967). Ecological studies of reproduction in Japanese monkeys (*Macaca fuscata*). I. Problems of the birth season. *Primates*, **8**, 35–73.

Kay, R. F. (1984). On the use of anatomical features to infer foraging behavior in extinct primates. In *Adaptations for Foraging in Nonhuman Primates: Contributions to an Organismal Biology of Prosimians, Monkeys, and Apes*, ed. P. S. Rodman & J. G. H. Cant. New York: Columbia University Press, pp. 21–53.

Klein, L. L. (1971). Observations on copulation and seasonal reproduction of two species of spider monkeys, *Ateles belzebuth* and *A. geoffroyi*. *Folia Primatologica*, **15**, 233–48.

Koenig, A., Borries, C., Chalise, M. K., & Winkler, P. (1997). Ecology, nutrition, and timing of reproductive events in an Asian primate, the hanuman langur (*Presbytis entellus*). *Journal of Zoology, London*, **243**, 215–35.

Korstjens, A. H. (2001). The mob, the secret sorority, and the phantoms: an analysis of the socio-ecological strategies of the three colobines of Taï. Ph. D. thesis, Utrecht University.

Kumar, A. & Kurup, G. U. (1985). Sexual behavior of the lion-tailed macaque, *Macaca silenus*. In *The Lion-Tailed Macaque: Status and Conservation*, ed. P. G. Heltne. New York: Liss, pp. 109–30.

Lancaster, J. B. & Lee, R. B. (1965). The annual reproductive cycle in monkeys and apes. In *Primate Behavior: Field Studies of Monkeys and Apes*, ed. I. De Vore. New York: Holt Rinehart and Winston, pp. 486–513.

Lindburg, D. G. (1987). Seasonality of reproduction in primates. In *Comparative Primate Biology*, Vol. 2B, ed. E. J. Mitchell. New York: Alan R. Liss, pp. 167–218.

Lycett, J. E., Weingrill, T., & Henzi, S. P. (1999). Birth patterns in the Drakensberg mountain baboons (*Papio cynocephalus ursinus*). *South African Journal of Science*, **95**, 354–6.

MacRoberts, M. H. & MacRoberts, B. R. (1966). Annual reproductive cycle of barbary ape (*Macaca sylvana*) in Gibraltar. *American Journal of Physical Anthropology*, **25**, 299–304.

Menard, N. & Vallet, D. (1993). Population dynamics of *Macaca sylvanus* in Algeria: a 8-year study. *American Journal of Primatology*, **30**, 101–18.

Nishida, T., Takasaki, H., & Takahata, Y. (1990). Demography and reproductive profiles. In *The Chimpanzees of the Mahale Mountains: Sexual and Life*

History Strategies, ed. T. Nishida. Tokyo: University of Tokyo Press, pp. 63–97.

Oi, T. (1996). Sexual behaviour and mating system of the wild pig-tailed macaque in West Sumatra. In *Evolution and Ecology of Macaque Societies*, ed. J. E. Fa & D. G. Lindburg. New York: Cambridge University Press, pp. 342–68.

Okamoto, K., Matsumura, S., & Wantanabe, K. (2000). Life history and demography of wild moor macaques (*Macaca maurus*): summary of ten years of observations. *American Journal of Primatology*, **52**, 1–11.

Omar, A. & De Vos, A. (1971). The annual reproductive cycle of an African monkey (*Cercopithecus mitis kolbi* Neuman). *Folia Primatologica*, **16**, 206–15.

Pereira, M. E. (1993). Seasonal adjustment of growth rate and adult body weight in ringtailed lemurs. In *Lemur Social Systems and Their Ecological Basis*, ed. P. M. Kappeler & J. U. Ganzhorn. New York: Plenum Press, pp. 205–21.

Pullen, S. L., Bearder, S. K., & Dixson, A. F.(2000). Preliminary observations on sexual behavior and the mating system in free-ranging lesser galagos (*Galago moholi*). *American Journal of Primatology*, **51**, 79–88.

Rahaman, H. & Parthasarathy, M. D. (1969). Studies on the social behavior of bonnet monkeys. *Primates*, **10**, 149–62.

Richard, A. F., Dewar, R. E., Schwartz, M., & Ratsirarson, J. (2000). Mass change, environmental variability and female fertility in wild *Propithecus verreauxi*. *Journal of Human Evolution*, **39**, 381–91.

Ross, C. (1992). Basal metabolic rate, body weight and diet in primates: an evaluation of the evidence. *Folia Primatologica*, **58**, 7–23.

Ross, C. & Jones, K. E. (1999). Socioecology and the evolution of primate reproductive rates. In *Comparative Primate Socioecology*, ed. P. C. Lee. Cambridge: University of Cambridge Press, pp. 73–110.

Rudran, R. (1973). The reproductive cycles of two subspecies of purple-faced langurs (*Presbytis senes*) with relation to the environmental factors. *Folia Primatologica*, **19**, 41–60.

— (1978). Socioecology of the blue monkeys (*Cercopithecus mitis stuhlmanni*) of the Kibale Forest, Uganda. *Smithsonian Contributions to Zoology*, **249**, 1–88.

Ruhiyat, Y. (1983). Socio-ecological study of *Presbytis aygula* in West Java. *Primates*, **24**, 344–59.

Siegel, S. (1956). *Nonparametric Statistics for the Behavioral Sciences*. New York: McGraw-Hill.

Sinclair, A. R. E., Mduma, S. A. R., & Arcese, P. (2000). What determines phenology and synchrony of ungulate breeding in Serengeti? *Ecology*, **81**, 2100–11.

Stanford, C. B. (1991). *The Capped Langur in Bangladesh: Behavioral Ecology and Reproductive Tactics*. Basel, Switzerland: S. Karger.

Strier, K. B., Mendes, S. L., & Santos, R. R. (2001). Timing of births in sympatric brown howler monkeys (*Alouatta fusca clamitans*) and northern muriquis (*Brachyteles arachnoides hypoxanthus*). *American Journal of Primatology*, **55**, 87–100.

Struhsaker, T. T. (1967). Ecology of vervet monkeys (*Cercopithecus aethiops*) in the Masai-Amboseli Game Reserve, Kenya. *Ecology*, **48**, 891–904.

Struhsaker, T. T. (1975). *The Red Colobus Monkey*. Chicago: University of Chicago Press.

— (1977). Infanticide and social organization in the redtail monkey (*Cercopithecus ascanius schmidti*) in the Kibale Forest, Uganda. *Zeitschrift fur Tierpsychologie*, **45**, 75–84.

Van Schaik, C. P. & van Noordwijk, M. A. (1985). Interannual variability in fruit abundance and the reproductive seasonality in Sumatran long-tailed macaques (*Macaca fascicularis*). *Journal of Zoology, London*, **206**, 533–49.

Van Schaik, C. P., Terborgh, J. W., & Wright, S. J. (1993). The phenology of tropical forests: adaptive significance and consequences for primary consumers. *Annual Review of Ecology and Systematics*, **24**, 353–77.

Watts, D. P. (1998). Seasonality in the ecology and life histories of mountain gorillas (*Gorilla gorilla beringei*). *International Journal of Primatology*, **19**, 929–48.

Wright, P. C. (1985). The costs and benefits of nocturnality for *Aotus trivirgatus* (the night monkeys). *Anthropology*. New York: City University of New York.

— (1999). Lemur traits and Madagascar ecology: coping with an island environment. *Yearbook of Physical Anthropology*, **42**, 31–72.

Wu, H. & Lin, Y. (1992). Life history varibles of wild troops of Formosan macaques (*Macaca cyclopis*) in Kentung, Taiwan. *Primates*, **33**, 85–97.

Zhao, Q. & Deng, Z. (1988). *Macaca thibetana* at Mt. Emei, China: II. Birth seasonality. *American Journal of Primatology*, **16**, 261–8.

12 *Energetic responses to food availability in the great apes: implications for hominin evolution*

CHERYL D. KNOTT
Department of Anthropology, Harvard University,
Cambridge MA 02138, USA

Introduction

The past 40 years of great-ape field research have seen the accumulation of a wealth of data that can be used to make predictions about the behavior of early hominins. We have only just begun, however, to undertake great-ape field studies that incorporate a physiological component in the wild. Understanding how the energetic and reproductive systems of living hominoids respond to environmental variation allows us to build more informed models with which to reconstruct the behavior, morphology, and responses to ecological pressures that were present in great-ape and human ancestors.

Fruit is the favored, although not always the most common, food of the great apes, and all ape populations that rely on fruit experience fluctuations in its availability. Given the importance of fruit in ape diets, such fluctuations are likely to have a major impact on these animals' biology and behavior. The primary goal of this chapter is to discuss how differences in overall fruit availability and fallback foods play a central role in shaping energetic and reproductive responses of the great apes, and then to examine the implications of this for hominin evolution.

A review of the literature reveals that relatively few studies present quantitative data on changes in diet or energetics over time. Descriptions are more likely to be statements about how diet or behavior changes in response to fluctuations in food availability. Thus, a secondary aim is to suggest ways in which data from both previous and future research can be presented in order to facilitate truly comparative studies of food availability and energetics.

Seasonality in Primates: Studies of Living and Extinct Human and Non-Human Primates, ed. Diane K. Brockman and Carel P. van Schaik. Published by Cambridge University Press.

Fluctuations in fruit availability: continental differences

Despite popular misconceptions, the rainforests of the world do not produce a constant cornucopia of fruit. The availability of food, especially fruit, fluctuates in all tropical forests where apes live. However, the extent and severity of fluctuations experienced by different species and populations vary. There appear to be some fundamental differences in forest fruit production patterns that may have affected ape evolution in each region.

In Southeast Asian forests, three major features characterize the pattern of fruiting. First, there are dramatic supra-annual peaks in fruit availability (Medway 1972; Appanah 1985; van Schaik 1986) known as mast fruitings. Many forests in this region are dominated by trees in the Dipterocarpaceae family, which is distinguished by its unusual reproductive pattern. Triggered by climatic events, these trees fruit and flower on an irregular basis, once every two to ten years (Ashton *et al.* 1988). They are joined by other tree families, resulting in a phenomenon in which more than 90% of trees produce fruit during these "mast" fruiting events (Medway 1972; Appanah 1981; van Schaik 1986). This provides an overabundance of fruit for animals, such as orangutans, during these periods. Second, Dipterocarpaceae dominance means that a significant portion of the plant biomass is non-reproductive during most years, lowering baseline fruit productivity. Appanah (1985) demonstrated that this results in lower overall productivity of these forests, meaning that they can support substantially less animal biomass than do other tropical rainforests. Due to the concentration of plant reproductive effort during masts, these synchronized fruiting events are often followed by long periods of low fruit availability (Appanah 1985; Leighton 1993; Knott 1998). Third, these supra-annual peaks, and the subsequent extended low-fruit periods, are irregular and unpredictable (Ashton *et al.* 1988).

African forests do not experience this kind of unpredictable community-wide mast fruiting (see Chapter 2). Instead, the predominant pattern is within-year seasonal variation in fruit availability in response to the more pronounced wet and dry seasons compared with Southeast Asia. Fruit availability, of course, also varies from year to year in Africa, but as pointed out in Chapter 2, the pattern is one of more constant higher fruit production with occasional bust years compared with the opposite pattern of generally low fruit production with occasional boom years in Southeast Asia.

Thus, while annual and supra-annual variation in fruit production does occur in African forests, the interannual variation in fruit production is significantly greater in Asia compared with the other tropical regions (see Chapter 2). It is not only that the mast fruitings provide such a superabundance of fruits when they occur, but also that the intervening periods of low fruit availability often tend to be much longer and more severe than those in Africa. Fleming *et al.* (1987) conclude that this fruiting pattern in Southeast Asia leads to forests in which fruit availability is more temporally and spatially patchy than in other tropical forests. Another important continental difference may be that African apes have higher-quality fallback foods to rely on during low-fruit periods, as discussed later in this chapter. This could temper the effects of low-fruit periods on apes in Africa. These differences between continents may help us to understand some of the unique features of orangutans compared with the African apes and may be particularly important for understanding reproductive patterning (van Schaik & van Noordwijk 1985; Knott 1999, 2001).

Energetic adaptations to fruit availability

Fluctuations in the availability of fruit are important because of the responses that they may elicit in their consumers (see also Chapter 3). Given that the focus here is on the energetic and physiological ramifications of such fluctuations, we can predict a number of energetically sensitive responses. Many factors are involved, including the magnitude and duration of the period of fruit scarcity or abundance, the availability of fallback foods, the social structure, and physiological adaptations and limitations. The quality, in addition to the quantity, of the food, is also important. If less fruit is available but it is of higher quality, then foraging time may not need to increase to maintain dietary adequacy. Below is a survey of the responses that have been observed in great apes, focusing particularly on energetic shifts and then moving on to reproductive ramifications.

Changes in diet composition and energy intake, and the impact of fallback foods

Most ape populations incorporate a high percentage of fruit in their diet when it is available. The fruits that are preferred, however, vary between species. Chimpanzees prefer succulent fruit (Wrangham 1977; Basabose 2002). Orangutans also show a high preference for succulent fruit, but they particularly value energy-rich fruits (Leighton 1993) and thus may prefer lipid-dense seeds over succulent fruits if they are higher in calories. Another contrast is that Western lowland gorillas always seem to incorporate large amounts of foliage in their diet, even when sympatric chimpanzees are eating an entirely frugivorous diet (Tutin *et al.* 1991).

When preferred fruit is scarce, apes incorporate fallback foods into their diet. Fallback foods are those that are permanently or frequently available but usually are ignored (Tutin *et al.* 1991) (see also Chapter 3). For apes, such foods include leaves, pith, terrestrial herbaceous vegetation (THV), bark, some insects, and less preferred fruits such as figs (see Table 12.1). These fallback foods are normally lower in quality than preferred foods, being more fibrous (Knott 1998; Tutin *et al.* 1991; Doran *et al.* 2002), lower in carbohydrates (Knott 1998), and/or lower in lipids (Knott 1998). Staple foods are those that are frequently available but form a more regular part of the diet and are adequate nutritionally to allow the animal to subsist on them. In some cases a food such as figs may be a staple at some sites but used as a fallback food at other sites.

It is important to note that some ape populations show little reliance on fallback foods, especially when there is a high herbaceous component to the diet. In particular, mountain gorilla diet does not vary significantly over time, except for increased bamboo eating by some groups when new shoots are put out (Watts 1984, 1998). Bonobos at Lomako (White 1998) do not show significant differences between months in their time spent feeding on fruits, figs, pith, leaves and flowers, and meat. Thus, despite fluctuations in fruit availability, bonobos do not appear to need to rely on certain foods as fallback resources. Taï chimpanzees also seem to be well buffered against seasonality because they have a more diverse set of high-quality preferred foods, including nuts (Boesch & Boesch-Achermann 2000). However, Doran (1997) found that when fruit abundance was unusually poor, Taï chimpanzees showed similar responses to those seen in east African populations.

Data on dietary composition, however, provide limited information on variation in the quality of the diet. We need nutritional analyses of the foods eaten to be able to report quantitatively on differences and assess their significance. This work has been done at a small number of ape study sites. Dietary quality (or, the amount of energy that a diet provides, and how easily it can be extracted) is measured by the relative percentages of nutrients in the diet, with high-quality diets being low in fiber and high in carbohydrates and lipids (Conklin-Brittain *et al.* 1998; Knott 1998).

Complete macronutrient analysis of the diet by season has been done for chimpanzees (Conklin-Britain *et al.* 1998) and for orangutans (Knott 1998, 1999). Four gorilla studies (Rogers *et al.* 1990; Remis *et al.* 2001;

Table 12.1 *Fallback foods eaten by great apes*

Species/study site	Fruits (including seeds)	Leaves and Stems	Pith	Bark	Insects
Orangutans					
Borneo					
Gunung Palung, Indonesia	Some figs	Yes	Yes	Yes	Yes
Kutai, Indonesia	Figs important	Yes	Yes	Yes	Sometimes
Tanjung Putting, Indonesia	No (figs rare)	Yes	Yes	Yes	Yes
Ulu Segama, Malaysia	Some figs	Yes	Yes	Yes	Sometimes
Sumatra					
Ketambe, Indonesia	Figs important	Yes	–	Yes	Yes
Gorillas					
Mountain gorillas					
Karisoke, Rwanda	No fallback foods used	–	–	–	–
Lowland gorillas					
Bai Hokou, CAR	*Duboscia*	Yes	Yes	Yes	No
Lopé, Gabon	Fibrous fruit *Duboscia*	*Zingiberaceae* *Marantaceae*	Yes	Yes	No
Mondika, CAR and/or RC	*Duboscia, Tetrapleura tetraptwra Klainedoxa*	Yes	Yes	Yes	No
Ndoki, Congo	*Duboscia*, figs	Yes	Yes	Yes	No
Kahuzi-Biega, DRC	No fallback foods	–	–	–	–
Bonobos					
Lomako, DRC	No fallback foods used	No	No	No	No
Wamba, DRC	None	Yes	Yes	–	No
Yalosidi, DRC	–	Yes (stems, shoots, sedges)	Yes	–	No
Chimpanzees					
East Africa					
Budongo, Uganda	Figs eaten as staple – no period of scarcity seen				

Table 12.1 (*cont.*)

Species/study site	Fruits (including seeds)	Leaves and Stems	Pith	Bark	Insects
Gombe, Tanzania	–	Yes	Oil-palm pith Some THV	–	–
Kalinzu, Uganda	*Musanga leo-errerae* Figs eaten as staple	Some	No THV eaten as staple	No	–
Kanyawara, Uganda	Figs	No	Yes (THV)	No	–
Mahale, Tanzania	–	Yes	–	Yes	–
Central Africa					
Ipassa, Gabon	–	Yes	Yes	Yes	Yes
Kahuzi-Biega, DRC	Figs eaten as staple	Yes (herbaceous)	Yes (THV)	No	Yes
Lopé, Gabon	*Duboscia macrocarpa* *Elaeis guineensis*	*Zingiberaceae* *Marantaceae*	Yes	Yes	Yes
Ndoki, Congo	*Duboscia, Ficus*	Yes	No	No	No
West Africa					
Bossou, Guinea	*Musanga cecropioides* Oil-palm (*Elaeis guineensis*) nuts	No	Oil-palm pith	Yes	No
Taï, Ivory Coast	Fibrous fruit, *Duboscia viridifolia, Klainedox gabonensis*	Yes	–	–	–

CAR, Central African Republic; DRC, Democratic Republic of Congo; RC, Republic of Congo; THV, terrestrial herbaceous vegtation.

References: Bai Hokou (Goldsmith 1999; Remis *et al.* 2001); Bossou (Yamakoshi 1998; Sugiyama & Koman 1992); Budongo (Newton-Fisher 1999; Newton-Fisher *et al.* 2000); Gombe (Wrangham 1977; Hladik 1977; Goodall 1986); Gunung Palung (Knott 1998, 1999); Ipassa (Hladik 1977); Kahuzi-Biega, chimpanzees (Basabose 2002); Kahuzi-Biega, gorillas (Goodall 1977); Kalinzu (Furuichi *et al.* 2001); Karisoke (Fossey & Harcourt 1977; Vedder 1984; Watts 1984, 1988, 1998); Kanyawara (Wrangham *et al.* 1991, 1993, 1996; Conklin-Brittain *et al.* 1998); Ketambe (Sugardjito *et al.* 1987; Rijksen 1978; Utami Atmoko 2000); Kutai (Rodman 1977; Leighton 1993); Lopé (Tutin *et al.* 1991; Tutin & Fernandez 1993; Rogers *et al.* 1990; Williamson *et al.* 1990); Lomako (Badrian *et al.* 1981; Badrian & Malenky 1984; White 1998); Mahale (Nishida 1974); Mondika (Doran *et al.* 2002); Ndoki (Nishihara 1995; Kuroda *et al.* 1996); Taï (Doran 1997; Boesch & Boesch-Achermann 2000); Tanjung Putting (Galdikas 1979, 1988); Ulu Segama (MacKinnon 1971, 1974); Wamba (Kuroda 1989; Kano 1992; Kano & Mulavwa 1984); Yalosidi (Kano 1983).

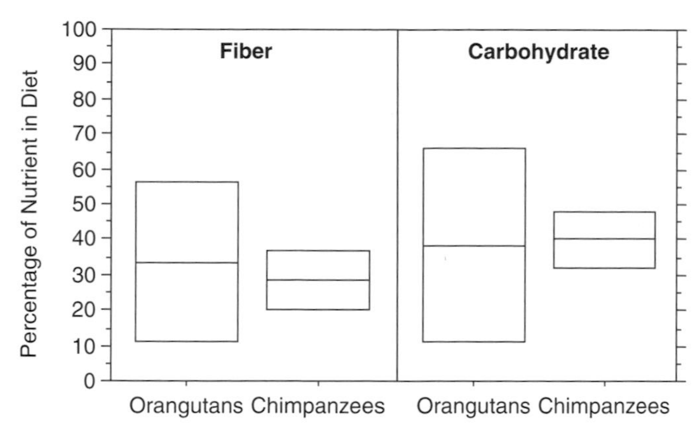

Figure 12.1 Box-plot comparisons of range and mean of percentages of fiber and carbohydrate in the diet of Gunung Palung orangutans (Knott 1999) and Kanyawara chimpanzees (Conklin-Brittain *et al.* 1998).

Popovich *et al.* 1997; Calvert 1985) and two other orangutan studies (Leighton 1993; Hamilton & Galdikas 1994) have examined some of the macronutrients but not seasonal differences. For chimpanzees at Kanyawara, ripe fruit availability was associated with a significant increase in consumption of lipids, free simple sugars, and total non-structural carbohydrates (Conlin-Brittain *et al.* 1998). Protein content of the diet did not vary seasonally. Of the fiber portion of the diet, neutral detergent fiber (NDF) did not vary significantly over the year. During the high-fruit period, the orangutan diet was significantly higher in lipid and carbohydrate and lower in fiber (NDF) compared with in the low-fruit period (Knott 1998). Similar to the diet of chimpanzees, protein content did not significantly vary.

Thus, the general nutrient composition of the diet changes in similar ways in these two ape populations. However, the degree of variability in nutrient composition differs between the Kanyawara chimpanzees and the Gunung Palung orangutans. The percentage of carbohydrate and fiber showed a much wider range of variation in the orangutans, although the means are similar (Fig. 12.1). Thus, compared with chimpanzees, orangutans ate a much higher-quality diet during some months and a lower-quality diet during other months. Caloric intake has been extrapolated for orangutans, and the data show a dramatic response, with five times or more the calories consumed during the high-fruit period compared with in the fruit-poor period (Knott 1998, 1999). It is unknown whether chimpanzees do the same and, if so, to what extent.

Thus, the nature of the fallback food helps explain energetic differences between orangutans and chimpanzees. In this case, the caloric content of the fallback foods and the severity and length of the fruit-shortage period are particularly important. Wrangham *et al.* (1991) show that chimpanzees in Kanyawara, and at many other sites, rely on THV or pith as an important fallback food. Pith appears to be a calorie-rich fallback food resource (Wrangham *et al.* 1991). Orangutans in Borneo do consume pith, but this is not a prime backup food; instead, leaves and bark are the most important during fruit-scarce periods. These foods have about half the calories, and are much higher in fiber and lower in carbohydrate, than their preferred foods (Knott 1998). It may be that the greater availability of pith in African rainforests is an important difference between the two ape habitats, allowing chimpanzees to enjoy the energetic benefits of a better fallback food.

Energy expenditure and energy balance

Studies of the relationship between primate diet and foraging effort suggest that primates that eat high-energy foods with high spatiotemporal patchiness, such as fruit, have longer day ranges than do primates that eat more evenly distributed low-quality foods, such as leaves (Milton & May 1976; Clutton-Brock & Harvey 1977). Thus, we would predict that when diets become more folivorous, apes would decrease their range use, and this does appear, in general, to be the case. Goldsmith (1999) found that Western lowland gorillas responded to decreasing fruit availability by increasing their consumption of leaves, stems, and bark and by decreasing daily path length. Orangutans also conformed to this pattern (Galdikas 1979; Knott 1999). At Gunung Palung, male orangutans significantly decreased their travel time, and both sexes decreased their day range during the period of lowest fruit availability (Knott 1999). This was possible because they shifted to a more folivorous and evenly distributed diet during the fruit-poor period. When fruits are scarce, male chimpanzees also reduce their range size (Wrangham 1977). When food was plentiful, chimpanzee females at Mahale increased their day range (Hasegawa 1990). During the months with highest fruit production, Taï chimpanzees ranged the farthest, whereas they ranged the least during the month of highest *Coula* nut production (Boesch & Boesch-Achermann 2000) and during periods of particular fruit scarcity or when figs with large fruit crops were fruiting (Doran 1997).

How does general activity level, as opposed to travel, respond to changes in food availability? Surprisingly, this has been little studied in apes. Doran (1997) examined the question in chimpanzees at Taï and found that these apes spent more time feeding during the period of fruit scarcity. Data from a study of wild orangutans in Borneo (Knott 1998, 1999) show that orangutans expended more calories during the fruit-rich period. This is explained by a significant increase in caloric expenditure due to travel. Conversely, relying on more abundant leaves and bark and shortening their day range were effective energy-conservation strategies for orangutans during fruit-poor periods.

Energy balance is the difference between energy intake and energy expenditure and indicates whether energy intake during a given period is adequate to meet an organism's needs. Weight data on chimpanzees from Mahale and Gombe show that these animals lose weight when fruit is scarce (Wrangham 1977; Uehara & Nishida 1987), although provisioning may have affected these observations. No weight data are available for wild gorillas or orangutans.

Ketones, excreted in urine, are produced when an organism is metabolizing its own fat deposits in order to meet its energy needs. Chimpanzee researchers at Kanyawara (Wrangham, unpublished data) and Mahale (Huffman, unpublished data) have never detected systematic ketone production. While chimpanzees in these forests are likely experiencing decreases in energetic status during fruit-poor periods, it appears that their energy deficits do not reach the "starvation level" indicated by ketosis. In contrast, during months of severe fruit shortage, ketones do appear in a large proportion of orangutan urine samples at Gunung Palung (Knott 1998). They are totally absent during periods of high fruit availability. These results indicate that during fruit-poor periods, orangutans are metabolizing their fat reserves to make up for insufficient energy intake. However, because the presence of ketones is indicative of starvation, they are unlikely to reveal small weight losses and it is noteworthy that we have not detected ketones during less severe periods of fruit shortage.

When kilocalories consumed were compared with those expended in orangutans, the mast fruiting was shown to lead to a daily caloric intake that exceeded energy requirements by several thousand calories (Knott 1999). This excess indicates that the orangutans were gaining weight during these periods. Correspondingly, during this period, energy balance was negative, indicating that the orangutans were losing weight, a conclusion supported independently by the ketone analyses.

Reproductive responses to energetic status

We know from the considerable body of work on human reproduction (e.g. see Ellison *et al.* [1993]) that human ovarian function is best understood within an ecological context in which female reproductive hormones respond to changes in nutritional intake, energy expenditure, and energy balance. Explicit studies of primate reproductive ecology are only just beginning, but these suggest that the same factors are important in understanding non-human primate, and particularly ape, reproductive functioning as well (Bentley 1999; Knott 2001).

Like humans, apes have long interbirth intervals and exhibit high investment in single offspring. Furthermore, because of the unpredictability of fruit resources, apes such as the orangutan cannot time their births to coincide with peaks of fruit availability. Thus, Knott (2001) has argued that apes have evolved, as did humans, to time reproduction so that conception is more likely during periods of positive energy balance. Van Schaik & van Noordwijk (1985) make the argument that non-seasonally breeding primates, such as great apes, should time reproduction to coincide with periods of high food availability in order to store reserves that can be drawn on later during pregnancy. As suggested by Ellison (1990), if the initiation of reproduction during a state of positive energy balance is associated with improved reproductive outcome, then it should be favored by natural selection. Thus, we should expect female reproductive physiology in the apes to be influenced strongly by energetic factors resulting from fluctuations in food availability. This is in contrast to many seasonally breeding monkeys, where increased food supply is tied to the season of lactation rather than conception (Van Schaik & van Noordwijk 1985; Lindburg 1987).

Evidence of this ecological influence on reproductive timing – the "Ecological Energetics Hypothesis" – is mounting (Knott 1999, 2001) (see Table 12.2) (see also Knott [2001] for a discussion of alternative hypotheses). At Gunung Palung, in Borneo, Knott (1999) found that estrone conjugate (E1C) levels in orangutan urine were significantly lower during the fruit-poor period as opposed to the mast fruiting period. Additionally, matings occurred only during the high-fruit period. This suggests that orangutans responded to the low-fruit period physiologically as well as behaviorally by not investing in reproduction. In captivity, Masters and Markham (1991) showed that E1C levels in orangutans are also associated with energy balance. As in humans, orangutans appear to modulate the probability of conception so that it will be more likely to occur when there are sufficient energy reserves for gestation and lactation.

Several lines of evidence suggest that nutrition is also an important factor in regulating chimpanzee reproduction. First, chimpanzees in captivity – where, presumably, access to energy is more than adequate at all times – that are allowed to keep their babies with them have dramatically shorter interbirth intervals, earlier age at menarche, and earlier age at first birth compared with their wild counterparts (Tutin 1994) (reviewed in

Table 12.2 *Reproductive and ovarian responses to changes in fruit availability from sites where data are available.*

	Ovarian response		Other responses associated with fruit/food availability
Species and site	Fruit-rich period	Fruit-poor period	
Orangutans, Borneo			
Gunung Palung, Indonesia	Higher estrone conjugates All conceptions	Lower estrone conjugates No conceptions	
Tanjung Putting, Indonesia	More conceptions		
Gorillas			
Karisoke, Rwanda	None	None	
Chimpanzees			
Gombe, Tanzania	Increased overall conceptions All postpartum conceptions First full swellings in adolescents Maximal swelling in cycling females Resumption of cycling after weaning Peak number of swellings		High-ranking females with access to best foraging areas reach sexual maturity sooner, and have higher offspring survival, higher annual production of offspring, and shorter interbirth intervals
Kanyawara, Uganda	Increased conceptions		
Mahale, Tanzania	First postpartum swelling One conception peak (of two)		

References: Gombe (Tutin 1975; Goodall 1986; Wallis 1995, 1997; Pusey *et al.* 1997); Gunung Palung (Knott 1999, 2001); Karisoke (Watts 1998); Kanyawara (Sherry 2002); Mahale (Nishida *et al.* 1990); Tanjung Putting (Galdikas 1979, 1988).

Fruit availability was not measured directly at Gombe or Mahale.

Knott [2001]). Second, Wallis (1997) points out in her review of reproductive parameters from chimpanzees at Gombe that the individuals with the shortest interbirth intervals were all descendants of Flo. These family members were the most frequent visitors to the banana feeding station and Wallis speculates that they may have received better nourishment. Additionally, Pusey *et al.* (1997) show that Flo and her family occupied the most productive home range in Gombe.

Finally, seasonality in conceptions has also been reported at Gombe, with the majority of conceptions occurring during the dry season, which is associated with higher fruit availability (Wallis 1997), a finding that Wallis (1997) speculates may have been due to changes in diet. Sherry (2002) (reviewed in Chapter 13) has shown a significant relationship between increased food supply and conception in Kanyawara chimpanzees.

The influence of energetic status on reproduction in gorillas is less clear. Gorillas in captivity often have problems reproducing, and thus any influence of the improved energetic status of captivity on reproduction is obscured. Mountain gorillas have been studied most extensively, but they do not show seasonality in their reproductive parameters, which is not surprising given their unusually stable food supply. Mountain gorillas may maintain reasonably good energetic status year round and thus be reproducing at their full potential (Knott 2001). Future studies of lowland gorillas that focus on the relationship between energetics and reproduction in more seasonal environments should help to clarify this.

Summary of energetic and reproductive responses

The review presented above indicates that the nature of the food supply is one of the key variables determining the energetic and reproductive responses of the great apes to fluctuations in food abundance. Thus, if sufficient variables are considered, then apes can be seen to respond in logical and predictable ways to changes in the food supply. In evaluating why energetic and reproductive responses may differ between ape species and populations, several key conclusions can be drawn: (i) the temporal availability, distribution, and patch size of preferred and fallback foods play a critical role in ranging and grouping patterns; (ii) the caloric content of the diet during high-fruit periods is critical in determining overall energetic adequacy as well as selection for the ability to store excess energy; (iii) the nutritional value of the fallback foods and the efficiency with which nutrients can be extracted from them determine the extent to which energy is suboptimal during fruit-poor periods; (iv) the proportion of the diet that is made up of fallback foods influences the extent to which the fallback diet will be adequate; and (v) the length and severity of the period during which fallback foods are relied upon plays a key role in determining the nature and extent of physiological responses. The energetic responses elicited by apes to these conditions range from modifying ranging patterns and activity budgets to selecting superior fallback foods to storing up energy as fat and then catabolizing it. The reproductive responses mirror the severity of the energy deficits that the population experiences, with interbirth intervals growing as populations experience longer and more extreme periods of scarcity.

Comparison of the Bossou chimpanzees with other east African chimpanzees is particularly illustrative of these points. The three fallback foods eaten by chimpanzees at Bossou – Musanga fruits, oil palm nuts and pith – appear to be very calorie-rich (Hladik 1977; Yamakoshi 1998). Bossou chimpanzees also have much smaller home ranges and thus can, presumably, meet their caloric requirements while expending much less on travel. Thus, the nature of the Bossou food supply, and the animals' ability to access very high-quality fallback foods, means that they should be able to maintain a particularly favorable energy balance compared with other chimp populations. This very high energetic status is also correlated with the shortest interbirth interval of any chimpanzee population and is consistent with the "ecological energetics" model of primate reproductive ecology (Knott 2001).

At the other end of the spectrum sit Borneo orangutans, whose fallback foods are so energy-poor that ketonic starvation is sometimes widespread in a population. The energetic inadequacy of the fallback resources, coupled with the temporal extent of fruit scarcity, is such that these apes have had to evolve, or at least call in to action, other physiological responses to weather such crises. Their ability to store fat during seasons of abundance and then catabolize these stores is essential to their survival. The other important part of this physiological response lies in the fact that the fruit-rich periods allow excess energy consumption; orangutans spend more time feeding than would be predicted just to satisfy their subsistence needs (Knott 1999). This extreme variability in the food supply experienced by orangutans is also associated with the longest interbirth interval of the primates (Galdikas & Wood 1990; Knott 2001) (see also Table 12.3).

Table 12.3 compares the general relationships between fallback foods, energetics, and reproduction in ape populations. The contrast between Bornean orangutan and African ape populations is striking. Whereas orangutans, at least in Borneo, increase their feeding time when fruit is abundant, the opposite finding or more complex results often are reported for chimpanzee and gorilla populations. Orangutans seem to be taking advantage of periods of high fruit abundance to eat as much as possible and put on fat reserves. This is necessitated by the irregular nature of the fruiting peaks, the long periods of very low food availability, and their need to rely on very low-quality fallback foods. Gorillas and chimpanzees, in contrast, may sometimes decrease time spent feeding when fruit, or other preferred food, is abundant. For example, Watts (1988) shows that gorillas at Karioske

Table 12.3 *Relative fallback food quality with energetic responses and associated interbirth intervals*

Species/study site	Quality of fallback foods	Total time feeding		Daily distance traveled		Interbirth interval (years)
		High fruit	Low fruit	High fruit	Low fruit	
Orangutans						
Borneo						
Gunung Palung, Indonesia	Poor	Increase	Decrease	Increase	Decrease	7.0
Kutai, Indonesia	Poor	Increase (variable)	Decrease (variable)	Increase (variable)	–	–
Tanjung Puting, Indonesia	Poor	Increase	Decrease	–	Decrease	7.7
Ulu Segama, Malaysia	Poor	Increase	Decrease	–	–	–
Sumatra						
Ketambe, Indonesia	–	Increase[a]	Decrease[a]	Decrease[a]	Increase[a]	9.3[d]
West Langkat, Indonesia	–	–	Decrease	–	Increase	–
Gorillas						
Mountain gorillas						
Karisoke, Rwanda	None	Decrease (when bamboo shoots abundant)	–	Decrease (when bamboo shoots abundant)	–	3.9
Lowland gorillas						
Bai Hokou, CAR	Good	–	–	–	Decrease	–
Kahuzi-Biega, DRC	None	–	–	Increase (when eating *Myrianthus holstii*)	Decrease (when bamboo shoots abundant)	–
Lopé, Gabon	Good	–	–	Increase	Decrease	–
Mondika, CAR and RC	Good	–	–	Increase (prediction)	Decrease (prediction)	–

Lomako, DRC	Good/excellent	–	–	–	–	8.0
Wamba, DRC	Good/excellent	–	–	–	–	4.5
Chimpanzees						
East Africa						
Budongo, Uganda	None	–	–	–	–	–
Gombe, Tanzania	Good	Increase when fruit high and low; decrease when intermediate (data from males)		Increase	Decrease	5.5
Kanyawara, Uganda	Good	No relationship with time spent eating non-fig fruit[c]		–	–	6.2
Mahale, Tanzania	Good	–	–	Increase (females)	Decrease (females)	6.0
West Africa						
Bossou, Guinea	Excellent	Increase	Decrease	–	–	5.1
Taï, Ivory Coast	Excellent	Decrease	Increase	Increase (except when eating *Coula* nuts or large fig crops)	Decrease	5.8

CAR, Central African Republic; DRC, Democratic Republic of Congo; RC, Republic of Congo.

References: Bai Hokou (Goldsmith 1999; Remis *et al.* 2001); Bossou: (Yamakoshi 1998; Sugiyama & Koman 1992); Budongo (Newton-Fisher 1999); Gombe (Wrangham 1977; Goodall 1986); Gunung Palung: (Knott 1998, 1999); Kanyawara (Wrangham *et al.* 1991, 1993, 1996; Conklin-Brittain *et al.* 1998); Karisoke (Vedder 1984; Watts 1984, 1988); Ketambe (Utami 2000; van Schaik 1986; te Boekhorst *et al.* 1990); Kahuzi-Biega (Goodall 1977); Kutai (Rodman 1977; Mitani & Rodman 1979); Lomako (Badrian *et al.* 1981; Badrian & Malenky 1984; White 1998); Lopé (Tutin 1996; Tutin *et al.* 1991; Tutin & Fernandez 1993; Tutin 1996; Rogers *et al.* 1990; Williamson *et al.* 1990); Mahale (Hasegawa 1990; Nishida 1974); Mondika (Doran *et al.* 2002); Taï (Doran 1997; Boesch & Boesch-Achermann 2000); Tanjung Putting (Galdikas 1979, 1988); Ulu Segama (MacKinnon 1971, 1974); Wamba (Kuroda 1979; Kano 1992); West Langkat (MacKinnon 1974). Reproductive parameters from Knott 2001, except as noted. Relative fruit variability and quality of fallback foods are based on a qualitative cross-site comparison of authors' reports.

[a] Flanged and unflanged males; no data from females.

[b] Fruth unpublished data.

[c] Wrangham personal communication.

[d] Wich *et al.* (2004).

decreased total time spent feeding in areas where food abundance was high because they were able to satisfy their subsistence needs quickly.

Why would all apes not eat as much as they possibly can when preferred food is abundant? This has not been measured directly for most populations, but we would predict that all apes likely *do* increase their caloric intake when fruit (or preferred food) is abundant, even if they decrease total time spent feeding. But, there are tradeoffs between eating and other activities that animals can engage in. More time spent feeding means less time for mating, grooming, hunting, fighting, and other social activities. Goodall (1986), for example, reports that party size, competition for estrus females, and patrol activities tend to decrease the time spent feeding for chimpanzees at Gombe. Costs such as toxin load and limitations on gut and digestive capacity also must be figured in. Thus, how long an animal should keep eating beyond the point when it has met its daily energetic needs may depend largely on the probability that its energetic requirements can be met readily in the future. If fruit availability is unpredictable, if fruit scarce seasons can extend for many months, and if fallback foods are inadequate or too low in nutritional quality to survive on for the long term, then maximization of energy intake, such as that seen in orangutans, makes sense. In contrast, if the availability and distribution of food are fairly consistent and predictable, such as observed in mountain gorillas, then we would not expect huge increases in feeding time during food-rich periods. Large-scale differences in the patterning of resource availability between continents may indeed have selected for some of the differences in feeding behavior seen between the Asian and African apes.

A second major finding from this review is that in most ape populations, daily travel distance is increased when fruit availability is high and decreased during fruit-poor periods. However, this is not the case when high-fruit periods are associated with especially dense or concentrated resources that do not require much travel. For example, travel distance in Taï chimpanzees is decreased when eating rich Coula nuts and when feeding in very large fig crops (Boesch & Boesch-Achermann 2000; Doran 1997). This also may be the explanation for the decrease in travel during high-fruit periods at Ketambe, where huge fig trees form an important part of the diet (Sugardjito *et al.* 1987; Utami *et al.* 1997).

Several additional factors may obscure the above relationships in some ape populations. Party composition in chimpanzees has been shown to fluctuate with resource distribution (Chapman *et al.* 1995). Thus, chimpanzees can use their fission–fusion social behavior to cope with varying resource conditions and may not need to modify other energetic variables. Doran *et al.* (2002) predict that lowland gorillas should modify their group composition based on resource distribution. Second, Goodall (1986) has shown that differences in processing times between different foods are a major determinant of time spent feeding for Gombe chimpanzees. This may be particularly important in ape populations that habitually use tools. Third, competing social factors may influence energetic variables (Goodall 1986). Additionally, a high-versus-low-fruit comparison may be too simplistic at many sites. The temporal and spatial distribution of food, the size of food patches, and the caloric and nutrient content of preferred and fallback foods may not co-vary in the same way between different periods. This is well described for lowland gorillas at Mondika by Doran *et al.* (2002), who divided the year into three distinct periods based on the size, density, and distribution of gorillas foods and predicted differences in ranging and group cohesion based on these differences.

Interbirth intervals also appear to be associated with the degree of fruit fluctuation and the quality of the fallback foods. Orangutans have the longest interbirth intervals, and the data reviewed above suggest that this may be linked to very high variability and unpredictability in fruit production, long periods of low fruit availability due to mast fruiting, and a lack of high-quality fallback foods during these periods. The extent to which conceptions at a given site are concentrated during fruit-rich periods also may prove to be determined by the degree of dietary fluctuation.

Our ability to assess energetic variables in comparable units across sites will make significant progress towards explaining interpopulation and interspecies variation. The importance of continuing long-term studies of individual populations is clear, as data sets will need to be analyzed over multiyear periods to capture the range of supra-annual variation experienced at each site.

Recommendations for comparative studies

This review reveals that our ability to make direct quantitative comparisons of ape food availability between sites is limited due to a lack of comparable data. There are a number of reasons for this. First, ape field studies normally are carried out by primatologists or anthropologists who are not trained botanists or plant ecologists. Thus, the plant phenological component of a study is usually a supplement to the main behavioral data collected and, given limited funds, it is often not feasible to gather the depth of botanical and phenological information needed to make studies comparable. Secondly, tropical rainforest plant communities are extremely complex. To identify and monitor all species eaten by apes within a single locale may not be feasible. This is particularly true of Southeast Asia rainforests, where one study site can have thousands of tree species. The unfortunate consequence of these factors is that although it is now routine practice to gather phenological data, those data usually are only relative measures that can be used to demonstrate fluctuations in fruit availability at a single locale but do not provide quantitative data necessary for comparisons between sites. Although relative measures tell us how apes in a particular population are responding, comparisons among sites and between ape species are essential for understanding the differences observed.

A literal "common currency" of fruit availability is needed to compare between sites. Such a measure needs to take into account not only the quantity but also the quality (value) of fruit available at a given time. We suggest that *kilocalories of ape fruit available/hectare* could be such a comparative measure. To calculate this, we need to be able to do the following: (i) identify accurately the species eaten using long-term data; (ii) monitor the fruiting pattern of a representative sample of fruit species eaten by apes; (iii) count or estimate the absolute crop size of monitored trees as they fruit; (iv) determine the density of each of the species eaten within the home range(s) of the apes in question; (v) weigh each fruit type to determine the grams/fruit; and (vi) analyze the nutrient and caloric content of the fruits that are eaten. This would allow us to compare directly the temporal variability and the quality of ape fruits. Other factors such as patch size, distances between patches, toxin levels, foraging efficiency, and digestibility also could be factored in, but the above comparison would be a good first step.

Finally, it is important that researchers do not use "wet" or "dry" seasons as proxies for fruit availability, since fruit production is not associated consistently with one or another of these seasons but varies markedly from site to site (see Chapter 2). Rainfall may have no relationship to fruit abundance, as White (1998) found at Lomako. Year-to-year variation may be such that "dry-season" months in one year may have more rain than "wet-season" months in other years (White 1998).

Additionally, great potential exists for future studies of reproductive ecology in the apes and other primates. Most of the existing data were not gathered with this question in mind and thus do not have the precision needed to address this issue fully. In particular, we need more studies that examine the caloric content of the diet and its effect on reproductive hormones. Quantitative changes in food supply, or a simple dry season/ wet season analysis, may not be a precise enough indication of the actual energetic intake that we expect reproductive physiology to respond to.

Ovarian function needs to be compared with actual feeding behavior, ideally caloric intake, to properly gauge the effect of nutritional status on reproductive functioning. Future analysis of data on seasonal fluctuations in phenology, caloric intake, energetic patterns, and reproduction should yield further insight into these relationships among the apes.

Implications for hominin evolution

Changes in environmental conditions have been proposed as catalyzing agents during critical junctures of human evolution (Vrba 1985a, 1985b, 1989; Rogers *et al.* 1994). Environmental fluctuations have become more and more extreme since the Miocene Epoch (Potts 1998). Starting at six million years ago, deep-sea cores reveal an isotopic record of oxygen enrichment, indicating a marked global cooling trend with accompanied drying (deMenocal 1995). This general trend, however, is characterized by a two- to threefold increase in the degree of environmental fluctuations during the period of most recent hominin evolution (Potts 1998). The most extreme climatic oscillations seem to have occurred around 2.5, 1.7, and 1.0 Ma (deMenocal 1995; deMenocal & Bloemendal 1995). The implication of this is that early hominins probably experienced increasingly dramatic seasonal fluctuations in food availability. Using our knowledge from great apes, we can now look at the adaptations of early hominins in light of the energetic challenges that such seasonality would have posed.

Dental microwear studies suggest that fruits represented a significant component of the australopithecine diet (Teaford *et al.* 2002). The importance of fruit likely varied from site to site and through time, but, with the possible exception of *Paranthropus*, the austrolipithecine-grade hominins appear to have relied heavily on fruit and therefore would have been faced with very real energetic challenges as the availability of these resources became more patchy in both time and space. Thus, the fallback foods of these hominins would have played a central role in shaping these species' adaptations and evolution.

The dental anatomy of the australopithecines suggests that their fallback foods were significantly harder, tougher, or more gritty than fruit (Teaford *et al.* 2002). These hominins show a consistent and increasingly marked posterior megadontia accompanied by thickened molar enamel. The exact nature of the foods to which these teeth were designed has been the subject of considerable debate in the literature and remains enigmatic. However, it seems unlikely that these fallback resources were the same as those exploited by extant apes, since the thickened enamel and large flat molar crowns displayed by these hominins are very different from the thinner-enameled, more sharply cusped molars of the living apes. There is increasing evidence that underground storage organs (USOs), such as roots, tubers, and rhizomes, may have been selected as fallback foods by australopithecines and paranthopines (Hatley & Kappelman 1980; Wrangham *et al.* 1999; Conklin-Brittain *et al.* 2002). Interestingly, Conklin-Brittain *et al.* (2002) argue that the substitution of USOs for THV and other ape piths as fallback foods would have substantially increased the overall quality of the hominin diet over that of the chimpanzee's. Energetically, if early hominins did indeed adopt a more calorically adequate, abundant, and spatiotemporally reliable fallback food, then the impact of fruit fluctuations would have been reduced significantly. Thus, even though periods of food scarcity were likely longer and more severe than those experienced by extant apes, these early, hominins may have uncovered – literally – a way to moderate the fluctuations in the quality and energetic adequacy of their diet.

As in the apes, a second response of hominins may have been to modify their activity budgets and ranging behavior. This is especially interesting given the fact that one of the major morphological shifts characterizing the australopithecines is a switch to bipedal locomotion. As a way of saving energy, the adoption of bipedality may have represented a major energetic victory in coping with a changed resource base; movement between more distant energy-rich fruit patches would have been increasingly feasible (Rodman & McHenry 1980; Leonard & Robertson 1995, 1997a, 1997b) (see Steudel-Numbers [2003] for an alternative view).

The advent of the genus *Homo* might instructively be seen here as the time when hominins developed even more effective means of buffering themselves from perturbations in their food supply. Unlike the poor-quality, calorically marginal fallback foods exploited by most great apes, members of the genus *Homo* increasingly exploited high-quality, energy-rich alternatives to seasonal fruit, including both difficult-to-process plant foods, such as nuts and tubers, and meat (Milton 1999; Kaplan *et al.* 2000). Other technological advances in food processing, such as tool use and cooking (Wrangham *et al.* 1999; Knott 2001), would have further enhanced this pattern of improved resource quality and better access to energy at all times of the year.

Adopting new resources and restructuring energy budgets and ranging costs might not have been the only ways in which *Homo* responded to periods of low fruit availability. It is possible that they, like orangutans, were able to store fat during periods of abundance, which could then be drawn upon during lean periods. While there is no direct evidence for fat-storage abilities in early hominins, modern humans clearly possess this trait. However, since chimpanzees and gorillas do not tend to store large amounts of fat during seasons of plenty, the time at which fat storage evolved is debatable.

Although they had adopted and refined a suite of behaviors that improved access to high-quality fallback foods, this does not mean that these hominins were never subjected to periods of caloric scarcity. However, the extent to which the lean seasons caused energetic shortfalls is unclear. Indeed, the evolution of increasingly larger brains within the human lineage may signal that energy availability was no longer limited or unreliable in the way that it had been in early hominins. Brains are costly to grow and to maintain, and thus buffering mechanisms – of a dietary, technological, or physiological nature – must have been in place to ensure that this critical organ had reliable access to the energy it constantly demands. Fat storage may have been one of those brain-buffering mechanisms.

Reconstructions of hominin evolution tend to focus on food acquisition, ranging patterns, tool use, etc. These are assuredly important variables. Leonard and Robertson (1997a, 1997b), for example, argue that hominin energetic needs increased dramatically with the advent of the genus *Homo*. However, lacking from these descriptions is a discussion of how reproduction, particularly female reproduction, would have been affected by changing energetic status caused by fluctuations in the food supply. Data from great apes and modern humans (see Chapter 13) tell us that maintaining positive energy balance in order to achieve conception would have been an important part of the equation (Knott 2001).

The interaction between food, food acquisition, and reproduction may have been particularly important in changing the dynamics of the hominin interbirth interval. If *Homo* females were able to buffer more effectively against food seasonality and energy fluctuations and thus maintain a higher energetic status, this could have been one important factor in the shortening of the interbirth interval in humans as compared with great apes (Knott 2001). Changes in the food supply, food-acquisition techniques, offspring provisioning, and social structure would have made shorter interbirth intervals possible through allowing for the production of overlapping nutritionally dependent offspring for the first time in hominoid history (Lancaster & Lancaster 1983; Knott 2001). These adaptations would have allowed our ancestors to reproduce more rapidly and perhaps less seasonally. Additionally, if part of the human adaptation that allowed us to evolve large brains is due to the ability to switch to a lower-fiber, higher-quality diet (Aiello & Wheeler 1995), then it also may have increased the need to maintain positive energy balance in females to nourish developing human brains. Thus, even in the face of an increasingly variable environment, hominins appear to have circumvented the "energy crisis" that these environmental fluctuations would have posed for other apes.

Acknowledgments

I would like to thank Diane Brockman and Carel van Schaik for asking me to be part of this book and for their constructive comments on the text. Tim Laman and Nancy Conklin-Brittain provided useful comments and discussion. I wish to particularly thank Catherine Smith for her assistance in reviewing the literature and editing the manuscript.

References

Aiello, L. C. & Wheeler, P. (1995). The expensive tissue hypothesis: the brain and the digestive system in human and primate evolution. *Current Anthropology*, **36**, 199–221.

Appanah, S. (1981). Pollination in Malaysian primary forests. *Malay Forester*, **44**, 37–42.

— (1985). General flowering in the climax rain forests of South-east Asia. *Journal of Tropical Ecology*, **1**, 225–50.

Ashton, P. S., Givnish, T. J., & Appanah, S. (1988). Staggered flowering in the Dipterocarpaceae: new insights into floral induction and the evolution of mast fruiting. *American Naturalist*, **132**, 44–66.

Badrian, B. & Malenky, R. K. (1984). Feeding ecology of *Pan paniscus* in the Lomako Forest, Zaire. In *The Pygmy Chimpanzee*, ed. R. L. Susman. New York: Plenum Press, pp. 233–74.

Badrian, A., Badrian, N., & Susman, R. L. (1981). Preliminary observations on the feeding behavior of *Pan paniscus* in the Lomako Forest of central Zaire. *Primates*, **22**, 173–81.

Basabose, A. K. (2002). Diet composition of chimpanzees inhabiting the montane forest of Kahuzi, Democratic Republic of Congo. *American Journal of Primatology*, **58**, 1–21.

Bentley, G. R. (1999). Aping our ancestors: comparative aspects of reproductive ecology. *Evolutionary Anthropology*, **7**, 175–85.

Boesch, C. & Boesch-Achermann, H. (2000). *The Chimpanzees of the Taï Forest: Behavioural Ecology and Evolution*. New York: Oxford University Press.

Chapman, C. A., Wrangham, R. W., & Chapman, L. J. (1995). Ecological constraints on group size: an analysis of spider monkey and chimpanzee subgroups. *Behavioral Ecology and Sociobiology*, **36**, 59–70.

Clutton-Brock, T. H. & Harvey, P. H. (1977). Primate ecology and social organization. *Journal of Zoology, London*, **183**, 1–39.

Conklin-Brittain, N. L., Wrangham, R., & Hunt, K. D. (1998). Dietary response of chimpanzees and cercopithecines to seasonal variation in fruit abundance: II. Macronutrients. *International Journal of Primatology*, **19**, 971–98.

Conklin-Brittain, N. L., Wrangham, R. W., & Smith, C. C. (2002). A two-stage model of increased dietary quality in early hominid evolution: the role of fiber. In *Human Diet: Its Origin and Evolution*, ed. P. S. Ungar & M. F. Teaford. Westport, CT: Bergin and Garvey, pp. 61–76.

DeMenocal, P. B. (1995). Plio-Pleistocene African climate. *Science*, **270**, 53–9.

DeMenocal, P. B. & Bloemendal, J. (1995). Plio-Pleistocene climatic variability in subtropical Africa and the paleoenvironment of hominid evolution. In *Paleoclimate and Evolution with Emphasis on Human Origins*, ed. E. A. Vrba, G. H. Denton, T. C. Partridge, & L. H. Burckle. New Haven: Yale University Press, pp. 262–88.

Doran, D. (1997). Influences of seasonality on activity patterns, feeding behavior, ranging, and grouping patterns in Taï chimpanzees. *International Journal of Primatology*, **18**, 183–206.

Doran, D., McNeilage, A., Greer, D., *et al.* (2002). Western lowland gorilla diet and resource availability: new evidence, cross-site comparisons, and reflections on indirect sampling methods. *American Journal of Primatology*, **58**, 91–116.

Ellison, P. T. (1990). Human ovarian function and reproductive ecology: new hypotheses. *American Anthropologist*, **92**, 952–93.

— (2001). *Reproductive Ecology and Human Evolution.* New York: Walter de Gruyter.

Ellison, P. T., Panter-Brick, C., Lipson, S. F., & O'Rourke, M. T. (1993). The ecological context of human ovarian function. *Human Reproduction*, **8**, 2248–58.

Fleming, T. H., Breitwisch, R., & Whitesides, G. H. (1987). Patterns of tropical vertebrate frugivore diversity. *Annual Review of Ecology and Systematics*, **18**, 91–109.

Fossey, D. & Harcourt, A. H. (1977). Feeding ecology of free ranging mountain gorillas (*Gorilla gorilla beringei*). In *Primate Ecology: Feeding and Ranging Behavior of Monkeys, Lemurs, and Apes*, ed. T. H. Clutton-Brock. New York: Academic Press, pp. 415–47.

Furuichi, T., Hashimoto, C., & Tashiro, Y. (2001). Fruit availability and habitat use by chimpanzees in the Kalinzu forest, Uganda: examination of fallback foods. *International Journal of Primatology*, **22**, 929–46.

Galdikas, B. M. F. (1979). Orangutan adaptation at Tanjung Puting Reserve: mating and ecology. In *The Great Apes*, ed. D. L. Hamburg & E. R. McCown. London: W. A. Benjamin, pp. 195–233.

— (1988). Orangutan diet, range, and activity at Tanjung Puting, Central Borneo. *International Journal of Primatology*, **9**, 1–35.

Galdikas, B. M. F. & Wood, J. W. (1990). Birth spacing patterns in humans and apes. *American Journal of Physical Anthropology*, **83**, 185–91.

Goldsmith, M. L. (1999). Ecological constraints on the foraging effort of western gorillas (*Gorilla gorilla gorilla*) at Bai Hokou, Central African Republic. *International Journal of Primatology*, **20**, 1–23.

Goodall, A. (1977). Feeding and ranging behaviour of a mountain gorilla group (*Gorilla gorilla beringei*) in the Tshibinda-Kahuzi region, Zaire. In *Primate Ecology: Feeding and Ranging Behavior of Monkeys, Lemurs, and Apes*, ed. T. H. Clutton-Brock. New York: Academic Press, pp. 415–47.

Goodall, J. (1986). *The Chimpanzees of Gombe: Patterns of Behavior*. Cambridge: Harvard University Press.

Hamilton, R. A. & Galdikas, B. M. F. (1994). A preliminary study of food selection by the orangutan in relation to plant quality. *Primates*, **35**, 255–63.

Hasegawa, T. (1990). Sex differences in ranging patterns. *In The Chimpanzees of the Mahale Mountains: Sexual and Life History Strategies*, ed. T. Nishida. Tokyo: University of Tokyo Press, pp. 99–114.

Hatley, T. & Kappelman, J. (1980). Bears, Pigs, and Plio-Pleistocene hominids: a case for the exploitation of below ground food resources. *Human Ecology*, **8**, 371–87.

Hladik, C. M. (1977). Chimpanzees of Gabon and chimpanzees of Gombe: some comparative data on the diet. In *Primate Ecology*, ed. T. H. Clutton-Brock. London: Academic Press, pp. 481–501.

Kano, T. (1983). An ecological study of the pygmy chimpanzees (*Pan paniscus*) of Yalosidi, Republic of Zaire. *International Journal of Primatology*, **4**, 1–31.

— (1992). *The Last Ape: Pygmy Chimpanzee Behavior and Ecology*. Stanford: Stanford University Press.

Kano, T. & Mulavwa, M. (1984). Feeding ecology of the pygmy chimpanzees (*Pan paniscus*) of Wamba. In *The Pygmy Chimpanzee*, ed. R. L. Susman. New York: Plenum Press, pp. 233–74.

Kaplan, H., Hill, K., Lancaster, J., & Hurtado, A. M. (2000). A theory of human life history evolution: diet, intelligence, and longevity. *Evolutionary Anthropology*, **9**, 156–84.

Knott, C. D. (1998). Changes in orangutan diet, caloric intake and ketones in response to fluctuating fruit availability. *International Journal of Primatology*, **19**, 1061–79.

— (1999). Reproductive, physiological and behavioral responses of orangutans in Borneo to fluctuations in food availability. Ph. D. thesis, Harvard University.

— (2001). Female reproductive ecology of the apes: implications for human evolution. In *Reproductive Ecology and Human Evolution*, ed. P. Ellison. New York: Aldine de Gruyter, pp. 429–63.

Kuroda, S. (1979). Grouping of the pygmy chimpanzee. *Primates*, **20**, 161–83.

— (1989). Developmental retardation and behavioral characteristics of pygmy chimpanzees. In *Understanding Chimpanzees*, ed. P. G. Heltne & L. A. Marquardst. Cambridge: Harvard University Press, pp. 184–93.

Kuroda, S., Nishihara, T., Suzuki, S., & Oko, R. A. (1996). Sympatric chimpanzees and gorillas in the Ndoki Forest, Congo. In *Great Ape Societies*, ed. W. C. McGrew, L. F. Marchant, & T. Nishida. Cambridge: Cambridge University Press, pp. 71–81.

Lancaster, J. B. & Lancaster, C. S. (1983). Parental investment: the hominid adaptation. In *How Humans Adapt*, ed. D. Ortner. Washington, DC: Smithsonian Institution Press, pp. 33–58.

Leighton, M. (1993). Modeling diet selectivity by Bornean orangutans: evidence for integration of multiple criteria for fruit selection. *International Journal of Primatology*, **14**, 257–313.

Leonard, W. R. & Robertson, M. L. (1995). Energetic efficiency of human bipedality. *American Journal of Physical Anthropology*, **97**, 335–8.

— (1997a). Comparative primate energetics and human evolution. *American Journal of Physical Anthropology*, **102**, 265–81.

— (1997b). Rethinking the energetics of bipedality. *Current Anthropology*, **38**, 304–9.

Lindburg, D. G. (1987). Seasonality of reproduction in primates. In *Behavior, Cognition, and Motivation*, ed. G. Mitchell & J. Erwin. New York: Alan R. Liss. pp. 167–218.

MacKinnon, J. (1971). The orangutan in Sabah today. *Oryx*, **11**, 141–91.

MacKinnon, J. R. (1974). The behaviour and ecology of wild orang-utans (*Pongo pygmaeus*). *Animal Behaviour*, **22**, 3–74.

Masters, A. & Markham, R. J. (1991). Assessing reproductive status in orangutans by using urinary estrone. *Zoo Biology*, **10**, 197–208.

Medway, L. (1972). Phenology of a tropical rain forest in Malaya. *Biological Journal of the Linnean Society*, **4**, 117–46.

Milton, K. (1999). A hypothesis to explain the role of meat-eating in human evolution. *Evolutionary Anthropology*, **8**, 11–21.

Milton, K. & May, M. L. (1976). Body weight, diet and home range area in primates. *Nature*, **259**, 459–62.

Mitani, J. C. & Rodman, P. S. (1979). Territoriality: the relation of ranging pattern and home range size to defendability, with an analysis of territoriality among primate species. *Behavioral Ecology and Sociobiology*, **5**, 241–51.

Newton-Fisher, N. E. (1999). The diet of chimpanzees in the Budongo Forest Reserve, Uganda. *African Journal of Ecology*, **37**, 344–54.

Newton-Fisher, N. E., Reynolds, V., & Plumptre, A. J. (2000). Food supply and chimpanzee (*Pan troglodytes schweinfurthii*) party size in the Budongo Forest Reserve, Uganda. *International Journal of Primatology*, **21**, 613–28.

Nishida, T. (1974). Ecology of wild chimpanzees. In *Human Ecology*, ed. R. Ohtsuka, J. Tanaka, & T. Nishida. Tokyo: Kyoritsu-suppan, pp. 15–60.

Nishida, T., Takasaki, H., & Takahata, Y. (1990). Demography and reproductive profiles. In *The Chimpanzees of the Mahale Mountains: Sexual and Life History Strategies*, ed. T. Nishida. Tokyo: University of Tokyo Press, pp. 63–97.

Nishihara, T. (1995). Feeding ecology of western lowland gorillas in the Nouabale-Ndoki National Park Park, Congo. *Primates*, **36**, 151–68.

Potts, R. (1998). Variability selection in hominid evolution. *Evolutionary Anthropology*, **7**, 81–96.

Pusey, A., Williams, J., & Goodall, J. (1997). The influence of dominance rank on the reproductive success of female chimpanzees. *Science*, **277**, 828–31.

Remis, M. J., Dierenfeld, E. S., Mowry, C. B., & Carroll, R. W. (2001). Nutritional aspects of Western lowland gorilla (*Gorilla gorilla gorilla*) diet during seasons of fruit scarcity at Bai Hokou, Central African Republic. *International Journal of Primatology*, **22**, 807–36.

Rijksen, H. D. (1978). *A Field Study on Sumatran Orang-utans (Pongo pygmaeus abelii, Lesson 1827): Ecology, Behavior, and Conservation*. Wageningen, The Netherlands: H. Veenman and Zonen.

Rodman, P. S. (1977). Feeding behavior of orangutans in the Kutai Reserve, East Kalimantan. In *Primate Ecology*, ed. T. H. Clutton-Brock. London: Academic Press, pp. 383–413.

Rodman, P. S. & McHenry, H. M. (1980). Bioenergetics and the origin of hominid bipedalism. *American Journal of Physical Anthropology*, **52**, 103–6.

Rogers, M. E., Maisels, F., Williamson, E. A., Fernandez, M., & Tutin, C. E. G. (1990). Gorilla diet in the Lopé reserve, Gabon: a nutritional analysis. *Oecologia*, **84**, 326–39.

Rogers, M. J., Harris, J. W. K., & Feibel, C. S. S. (1994). Changing patterns of land use by Plio-Pleistocene hominids at the Lake Turkana basin. *Journal of Human Evolution*, **27**, 139–58.

Sherry, D. (2002). Reproductive seasonality in chimpanzees and humans: ultimate and proximate factors. Ph. D. thesis, Harvard University.

Steudel-Numbers, K. L. (2003). The energetic cost of locomotion: humans and primates compared to generalized endotherms. *Journal of Human Evolution*, **44**, 255–62.

Sugardjito, J., te Boekhorst, I. J. A., & Van Hooff, J. A. R. A. M. (1987). Ecological constraints on the grouping of wild orang-utans (*Pongo pygmaeus*) in the Gunung Leuser National Park, Sumatra, Indonesia. *International Journal of Primatology*, **8**, 17–41.

Sugiyama, Y. & Koman, J. (1992). The flora of Bossou: its utilization by chimpanzees and humans. *African Study Monograph*, **13**, 127–69.

Teaford, M. A., Ungar, P. S., & Grine, F. E. (2002). Paleontological evidence for the diets of African Plio-Pleistocene hominins with special reference to early *Homo*. In *Human Diet: Its Origin and Evolution*, ed. P. S. Ungar & M. F. Teaford. Westport, CT: Bergin and Garvey, pp. 143–66.

Te Boekhorst, I. J. A., Schurmann, C. L., & Sugardjito, J. (1990). Residential status and seasonal movements of wild orang-utans in the Gunung Leuser Reserve (Sumatera, Indonesia). *Animal Behaviour*, **39**, 1098–109.

Tutin, C. E. G. (1975). Sexual behaviour and mating patterns in a community of wild chimpanzees. Ph. D. thesis, University of Edinburgh.

— (1994). Reproductive success story: variability among chimpanzees and comparisons with gorillas. In *Chimpanzee Cultures*, ed. R. W. Wrangham, W. C. McGrew, F. B. M. de Waal, & P. G. Heltne. Cambridge: Harvard University Press, pp. 181–93.

— (1996). Ranging and social structure of lowland gorillas in the Lope Reserve, Gabon. In *Great Ape Societies*, ed. W. C. McGrew, L. F. Marchant, & T. Nishida. Cambridge: Cambridge University Press, pp. 58–70.

Tutin, C. E. G. & Fernandez, M. (1993). Composition of the diet of chimpanzees and comparisons with that of sympatric lowland gorillas in the Lopé Reserve, Gabon. *American Journal of Primatology*, **30**, 195–211.

Tutin, C. E. G., Fernandez, M., Rogers, M. E., Williamson, E. A., & McGrew, W. C. (1991). Foraging profiles of sympatric lowland gorillas and chimpanzees

in the Lopé Reserve, Gabon. *Philosophical Transactions of the Royal Society of London, Series B*, **334**, 179–86.
Uehara, S. & Nishida. (1987). Body weights of wild chimpanzees (*Pan troglodytes schweinfurthii*) of the Mahale Mountains National Park, Tanzania. *American Journal of Physical Anthropology*, **72**, 315–21.
Utami Atmoko, S. S. (2000). Bimaturism in orang-utan males: reproductive and ecological strategies. Ph. D. thesis, Utrecht University.
Utami, S., Wich, S. A., Sterck, E. H. M., & van Hooff, J. A. R. A. M. (1997). Food competition between wild orangutans in large fig trees. *International Journal of Primatology*, **18**, 909–27.
Van Schaik, C. P. (1986). Phenological changes in a Sumatran rainforest. *Journal of Tropical Ecology*, **2**, 327–47.
Van Schaik, C. P. & van Noordwijk, M. A. (1985). Interannual variability in fruit abundance and the reproductive seasonality in Sumatran long-tailed macaques (*Macaca fascicularis*). *Journal of Zoology, London (A)*, **206**, 533–49.
Vedder, A. L. (1984). Movement patterns of free-ranging mountain gorillas (*Gorilla gorilla beringei*) and their relation to food availability. *American Journal of Primatology*, **7**, 73–88.
Vrba, E. S. (1985a). Ecological and adaptive changes associated with early hominid evolution. In *Ancestors: The Hard Evidence*, ed. E. Delson. New York: Alan R. Liss, pp. 63–71.
— (1985b). Environment and evolution: alternative causes of the temporal distribution of evolutionary events. *South African Journal of Science*, **81**, 229–36.
— (1989). Late Pliocene climatic events and hominid evolution. In *Evolutionary History of the "Robust" Australopithecines*, ed. F. Grine. Chicago: Aldine, pp. 405–26.
Wallis, J. (1995). Seasonal influence on reproduction in chimpanzees of Gombe National Park. *International Journal of Primatology*, **16**, 435–51.
— (1997). A survey of reproductive parameters in the free-ranging chimpanzees of Gombe National Park. *Journal of Reproduction and Fertility*, **109**, 121–54.
Watts, D. P. (1984). Composition and variability of mountain gorilla diets in the central Virungas. *American Journal of Primatology*, **7**, 323–56.
— (1988). Environmental influences on mountain gorilla time budgets. *American Journal of Primatology*, **15**, 195–211.
— (1998). Long-term habitat use by mountain gorillas (*Gorilla gorilla beringei*). 2. Reuse of foraging areas in relation to resource abundance, quality and depletion. *International Journal of Primatology*, **19**, 681–702.
White, F. J. (1998). Seasonality and socioecology: the importance of variation in fruit abundance to Bonobo sociality. *International Journal of Primatology*, **19**, 1013–27.
Wich, W. A., Utami-Atmoko, S. S., Setia, T. M., *et al*. (2004). Life history of wild Sumatran orangutans (*Pongo abelii*). *Journal of Human Evolution*, **47**, 385–98.
Williamson, E. A., Tutin, C. E. G., Rogers, M. E., & Fernandez, M. (1990). Composition of the diet of lowland gorillas at Lopé in Gabon. *American Journal of Primatology*, **21**, 265–77.

Wrangham, R. (1977). Feeding behaviour of chimpanzees in Gombe National Park, Tanzania. In *Primate Ecology*, ed. T. H. Clutton-Brock. London: Academic Press, pp. 503–38.

Wrangham, R. W., Conklin, N. L., Chapman, C. A., & Hunt, K. D. (1991). The significance of fibrous foods for Kibale Forest chimpanzees. *Philosophical Transactions of the Royal Society of London, Series B*, **334**, 171–8.

Wrangham, R. W., Conklin, N. L., Etot, G., *et al.* (1993). The value of figs to chimpanzees. *International Journal of Primatology*, **14**, 243–56.

Wrangham, R. W., Chapman, C., Clark-Arcadi, A. P., & Isabirye-Basuta, G. (1996). Social ecology of Kanywara chimpanzees: implications for understanding the costs of great ape groups. In *Great Ape Societies*, ed. W. C. McGrew, L. F. Marchant, & T. Nishida. Cambridge: Cambridge University Press, pp. 45–57.

Wrangham, R. W., Jones, J. H., Laden, G., Pilbeam, D., & Conklin-Brittain, N. (1999). The raw and the stolen: cooking and the ecology of human origins. *Current Anthropology*, **40**, 567–94.

Yamakoshi, G. (1998). Dietary response to fruit scarcity of wild chimpanzees at Bossou, Guinea: possible implications for ecological importance of tool use. *American Journal of Physical Anthropology*, **106**, 283–96.

13 *Human birth seasonality*

PETER T. ELLISON, CLAUDIA R. VALEGGIA*
& DIANA S. SHERRY
Department of Anthropology Harvard University
Cambridge MA 02138, USA
* and Consejo Nacional de Investigaciones
Científicas y Tecnológicas (CONICET) Argentina

Introduction

Seasonal variation in the frequency of births is a nearly universal phenomenon in human populations (Cowgill 1965; Lam & Miron 1991; Bronson 1995). Indeed, the absence of birth seasonality in any particular population can be considered a remarkable observation (Brewis *et al.* 1996; Pascual *et al.* 2000). However, the broad prevalence of human birth seasonality does not imply simple causation. Several mechanisms have been proposed to account for the seasonality of births in different specific cases, and most investigators acknowledge that multiple causes are almost certainly involved. Nevertheless, some of the causes of human birth seasonality are likely to have deeper roots in our biology as hominoid primates than others. It will not be possible to review all the hypotheses that have been put forth or to survey the extensive empirical literature on human birth seasonality in this chapter. Rather, the purpose of this chapter is to discuss some of the major hypothesized causes of human birth seasonality in a way that highlights their relevance to the evolutionary ecology of our species and their relationship to the ecology of other primates.

The major hypotheses regarding human birth seasonality can be grouped under three headings: seasonality due to social factors that influence the frequency of intercourse; seasonality due to climatological factors that directly affect human fecundity; and seasonality due to energetic factors that principally affect female fecundity. The first group of hypotheses does not necessarily posit any change in reproductive physiology underlying human birth seasonality, placing primary emphasis on behavior. The second group of hypotheses posits changes in reproductive physiology tied to climatological variation, but often without implying any clear adaptive significance to the resulting pattern. The third group of hypotheses both posits a physiological basis for birth seasonality and argues for the adaptive significance of the underlying mechanism.

Seasonality in Primates: Studies of Living and Extinct Human and Non-Human Primates, ed. Diane K. Brockman and Carel P. van Schaik. Published by Cambridge University Press.

It is important to acknowledge at the outset, however, that these different hypotheses are not mutually exclusive and to reiterate that more than one cause can be operating simultaneously on a given population. Given this fact and the inherently cyclical nature of seasonal phenomena, distinguishing between competing hypotheses can be extremely difficult. It is also important to note that several of the major hypotheses regarding human birth seasonality actually view seasonality in the strict sense as an epiphenomenon and not as a consciously desired or adaptively designed outcome. That is, some hypotheses posit that a given aspect of human reproduction is contingent on an environmental variable for reasons that are independent of any seasonal pattern in that variable. When that particular environmental variable varies in a strongly seasonal pattern, reproductive seasonality may result as an epiphenomenon. For example, a relationship between high ambient temperature and sperm production might produce reproductive seasonality in an environment with seasonally high temperatures. The relationship between temperature and sperm production, however, may have an ultimate explanation that is totally unrelated to environmental seasonality. This differs from mechanisms that are hypothesized to have evolved in order to synchronize reproductive patterns with a seasonally varying environment.

Social factors influencing human birth seasonality

Until relatively recently, the leading hypothesis for human birth seasonality in the demographic literature has been that social factors influencing the probability of intercourse often lead to seasonal clustering of conceptions (Huntington 1938; Udry & Morris 1967). Conception peaks often can be identified with major secular and religious holidays, such as the Christmas–New Year season in the USA and the August vacation period in France (National Center for Health Statistics 1966; Lam & Miron 1994). Differences in conception peaks between different religious groups are sometimes observed to correlate with differences in the occurrence of major religious festivals or other aspects of the religious calendar (Rajan & James 2000).

Socially determined seasonal patterns of conception in developed countries often are viewed as divorced from any important ecological causation. In subsistence societies, however, seasonal cycles of social activity and interaction often are shaped and constrained by the annual subsistence cycle. In many agricultural populations, for example, marriages tend to be clustered either before or after the major season of agricultural labor

(Danubio & Amicone 2001). In migratory populations, including circumstances of seasonal labor migration, that undergo seasonal dispersal and aggregation, marriage and conception peaks often are observed in the season of aggregation (Menken 1979; Condon 1982; Huss-Ashmore 1988). Major cultural and religious festivals also tend to be coordinated with the calendar of annual subsistence activities. It has been argued that the cultural patterning of conceptions in subsistence societies is thus partly a result of ecological constraints, producing the most pronounced birth seasonality in populations where the subsistence regime is itself most seasonally constrained (Condon & Scaglion 1982).

The seasonality of human subsistence activities in non-industrial societies usually correlates with seasonality of weather conditions and with seasonality of food intake and workload. Because of this, there is a certain degree of overlap between hypotheses of human birth seasonality based on social factors and those based on climatological and energetic factors. For instance, both the heat of the summer months and the exhaustion of the agricultural work season have been suggested as causes of reduced frequency of intercourse in agricultural societies, as has anxiety for the coming harvest (Stoeckel & Chowdhury 1972; Thompson & Robbins 1973; Malina & Himes 1977; Ayeni 1986). As noted above, it is not necessary that we treat these as mutually exclusive possibilities. Rather, in this context, we can include them as factors related to subsistence ecology that may influence the seasonality of conception.

If we view the social patterning of conception frequency as having a historical basis in the constraints of human subsistence ecology, then we may be able to see some phylogenetic continuity as well. Group size can influence mating opportunity in some ape species, and where group size is influenced by ecological factors, seasonality of mating opportunity can result. The occurrence of large fruiting patches can, for instance, lead to increased mating frequency in chimpanzees (Wrangham 2000; Knott 2001). Orangutans provide a particularly extreme example of the influence of ecological seasonality on mating frequency. Synchronous masting patterns or major food trees results in increasing density of both female and male orangutans, higher encounter rates between potential mating partners, and higher mating frequencies than during non-mast periods (Knott 1997a, 1999).

Climatological factors influencing human birth seasonality

As noted above, climatological factors sometimes have been invoked as causes of seasonal variation in the frequency of intercourse within human populations. High ambient temperatures often are cited as having a negative influence, while heavy rainfall sometimes is cited as having a positive influence, on the assumption that it leads to couples spending more leisure time indoors (Thompson & Robbins 1973; Dyson & Crook 1981). More recently, however, increasing attention has been paid to the possibility that climatological factors have a direct influence on human reproductive physiology and thereby on human fecundity, or the biological capacity to conceive and bear offspring. The two factors most often discussed in this context are temperature and photoperiod.

Sperm production in mammals is temperature-sensitive and optimized at temperatures below the core body temperature of many species (Bedford 1991). As a result, testes are located in external scrotal sacs in most mammals, including primates (Bedford 1977). Sperm production in humans similarly has been observed to be temperature-sensitive, even leading to suggestions for contraceptive manipulation (Appell & Evans 1977; Kandeel & Swerdloff 1988).

This well-known temperature sensitivity of sperm production has led to interest in whether high ambient temperatures could compromise male fecundity by affecting sperm quantity or quality. Certain geographical patterns tend to support such a hypothesis. A north–south gradient in the amplitude of birth seasonality in the USA has been noted, with higher amplitude at lower latitude (Lam & Miron 1994, 1996). It has also been suggested that the amplitude of the birth peak in the southern states has declined in recent years with the wider prevalence of air-conditioning (Seiver 1989). Initial studies in the southern USA produced evidence of significant seasonal variation in sperm production among men working outdoors, with suppressed values being found in the hot months of the year. However, subsequent studies found similar seasonal patterns in the indices of sperm production among men who worked in air-conditioned indoor environments and among those who worked outdoors at high ambient temperatures, weakening the case for high ambient temperature exposure as a cause of the pattern (Levine *et al.* 1988; Levine 1994).

The temperature sensitivity of sperm production clearly has been a selective force in mammalian evolution resulting in external testes as a common anatomical characteristic. Given the ancient nature of this adaptation, however, and the evolutionary origin of our species in tropical latitudes, it is difficult to imagine that humans would be less well adapted to ambient temperature effects than other tropical primate species. If temperature effects on male fecundity do occur in modern human populations, then they are likely to result from temperature exposures that would be uncharacteristic of our evolutionary past, as a result of either the colonization of extreme habitats or the adoption of clothing that inhibits natural scrotal heat loss.

Photoperiod also has been suggested to influence human fecundity (Wehr 2001). The argument in this case stems primarily from the observation of photosensitive secretion of melatonin by the pineal gland. Pineal melatonin secretion is known to be important in the maintenance of both circadian cycles and seasonal patterns of reproduction in many mammals (Karsch *et al.* 1984; Tamarkin *et al.* 1985; Bronson 1989).

Melatonin secretion in humans is photosensitive, occurring primarily at night (Bojkowski *et al.* 1987; Wehr *et al.* 1993). Evidence of seasonal variation in the period or magnitude of melatonin secretion in human populations is equivocal, however (Illnerova *et al.* 1985; Kauppila *et al.* 1987; Matthews *et al.* 1991; Wehr 1991). It is possible that artificial light now buffers natural photoperiodicity in many populations and so attenuates any seasonal effect. However, it is not clear whether variation in melatonin levels or periods of secretion within the normal range has any effect on the human reproductive axis. Most clinical reviews of the subject have concluded that the evidence for such an effect is weak (Brzezinski & Wurtman 1988).

There is no question that the physiological substrate for sensitivity to photoperiod is intact in most mammals, including primates, even if it does not have a direct influence on reproductive physiology. Given the presence of such an endogenous signal of day length, it is easy to imagine selection acting to make use of that information to coordinate reproduction when circumstances render such coordination advantageous. Photoperiod sensitivity has been documented in rhesus monkeys related to patterns of seasonal reproduction observed both in the wild and in captivity (Wilson & Gordon 1989; Chik *et al.* 1992; Wehr 2001). As human populations spread out of the tropics to higher latitudes, they may have become subject to new selective pressures for the coordination of reproduction, with day length as a reliable predictor of environmental conditions. The best use of that information is not intuitively obvious, however. It would depend on there being a single critical period in the human reproductive cycle that could be optimized by seasonal coordination; whether that critical period is conception, gestation, birth, early lactation, or weaning is debatable. Nor is it clear whether the critical feature of the environment is temperature, food availability, or disease prevalence.

Thus, while an effect of photoperiod on human fecundity may be reasonable from a phylogenetic and evolutionary perspective, clear evidence of such an effect in contemporary human populations is lacking. Nor are the specific selective pressures that would have activated a latent capacity to respond to photoperiod understood well.

Energetic factors influencing female fecundity

Reproduction is an energetically demanding process for most organisms. Among placental mammals, females bear a particularly heavy energetic burden due to the metabolic requirements of gestation and lactation. Humans are no exception. At the end of gestation and during early lactation, the metabolic requirements of an offspring can equal as much as a third of its mother's non-pregnant metabolic budget. A large part of the offspring's metabolic demand derives from the requirements of its large and rapidly growing brain, a characteristic that, while shared with other Old World primates, is particularly accentuated in humans.

A number of features of human reproductive biology help to meet the energetic demands of reproduction for females, many of these being shared with other primates (Prentice & Whitehead 1987; Martin 1990; Lee *et al.* 1991). Fetal and infant growth rate tend to be slow relative to maternal mass, and the gestation period is relatively long compared with that of other mammals, resulting in a relatively low daily energy demand. Maternal metabolism favors fat storage early in gestation, when the energetic demands of the pregnancy are low. These energy reserves are then mobilized to help meet the high demands of late gestation and early lactation. Under conditions of constrained energy availability, human females can lower their own basal metabolic rates to free up energy for gestation and lactation, even at a cost to their own wellbeing.

In addition to these general features of energetic management, there is substantial evidence that human ovarian function is directly sensitive to female energetic conditions (Ellison 2001). Three aspects of female energetics need to be distinguished in this context: (i) energy status, or the amount of stored somatic energy that can be mobilized to meet the demands of reproduction, primarily in the form of fat and glycogen; (ii) energy balance, or the net of energy intake over expenditure; and (iii) energy flux, or the total rate of energy turnover. Although often correlated, these aspects of energetics are logically separable and often vary independently. Low energy status, for example, is often associated with a recent history of negative energy balance. But two individuals can have the same energy status (level of stored fat) while being in opposite states of energy balance, one losing weight and fat while the other is gaining weight and fat. Similarly, both high energy flux, as in a trained endurance athlete, and low energy flux, as in a famine victim, can be associated with low energy status and neutral energy balance.

Considerable evidence now exists indicating that human ovarian function is sensitive to energy balance. Even modest changes in weight, in the range of 2 kg in a month, are associated with significant changes in the production of the principal ovarian hormones, estradiol and progesterone (Pirke *et al.* 1985; Lager & Ellison 1990; Lipson & Ellison 1996). These changes in turn are associated with changes in the probability of conception in a given cycle. In addition, there is evidence that high energy flux is associated with suppressed ovarian function, even when energy balance is neutral and energy status is normal. Moderate recreational exercise, for example, has been associated with lower ovarian steroid levels in women of constant weight (Ellison & Lager 1986). Similarly, heavy agricultural work has been associated with suppressed ovarian steroid profiles, even when energy intake is increased to result in neutral energy balance (Jasienska & Ellison 1998). Evidence for an independent effect of energy status on ovarian function is lacking, however, at least within the normal range of healthy human variation in energy status (Ellison 2001). As noted above, extremes of energy status are difficult to dissociate from the confounded effects of energy balance and energy flux.

Several physiological pathways exist by which energetic variables can affect female ovarian function. Both high and low energy flux have been shown to reduce the pulsatile release of gonadotropin hormones by the pituitary, suggesting an effect on the hypothalamic-pituitary (H–P) axis (Boyar *et al.* 1974; Veldhuis *et al.* 1985; Vigersky *et al.* 1977; Dixon *et al.* 1984). Both endogenous opioids and corticotropin-releasing hormone, which are increased under conditions of energetic stress, may contribute to this suppression of pituitary gonadotropin release (Loucks *et al.* 1989; Loucks 1990; Ferin 1999). Leptin, a protein hormone secreted by adipose tissue, may affect the H–P axis, in this case supporting pituitary gonadotropin release (Flier 1998). Leptin also may affect ovarian function directly (Spicer & Francisco 1997; Brannian *et al.* 1999). Leptin levels are affected not only by the amount of adipose tissue that an individual has but also by energy balance and energy flux (Rosenbaum *et al.* 1997; Warden *et al.* 1998). Leptin production is also influenced strongly by both insulin, a major regulator of energy metabolism, and gonadal steroids (Rosenbaum & Leibel 1999; Carmina *et al.* 1999). Hence, the information carried by leptin levels is a complex signal reflecting energetic variables, gender, and reproductive state. Finally, there is increasing evidence that ovarian function is subject to regulation by major metabolic hormones, including insulin and growth hormone (Poretsky *et al.* 1999; Yoshimura *et al.* 1993). Insulin in particular stimulates the ovarian response to gonadotropins (Willis *et al.* 1996). Insulin levels increase under conditions that favor energy storage and therefore help to promote ovarian function at such times.

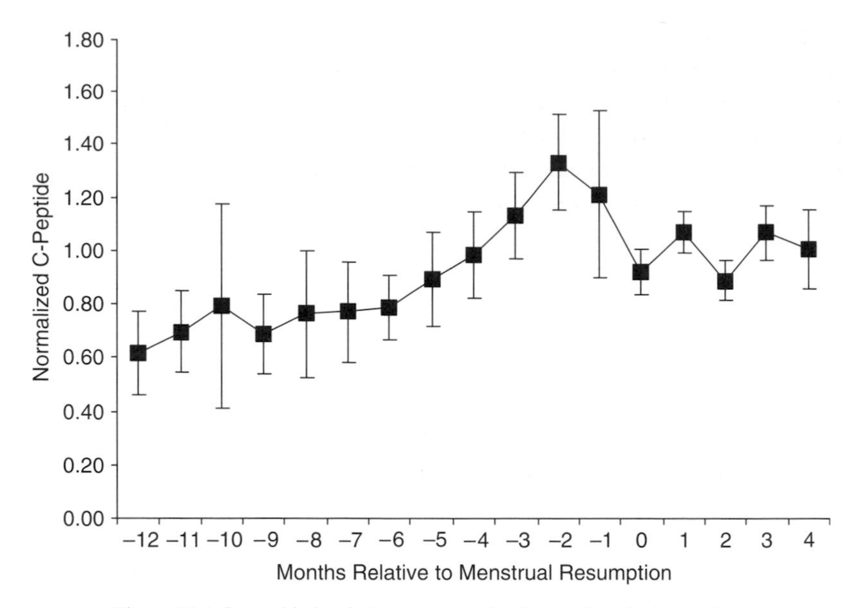

Figure 13.1 C-peptide levels (mean ± standard error) in relation to the resumption of menstrual cycles postpartum in 70 lactating Toba women from northern Argentina, expressed as a fraction of each individual's average level when menstruating regularly. C-peptide is a marker of insulin production.

The resumption of ovarian function postpartum is also sensitive to energetic variables (Valeggia & Ellison 2001). Intensive lactation represents a particularly high state of energy flux that is usually associated with suppressed ovarian activity (Ellison 1995). It appears that this low level of ovarian activity is more a function of suppressed responsiveness to gonadotropin stimulation than a function of low gonadotropin stimulation itself (McNeilly *et al.* 1994). As the metabolic demand of lactation declines, particularly with the introduction of supplementary foods to the infant's diet, ovarian function resumes. Resumption of ovarian function is correlated closely with a shift in energy balance reflected in rising insulin levels (Valeggia & Ellison 2001; Ellison & Valeggia 2003). The rise in insulin levels is, in turn, implicated in the restoration of ovarian responsiveness to gonadotropin stimulation (Fig. 13.1).

The sensitivity of ovarian function to energetic variables provides a mechanism by which ecological constraints on energy availability for reproduction can affect the probability of conception. It is likely that this sensitivity is adaptive and reflects the action of natural selection (Ellison 2001). Given that reproduction represents a sizable metabolic commitment for a female over and above her own maintenance requirements, success will be dependent on the ability of the female to divert energy to reproduction. Positive energy balance is the single best predictor of such excess energy availability, and negative energy balance is a strong predictor of constraint. High and low energy flux ordinarily also indicate a situation of constrained energy availability for humans, the former representing high workloads and the latter inadequate intakes (Jasienska 2001). Energy status, perhaps counterintuitively, is often a poor indicator of current energetic constraints, being more reflective of past history than of current conditions. The sensitivity of ovarian function to these signals of energetic constraint may well have evolved because energetic constraints have such an immediate impact on the ability of a female to meet the metabolic demands of reproduction.

When ecological constraints on energy availability vary seasonally, they can result in a seasonal pattern of conceptions and births. This pathway has been traced in detail for the Lese, a group of slash-and-burn horticulturalists living in the Ituri Forest of the Democratic Republic of the Congo (Ellison *et al.* 1989; Bailey *et al.* 1992). The subsistence ecology, reproductive ecology, and demography of the Lese were studied for over a decade, from the early 1980s to the early 1990s. Long-term data revealed a typical "hunger" season between January and June, when food stores from the previous year's harvest become progressively depleted. During the hunger season, most individuals in the population lose weight, indicating negative energy balance. Indices of ovarian function, including levels of estradiol and progesterone, frequency of ovulation, and duration of menstruation all decline in parallel with weight during this period, while the average intermenstrual interval increases. With the onset of the new harvest season in June, body weights begin to increase and the indices of ovarian function improve. As one would predict from this pattern of ovarian function, the distribution of births and backdated conceptions over the decade of observation indicates a dearth of conceptions during the hunger season and a peak in conceptions during the period of positive energy balance following the new harvests. Variation in the severity of the hunger season between years was also correlated with variation in the degree of conception clustering.

Birth seasonality in the western Toba community of Vaca Perdida

Other agricultural populations show similar patterns of birth seasonality correlated with harvests and either observed or imputed variation in energy balance. Foraging populations are often but not always, buffered from extreme seasonality by their ability to shift between food resources, (Hurtado & Hill 1990). One example of birth seasonality associated with a foraging subsistence ecology is presented by our current work with the indigenous populations of northern Argentina.

The Toba, who belong to the Guaycuruan linguistic family, are one of the many indigenous peoples inhabiting the Gran Chaco of South America (Miller 1999). Historically, these populations have been semi-nomadic, hunter-gatherers. Nowadays, Chacoan Indians can be found in sedentary villages with different degrees of acculturation, from rural communities relying on the forest and river for most of their subsistence to urban communities with cash economy.

Vaca Perdida is a Toba village of approximately 250 people located in western Formosa, northern Argentina. The western region of the Gran Chaco is characterized by low xeric vegetation, patches of thorny bushes, bromeliads, and cacti. Seasonal changes in temperature and resources are pronounced. The first rains of September mark the beginning of the rainy and "bountiful" season. During the summer, western Toba have abundant food from the forest, such as game, fruits, and honey. The highlight of the season is the collection of algarroba (*Prosopis alba*) fruits, which reaches its peak in December (Mendoza 1999). Riverine communities enjoy a second bountiful season during the winter (June–July), as this is when fish in the Pilcomayo River are most abundant (Mendoza 1999).

The western Toba have been experiencing a profound ecological and cultural transition, which started in the 1930s with the arrival of the Anglican church missionaries (Gordillo 1994). One of the most important changes in their lifestyle has been the sedentarization of their communities, including Vaca Perdida. However, until the late 1980s, most villagers were still relying heavily on hunting, fishing, and gathering for their subsistence. During the late 1980s and early 1990s, the subsistence economy changed from mostly hunting/gathering to a newly acquired cash economy. However, during the last half of the 1990s, many men lost their cash-paid jobs (Gordillo 2002). For the past five years, at the time of writing, many families in Vaca Perdida have relied on a combination of foraging and temporary jobs.

We analyzed a total of 256 births from 105 women, corresponding to all births occurring in Vaca Perdida between 1980 and 2001, to determine whether the changes in subsistence ecology were associated with changes in birth seasonality. We divided the sample into three groups: births occurring before the transition to market economy (1980–86, $n = 78$), births occurring during the increase in male waged labor (1987–1993, $n = 78$), and births occurring after the transition (1994–2001, $n = 100$). There was a marked

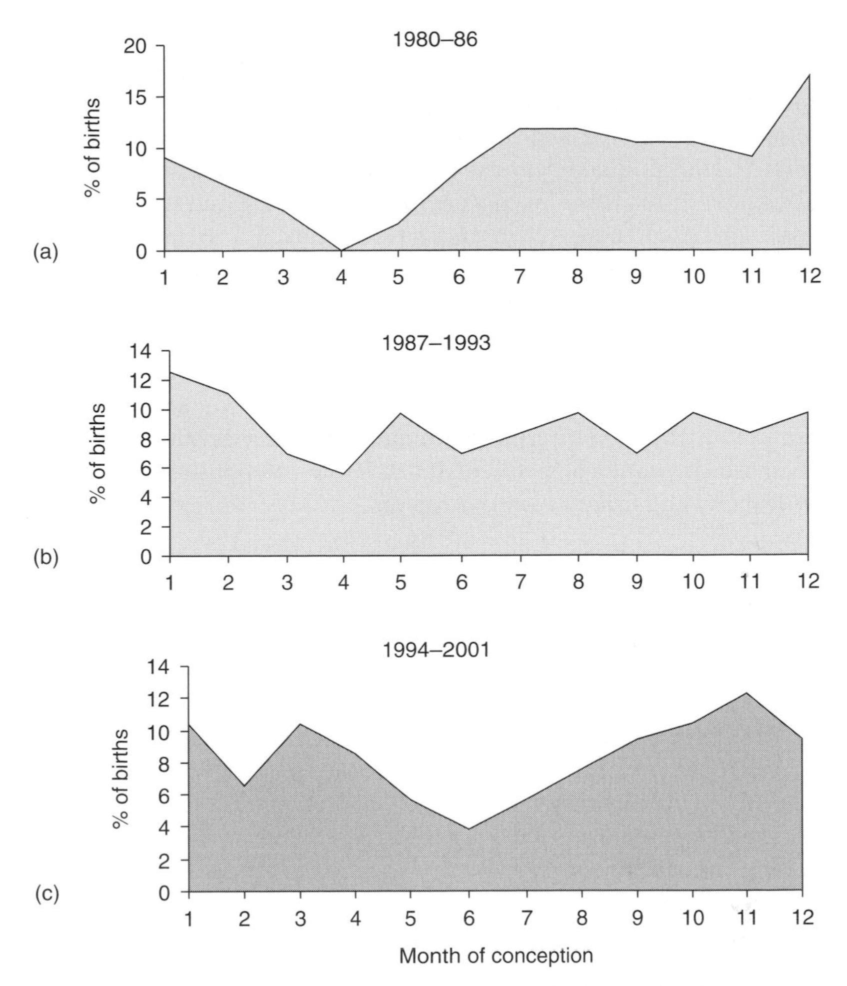

Figure 13.2 Distribution of births by inferred month of conception for the women of Vaca Perdida, Argentina, during three time periods. Seasonality of the 1980–86 period is significant at the $P < 0.05$ level.

birth seasonality in the first group (Rayleigh's $z = 7.41$, $P < 0.001$). Births occurred more frequently in late fall and winter (May–September) than during the summer (December–March) (Fig. 13.2a). Conception frequency rises from the late winter, after the fishing season, and continues to be high through the spring and summer periods of resource abundance, declining in the fall after the algarroba season has passed. Births occurring between 1987 and 1993 are distributed more uniformly across the year (Rayleigh's $z = 0.89$, $P > 0.2$) (Fig. 13.2b). The availability of cash in most households buffered the seasonal variation in food availability and reduced the seasonal energy expenditure associated with foraging (Gordillo 1994). Although not as seasonal as the 1980–86 period, conceptions occurring during the socioeconomically unstable mid and late 1990s are less frequent during the winter (Fig. 13.2c). Informal interviews with Vaca Perdida villagers indicated that the winter is a "bad time for *changas* [temporary jobs]" and that younger men "have forgotten" how to fish and have developed a preference for store-bought food. Therefore, they say, the winter has become their hunger season.

These data are preliminary and should be taken with caution due to the small sample size. However, they do suggest that changes in the subsistence ecology of the Vaca Perdida community are associated with changes in the degree of birth seasonality that they experience. We are currently collecting reproductive histories and hormonal data on seven other villages in the area, which should help to further test the relationship between food availability and fecundity in this region.

Energetics and birth seasonality in other great apes

Although it has been known for some time that apes do not exhibit restricted birth seasons, seasonal distributions in the frequency of births have become evident as a result of long-term field studies. Early reports suggested the possibility of birth peaks for mountain gorillas (Lancaster & Lee 1965), gibbons (Lindburgh 1987), and chimpanzees (Goodall 1986), but small sample sizes and long interbirth intervals made it difficult to observe and document reproductive seasonality for apes in the wild. Over the past decade or so, a clearer picture has emerged with respect to chimpanzees as data continue to accumulate from long-term field studies in a diverse set of habitats and locations.

Chimpanzee communities are being studied in semideciduous woodlands at Gombe, Tanzania, in lakeshore forest at Mahale, Tanzania, in mid-altitude rainforest at Kibale, Uganda, in lowland rainforest at Taï, Ivory Coast, and in highland rainfall/secondary forest at Bossou, Guinea. Reports from Gombe (Wallis 1995, 1997), Mahale (Nishida 1990), Taï (Boesch & Boesch-Achermann 2000), and Kibale (Sherry 2002) indicate that seasonal birth peaks are a typical feature of chimpanzee reproduction in the wild. Like natural fertility populations of humans, births for chimpanzees generally take place throughout the year, but with notable peaks or clusters in certain seasons. For chimpanzees, the pattern of reproductive seasonality appears to be related to seasonal rainfall and changes in food supply (Wallis 1995; Nishida 1990; Boesch & Boesch-Achermann 2000).

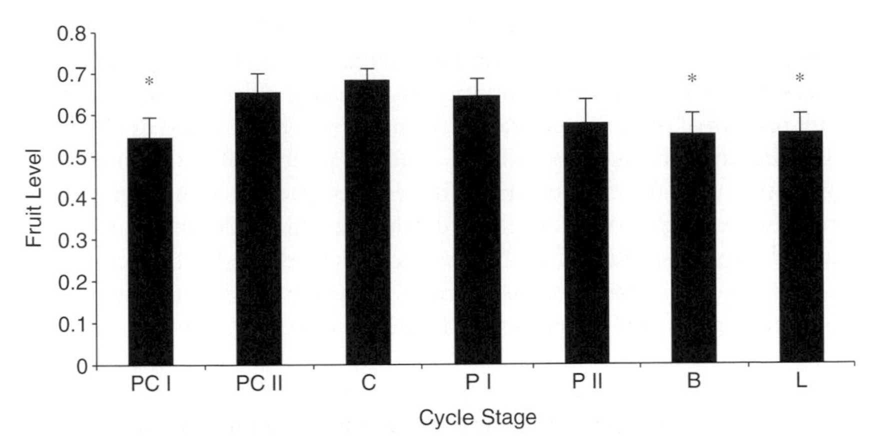

Figure 13.3 The fraction of observed feeding episodes (mean ± standard error) per month that involved ripe fruit among reproductive female chimpanzees in Kibale, Uganda, by stage of the reproductive cycle. PCI, 3–4 months preconception; PCII, 1–2 months preconception; C, month of conception; PI, months 2–4 of pregnancy; PII months 5–7 of pregnancy; B, month of birth; L, first three months of lactation.

*Significantly different from C at the $P < 0.05$ level.

Recently, research at Kibale has focused on a more detailed, functional explanation of chimpanzee reproductive seasonality and has found that conceptions tend to occur during periods of increased food supply. Based on long-term feeding records and phenological data, Sherry (2002) has evaluated food availability and dietary quality across the various stages of the chimpanzee reproductive cycle. The most favorable feeding conditions were associated with the conception period rather than with birth or lactation. Conceptions occurred in association with periods in which high-quality food in the diet (ripe fruit and preferred succulent fruits) was both more abundant in the environment and recorded more frequently in feeding observations compared with both preceding and subsequent stages of the reproductive cycle (Fig. 13.3). The least favorable feeding conditions were associated with the birth window and early lactation phase of the reproductive cycle, both of which contained significantly lower proportions of ripe and succulent fruits in feeding observations compared with the conception period (analysis of variance [ANOVA], $P < 0.05$). The early lactation phase also showed the highest proportion of figs in feeding observations. Figs are considered a less preferred fallback food for chimpanzees at Kibale (Wrangham *et al.* 1993).

Additional evidence of an effect of energetics on the reproductive ecology of wild apes comes from orangutans in Borneo. Knott (1997b, 1998,

1999) (see also Chapter 12) has described the marked variation in food availability and diet quality experienced by orangutans at Gunung Palung, Borneo, as a result of the dramatic mast fruiting episodes that alternate with periods of pronounced scarcity of preferred fruits. Patterns of ketone excretion in urine samples indicate that the periods of fruit scarcity are associated with the catabolism of fat reserves. Steroid profiles measured from urine samples indicate that female ovarian function is suppressed during these periods, while higher steroid levels indicative of higher fecundity occur during the mast periods. Conceptions also are more likely during the periods of dramatic food abundance.

Adaptive significance of the connection between energetics and female fecundity

Ovarian function in human females appears to be sensitive to the availability of metabolic energy to support reproduction, with a result that conception probability is increased when energy balance is positive and decreased when energy balance is negative. Both high and low energy flux appear to suppress ovarian function. However, high and low energy flux represent conditions under which energy availability for reproduction is constrained, either by high energy expenditure on other functions or by low energy intake. In either case, the potential for allocating additional energy to reproduction is reduced. Hence, positive energy balance and moderate energy flux are conditions that favor a female's ability to allocate energy to reproduction. Energy status is not a particularly good marker of this ability, since it is more reflective of history than current state.

The ability to allocate energy to reproduction is particularly important in early pregnancy when fat is stored to offset the high energy demands of later pregnancy and early lactation. Pre-pregnancy weight and early gestational weight gain are both important predictors of birth weight, which is in turn a powerful predictor of infant survivorship (Villar *et al.* 1992). Other periods in the human reproductive cycle are more demanding of energy, however, than the period immediately after conception. Would it not be more adaptive to have a mechanism that would synchronize one of these other stages with periods of high energy availability?

The fact that humans share the pattern of increased conception probability during periods of high energy availability with other apes suggests that the adaptive significance of this pattern may lie in shared aspects of our ecology. One important factor may be the degree to which periods of high energy availability are predictable in advance. Preferred chimpanzee food resources are notoriously patchy in space and time, and the masting episodes that characterize the Bornean rainforest are particularly irregular. It may well be that food resources in formative human environments were similarly unpredictable. Other features of our biology, such as our highly developed subcutaneous adipose deposits, suggest adaptation to unpredictable resources. In an unpredictable environment, optimizing energetic conditions at conception may have had a more positive effect on fitness than strategies dependent on predicting conditions in the future.

A second biological characteristic shared by humans and other apes is our relatively slow life history, with long periods of gestation and lactation periods that are both longer and lower in demand at their peaks than those of other mammals of comparable body size (Martin 1990; Oftedal 1984; Lee *et al.* 1991). This low but prolonged arc of energetic investment during reproduction effectively eliminates the pronounced peaks of energetic investment seen in most other mammals. For female apes, a low level of sustained energetic investment may have allowed a shift away from coordinating any one particular stage of the reproductive cycle with seasonal food availability and may have placed emphasis instead on female condition going into a reproductive bout.

These hypothesized scenarios for the selective advantage derived from coordinating conception with favorable energetic conditions are not mutually exclusive. It has been argued elsewhere (Ellison 2001) that unpredictable high-amplitude variation in energy availability in the environment may have been a particular feature of early hominid evolution, selecting for increased ability to sequester energy reserves as fat primarily in the service of survival rather than reproduction. Later in human evolution, this same ability may have been co-opted to support the development of an increasingly large-brained fetus. An enhanced sensitivity of female reproductive function to environmental energy constraints may have been a result. Nevertheless, it is likely that the general pattern of female reproduction being sensitive to environmental energy availability is much more ancient and typical of mammalian species with slow life histories and/or unpredictable environments. There is increasing evidence that this pattern is shared with other great apes at the very least.

Summary

The pattern of birth seasonality displayed by human populations around the world probably derives from multiple causes, with the most important factors varying between populations and through time. Three groups of factors have been suggested most often as playing important roles: social factors affecting the frequency of intercourse; climatological factors affecting male and female fecundity; and energetic factors principally affecting female ovarian function. Continuity with broader primate patterns can be seen in all three. However, evidence is particularly compelling for phylogenetic continuity in the importance of energetic factors on female fecundity leading to a coordination of conception with periods of enhanced energy availability. The adaptive significance of a link between energetics and female fecundity may lie in optimizing female reproductive effort under conditions of unpredictable resources and/or under a prolonged pattern of heavily reproductive investment.

References

Appell, R. A. & Evans, P. R. (1977). The effect of temperature on sperm motility and viability. *Fertility and Sterility*, **28**, 1329–32.

Ayeni, O. (1986). Seasonal variation in births in rural southwestern Nigeria. *International Journal of Epidemiology*, **15**, 91–4.

Bailey, R. C., Jenike, M. R., Ellison, P. T., *et al.* (1992). The ecology of birth seasonality among agriculturalists in central Africa. *Journal of Biosocial Science*, **24**, 393–412.

Bedford, J. M. (1977). Evolution of the scrotum: the epididymis as prime mover? In *Reproduction and Evolution*, ed. J. H. Calaby & P. Tynsdale-Biscoe. Canberra: Australian Academy of Science, pp. 171–82.

— (1991). Effects of elevated temperature on the epididymis and the testis: experimental studies. In *Advances in Experimental Biology and Medicine: Temperature and Environmental Effects on the Testis*, ed. A. W. Zorgniotti. New York: Plenum, pp. 19–32.

Boesch, C. & Beosch-Achermann, H. (2000). *The Chimpanzees of the Taï Forest: Behavioral Ecology and Evolution*. Oxford: Oxford University Press.

Bojkowski, C. J., Aldhous, M. E., English, J., *et al.* (1987). Suppression of nocturnal plasma melatonin and 6-sulphatoxy melatonin by bright and dim light in man. *Hormone and Metabolic Research*, **19**, 437–40.

Boyar, R. M., Katz, J., Finkelstein, J. W., *et al.* (1974). Anorexia nervosa: immaturity of the 24-hour luteinizing hormone secretory pattern. *New England Journal of Medicine*, **297**, 861–5.

Brannian, J. D., Zhao, Y., & McElroy, M. (1999). Leptin inhibits gonadotrophin-stimulated granulosa cell progesterone production by antagonizing insulin action. *Human Reproduction*, **14**, 1445–8.

Brewis, A., Laylock, J., & Huntsman, J. (1996). Birth non-seasonality on the Pacific equator. *Current Anthropology*, **87**, 842–51.

Bronson, F. H. (1989). *Mammalian Reproductive Biology*. Chicago: University of Chicago Press.

— (1995). Seasonal variation in human reproduction: environmental factors. *Quarterly Review of Biology*, **70**, 141–64.

Brzezinski, A. & Wurtman, R. J. (1988). The pineal gland: its possible roles in human reproduction. *Obstetric and Gynecological Survey*, **43**, 197–207.

Carmina, E., Ferin, M., Gonzalea, F. & Lobo, R. (1999). Evidence that insulin and androgens may participate in the regulation of serum leptin levels in women. *Fertility and Sterility*, **72**, 926–31.

Chik, C. L., Almeida, O. F. X., Libre, E. A., *et al.* (1992). Photoperiod-driven changes in reproductive function in male rhesus monkeys. *Journal of Clinical Endocrinology and Metabolism*, **74**, 1068–74.

Condon, R. G. (1982). Inuit natality rhythms in the central Canadian arctic. *Journal of Biosocial Science*, **14**, 167–77.

Condon, R. G. & Scaglion, R. (1982). The ecology of human birth seasonality. *Human Ecology*, **10**, 495–510.

Cowgill, U. M. (1965). Season of birth in man, contemporary situation with special reference to Europe and the Southern Hemisphere. *Ecology*, **47**, 614–23.

Danubio, M. E. & Amicone, E. (2001). Biodemographic study of a central Apennine area (Italy) in the 19th and 20th centuries: marriage seasonality and reproductive isolation. *Journal of Biosocial Science*, **33**, 427–49.

Dixon, G., Eurman, P., Stern, B. E., Schwartz, B., & Rebar, R. W. (1984). Hypothalamic function in amenorrheic runners. *Fertility and Sterility*, **42**, 377–83.

Dyson, T. & Crook, N. (1981). Seasonal patterns of births and deaths. In *Seasonal Dimensions to Rural Poverty*, ed. R. Chambers, R. Longhurst, & A. Pacey. London: Pinter, pp. 135–62.

Ellison, P. T. (1995). Breastfeeding, fertility, and maternal condition. In *Breastfeeding: Biocultural Perspectives*, ed. K. A. Dettwyler & P. Stuart-Macadam. Hawthorne, NY: Aldine de Gruyter, pp. 305–45.

— (2001). *On Fertile Ground.* Cambridge, MA: Harvard University Press.

Ellison, P. T. & Lager, C. (1986). Moderate recreational running is associated with lowered salivary progesterone profiles in women. *American Journal of Obstetrics and Gynecology*, **154**, 1000–1003.

Ellison, P. T. & Valeggia, C. R. (2003). C-peptide levels and the duration of lactational amenorrhea. *Fertility and Sterility*, **80**, 1279–80.

Ellison, P. T., Peacock, N. R. & Lager, C. (1989). Ecology and ovarian function among Lese females of the Ituri Forest, Zaire. *American Journal of Physical Anthropology*, **78**, 519–26.

Ferin, M. (1999). Stress and the reproductive cycle. *Journal of Clinical Endocrinology and Metabolism*, **84**, 1768–74.

Flier, J. S. (1998). What's in a name? In search of leptin's physiologic role. *Journal of Clinical Endocrinology and Metabolism*, **83**, 1407–13.

Goodall, J. (1986). *The Chimpanzees of Gombe: Patterns of Behavior*. Cambridge, MA: Harvard University Press.

Gordillo, G. (1994). La presión de los más pobres: reciprocidad, diferenciación social y conflicto entre los Tobas del oeste de Formosa. *Cuadernos del Instituto Nacional de Antropología y Pensamiento Latinoamericano*, **15**, 53–82.

— (2002). Locations of hegemony: the making of places in the Toba's struggle for *La Comuna*, 1989–99. *American Anthropologist*, **104**, 262–77.

Huntington, E. (1938). *Season of Birth*. New York: John Wiley & Sons.

Hurtado, A. M. & Hill, K. R. (1990). Seasonality in a foraging society: variation in diet, work effort, fertility, and sexual division of labor among the Hiwi of Venezuela. *Journal of Anthropological Research*, **46**, 293–346.

Huss-Ashmore R. (1988). Seasonal patterns of birth and conception in rural highland Lethoto. *Human Biology* **60**, 493–506.

Illnerova, H., Zvolsky, P., & Vanecek, J. (1985). The circadian rhythm in plasma melatonin concentration of the urbanized man: the effect of summer and winter time. *Brain Research*, **328**, 186–9.

Jasienska, G. (2001). Why energy expenditure causes reproductive suppression in women: an evolutionary and bioenergetic perspective. In *Reproductive Ecology and Human Evolution*. ed. P. T. Ellison. New York: Aldine de Gruyter, pp. 59–84.

Jasienska, G. & Ellison, P. T. (1998). Physical work causes suppression of ovarian function in women. *Proceedings of the Royal Society of London B*, **265**, 1847–51.

Kandeel, F. R. & Swerdloff, R. S. (1988). Role of temperature in regulation of spermatogenesis and the use of heating as a method for contraception. *Fertility and Sterility*, **49**, 1–23.

Karsch, F. J., Bittman, E. L., Foster, D. L., *et al.* (1984). Neuroendocrine basis of seasonal reproduction. *Recent Progress in Hormonal Research*, **40**, 185–232.

Kauppila, A., Kivelä, A., Pakarinen, A., & Vakkuri, O. (1987). Inverse seasonal relationship between melatonin and ovarian activity in humans in a region with a strong seasonal contrast in luminosity. *Journal of Clinical Endocrinology and Metabolism*, **65**, 823–8.

Knott, C. D. (1997a). Interactions between energy balance, hormonal patterns and mating behavior in wild Bornean orangutans (*Pongo pygmaeus*). *American Journal of Physical Anthropology*, **24**, 145.

— (1997b). Interactions between energy balance, hormonal patterns and mating behavior in wild Bornean orangutans (*Pongo pygmaeus*). *American Journal of Primatology*, **42**, 124.

— (1998). Changes in orangutan diet, caloric intake and ketones in response to fluctuating fruit availability. *International Journal of Primatology*, **19**, 1061–79.

— (1999). Reproductive, physiological, and behavioral responses of orangutans in Borneo to fluctuations in food availability. Ph. D. thesis, Harvard University.

— (2001). Female reproductive ecology of the apes: implications for human evolution. In *Reproductive Ecology and Human Evolution*, ed. P. T. Ellison. New York: Aldine de Gruyter, pp. 429–64.

Lager, C. & Ellison, P. T. (1990). Effect of moderate weight loss on ovarian function assessed by salivary progesterone measurements. *American Journal of Human Biology*, **2**, 303–12.

Lam, D. & Miron, J. (1991). Seasonality of births in human populations. *Social Biology*, **38**, 51–78.

— (1994). Global patterns of seasonal variation in human fertility. *Annals of the New York Academy of Sciences*, **709**, 9–28.

— (1996). The effect of temperature on human fertility. *Demography*, **33**, 291–305.

Lancaster, J. B. & Lee, R. B. (1965). The annual reproductive cycle in monkeys and apes. In *Primate Behavior: Field Studies of Monkeys and Apes*, ed. I. DeVore. New York: Holt, Rinehart, & Winston, pp. 486–513.

Lee, P. C., Majluf, P., & Gordon, I. J. (1991). Growth, weaning and maternal investment from a comparative perspective. *Journal of Zoology*, **225**, 99–114.

Levine, R. J. (1994). Male factors contributing to seasonality of human reproduction. *Annals of the New York Academy of Sciences*, **709**, 29–45.

Levine, R. J., Bordson, B. L., Mathew, R. M., *et al.* (1988). Deterioration of semen quality during summer in New Orleans. *Fertility and Sterility*, **49**, 900–907.

Lindburgh, W. G. (1987). Seasonality of reproduction in primates. In *Comparative Primate Biology*, Vol. 2B. Behavior, Cognition, and Motivation, ed. W. R. Dukelow & J. Erwin. New York: Alan R. Liss, pp. 167–218.

Lipson, S. F. & Ellison, P. T. (1996). Comparison of salivary steroid profiles in naturally occurring conception and non-conception cycles. *Human Reproduction*, **11**, 2090–96.

Loucks, A. B. (1990). Effects of exercise training on the menstrual cycle: existence and mechanisms. *Medicine and Science in Sports and Exercise*, **22**, 275–80.

Loucks, A. B., Mortola, J., Girton, L., & Yen, S. (1989). Alterations in the hypothalamic-pituitary-ovarian and hypothalamic-pituitary-adrenal axes in athletic women. *Journal of Clinical Endocrinology and Metabolism*, **68**, 402–11.

Malina, R. M. & Himes, J. H. 1977. Seasonality of births in a rural Zapotec Municipio, 1945–1970. *Human Biology*, **49**, 125–37.

Martin, R. D. (1990). Primate reproductive biology. In *Primate Origins and Evolution: A Phylogenetic Reconstruction*, ed. R. D. Martin. Princeton: Princeton University Press, pp. 427–75.

Matthews, C. D., Guerin, M. V., & Wang, X. (1991). Human plasma melatonin and urinary 6-sulphatoxy melatonin: studies in natural annual photoperiod and extended darkness. *Clinical Endocrinology*, **35**, 21–7.

McNeilly, A. S., Tay, C. C. K., & Glasier, A. (1994). Physiological mechanisms underlying lactational amenorrhea. *Annals of the New York Academy of Sciences*, **709**, 145–55.

Mendoza, M. (1999). The western Toba: family life and subsistence of a former hunter-gatherer society. In *Peoples of the Gran Chaco: Native Peoples of the Americas*, ed. E. S. Miller. Westport, CT: Bergin & Garvey, pp. 81–108.

Menken, J. (1979). Seasonal migration and seasonal variation in fecundability: effects on birth rates and birth intervals. *Demography*, **16**, 103–19.

Miller, E. (1999). *Peoples of the Gran Chaco: Native Peoples of the Americas.* Westport, CT: Bergin & Garvey.

National Center for Health Statistics. (1966). Seasonal variation of births: United States, 1933–1963. Series 21, no. 9. Washington, DC: US Government Printing Office.

Nishida, T. (1990). Demography and reproductive profiles. In *The Chimpanzees of the Mahale Mountains*, ed. T. Nishida. Tokyo: University of Tokyo Press, pp. 63–97.

Oftedal, O. T. (1984). Milk composition, milk yield and energy output at peak lactation: a comparative review. *Symposia of the Zoological Society of London*, **51**, 33–85.

Pascual, J., Gardia-Moro, C., & Hernandez, M. (2000). Non-seasonality of births in Tierra del Fuego (Chile). *Annals or Human Biology*, **27**, 517–24.

Pirke, K. M., Schweiger, U., Lemmel, W., Krieg, J. C., & Berger, M. (1985). The influence of dieting on the menstrual cycle of healthy young women. *Journal of Clinical Endocrinology and Metabolism*, **60**, 1174–9.

Poretsky, L., Cataldo, N. A., Rosenwaks, Z., & Giudice, L. C. (1999). The insulin-related ovarian regulatory system in health and disease. *Endocrine Reviews*, **20**, 535–82.

Prentice, A. M. & Whitehead, R. G. (1987). The energetics of human reproduction. *Proceedings of the Royal Society of London*, **57**, 275–304.

Rajan, S. I. & James, K. S. (2000). The interdependence of vital events: twentieth-century Indian Kerala. *Journal of Interdisciplinary History*, **31**, 21–41.

Rosenbaum, M. & Leibel, R.L. (1999). Role of gonadal steroids in the sexual dimorphisms in body composition and circulating concentrations of leptin. *Journal of Clinical Endocrinology and Metabolism*, **84**, 1784–9.

Rosenbaum, M., Nicolson, M., Hirsch, J., *et al.* (1997). Effects of weight change on plasma leptin concentrations and energy expenditure. *Journal of Clinical Endocrinology and Metabolism*, **82**, 3647–54.

Seiver, D. A. (1989). Seasonality of fertility: new evidence. *Population and Environment*, **10**, 245–57.

Sherry, D. S. (2002). Reproductive seasonality in chimpanzees and humans: ultimate and proximate factors. Ph. D. thesis, Harvard University.

Spicer, J. L. & Francisco, C. C. (1997). The adipose gene product, leptin: evidence of a direct inhibitory role in ovarian function. *Endocrinology*, **138**, 3374–9.

Stoeckel, J. & Chowdhury, A. (1972). Seasonal variation in births in rural East Pakistan. *Journal of Biosocial Science*, **4**, 107–16.

Tamarkin, L., Baird, C. J., & Almeida, O. F. X. (1985). Melatonin: a coordinating signal for mammalian reproduction? *Science*, **263**, 1118–21.

Thompson, R. W. & Robbins, M. S. (1973). Seasonal variation in conception in rural Uganda and Mexico. *American Anthropologist*, **75**, 676–81.

Udry, J. R. & Morris, N. M. (1967). Seasonality of coitus and seasonality of birth. *Demography* **4**, 673–81.

Valeggia, C. R. & Ellison, P. T. (2001). Lactation, energetics, and postpartum fecundity. In *Reproductive Ecology and Human Evolution*, ed. P. T. Ellison. New York: Aldine de Gruyter, pp. 85–105.

Veldhuis, J. D., Evans, W. S., Demers, L. M., *et al.* (1985). Altered neuroendocrine regulation of gonodotropin secretion in women distance runners. *Journal of Clinical Endocrinology and Metabolism*, **60**, 557–63.

Vigersky, R. A., Anderson, A. E., Thompson, R. H., & Loriaux, D. L. (1977). Hypothalamic dysfunction in secondary amenorrhea associated with simple weight loss. *New England Journal of Medicine*, **297**, 1141–5.

Villar, J., Cagswell, M., Kestler, E., *et al.* (1992). Effect of fat and fat-free mass deposition during pregnancy on birth weight. *American Journal of Obstetrics and Gynecology*, **167**, 1344–52.

Wallis, J. (1995). Seasonal influence on reproduction in chimpanzees of Gombe National Park. *International Journal of Primatology*, **16**, 533–49.

— (1997). A survey of reproductive parameters in the free-ranging chimpanzees of Gombe National Park. *Journal of Reproduction and Fertility*, **109**, 297–307.

Warden, T. A., Considine, R. V., Foster, G. D., *et al.* (1998). Short- and long-term changes in serum leptin in dieting and obese women: effects of caloric restriction and weight loss. *Journal of Clinical Endocrinology and Metabolism*, **83**, 214–18.

Wehr, T. A. (1991). The durations of human melatonin secretion and sleep respond to changes in daylength (photoperiod). *Journal of Clinical Endocrinology and Metabolism*, **73**, 1276–80.

— (2001). Photoperiodism in humans and other primates: evidence and implications. *Journal of Biological Rhythms*, **16**, 348–64.

Wehr, T. A., Moul, D. E., Barbato, G., *et al.* (1993). Conservation of photoperiod-responsive mechanisms in humans. *American Journal of Physiology*, **265**, R846–57.

Willis, D., Mason, H., Gilling-Smith, C., & Franks, S. (1996). Modulation by insulin of follicle-stimulating hormone and luteinizing hormone actions in human granulosa cells of normal and polycystic ovaries. *Journal of Clinical Endocrinology and Metabolism*, **81**, 302–9.

Wilson, M. E. & Gordon, T. P. (1989). Short-day melatonin pattern advances puberty in seasonally breeding rhesus monkeys (*Macaca mulatta*). *Journal of Reproduction and Fertility*, **86**, 435–44.

Wrangham, R. W. (2000). Why are male chimpanzees more gregarious than mothers? A scramble competition hypothesis. In *Great Ape Societies*, ed. P. Kappeler. Cambridge: Cambridge University Press, pp. 248–58.

Wrangham, R. W., Conklin, N. L., Etot, G., *et al.* (1993). The value of figs to chimpanzees. *International Journal of Primatology*, **14**, 243–56.

Yoshimura, Y., Makamura, Y., Koyama, N., *et al.* (1993). Effects of growth hormone on follicle growth, oocyte maturation, and ovarian steroidogenesis. *Fertility and Sterility*, **59**, 917–23.

14 *Seasonality, social organization, and sexual dimorphism in primates*

J. MICHAEL PLAVCAN
Department of Anthropology, University of Arkansas, Fayetteville AR 72701, USA

CAREL P. VAN SCHAIK
Anthropologisches Institut, University of Zurich, Winterthurerstrasse 190, CH-8057, Zurich, Switzerland

W. SCOTT MCGRAW
Department of Anthropology, Ohio State University, Columbus OH 43210, USA

Introduction

Primates live in habitats in which food abundance and other resources fluctuate over time, usually on a seasonal basis, and space. This variation affects the lives of primates in many ways, from behavioral ecology to reproduction (see Chapters 3 and 11). In this chapter, we explore how environmental and behavioral seasonality affect sexual dimorphism.

Sexual dimorphism in body and canine size among primates generally is viewed as primarily a consequence of sexual selection operating through the mechanism of male–male competition for mates (Leutenegger & Kelly 1977; Clutton-Brock *et al.* 1977; Kay *et al.* 1988; Plavcan & van Schaik 1992, 1997; Ford 1994; Lindenfors & Tullberg 1998) and modified by female choice for male traits (Plavcan 2004). Sexual dimorphism can be affected by environmental seasonality in two independent ways (see Fig. 14.1): first through the indirect impact of seasonality on the potential for mate monopolization (Mitani *et al.* 1996a; Nunn 1999; Pereira *et al.* 2000), and second through the direct impact of seasonality on male and female body size (Albrecht 1978; Turner *et al.* 1997). In the first case, phenological or climatic seasonality brings about the simultaneous presence of multiple cycling females due to reproductive seasonality and also may favor larger female group size. These effects in turn should affect the number of males present in a group, and patterns of male–male

Seasonality in Primates: Studies of Living and Extinct Human and Non-Human Primates, ed. Diane K. Brockman and Carel P. van Schaik. Published by Cambridge University Press.

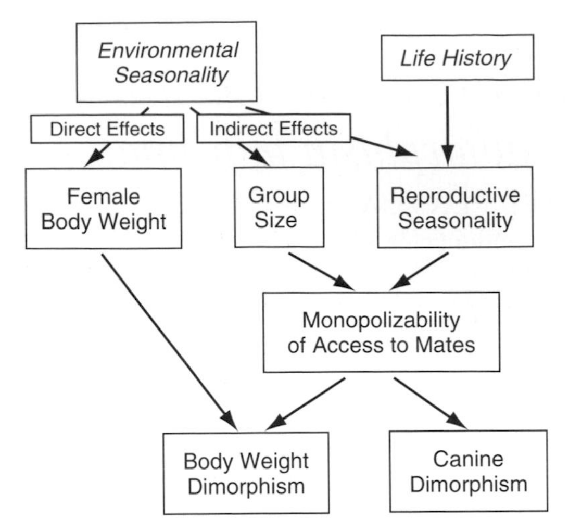

Figure 14.1 Path diagram of hypothetical relationships between seasonality and dimorphism via direct and indirect effects.

competition and resulting reproductive skew – in other words, several aspects of social organization and the mating system, all of which are tied to sexual dimorphism. In the second case, seasonality may affect optimum female body size, or perhaps limit male body size, thereby constraining or altering the possible range of dimorphism (Albrecht 1980; Leigh & Terranova 1998).

The possible effect of seasonality on dimorphism has received little attention in the literature, although some studies provide evidence suggesting a general relationship. Mitani *et al.* (1996b) directly incorporate data on birth seasonality in their estimates of operational sex ratios, indicating that male reproductive skew and size dimorphism are linked to variation in mating seasonality. In a different analysis, Nunn (1999) demonstrates that the number of males in primate groups is contingent on not only the number of females but also the likelihood of estrous overlap, which is in turn associated with female group size and breeding seasonality. This should affect operational sex ratios, which should be correlated with sexual selection and dimorphism. Several studies have found an association between environmental seasonality and/or productivity and dimorphism among subspecies or congeneric species (Albrecht 1978, 1980; Albrecht *et al.* 1990; Dunbar 1990; Albrecht & Miller 1993; Fooden & Albrecht 1993; Turner *et al.* 1997). However, there have been no broadly based studies comparing seasonality and dimorphism directly in primates.

Here, we first outline how we expect dimorphism to be related to breeding seasonality, environmental seasonality, and productivity. We then provide a series of comparative tests of the association between seasonality and dimorphism across and within species with available data. We emphasize that, as shown in Fig. 14.1, multiple ecological and social variables can act independently to affect sexual dimorphism, and they may interact in non-linear ways. Lacking sufficient empirical research and data to evaluate all of these factors simultaneously, our goal here is only to test the direct correlates between seasonality and dimorphism. We emphasize that localized comparisons may reveal more subtle relationships between any of these factors.

Indirect effects: seasonality and male competition

In primates, the strength of sexual selection should be proportional to the reproductive skew resulting from male–male competition (Andersson 1994; Mitani *et al.* 1996b; Plavcan 1999, 2004; Altmann 2000; Pereira *et al.* 2000). Reproductive skew will be proportional to the degree to which a dominant male is able to monopolize matings with females (Emlen & Oring 1977). To the degree that size or weaponry (canine teeth) assists males in winning fights for access to females, dimorphism in these traits should be proportional to the monopolization potential of matings by males (Plavcan & van Schaik 1992).

Environmental seasonality should impact this monopolization potential in two ways. First, seasonal breeding in primates is linked to phenological seasonality as a means of reducing maternal and infant mortality by timing key aspects of the reproductive cycle to coincide with peaks in food availability (see Chapter 10). In principle, monopolization of matings is contingent on the spatial and temporal distribution of female "estrous" periods (the periovulatory periods when fertilization is possible, usually coinciding with active mating). The priority-of-access model (Altmann 1962) postulates that a dominant male will obtain all the matings in a group as long as female estrus is asynchronous, unless females actively prevent monopolization attempts by males. In groups where two or more females are simultaneously sexually attractive, it is highly unlikely that a single male, no matter how strong, can monopolize all matings. Hence, where groups are not too small, we expect that seasonal breeding will reduce the potential for the dominant male to monopolize all fertile matings in a group by increasing the likelihood of overlap in female estrus (Emlen & Oring 1977; Clutton-Brock 1989; Mitani *et al.* 1996b; Nunn 1999; Altmann 2000; Dunbar 2000; Pereira *et al* 2000).

A second pathway to reduced monopolization potential is through larger female group size (Andelmann 1986; Mitani *et al.* 1996a; Nunn 1999; Cords 2000; Pereira *et al.* 2000). As female group size increases, the chances of more females becoming sexually attractive increases simultaneously, forcing mating competition to grade toward scramble and reducing reproductive skew and the fitness benefits of increased male strength and size. Groups of the same species or of geographically separated congeners tend to be larger in more seasonal habitats at similar latitudes or at higher latitudes, where seasonality is increased (Southwick *et al.* 1996). Numerous analyses have confirmed that female group size and both the observed and expected degree of estrus overlap determine the number of males in primate groups (Andelman 1986; Mitani *et al.* 1996a; Srivastava & Dunbar 1996; Nunn 1999; Cords 2000), suggesting that the spatiotemporal distribution of females should affect male reproductive skew.

Overall, then, increased seasonality should lead to reduced monopolization potential through a tendency for female estrous periods to overlap due to either seasonal breeding or larger group size, or both. Because the number of males in a group is contingent on the monopolization potential of females and female group size, these two pathways imply strongly that seasonality effects should be manifest especially across species showing multi-male groups and possibly also in those species showing variation in male numbers, but that no such effects should be expected in pair-living species or those always found in single-male groups. Therefore, we predict that increased breeding seasonality will be accompanied by reduced levels of sexual dimorphism in species living in multi-male groups but not necessarily in other social arrangements.

This prediction identifies breeding seasonality, rather than environmental seasonality, as the key variable, because the link between the two forms of seasonality is not uniform. In fact, there are numerous examples of species that breed non-seasonally in seasonal environments (see Chapter 11), perhaps because life history affects this response. Nonetheless, even if we limit the prediction to refer to breeding seasonality, then it is not very strong, because females may actively modify monopolization potential, thus overruling any influence of seasonality. On the one hand, even where mating is highly seasonal, females may nevertheless show non-overlapping estrous periods, as in *Lemur catta* (Pereira *et al.* 2000). On the other hand, if they benefit from the presence of multiple males (because it reduces the risk of predation or infanticide; e.g. van Schaik [2000]), then females may reduce the dominant male's monopolization potential by actively pursuing polyandrous mating behavior or by increasing the likelihood of estrous overlap over that expected by environmental forcing (cf. Nunn [1999]). Thus, the presence of these confounding effects dilutes the prediction.

Direct effects: seasonality and body size

Body mass is a key adaptation of an animal to its environment, in part constrained by the nutrition and health of the individual during its period of growth (cf. Stearns & Koella [1986]). If these constraints differ between the sexes, then dimorphism will be affected by divergent responses of male and female body mass along an environmental seasonality gradient. Consequently, we might expect that body mass dimorphism varies along this gradient as well, although, as we shall see, it is hard to predict exactly how. (In principle, canine tooth size should not be affected directly by environmental seasonality. Hence, we will ignore canine dimorphism in this particular analysis.)

Intraspecific comparisons generally indicate that female body size increases with increased annual productivity, usually as indexed by annual rainfall, which in turn is assumed to be a proxy measure of rainfall seasonality (Albrecht 1983; Albrecht *et al.* 1990; Dunbar 1990; Turner *et al.* 1997). These differences may reflect adaptive phenotypic responses. In most species, greater resource availability leads to better nutrition, favoring faster growth in males, females, or both sexes, and, hence, larger adult size (Stearns & Koella 1986). This explains why zoo animals and provisioned animals (either experimentally or through crop raiding) tend to be heavier than their non-provisioned wild counterparts (Dunbar 1990; Leigh 1994).

Alternatively, the differences may be genetic. In populations where lean seasons are long and, hence, cannot be bridged by storing reserves and mobilizing them during the lean period, smaller individuals would be favored selectively because they have lower absolute energy requirements, making it easier to satisfy nutritional demands. However, this prediction assumes that competition for the fallback foods used to bridge the lean season (see Chapter 3) is by scramble. If these fallback resources occur in defendable clumps, then contest competition for resources may increase during the lean season, and the sex competing most severely for access to food (females) may show a less drastic reduction in body size, or no reduction at all, compared with in the region with less seasonality.

Whether these changes in body size due to seasonality alter size dimorphism depends on whether male or female sizes change in parallel or differentially. The simplest expectation is that resource abundance should impact male and female size equally, and no effect of seasonality on body size dimorphism is expected. However, one could also argue that females, being more constrained by reproductive pressures, may have less leeway either to delay maturation or to grow faster with greater resource abundance. In either case, body size dimorphism would increase with lesser seasonality through a relatively greater increase in male size compared with that of females. Alternatively, female mass might be more affected by seasonality if females need to shunt energy away from growth to reproduction. In this case, body size dimorphism should increase with greater seasonality through a (stronger) reduction in female body size. The exception would be where a lean season leads to increased contest competition among females. Leigh and Shea (1995) use this idea to explain the lower dimorphism in chimpanzees compared with gorillas: female chimpanzees experience greater within-group contest competition than female gorillas, and hence delayed female growth and larger female size are favored.

It is impossible to predict which, if any, of these countervailing processes might predominate among primates. Lacking precise data for fine-grained comparisons, we are limited to evaluating empirically whether there is a correlation between body-size dimorphism and broad measures of environmental seasonality.

Methods

Morphological data

Morphological, behavioral, and ecological data were assembled from previously published data sets. Canine tooth size data were taken from Plavcan (1990). Body mass data were gleaned from Smith and Jungers (1997) and Plavcan and van Schaik (1997) and from museum records of specimens represented in the craniometric and dental data sets, where available. Craniometric data were gathered by one author (JMP) and are available from the author on request. Appendix 14.1 summarizes the morphological data used in the analyses. Data from restricted geographic samples were preferred whenever possible.

Dimorphism was estimated as the natural log of the ratio of the mean male to the mean female trait values. This transformation preserves linearity in estimates of dimorphism and preserves proportionality in the ratio when male values are less than those of females (Smith 1999).

We used maxillary canine crown height to estimate canine dimorphism because this measure of canine size is, theoretically, the most responsive to selection, shows the greatest variation in dimorphism, and is the strongest correlate of various measures of sexual selection and male–male competition (Greenfield & Washburn 1991; Greenfield 1992; Plavcan & van Schaik 1992; Plavcan *et al.* 1995; Plavcan 2000, 2001, 2004).

We evaluated size dimorphism using both body mass data and craniometric data. It is accepted widely that body size is a target of sexual selection in primates (Leutenegger & Kelly 1977; Clutton-Brock *et al.* 1977; Ford 1994; Plavcan & van Schaik 1997; Lindenfors & Tullberg 1998; Plavcan 2000, 2001, 2004; Lindenfors 2002), and therefore it is appropriate to estimate body mass dimorphism. However, there are well-known problems with body mass data from the literature (Albrecht *et al.* 1990; Smith & Jungers 1997). Sample sizes for many species are small. Often, data are reported without identifying the geographic origin of the specimens, opening the possibility that geographic variation may significantly affect estimates of dimorphism. It is clear from comparing museum records with skeletons that individuals reported as "adult" often have not achieved their full adult body mass. Field data often fail to note whether females were pregnant or lactating, whether specimens showed significant blood loss, and whether specimens defecated or vomited before being weighed, and often round numbers to the nearest pound or kilogram – a significant source of error when dealing with specimens that, for the most part, weigh less than 10 kg.

Consequently, we also estimated male and female size using craniometric data. For each specimen, we calculated the geometric mean of nine non-overlapping dimensions of the skull that characterize overall skull shape in the sagittal and coronal planes and show low measurement error (porion–inion, bimastoid breadth, vomer–basion, opisthion–inion, bizygomatic breadth, palate length, palate breadth at M1, facial height, biorbital breadth; see Plavcan [2003] for definitions). These data have the advantage of being available for relatively large numbers of individuals from populations that are relatively restricted geographically (Albrecht 1990). However, skull size is clearly not the exact equivalent of body mass, and it is well known that different species vary significantly in the relative size of the skull and its component parts. Even so, the issue of what constitutes body size is sometimes contentious, and most parties agree that there is no single "true" measure of body size. Geometric mean skull size is correlated strongly with body mass across all taxa (males: $n = 127$, $r = 0.979$, $P << 0.001$; females: $n = 127$, $r = 0.974$, $P << 0.001$; based on natural-log-transformed data) and within species for which body weights are available for individual specimens. Wherever results differ for analyses using skull size and body size, we underscore the potential differences in the measures as estimates of size. Geometric means were subsequently averaged for each sex. Ratios of male and female geometric means were log-transformed as for estimates of body mass and canine tooth size.

Behavioral/ecological data

For interspecific comparative analyses, group size data and standard classifications of primate mating and breeding systems were gathered from the literature and are summarized in Plavcan (2004). Additional group size data for intraspecific analyses were gathered from the literature (see Appendix 14.2). Annual rainfall data for localities corresponding to the group size data were used.

For interspecific analyses, we used data on birth seasonality and environmental seasonality gathered from the literature. Standardized interspecific data on both of these variables are particularly difficult to come by. We relied primarily on estimates of seasonality in rainfall, temperature, and birth seasonality using the concentration measures (r) based on circular statistics presented in Janson and Verdolin (Chapter 11). This appears to be the most comprehensive data set on these variables currently available. We used data only for those species represented in the morphological data sets. Where more than one subspecies was available for either data set, we matched together the morphological and seasonality data. We also used data on breeding seasonality, observed estrous overlap and expected estrous overlap reported in Nunn (1999). We used both the raw estrous overlap data, and estrous overlap data corrected for group size following the phylogenetic correction procedure detailed in Nunn (1999), in which residual overlap estimates are calculated from a regression generated using phylogenetic contrast analysis (see below). In calculating residual expected overlap, we used the same group size data reported by Nunn (1999) that was used to calculate the expected overlap in the first place.

For intraspecific analyses, monthly rainfall data were not available for many localities. Hence, we used annual rainfall estimates with the understanding that these are only approximations of annual productivity, and not seasonality per se, following the recommendation of Srivastava and Dunbar (1996). It is important to stress that most sources for group size that we evaluated noted that primates live in seasonal habitats, regardless of the actual annual rainfall. For example, both Tana River (annual rainfall 466 mm [Mowry *et al.* 1996]) and Parc National Tinigua, Colombia (annual rainfall 2623 mm [Stevenson 2000]) are characterized as markedly seasonal. What becomes clear very quickly is that some authors think of seasonality in terms of marked intra-annual variation of rainfall, regardless of the year's total, while others refer to temperature or year-to-year variation (Herbinger *et al.* 2001; Reed *et al.* 1997; DiFiore & Rodman 2001). Clearly, standardization is necessary. Apart from using annual rainfall, however, this task is beyond the scope of this analysis.

Finally, even though seasonality data are available for many species, it is important to note that in most cases, we do not have morphological data on the populations for which the behavioral/ecological data were derived. This severely limits our ability to test "fine-grained" hypotheses about intraspecific covariation between environmental seasonality and dimorphism and may have a confounding effect on interspecific comparisons.

Analyses

A series of direct comparisons between dimorphism and the seasonality estimates were carried out using both species values and phylogenetic contrast analysis. The latter was carried out using the phylogeny of Smith and Cheverud (2002), modified with the Purvis and Webster phylogeny (1999) where necessary. We used Garland's PDTREE program (Garland *et al.* 1993; Garland *et al.* 1999). Branch lengths were set initially using the estimates provided by the above papers. However, subsequent analysis of the correlation between branch lengths and contrasts revealed that the "Pagel" branch transformation (setting terminal branch lengths equal to 1 and then adjusting intermediate branches so that all species are equidistant from the root node) consistently yielded the lowest correlations between branch lengths and contrasts. Consequently, we report results for contrasts only for branch lengths manipulated in this way, following the recommendation of Garland *et al.* (1999). In practice, alternative branch length schemes (using the original branch lengths, constant branch lengths, natural-log-transformed branch lengths, and those transformed using the "Grafen" and "Nee" procedures) made little difference to the outcome of most of the analyses.

As noted above, we predict specifically that there should be a stronger relationship between dimorphism and seasonality in multi-male species as opposed to single-male species. We therefore repeated analyses within these social system measures. Because *Theropithecus gelada*, *Papio hamadryas*, and possibly *Nasalis larvatus* show single-male breeding units within multi-male social units, we repeated analyses classifying these species as either single-male or multi-male. Classifications made little difference to the results. We report results classifying these species as multi-male (Dunbar 2000; Plavcan 2004).

We carried out a series of regressions of male versus female body mass and skull size among congeneric species, and among conspecific subspecies when available. Sample-specific averages for males and females were used, and all data were natural-log-transformed. Reduced major axis regression lines were fitted to the data to check whether the relationship differed from isometry. These analyses allowed us to identify samples in which either male or female body size was substantially larger or smaller than that of close relatives. Data were compared directly with a phylogeny to infer whether changes in dimorphism were due to changes in male or female size.

Results

Environmental seasonality and female body size

Here we review studies that examine geographic variation in body size within species or groups of allopatric congeners and compare them directly with data available for this analysis.

Both living and fossil Malagasy strepsirrhines tend to show larger skull size in more productive, less seasonal eastern forests than in western forests (Albrecht *et al.* 1990). The body mass data available for this study partly corroborate this: western populations, subspecies, or species are smaller in the genera *Avahi*, *Cheirogaleus*, and *Propithecus* (quite dramatically in the latter genus), although no such pattern is apparent in the genera *Eulemur* and *Hapalemur*. In *Lepilemur*, southern populations tend to be smaller than northern populations. The largest extant lemurs, *Indri*, are found in rainforest. Likewise, among the related genera *Lemur* and *Varecia*, the smaller *Lemur* is found in the arid southwest, whereas the larger *Varecia* is found in eastern rainforests. Thus, the (monomorphic) lemurs show a clear trend toward larger body sizes in less seasonal habitats. Ravosa (1998) demonstrates that *Nycticebus* increases size with increasing distance (north or south) from the Equator.

Among platyrrhines, body size in *Callithrix jacchus* increases in a latitudinal gradient associated with rainfall and humidity, such that populations living in more humid, productive environments are larger (Albrecht 1983). Both Central America and the more southern areas of South America tend to be more seasonal in climate than Equatorial South America (Kay *et al.* 1997). In the data available here, we note that more southern howler monkeys (*Alouatta fusca*, *A. caraya*) tend to be smaller than the more Equatorial species. In contrast, those living in Central America (*A. palliata*, *A. villosa*) tend to be somewhat larger. Among

Cebus and *Ateles*, we find no consistent trend for more northern or southern forms to be larger or smaller (cf. Hershkovitz [1977]).

Among Old World monkeys, latitudinal variation in body size has been demonstrated for *Macaca* (within species in *nemestrina*, *mulatta*, *arctoides*, *assamensis*, *radiata*, and across the *sinica* and *mulatta* species groups) (Albrecht *et al.* 1990; Fooden & Albrecht 1993; Albrecht 1978, 1980) and *Semnopithecus entellus* (Albrecht & Miller 1993). Each of these cases conforms to Bergmann's rule, with body mass increasing with increasing distance from the Equator. Skull size data available for this study confirm these results clearly for *Semnopithecus entellus* and *Macaca mulatta* (notably, we use data from many of the same specimens as Albrecht). The two macaques living in the most temperate seasonal environments (*M. fuscata* and *M. sylvanus*) are among the largest.

Two detailed studies of the relationship between body size and rainfall (and/or temperature) are available. Dunbar (1990) demonstrates that body size in *Papio* increases in areas of higher rainfall, with males increasing in size faster than females. Turner *et al.* (1997) demonstrate that females in *Cercopithecus aethiops* are larger in areas of higher rainfall, which also are areas where crop cultivation is more widespread. These studies link increases in body size not to seasonality per se but rather to productivity associated with increased annual rainfall.

Thus, the few detailed studies relating geographic variation in primates to environmental variation reviewed here note two independent trends: (i) species or species groups spanning large latitudinal gradients tend to follow Bergmann's rule, with larger animals occurring farther from the Equator; and (ii) animals living in more productive habitats (higher rainfall and less seasonality) tend to be larger.

Not surprisingly, given the large range of body sizes and niches occupied by different primate species, these patterns do not hold when examined across all species. There is no correlation between female body size or skull size (as a measure of overall body size) and either of the measures of environmental variation (*r*-temp and *r*-rain) using either species values or independent contrasts (Table 14.1).

Environmental seasonality and sexual dimorphism

Published studies show no consistent intraspecific relationship between environmental seasonality and sexual dimorphism in primates. Turner *et al.* (1997) find that sexual dimorphism in *Cercopithecus aethiops* decreases with increasing rainfall and cultivated food abundance,

Table 14.1 *Analyses of female skull size and body mass versus circular measures of rainfall and temperature seasonality.*

	Species analyses				Independent contrasts			
Dimorphism measure	*n*	*r*	*k*	*P*	DF	*r*	*k*	*P*
Body mass								
r-temp	41	0.028	0.342	NS	38	0.294	1.636	NS
r-rain	41	0.119	0.746	NS	38	0.117	0.329	NS
Skull Size								
r-temp	40	0.029	0.086	NS	38	0.007	0.044	NS
r-rain	40	0.095	0.148	NS	38	0.117	0.329	NS

DF, degrees of freedom; NS, not significant.

suggesting that female body mass increases faster than that of males as productivity increases. In contrast, Dunbar (1990) finds that dimorphism in *Papio* increases with increasing rainfall. Albrecht (1980) demonstrates that in *Macaca nemestrina* (both subspecies), sexual dimorphism in skull size increases with increasing distance from the Equator, due to male size increasing relatively faster than female size.

The relationship between male and female size within a series of closely related species and subspecies can be examined for consistent changes in dimorphism with changes in female size, which would give us a partial test of the hypothesis that changes in seasonality will produce changes in dimorphism. Table 14.2 lists results for regressions of male skull and body size against female skull and body size. Though sample sizes are small, there is a general trend for scaling that is not significantly different from isometry in the majority of colobine and platyrrhine comparisons. In cercopithecines and hominoids, there is a tendency to scale with positive allometry.

Unfortunately, the positively allometric trends shown among cercopithecines and great apes are not attributable to differences in latitude or seasonality. Likewise, among South American monkeys, there is no indication that latitudinal differences are associated consistently with changes in either female or male size. Particularly telling is that within *Semnopithecus entellus*, dimorphism changes isometrically with the dramatic changes in body size associated with latitudinal variation (cf. Albrecht & Miller [1993]). Only among the macaques is there any indication that seasonality may affect dimorphism (Fig. 14.2). The two highest-latitude species, *Macaca sylvanus* and *M. fuscata*, both show lower levels of dimorphism than other macaques (removal of these two species from the regression of male on female skull size produces a near-linear, positively allometric relationship among the

Table 14.2 *Statistics for least squares (LS) regressions, including reduced major axis (RMA) of male versus female natural-log-transformed geometric mean skull size and body mass*

Group[a]		n	r	Intercept	LS slope	RMA slope	P	Allometry
Alouatta	Skull	8	0.672	1.579	0.611	0.909	NS	–
	Mass	7	0.804	0.355	1.012	1.259	0.029	Isometric
Ateles	Skull	4	0.966	0.560	0.855	0.885	0.034	Isometric
	Mass	4	0.882	0.792	0.642	0.728	NS	–
Cebus	Skull	6	0.950	−2.333	1.679	1.767	0.004	Positive[b]
	Mass	6	0.650	0.966	0.275	0.423	NS	–
Semnopithecus	Skull	9	0.868	−0.122	1.063	1.224	0.002	Isometric
	Mass	–	–	–	–	–	–	–
Presbytis	Skull	11	0.915	−0.814	1.237	1.352	<0.001	Isometric
	Mass	11	0.956	−0.022	1.049	1.097	<0.001	Isometric
Colobus	Skull	10	0.967	−0.367	1.120	1.158	<0.001	Isometric
	Mass	8	0.911	−0.060	1.131	1.241	0.002	Isometric
Nasalis/ Pygathrix	Skull	6	0.823	−0.117	1.058	1.285	0.044	Isometric
	Mass	4	0.207	2.122	0.173	0.835	NS	–
Cercopitheciins	Skull	24	0.944	−0.635	1.211	1.283	<0.001	Positive
	Mass	17	0.982	0.134	1.236	1.259	<0.001	Positive
Papionins	Skull	11	0.946	−1.742	1.492	1.577	<0.001	Positive
	Mass	11	0.948	−0.079	1.268	1.338	<0.001	Positive[b]
Macaca	Skull	11	0.970	−0.110	1.067	1.100	<0.001	Isometric[c]
	Mass	9	0.956	0.556	0.948	0.992	<0.001	Isometric
Great apes	Skull	8	0.969	−2.837	1.685	1.739	<0.001	Positive
	Mass	8	0.893	−0.379	1.247	1.396	0.003	Isometric

NS, not significant.

[a] Composition of groups as follows, from Appendix 14.1: *Alouatta*, all; *Ateles*, all; *Cebus*, all taxa; *Semnopithecus*, all subspecies; *Colobus*, all *Colobus* and *Procolobus; Presbytis*, all *Presbytis sensu lato* excluding *Semnopithecus*; *Nasalis/Pygathrix*, all *Nasalis*, *Simias*, *Pygathrix*, and *Rhinopithecus*; Cercopitheciins, all *Cercopithecus*, *Erythrocebus*, and *Miopithecus*; Papionins, all *Papio*, *Mandrillus*, *Theropithecus*, *Cercocebus*, and *Lophocebus*; *Macaca*, all *Macaca*; great apes, all *Pan*, *Gorilla*, and *Pongo*.

[b] Positive allometry only for the RMA slope.

[c] Isometric only when *M. fuscata* and *M. mulatta* are included. Significant positive allometry for both LS and RMA slopes when these species are excluded ($n = 9$, $r = 0.996$, slope = 1.159, $P < 10.001$).

remaining macaques). Phylogenetic comparison (using matched-pair comparisons) suggests that female body size is relatively large in both species.

Analysis of both species values and independent contrasts reveals no significant interspecific relationship between measures of environmental

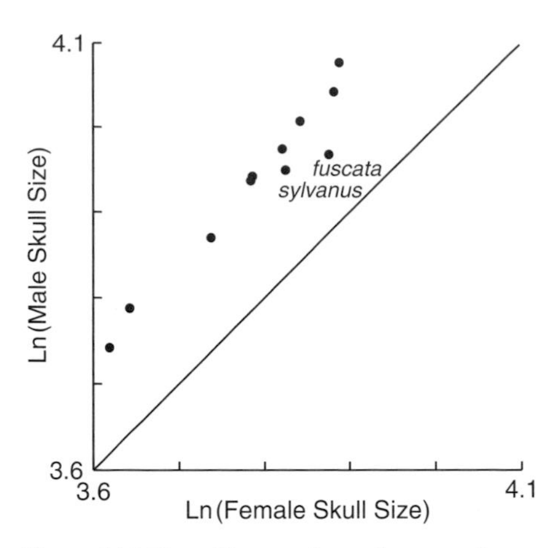

Figure 14.2 Plot of ln-transformed geometric means of male skull size versus female skull size in macaques. The line represents isometry and monomorphism. Note the near linear relationship for all macaques, except *M. sylvanus* and *M. fuscata*, both of which live in temperate seasonal environments. Phylogenetic comparison suggests that the deviations reflect larger female size rather than smaller male size.

seasonality (*r*-temp, *r*-rain) and sexual dimorphism in body mass, skull size, and canine tooth size, using either species values or phylogenetic contrasts. Repeating the analyses within mating systems yields the same result (Table 14.3). Hence, there is little evidence that sexual dimorphism in primates responds to increased environmental seasonality in a predictable way.

Environmental seasonality and group size

Using species values, female group size is related significantly to canine, body mass, and skull dimorphism in primates. However, phylogenetic contrasts demonstrate no significant relationship between group size and dimorphism. Nonetheless, it is of interest to see how group size responds to environmental seasonality.

Of 24 species for which we had annual rainfall data and group size data from more than one locality, 9 showed larger groups in areas with less annual rainfall and 15 showed the opposite pattern. Most of these comparisons involve too few localities to carry out statistical tests. The two species with the largest numbers of localities (*Alouatta palliata*, $n = 38$; *A. seniculus*, $n = 25$) both show a weakly significant correlation between rainfall and group size, but in opposite directions (Table 14.4). None of the

Table 14.3 *Analyses of dimorphism versus circular measures of birth, rainfall, and temperature seasonality*

	Species analyses				Independent contrasts			
Dimorphism measure	*n*	*r*	*k*	*P*	DF	*r*	*k*	*P*
All species								
Canines								
r-birth	27	0.106	−0.097	NS	24	0.100	0.057	NS
r-temp	27	0.185	0.517	NS	24	−0.073	−0.099	NS
r-rain	27	0.206	0.326	NS	24	0.075	0.064	NS
Body mass								
r-birth	41	0.154	−0.119	NS	36	0.066	0.050	NS
r-temp	41	0.018	−0.043	NS	36	0.045	0.074	NS
r-rain	41	0.110	−0.130	NS	36	−0.211	−0.173	NS
Skull Size								
r-birth	40	0.324	−0.071	NS	38	0.065	−0.014	NS
r-temp	40	0.017	−0.011	NS	38	−0.132	−0.061	NS
r-rain	40	0.010	−0.993	NS	38	−0.211	−0.049	NS
Multi-male species								
Canines								
r-birth	17	0.147	−0.127	NS	15	−0.212	−0.112	NS
r-temp	17	0.093	0.206	NS	15	−0.246	−0.212	NS
r-rain	17	0.152	0.247	NS	15	0.002	0.002	NS
Body mass								
r-birth	25	0.171	−0.098	NS	23	−0.256	−0.150	NS
r-temp	25	0.155	0.231	NS	23	−0.032	−0.029	NS

DF, degrees of freedom; NS, not significant.

Table 14.4 *Within-species correlations between group size and annual rainfall*

Species	*n*	*r*	*P*
Alouatta paliatta	38	0.349	0.034
Alouatta seniculus	25	−0.477	0.021
Colobus guereza	15	0.226	NS
Cercopithecus aethiops	8	0.532	NS
Semnopithecus entellus	8	−0.166	NS
Alouatta pigra	7	0.143	NS
Procolobus badius	7	0.265	NS
Colobus satanas	5	0.163	NS

NS, not significant.

Table 14.5 *Interspecific analysis of group size versus measures of birth, temperature, and rainfall seasonality*

	Species analyses				Independent contrasts			
Seasonality measure	*n*	*r*	*k*	*P*	*n*	*r*	*k*	*P*
All species								
Breeding	51	0.098	−0.124	NS	39	−0.326	−0.310	NS
r-birth	37	0.151	−0.509	NS	34	0.013	0.039	NS
r-temp	37	0.039	0.384	NS	34	0.186	0.120	NS
r-rain	37	0.136	0.738	NS	34	0.203	0.686	NS
Multi-male species								
Breeding	39	0.126	−0.128	NS	27	−0.153	−0.131	NS
r-birth	24	0.167	−0.485	NS	21	−0.095	−0.346	NS
r-temp	24	0.019	0.140	NS	21	0.176	0.944	NS
r-rain	24	0.112	0.492	NS	21	0.100	0.375	NS
Single-male species								
Breeding	9	0.307	−0.327	NS	7	−0.128	−0.125	NS
r-birth	9	0.485	0.833	NS	7	0.193	0.195	NS
r-temp	9	0.044	−0.496	NS	7	0.334	2.793	NS
r-rain	9	0.351	−1.137	NS	7	−0.211	−0.331	NS

NS, not significant.

other species with sample sizes greater than five show the predicted relationship between group size and rainfall.

Interspecific analysis shows no relationship between group size and either environmental seasonality (*r*-temp, *r*-rain) or birth seasonality (*r*-birth, breeding season) using either species values or phylogenetic contrasts (Table 14.5). Reanalysis within mating system classifications likewise fails to yield any significant correlation.

Hence, both intraspecific and interspecific analyses reveal no consistent relationship between annual rainfall and group size. Accordingly, there is no evidence for an indirect effect of environmental seasonality on group size and, therefore, no indication for a possible effect of group size, through monopolization potential, on sexual dimorphism.

Reproductive seasonality, group size, and dimorphism

Using all primates, no measure of birth seasonality (*r*-birth, breeding season) or estrous overlap (observed or expected, raw or corrected for group size) is related to variation in sexual dimorphism in canine size, body mass, or skull size for either species values or independent contrasts (Table 14.6). However, there is a significant relationship between *r*-birth and skull dimorphism in multi-male species, both when using species values and when using independent contrasts (Table 14.3). This finding agrees with the prediction that birth seasonality should temper male monopolization potential in large female groups. Likewise, independent contrasts also show a weakly significant negative correlation between expected overlap and body mass dimorphism in multi-male species. Independent contrast analysis also indicates a weak relationship between observed estrous overlap and canine dimorphism in single-male species. These results are robust to variation in branch lengths and to variation in classifications of multi-level society species as single-male or multi-male.

Discussion

Interspecific comparative analyses demonstrate that sexual dimorphism across primates is not associated directly with variation in either birth seasonality or environmental seasonality. These results stand in contradiction to the accumulating evidence that changes in male and female body size and skull size within species are associated with variation in environmental seasonality and resource productivity (see above) and to generalizations that increased mating seasonality is associated with lower levels of intrasexual selection among males because of a decrease in mate monopolization potential (e.g. Nunn [1999]). The contrast between these results is important both for stressing the caution that needs to be used in extrapolating toward broad interspecific patterns from intraspecific ones and in identifying a basic dissociation between the norm of reaction found within a species and effects on mean values arising from selection, both across populations within a species and, more commonly, among species.

Direct effects

Intraspecific and intrageneric comparisons between body size and seasonality and productivity clearly suggest that animals are larger in regions with increasing productivity, as measured by annual rainfall, and at higher latitudes, following Bergmann's rule. If more seasonal environments are less productive, leading to smaller body size, then animals should also get smaller with increasing distance from the Equator, since seasonality generally increases in the same direction. Hence, two conflicting mechanisms

Table 14.6 *Analysis of dimorphism versus measures of breeding season, expected estrous overlap, and observed estrous overlap (from Nunn [1999])*

	Species analyses				Independent contrasts			
Dimorphism measure	*n*	*r*	*k*	*P*	DF	*r*	*k*	*P*
Breeding season								
All Species								
Canines	33	0.001	0.00	NS	31	−0.048	−0.013	NS
Body mass	53	0.164	−0.047	NS	39	−0.033	−0.013	NS
Skull size	52	0.065	0.005	NS	38	0.097	0.120	NS
Multi-male								
Canines	25	0.038	0.011	NS	22	0.003	0.001	NS
Body mass	40	0.174	−0.039	NS	27	−0.232	−0.064	NS
Skull size	40	0.170	0.012	NS	27	−0.070	0.005	NS
Single-male								
Canines	7	0.618	−0.117	NS	6	−0.788	−0.205	< 0.05
Body mass	9	0.148	0.069	NS	7	−0.296	−0.116	NS
Skull size	9	0.079	−0.009	NS	7	−0.330	−0.032	NS
Expected estrous overlap								
All species								
Canines	29	0.210	0.031	NS	27	0.063	0.006	NS
Body mass	43	0.045	0.006	NS	31	−0.420	−0.062	< 0.02
Skull size	42	0.055	−0.002	NS	30	−0.334	−0.014	NS
Multi-male								
Canines	24	0.077	0.011	NS	22	0.105	0.010	NS
Body mass	35	0.016	−0.002	NS	25	−0.409	−0.051	< 0.05
Skull size	35	0.142	−0.004	NS	25	−0.198	−0.007	NS
Single-male								
Canines	4	0.676	0.072	NS	–	–	–	–
Body mass	5	0.432	−0.086	NS	–	–	–	–
Skull size	5	0.203	−0.010	NS	–	–	–	–
Observed estrous overlap								
All Species								
Canines	17	0.475	0.083	NS	15	0.095	0.011	NS
Body mass	27	0.235	0.038	NS	19	−0.326	−0.054	NS
Skull Size	26	0.245	0.011	NS	18	−0.295	−0.014	NS
Multi-male								
Canines	13	0.352	0.069	NS	–	–	–	–
Body mass	18	0.126	−0.018	NS	–	–	–	–
Skull size	18	0.024	−0.001	NS	–	–	–	–
Single-male								
Canines	–	–	–	–	–	–	–	–
Body Mass	5	0.323	−0.041	NS	–	–	–	–
Skull size	5	0.057	−0.002	NS	–	–	–	–

DF, degrees of freedom.

are probably at work. On the one hand, animals that eat better diets reach bigger adult size – a phenomenon that is both well known (Leigh 1994) and expected by evolutionary theory (Stearns & Koella 1986). On the other hand, Bergmann's rule commonly is attributed to local selection for increased body size in association with thermoregulation, though other factors may be involved (Freckleton *et al.* 2003). Hence, wherever seasonality involves both productivity and temperature, we should expect patterns to reflect a balance of these two processes.

Regardless, the most interesting observation from this study is that changes in male or female body size in closely related species fail to produce a consistent pattern of changes in sexual dimorphism. Three documented cases in the literature – *Papio*, *Cercopithecus aethiops*, and *Macaca nemestrina* – demonstrate clearly that sexual size dimorphism can change with changes in overall size. However, as we have seen, each case shows a different pattern. Similarly, the data presented here suggest that body size dimorphism is reduced in *M. fuscata* and *M. sylvanus* as a result of a disproportionate increase in female body size, while in *Semnopithecus entellus* dimorphism remains unchanged in spite of a large latitudinal gradient in body size.

The data compiled for this study also show that for the majority of closely related, non-cercopithecine species and subspecies, size dimorphism does not change substantially with changes in size. The stability of dimorphism in the face of changing size could have at least two basic explanations. First, female body size is optimized adaptively, whereas male size represents a compromise between this and sexual selection for larger size. Thus, changes in female size result in parallel changes in male size. In this sense, female size can be seen as the approximate baseline, while male size would be seen as modified by sexual selection to a maximum degree from this baseline, or ideal size. This idea is supported to some degree by the observation that in species lacking sexual selection for male size, size dimorphism is minimal or absent. Second, size dimorphism is constrained by correlated response, such that changes in size in one sex place a minimum or maximum size limit on the other sex (Lande 1980; Martin *et al.* 1994; Lindenfors 2002). This second possibility receives support from the observation that male and female size changes are correlated regardless of mass dimorphism. Of course, correlated response does not explain the clear positive allometry between male and female skull size and body mass seen among most cercopithecines (cf. Smith & Cheverud [2002], Lindenfors & Tullberg [1998]).

These observations and models are not mutually exclusive. Taken together, they suggest that changes in overall size can influence dimorphism, but perhaps because seasonality is merely one among many factors affecting size, its influence is not so pervasive that it produces a general effect across species.

Indirect effects

Seasonality is often linked to sexual dimorphism through its effect on the strength of sexual selection (Mitani *et al.* 1996b; Srivastava & Dunbar 1996; Nunn 1999; Strier 2003). Several studies demonstrate that the numbers of males in primate groups are associated with female group size, breeding seasonality, and observed and predicted estrous overlap (Andelman 1986; Mitani *et al.* 1996a; Srivastava & Dunbar 1996; Nunn 1999; Altmann 2000).

The results of our analysis partly confirm our prediction that breeding seasonality affects sexual dimorphism in a limited set of conditions. Interspecific analyses indicate that skull size dimorphism is correlated with birth seasonality in multi-male species. This is expected because in situations with large numbers of males and females, small changes in monopolization potential can alter male strategies on the scramble-contest continuum. This finding echoes that of Mitani *et al.* (1996b), who demonstrated a significant link between dimorphism and estimates of operational sex ratios that incorporated information on breeding seasonality (as well as estrous duration, number of cycles to conception, interbirth intervals, and adult sex ratios). However, we do not find the same pattern in canine dimorphism or body mass dimorphism. While one might argue reasonably that the results pertaining to body mass dimorphism are biased by the generally poor quality of body mass data for most species, this argument fails to explain why the canine data also are not correlated with seasonality, especially given that canine tooth size is thought to be governed by a less complex suite of factors than body size (Plavcan 2001, 2004). The near absence of an effect of seasonality on various kinds of dimorphism among single-male species is not unexpected because we expect only minor changes in dimorphism as long as males continue to monopolize access to females.

Overall, therefore, if there is a direct, general correlation between breeding seasonality and dimorphism in primates, then it is so weak as to be undetectable in comparative analysis with most current data. Perhaps reproductive seasonality is just one among several factors affecting operational sex ratios and male reproductive skew, such as within-group sex ratios, male tenure, and population-specific patterns of estrous overlap (Plavcan 2004; Altmann 2000; Pereira *et al.* 2000). It is also possible, however, that natural selection has molded female strategies to produce a degree of sexual dimorphism that is most tolerable to them as a class in spite of varying forces.

Group size is another factor that is, in theory, associated with potential estrous overlap and, thus, dimorphism (Nunn 1999). Unfortunately, phylogenetic analysis fails to find a correlation between group size and dimorphism across all species, although Plavcan (2003) demonstrated that female group size is associated with variation in female canine size and, weakly, with canine dimorphism, consistent with the Janson and Goldsmith (1995) model that female competition is more intense with smaller female group sizes.

Within species, variation in group size is associated inconsistently with variation in rainfall, which is a crude estimate of productivity (Srivastava & Dunbar 1996). Comparative analyses fail to demonstrate any interspecific association between female group size and any of the measures of seasonality based on circular statistics. Furthermore, our within-species analyses, while limited, also failed to show any consistent pattern of larger groups inhabiting either more seasonal or less productive habitats. This result is surprising, given reports that primate group size tends to increase along latitudinal gradients (e.g. Southwick *et al.* [1996]). Even though our samples are not exhaustive, we are nevertheless surprised to find virtually no consistent relationship at all in our data. This hypothesis clearly needs a comprehensive test. Regardless, given the clear association between group size and other measures related to operational sex ratios mentioned above, these results are consistent with the lack of association between dimorphism and measures of environmental seasonality.

Seasonality, dimorphism, and hominins

It is interesting to consider the implications of this analysis for hominin evolution. Hominin evolution is characterized by several trends that might be associated with seasonality: (i) pronounced sexual dimorphism in body size that appears to reduce dramatically in the evolution of *Homo*; (ii) an increase in body size with the origin of the genus *Homo*; and (iii) a move to more open and more seasonal habitats, with an increase in tool use and hunting, among early representatives of the genus *Homo* (see Chapter 17).

As a general rule in primates, dimorphism does not change in a predictable way in seasonal environments. However, our analyses have revealed two relevant trends. First, females (and, to some extent, males) living in more seasonal or less productive environments tend to be smaller than those living in less seasonal or more productive environments; second, animals living farther away from the Equator tend to be larger than those living near it. If hominins moved from a less to a more seasonal environment, then we should expect, if anything, a reduction in body size, although not necessarily a change in dimorphism. While early hominins tend to be small compared with modern *Homo*, there is a clear increase in body size in the evolution of *Homo*. Indeed, McHenry (1992) suggests specifically that dimorphism may have been reduced in *H. erectus* and *H. ergaster* through a disproportionate *increase* in female body size. It is tempting to attribute the increase in size to Bergmann's rule – indeed *H. erectus* is the first hominin to clearly expand its range northward out of Africa. However, it is also characteristic of African *H. erectus*, which does not differ from earlier hominins in geographic location.

Hence, the general pattern seen with the rise of the genus *Homo* is not consistent with the response to increased seasonality seen in the broad primate sample. A different explanation therefore is needed. The larger body size may reflect greater mobility and the ability to exploit multiple resources, including animal protein and vegetable material year round. With a shift to more elaborate tool use and hunting, *Homo* could have enjoyed a more reliable and productive diet, allowing an increase in body size. Thus, *Homo erectus*'s flexibility in resource utilization in both time and space might have freed it from the impact of increased seasonality found among primates generally.

However, because in most primates changes in female body size are accompanied by parallel changes in male body size, the reduction in dimorphism requires a separate explanation. It may indicate a fundamental shift in social organization, reflecting reduced male–male competition for access to females. The latter, in turn, could be due to increased male–female bonding and male care for offspring (perhaps linked to life history changes leading to increased dependence of immatures) or vastly increased group sizes, perhaps induced by savanna life, and a corresponding tendency toward scramble competition for mates.

Conclusions

Our analyses and review indicate that reduced body size in various primates is associated with more seasonal or less productive environments, often without affecting sexual dimorphism. While increased breeding seasonality is correlated with reduced dimorphism in species living in large multi-male groups, this trend is not present among primates living in other forms of social organization. Female group size is not correlated with seasonality measures. Thus, overall, there are remarkably poor direct links between seasonality and dimorphism. The increase in body size and reduction in dimorphism in early *Homo*, which are thought to have shifted to habitats with more extreme seasonality, therefore are not explained simply as responses to this increased seasonality. Fundamental changes in lifestyle, including the nature of resource exploitation and the breeding system, therefore are more likely candidate explanations for these changes.

Appendix 14.1 *Natural-log-transformed values of male and female body mass, male and female mean geometric skull size, and maxillary canine crown height dimorphism used in analyses*

Species	Male mass	Female mass	Male skull	Female skull	Canine dimorph
Alouatta belzebul discolor	–	–	3.695	3.841	–
Alouatta belzebul nigerrima	1.971	1.645	3.882	3.741	0.594
Alouatta caraya	1.859	1.466	3.841	3.698	0.398
Alouatta fusca	1.907	1.470	3.817	3.660	0.412
Alouatta palliata aequatorialis	1.950	1.735	3.808	3.696	0.411
Alouatta pigra	2.434	1.861	3.825	3.699	0.312
Alouatta seniculus insulanis	1.841	1.416	3.838	3.662	–
Alouatta seniculus seniculus	1.901	1.651	3.820	3.689	0.404
Aotus azarae	0.140	0.215	3.279	3.277	–
Aotus lemurinus	−0.086	−0.135	3.232	3.245	0.158
Aotus vociferans	−0.345	−0.360	3.243	3.223	–
Ateles belzebuth belzebuth	2.115	2.061	3.818	3.802	–
Ateles fuscata	2.185	2.215	3.793	3.794	–
Ateles geoffroyi vellerosus	2.052	1.987	3.735	3.712	0.421
Ateles paniscus chameck	2.209	2.133	3.806	3.784	0.450
Brachyteles arachnoides	2.263	2.126	3.822	3.806	0.288
Cacajao calvus	1.238	1.058	3.639	3.590	0.227
Cacajao melanocephalus	1.151	0.997	3.595	3.547	–
Callicebus moloch discolor	0.020	−0.045	3.247	3.225	0.076
Callicebus torquatus lugens	0.247	0.191	3.290	3.269	0.043
Callithrix argentata argentata	−1.109	−1.022	2.943	2.946	–
Callithrix humeralifer humeralifer	−0.744	−0.751	2.956	2.963	–
Callithrix jacchus jachus	−1.016	−0.965	–	–	–
Callithrix jacchus penicilata	−1.067	−1.181	2.916	2.918	–
Cebus albifrons cesarae	1.157	0.829	3.571	3.525	–
Cebus apella libidinosus	1.166	0.728	3.634	3.545	0.343
Cebus apella paraguayanus	1.141	0.593	3.624	3.546	–
Cebus capucinus capucinus	1.303	0.932	3.662	3.575	–
Cebus nigrivittatus apiculatus	1.191	0.924	3.662	3.569	–
Cebus nigrivitatus casteneus	1.191	0.924	3.653	3.571	–

Appendix 14.1 (*cont.*)

Species	Male mass	Female mass	Male skull	Female skull	Canine dimorph
Chiropotes albimanus	1.147	0.912	3.585	3.508	–
Chiropotes satanas chiropotes	1.065	0.948	3.515	3.499	0.179
Lagothrix lagothricha cana	2.322	2.035	3.762	3.685	–
Lagothrix lagothricha lugens	–	–	3.726	3.697	–
Lagothrix lagothricha poeppigii	1.898	1.712	3.743	3.689	0.570
Leontopithecus rosalia	−0.478	−0.514	–	–	–
Pithecia irrorata irrorata	0.811	0.728	3.494	3.433	–
Pithecia monachus	0.959	0.747	3.480	3.461	–
Pithecia pithecia	0.588	0.412	3.443	3.392	0.277
Saguinus bicolor ochraseus	−0.849	−0.844	3.057	3.067	–
Saguinus fuscicollis nigrifrons	−1.070	−1.027	2.978	2.980	0.034
Saguinus inustus	–	–	3.054	3.071	–
Saguinus midas niger	−0.664	−0.553	3.022	3.019	−0.007
Saguinus mystax mystax	−0.673	−0.618	3.060	3.061	–
Saguinus nigricollis graellsi	−0.759	−0.726	2.993	2.992	–
Saimiri oerstedii oerstedii	−0.288	−0.511	3.214	3.179	0.521
Saimiri sciureus boliviensis	0.262	−0.317	3.202	3.148	–
Saimiri sciureus macrodon	−0.250	−0.412	3.256	3.212	0.347
Saimiri ustus	−0.082	−0.224	3.229	3.191	–
Allenopithecus nigroviridis	1.813	1.157	3.758	3.574	–
Cercocebus agilis agilis	2.251	1.733	3.901	3.775	0.674
Cercocebus torquatus atys	2.398	1.825	3.927	3.793	0.904
Cercopithecus aethiops hilgerti	1.522	1.182	3.671	3.562	0.593
Cercopithecus aethiops ngamiensis	1.787	1.335	3.772	3.620	–
Cercopithecus aethiops sabaeus	1.668	1.194	3.777	3.647	–
Cercopithecus ascanius whitesidei	1.308	1.072	3.650	3.567	0.448
Cercopithecus cephus cephus	1.456	1.058	3.686	3.611	0.572
Cercopithecus denti	–	–	3.698	3.560	–
Cercopithecus diana diana	1.649	1.361	3.762	3.661	0.481
Cercopithecus diana roloway	1.649	1.361	3.795	3.674	–
Cercopithecus erythrotis camerunensis	–	–	3.720	3.548	–
Cercopithecus erythrogaster	–	–	3.675	3.546	–
Cercopithecus hamlyni	1.703	1.212	3.759	3.604	–
Cercopithecus lhoesti lhoesti	1.787	1.238	3.839	3.652	0.611
Cercopithecus mitis kolbi	1.953	1.399	3.791	3.628	0.678
Cercopithecus mitis stuhlmanni	1.766	1.369	3.787	3.681	–
Cercopithecus mona	–	–	3.742	3.569	0.645
Cercopithecus neglectus	1.844	1.376	3.802	3.667	0.542
Cercopithecus nictitans nictitans	1.907	1.454	3.794	3.678	0.478
Cercopithecus petaurista petaurista	1.500	1.065	3.671	3.574	–
Cercopithecus pogonias grayi	1.449	1.065	3.682	3.594	0.443

Appendix 14.1 (*cont.*)

Species	Male mass	Female mass	Male skull	Female skull	Canine dimorph
Cercopithecus preussi preussi	–	–	3.744	3.613	–
Cercopithecus wolfi wolfi	1.364	1.054	3.660	3.562	0.633
Colobus angolensis cottoni	2.264	2.050	3.860	3.776	0.613
Colobus guereza caudatus	2.233	2.058	3.900	3.789	0.383
Colobus guereza kikiyuensis	–	–	3.865	3.763	–
Colobus guereza occidentalis	2.233	2.058	3.875	3.788	–
Colobus polykomos polykomos	2.293	2.116	3.851	3.778	0.570
Colobus satanas	2.342	2.004	3.797	3.754	0.563
Erythrocebus patas	2.518	1.872	3.959	3.743	0.778
Gorilla gorilla beringei	5.091	4.580	4.700	4.491	–
Gorilla gorilla gorilla	5.138	4.270	4.672	4.453	0.553
Gorilla gorilla grauri	5.166	4.263	4.720	4.476	–
Hylobates concolor	2.053	2.031	3.750	3.723	0.148
Hylobates hoolock	1.927	1.929	3.803	3.769	0.038
Hylobates klossi	1.735	1.778	3.663	3.650	−0.056
Hylobates lar carpenteri	1.742	1.668	3.732	3.710	0.148
Hylobates syndactylus	2.384	2.361	3.930	3.882	0.173
Kasi johnii	2.485	2.416	3.775	3.722	–
Kasi vetulus	1.735	1.631	3.718	3.642	0.596
Lophocebus albigena aterrimus	2.059	1.751	3.842	3.761	0.640
Lophocebus albigena johnstoni	2.035	1.872	3.875	3.786	–
Macaca fascicularis fascicularis	1.571	1.115	3.788	3.643	0.814
Macaca fuscata fuscata	2.398	2.083	3.968	3.876	0.713
Macaca hecki	2.416	1.917	4.007	3.842	0.721
Macaca mulatta mulatta	1.917	1.645	3.870	3.737	0.736
Macaca mulatta villosa	–	–	3.942	3.785	–
Macaca nemestrina leonina	2.041	1.589	3.974	3.821	–
Macaca nemestrina nemestrina	2.416	1.872	4.076	3.888	0.859
Macaca nigra	2.292	1.699	3.937	3.784	0.961
Macaca sinica	1.737	1.163	3.742	3.619	0.681
Macaca sylvanus	–	–	3.949	3.825	–
Macaca tonkeanna	2.701	2.197	4.041	3.882	0.662
Mandrillus leucophaeus leucophaeus	2.833	2.303	4.256	3.965	1.411
Mandrillus sphinx	3.453	2.557	4.365	4.001	–
Miopithecus talapoin talapoin	0.322	0.113	3.388	3.321	0.527
Nasalis larvatus	3.016	2.284	3.928	3.758	0.790
Pan paniscus	3.807	3.503	4.201	4.178	0.325
Pan troglodytes scheweinfurthi	3.754	3.517	4.350	4.302	–
Pan troglodytes troglodytes	4.089	3.824	4.346	4.295	0.353
Papio anubis	3.186	2.761	4.254	4.051	–
Papio cynocephalus kindae	2.847	2.277	4.004	3.873	1.030
Papio hamadryas	3.045	2.434	4.179	4.018	1.009
Papio ursinus	3.394	2.695	4.314	4.092	–
Pongo pygmaeus abelli	4.355	3.572	4.500	4.317	–

Appendix 14.1 (*cont.*)

Species	Male mass	Female mass	Male skull	Female skull	Canine dimorph
Pongo pygmaeus pygmaeus	4.363	3.578	4.481	4.307	0.526
Presbytis comata	1.899	1.904	3.649	3.623	0.411
Presbytis melalophos chrysomelas	1.869	1.844	3.677	3.628	0.538
Presbytis potenziani	1.820	1.856	3.665	3.651	0.350
Presbytis rubicunda chrysomelas	1.813	1.822	3.607	3.593	0.407
Procolobus badius badius	2.123	2.105	3.737	3.682	0.636
Procolobus badius oustaletti	2.510	2.110	3.854	3.733	–
Procolobus verus	1.548	1.435	3.617	3.560	0.696
Pygathrix nemaeus nemaeus	–	–	3.858	3.813	–
Pygathrix nemaeus nigripes	2.398	2.133	3.805	3.725	0.664
Rhinopithecus avunculus	–	–	3.866	3.737	–
Rhinopithecus roxellana roxellana	2.885	2.451	3.908	3.801	–
Semnopithecus entellus achates	–	–	3.872	3.764	–
Semnopithecus entellus ajax	–	–	4.067	3.922	–
Semnopithecus entellus anchises	–	–	3.966	3.751	–
Semnopithecus entellus entellus	–	–	3.968	3.810	–
Semnopithecus entellus hector	–	–	4.030	3.905	–
Semnopithecus entellus iulus	–	–	3.763	3.746	–
Semnopithecus entellus priam	–	–	3.878	3.790	–
Semnopithecus entellus schistacea	2.955	2.695	4.053	3.946	–
Semnopithecus entellus thersites	2.365	1.902	3.821	3.702	0.568
Simias concolor	2.214	1.917	3.710	3.642	0.544
Theropithecus gelada	2.944	2.460	4.127	3.926	1.172
Trachypithecus cristata pyrrhus	1.889	1.751	3.770	3.679	0.529
Trachypithecus cristata ultima	1.924	1.759	3.697	3.607	–
Trachypithecus francoisi	2.041	1.995	3.769	3.725	–
Trachypithecus obscura obscura	1.992	1.899	3.700	3.650	0.574
Trachypithecus pileata shortridgei	2.485	2.288	3.855	3.752	0.635

Appendix 14.2 *Group size and rainfall data within species*

Species and group size	Annual rainfall (mm)	References
Alouatta caraya		
7.9	1400	Pope (1966, 1968)
2	1400	Pope (1966, 1968)
8.05	1400	Thorington *et al.* (1984)
4	1500	Wallace *et al.* (1998)
Alouatta palliata		
9.3	2730	Altmann (1959)
16.5	2730	Bernstein (1964)
17.3	2730	Carpenter (1934, 1962)
17.4	2730	Carpenter (1934, 1962)
18.2	2730	Carpenter (1934, 1962)
18.5	2730	Carpenter (1934, 1962)
14.7	2730	Chivers (1969)
15.5	1450	Clarke *et al.* (1986)
8	2730	Collias & Southwick (1952)
9.12	4953	Estrada (1982)
13.6	1531	Fedigan *et al.* (1985)
13.75	4015	Fishkind & Sussman (1987)
8.13	1531	Freese (1976)
8.9	1450	Heltne *et al.* (1976)
9.9	1450	Heltne *et al.* (1976)
10	1450	Heltne *et al.* (1976)
10	1450	Heltne *et al.* (1976)
10.4	1450	Heltne *et al.* (1976)
10.8	1450	Heltne *et al.* (1976)
12.5	1450	Heltne *et al.* (1976)
13.2	1450	Heltne *et al.* (1976)
13.5	1450	Heltne *et al.* (1976)
15.5	1450	Heltne *et al.* (1976)
13.1	1450	Heltne *et al.* (1976)
11.3	1450	Heltne *et al.* (1976)
18.7	2730	Milton (1982)
18.9	2730	Milton (1982)
18.9	2730	Milton (1982)
20.2	2730	Milton (1982)
20.4	2730	Milton (1982)
20.8	2730	Milton (1982)
21.3	2730	Milton (1982)
21.4	2730	Milton (1982)
23	2730	Milton (1982)
16.2	2730	Mittermeier (1973)
13.77	1878	Rodriguez (1985)
13.8	2730	Smith (1977)
14	3962	Stoner (1996)

Appendix 14.2 (*cont.*)

Species and group size	Annual rainfall (mm)	References
Alouatta pigra		
4.2	2032	Bolin (1981)
6.25	1350	Coelho *et al.* (1976)
6.22	2030	Horwich (1983)
5	1350	Horwich & Johnson (1986)
6.3	2032	Horwich & Johnson (1986)
5	1350	Schlicte (1978)
5.9	1955	Silver *et al.* (1998)
Alouatta seniculus		
6.3	1424	Braza *et al.* (1981)
7.7	1693	Crockett (1984)
10.5	1693	Crockett (1984)
7.8	1693	Crockett & Eisenberg (1987)
8.3	1693	Crockett & Eisenberg (1987)
9.1	1693	Crockett & Eisenberg (1987)
9.7	1693	Crockett & Eisenberg (1987)
5.75	2100	Defler (1981)
8.7	1693	Eisenberg (1979)
9	1947	Gaulin & Gaulin (1982)
3.3	2705	Green (1978)
3.3	2705	Green (1978)
6	3125	Julliot (1996)
5	3050	Neville (1976)
6.9	1693	Neville (1972, 1976)
7.9	1693	Neville (1972, 1976)
8.5	1693	Neville (1972, 1976)
8.8	1693	Neville (1972, 1976)
9	1693	Neville (1972, 1976)
9	1693	Neville (1972, 1976)
6	3836	Palacios & Rodriguez (2001)
8.9	1693	Rudran (1979)
8	3125	Simmen & Sabatier (1996)
3.3	1500	Wallace *et al.* (1998)
Ateles paniscus		
13	3125	Simmers & Sabatier (1996)
6.2	1500	Wallace *et al.* (1998)

Appendix 14.2 (*cont.*)

Species and group size	Annual rainfall (mm)	References
Cebus apella		
27	2000	DiBietti & Janson (2000)
13	3125	Simmers & Sabatier (1996)
9	1500	Wallace *et al.* (1998)
Cercopithecus aethiops		
12	874	Dunbar & Dunbar (1974)
42	867	Galat & Galat Luong (1976)
12	1150	Hall & Gartlan (1965)
25	1056	Henzi & Lucas (1980)
28	568	Isbell & Pruetz (1998)
76	1571	Kavanagh (1978)
18.5	497	Nakagawa (1999)
39	848	Whitten (1983)
Cercopithecus ascanius		
23	2000	Cords (1984)
14.7	1850	McGraw (1994)
32	1500	Struhsaker & Leland (1979)
Cercopithecus mitis		
35	2000	Cords (2000)
29	1744	Kaplin & Moermond (2000)
24	1500	Struhsaker & Leland (1979)
Colobus angolensis		
5	1850	McGraw (1994)
4.9	2510	McKey (1978)
Colobus guereza		
6	1500	Clutton-Brock (1975)
7.4	1230	Dunbar & Dunbar (1974)
7.8	945	Dunbar (1987)
7	2000	Fashing (2001)
10	2000	Fashing (2001)
11	2000	Fashing (2001)
11.9	2000	Fashing (2001)
15	2000	Fashing (2001)
21	2000	Fashing (2001)
5.4	1191	Groves (1973)
6.3	1050	Marler (1969)
8.4	1050	Marler (1969)

Appendix 14.2 (*cont.*)

Species and group size	Annual rainfall (mm)	References
6	1500	Oates (1977)
19	655	Rose (1977)
6.9	1050	Suzuki (1979)
Colobus satanas		
11	1775	Fleury & Gautier-Hion (1999)
17.5	1775	Fleury & Gautier-Hion (1999)
22	1775	Fleury & Gautier-Hion (1999)
15	4000	McKey & Waterman (1982)
12.1	1531	Tutin *et al.* (1997)
Erythrocebus patas		
39	568	Isbell & Pruetz (1998)
11	497	Nakagawa (1999)
Gorilla gorilla gorilla		
8.4	1412	Parnell (2002)
13	1400	Remis (1997)
5.3	1531	Tutin *et al.* (1997)
Lagothrix lagothricha		
24	3832	Deffler (1996)
18	2623	Stevenson (1998)
30	2623	Stevenson (2000)
Lophocebus albigena		
18	1600	Poulsen *et al.* (2001)
15	1500	Struhsaker & Leland (1979); Oluput *et al.* (1994)
18.9	1531	Tutin et al. (1997)
Nasalis larvatus		
9	2977	Tilson (1977); Watanabe (1981); Bennett & Sebastian (1988); Yeager (1989, 1991)
12	2400	Tilson (1977); Watanabe (1981); Bennett & Sebastian (1988); Yeager (1989, 1991)
21	3103	Boonratana (2000)
Pan troglodytes verus		
11	1850	Herbinger *et al.* (2001)
35	1850	Herbinger *et al.* (2001)

Appendix 14.2 (*cont.*)

Species and group size	Annual rainfall (mm)	References
63	1850	Herbinger *et al.* (2001)
18	2230	Yamakoshi (1998)
Presbytis melalophos		
6	2977	Bennett & Davies (1994)
14	2207	Bennett & Davies (1994)
15	2120	Bennett & Davies (1994)
Presbytis rubicunda		
6.1	2400	Bennett & Davies (1994)
7	2977	Bennett & Davies (1994)
Procolobus badius		
33	2708	Davies *et al.* (1999)
16	1511	Galat-Luong & Galat (1979)
50	1700	Maisels *et al.* (1994)
32	1504	Siex & Struhsaker (1999)
33.6	1504	Silkiluwasha (1981)
40	2963	Struhsaker (1975)
34	1500	Struhsaker (1975); Struhsaker & Leland (1987)
Semnopithecus entellus		
12.8	1600	Oppenheimer (1977); Newton (1987)
15.5	1298	Oppenheimer (1977); Newton (1987)
15.3	1298	Oppenheimer (1977); Newton (1987)
21.3	820	Oppenheimer (1977); Newton (1987)
35	450	Oppenheimer (1977); Newton (1987)
12	1332	Vogel (1971); Curtin (1975); Boggess (1976); Bishop (1979)
26	1200	Ziegler *et al.* (2000)
Trachypithecus auratus		
9	1523	Bernstein (1968); Wolf (1978); Kool (1989)
14.5	3215	Bernstein (1968); Wolf (1978); Kool (1989)
Trachypithecus cristata		
17	2855	Bernstein (1968) Wolf (1978); Kool (1989)
26.5	2506	Bernstein (1968); Wolf (1978); Kool (1989)

Appendix 14.2 (*cont.*)

Species and group size	Annual rainfall (mm)	References
Trachypithecus pileatus		
7.5	1633	Bennett & Davies (1994)
8.3	1633	Bennett & Davies (1994)
9.6	2218	Bennett & Davies (1994)

References

Albrecht, G. H. (1978). *The Craniofacial Morphology of the Sulawesi Macaques.* Basel: Karger.

— (1980). Latitudinal, taxonomic, sexual, and insular determinants of size variation in pigtail macaques, *Macaca nemestrina. International Journal of Primatology*, **1**, 141–52.

— (1983). Geographic variation in the skull of the crab-eating macaque, *Macaca fascicularis* (Primates: Cercopithecidae). *American Journal of Physical Anthropology*, **60**, 169.

Albrecht, G. H. & Miller, J. M. A. (1993). Geographic variation in primates: a review with implications for interpreting fossils. In *Species, Species Concepts, and Primate Evolution*, ed. W. H. Kimbel & L. B. Martin. New York: Plenum Press, pp. 211–37.

Albrecht, G. H., Jenkins, P. D., & Godfrey, L. R. (1990). Ecogeographic size variation among the living and subfossil prosimians of Madagascar. *American Journal of Primatology*, **22**, 1–50.

Altmann, J. (2000). Models of outcome and process: predicting the number of males in primate groups. In *Primate Males*, ed. P. M. Kappeler. Cambridge: Cambridge University Press, pp. 236–47.

Altmann, S. A. (1959). Field observations on a howling monkey society. *Journal of Mammalogy*, **40**, 317–30.

— (1962). A field study of the sociobiology of the rhesus monkey, *Macaca mulatta. Annals of the New York Academy of Sciences*, **102**, 338–435.

Andersson, M. (1994). *Sexual Selection.* Princeton: Princeton University Press.

Andelman, S. J. (1986). Ecological and social determinants of cercopithecine mating patterns. In *Ecological Aspects of Social Evolution: Birds and Mammals*, ed. D. I. Rubenstein & R. W. Wrangham. Princeton: Princeton University Press, pp. 201–16.

Bennett, E. L. & Davies, A. G. (1994). The ecology of Asian colobines. In *Colobine Monkeys*, ed. A. G. Davies & J. F. Oates. Cambridge: Cambridge University Press, pp. 129–72.

Bennett, E. L. & Sebastian, A. C. (1988). Social organization and ecology of proboscis monkeys (*Nasalis larvatus*) in mixed coastal forest in Sarawak. *International Journal of Primatology*, **9**, 233–55.

Bernstein, I. S. (1964). A field study of the activities of howler monkeys. *Animal Behavior*, **12**, 92–7.
— (1968). The lutong of Kuala Selangor. *Behavior*, **32**, 1–16.
Bishop, N. H. (1979). Himalayan langurs: temperate colobines. *Journal of Human Evolution*, **8**, 251–81.
Boggess, J. (1976). The social behavior of the Himalayan langur (*Presbytis entellus*) in Eeastern Nepal. Ph. D. thesis, University of California.
Bolin, I. (1981). Male parental behavior in black howler monkeys (*Alouatta palliata pigra*) in Belize and Guatemala. *Primates*, **22**, 349–60.
Boonratana, R. (2000). Ranging behavior of proboscis monkeys (*Nasalis larvatus*) in the lower Kinabatangan, Northern Borneo. *International Journal of Primatology*, **21**, 497–518.
Braza, A. A., Alvarez, A., & Azacarte, T. (1981). Behavior of the red howler monkey (*Alouatta seniculus*) in the llanos of Venezuela. *Primates*, **22**, 459–73.
Carpenter, C. R. (1934). A field study of the behavior and social relations of howling monkeys. *Comparative Psychology Monographs*, **10**, 1–168.
— (1962). Field studies of a primate population. In *Roots of Behavior*, ed. E. L Bliss. New York: Harper and Row, pp. 286–94.
Chivers, D. J. (1969). On the daily behavior and spacing of howling monkey groups. *Folia Primatologica*, **10**, 48–102.
Clarke, M. R., Zucker, E. L., & Scott, N. J. (1986). Population trends of the mantled howler groups of La Pacifica, Guanacaste, Costa Rica. *American Journal of Primatology*, **11**, 79–88.
Clutton-Brock, T. H. (1975). Feeding behavior of red colobus and black and white colobus in East Africa. *Folia Primatologica*, **23**, 165–207.
— (1989). Mammalian mating systems. *Proceedings of the Royal Society, London, Series B*, **236**, 339–72.
Clutton-Brock T. H., Harvey, P. H., & Rudder, B. (1977). Sexual dimorphism, socionomic sex ratio and body weight in primates. *Nature*, **269**, 191–5.
Coelho. A. M., Coelho, L. S., Bramblett, C. A., Bramblett, S. S., & Quick, L. B. (1976). Ecology, population characteristics and sympatric associations in primates: a bioenergetic analysis of howler and spider monkeys in Tikal. *Yearbook of Physical Anthropology*, **20**, 96–135.
Collias, N. & Southwick, C. (1952). A field study of population density and social organization in howling monkeys. *Proceedings of the American Philosophical Society*, **96**, 143–56.
Cords, M. (1984). Mating patterns and social structure in redtail monkeys (*Cercopithecus ascanius*). *Zeitschrift fur Tierpsychologie*, **64**, 313–29.
— (2000). The number of males in guenon groups. In *Primate Males*, ed. P. M. Kappeler. Cambridge: Cambridge University Press, pp. 84–96.
Crockett, C. M. (1984). Emigration by female red howler monkeys and the case for female competition. In *Female Primates: Studies by Female Primatologists*, ed. M. F. Small. New York: Alan R. Liss, pp. 159–73.
Crockett, C. M. & Eisenberg, J. F. (1987). Howlers: variation in group size and demography. In *Primate Societies*, ed. B. Smuts, D. Cheney, R. Wrangham, & T. T. Struhsaker. Chicago: University of Chicago Press, pp. 54–68.

Curtin, R. A. (1975). The socio-ecology of the common langur, *Presbytis entellus*, in the Nepal Himalaya. Ph. D. thesis, University of California.

Davies, A. G., Oates, J. F., & Dasilva, G. L. (1999). Patterns of frugivory in three West African colobine monkeys. *International Journal of Primatology*, **20**, 327–57.

Defler, T. R. (1981). The density of *Alouatta seniculus* in the eastern llanos of Colombia. *Primates*, **22**, 564–9.

— (1996). Aspects of the ranging pattern in a group of wild woolly monkeys (*Lagothrix lagothricha*), *American Journal of Primatology*, **38**, 289–302.

DiBietti, M. S., & Janson, C. H. (2000). Reproductive socioecology of tufted capuchins (*Cebus apella nigritus*) in Northeastern Argentina. *International Journal of Primatology*, **22**, 127–142.

DiFiore, A. & Rodman, P. S. (2001). Time allocation patterns of lowland woolly monkeys (*Lagothrix lagothricha peoppigii*) in a neotropical terra firma forest. *International Journal of Primatology*, **22**, 449–80.

Dunbar, R. I. M. (1987). Habitat quality, population dynamics, and group composition in colobus monkeys (*Colobus guereza*). *International Journal of Primatology*, **8**, 299–330.

— (1990). Environmental determinants of intraspecific variation in body weight in baboons (*Papio* spp.). *Journal of Zoology, London*, **220**, 157–69.

— (2000). Male mating strategies: a modeling approach. In *Primate Males*, ed. P. M. Kappeler. Cambridge: Cambridge University Press, pp. 259–68.

Dunbar, R. I. M. & Dunbar, E. P. (1974). Ecology and population dynamics of *Colobus guereza* in Ethiopia. *Folia Primatologica*, **21**, 188–208.

Eisenberg, J. F. (1979). Habitat, economy, and society: some correlations and hypotheses for the Neotropical primates. In *Primate Ecology and Human Origins*, ed. I. S. Bernstein & E. O. Smith. New York: Garland Press, pp. 215–62.

Emlen, S. T. & Oring T. (1977). Ecology, sexual selection, and the evolution of mating systems. *Science*, **191**, 215–33.

Estrada, A. (1982). Survey and census of howler monkeys (*Alouatta palliata*) in the rain forest of "Los Tuxtlas" Veracruz, Mexico. *American Journal of Primatology*, **2**, 363–72.

Fashing, P. J. (2001). Feeding ecology of guerezas in the Kakamega Forest Kenya: the importance of Moracaceae fruit in their diet. *International Journal of Primatology*, **22**, 579–610.

Fedigan, L. M., Fedigan, L., & Chapman, C. A. (1985). A census of *Alouatta palliata* and *Cebus capucinus* in Santa Rosa National Park, Costa Rica. *Brenesia*, **23**, 309–22.

Fishkind, A. S. & Sussman, R. W. (1987). Preliminary survey of the primates of the Zona Protectora and La Selva Biological Station, northeast Costa Rica. *Primate Conservation*, **8**, 63–6.

Fleury, M. & Gautier-Hion, A. (1999). Seminomadic ranging in a population of black colobus (*Colobus satanas*) in Gabon and its ecological correlates. *International Journal of Primatology*, **20**, 491–510.

Fooden, J. & Albrecht, G. H. (1993). Latitudinal and insular variation of skull size in crab-eating macaques (Primates, Cercopithecidae: *Macaca fascicularis*). *American Journal of Physical Anthropology*, **92**, 521–38.

Ford, S. M. (1994). Evolution of sexual dimorphism in body weight in platyrrhines. *American Journal of Primatology*, **34**, 221–4.

Freckleton, R. P., Harvey, P. H., & Pagel, M. (2003). Bergmann's rule and body size in mammals. *American Naturalist*, **161**, 821–5.

Freese, C. (1976). Censusing *Alouatta palliata* and *Cebus capucinus* in Costa Rican dry forest. In *Neotropical Primates: Field Studies and Conservation*, ed. R. W. Thorington & P. G. Heltne. Washington, DC: National Academy of Sciences, pp. 4–9.

Galat, G. & Galat-Luong, A. (1976). La colonization de la mangrove par *Cercopithecus aethiops sabaeus* au Senegal. *La Terre et Vie*, **30**, 3–30.

Galat-Luong, A. & Galat, G. (1979). Quelques observations sur l'ecologie de *Procolobus badius oustaleti* en Empire Centrafrician. *Mammalia*, **43**, 309–12.

Garland, T., Jr, Dickerman, A. W., Janis, C. M., & Jones, J. A. (1993). Phylogenetic analysis of covariance by computer simulation. *Systematic Biology*, **42**, 265–92.

Garland, T., Jr, Midford, P. E., & Ives, A. R. (1999). An introduction to phylogenetically based statistical methods, with a new method for confidence intervals on ancestral states. *American Zoologist*, **39**, 374–88.

Gaulin, S. J. C. & Gaulin, C. K. (1982). Behavioral ecology of *Alouatta seniculus* in Andean cloud forest. *International Journal of Primatology*, **3**, 1–32.

Green, S. (1978). Primate censusing in northern Colombia: a comparison of two techniques. *Primates*, **19**, 537–50.

Greenfield, L. O. (1992). Relative canine size, behavior, and diet in male ceboids. *Journal of Human Evolution*, **23**, 469–80.

Greenfield, L. O. & Washburn, A. (1991). Polymorphic aspects of male anthropoid canines. *American Journal of Physical Anthropology*, **84**, 17–34.

Groves, C. P. (1973). Notes on the ecology and behavior of the Angola colobus (*Colobus angolensis* P. L. Sclater 1860) in N.E. Tanzania. *Folia Primatologica*, **20**, 12–26.

Hall, K. R. L. & Gartlan, J. S. (1965). Ecology and behavior of the vervet monkey, *C. aethiops*, Lolui Island, Lake Victoria. *Proceedings of the Zoological Society, London*, **145**, 37–56.

Heltne, P. G., Turner, D. C., & Scott, N. J. (1976). Comparison of census data on *Alouatta palliata* from Costa Rica and Panama. In *Neotropical Primates: Field Studies and Conservation*, ed. R. W. Thorington & P. G. Heltne. Washington, DC: National Academy of Sciences, pp. 10–19.

Henzi, S. P., & Lucas, J. W. (1980). Observations on the inter-troop movement of adult vervet monkeys (*Cercopithecus aethiops*). *Folia Primatologica*, **33**, 220–35.

Herbinger, I., Boesch, C., & Rothe, H. (2001). Territory characteristics among three neighboring chimpanzee communities in the Tai National Park, Cote d'Ivoire. *International Journal of Primatology*, **22**, 143–68.

Hershkovitz P. (1977). *Living New World Monkeys* (Platyrrhini), Vol. 1. Chicago: University of Chicago Press.

Horwich, R. W. (1983). Species status of the black howler monkey, *Alouatta pigra*, of Belize. *Primates*, **27**, 53–62.

Horwich, R. H. & Johnson, E. D. (1986). Geographical distribution of the black howler (*Alouatta pigra*) in Central America. *Primates*, **27**, 53–62.

Isbell, L. A. & Pruetz, J. D. (1998). Differences between vervets (*Cercopithecus aethiops*) and patas monkeys (*Erythrocebus patas*) in agonistic interactions between adult females. *International Journal of Primatology*, **19**, 837–56.

Janson, C. H., & Goldsmith, M. L. (1995). Predicting group size in primates: foraging costs and predation risks. *Behavioural Ecology*, **6**, 326–36.

Julliot, C. (1996). Fruit choice by red howler monkeys (*Alouatta seniculus*) in a tropical rain forest. *American Journal of Primatology*, **40**, 261–82.

Kaplin, B. A. & Moermond, T. C. (2000). Foraging ecology of the mountain monkey (*Cercopithecus l'hoesti*): implications for its evolutionary history and use of disturbed forest. *American Journal of Primatology*, **50**, 227–46.

Kavanagh, M. (1978). The diet and feeding behavior of *Cercopithecus aethiops tantalus*. *Folia Primatologica*, **30**, 76–98.

Kay, R. F., Plavcan, J. M., Glander, K. E., & Wright, P. C. (1988). Sexual selection and canine dimorphism in New World monkeys. *American Journal of Physical Anthropology*, **77**, 385–97.

Kay, R. F., Madden, R. H., van Schaik, C. & Higdon, D. (1997). Primate species richness is determined by plant productivity: implications for conservation. *Proceedings of the National Academy of Sciences, USA*, **94**, 13023–7.

Kool, K. M. (1989). Behavioral ecology of the silver leaf monkey, *Trachypithecus auratus sondaicus*, in the Pangandaran Nature Reserve, West Java, Indonesia. Ph. D. thesis, University of New South Wales.

Lande R. (1980). Sexual dimorphism, sexual selection, and adaptation in polygenic characters. *Evolution*, **33**, 292–305.

Leigh, S. R. (1994). Relations between captive and noncaptive weights in anthropoid primates. *Zoo Biology*, **13**, 21–43.

Leigh, S. R., & Shea, B. T. (1995). Ontogeny and the evolution of adult body size dimorphism in apes. *American Journal of Primatology*, **36**, 37–60.

Leigh, S. R. & Terranova, C. J. (1998). Comparative perspectives on bimaturism, ontogeny, and dimorphism in lemurid primates. *International Journal of Primatology*, **19**, 723–49.

Leutenegger, W. & Kelly, J. T. (1977). Relationship of sexual dimorphism in canine size and body size to social, behavioral and ecological correlates in anthropoid primates. *Primates*, **18**, 117–36.

Lindenfors, P. (2002). Sexually antagonistic selection on primate size. *Journal of Evolutionary Biology*, **15**, 595–607.

Lindenfors, P. & Tullberg, B. S. (1998). Phylogenetic analysis of primate size evolution: the consequences of sexual selection. *Biological Journal of the Linnean Society*, **64**, 413–47.

Maisels, F., Gautier-Hion, A., & Gautier, J. P. (1994). Diets of two sympatric colobines in Zaire: more evidence on seed-eating in forests on poor soils. *International Journal of Primatology* **15**, 681–701.

Marler, P. (1969). *Colobus guereza*: territoriality and group composition. *Science*, **163**, 93–5.

Martin, R. D., Willner, L. A., & Dettling, A. (1994). The evolution of sexual size dimorphism in primates. In *The Differences Between the Sexes*, ed. R. V. Short & E. Balaban. Cambridge: Cambridge University Press, pp. 159–200.
McGraw, W. S. (1994). Census, habitat preference and polyspecific associations of six monkeys in the Lomako Forest, Zaire. *American Journal of Primatology*, **34**, 295–307.
McHenry, H. M. (1992). Body size and proportions in early hominids. *American Journal of Physical Anthropology*, **87**, 407–31.
McKey, D. B. (1978). Plant chemical defenses and the feeding and ranging behavior of colobus monkeys in African rain forests. Ph. D. thesis, University of Michigan.
McKey, D. B. & Watermann, P. G. (1982). Ranging behavior of a group of black colobus (*Colobus satanas*) in the Douala-Edea Reserve, Cameroon. *Folia Primatologica*, **39**, 264–304.
Milton, K. (1982). Dietary quality and demographic regulation in a howler monkey population. In *The Ecology of a Tropical Forest*, ed. E. G. Leigh, A. S. Rand, & D. M. Windsor. Washington, DC: Smithsonian Institution Press, pp. 273–89.
Mitani, J., Gros-Louis, J., & Richards, A. F. (1996a). Number of males in primate groups: comparative tests of competing hypotheses. *American Journal of Primatology*, **38**, 315–32.
— (1996b). Sexual dimorphism, the operational sex ratio, and the intensity of male competition in polygynous primates. *American Naturalist*, **147**, 966–80.
Mittermeier, R. A. (1973). Group activity and population dynamics of the howler monkey on Barro Colorado Island. *Primates*, **14**, 1–19.
Mowry, C. B., Decker, B. S., & Shure, D. J. (1996). The role of phytochemistry in dietary choices of Tana River red colobus monkeys (*Procolobus badius rufomitratus*). *International Journal of Primatology*, **17**, 63–84.
Nakagawa, N. (1999). Differential habitat utilization by patas monkeys *Erythrocebus patas*) and tantalus monkeys (*Cercopithecus aethiops tantalus*) living sympatrically in northern Cameroon. *American Journal of Primatology*, **49**, 243–64.
Neville, M. K. (1972). The population structure of red howler monkeys (*Alouatta seniculus*) in Trinidad and Venezuela. *Folia Primatologica*, **17**, 56–86.
— (1976). The population and conservation of howler monkeys in Venezuela and Trinidad. In *Neotropical Primates: Field Studies and Conservation*, ed. R. W. Thorington & P. G. Heltne. Washington, DC: National Academy of Sciences, pp. 101–9.
Newton, P. N. (1987). The social organization of forest hanuman langurs (*Presbytis entellus*). *International Journal of Primatology*, **8**, 199–232.
Nunn, C. L. (1999). The evolution of exaggerated sexual swellings in primates and the graded-signal hypothesis. *Animal Behavior*, **58**, 229–46.
Oates, J. F. (1977). The social life of black and white colobus monkeys, *Colobus guereza*. *Zeitschrift für Tierpsychologie*, **45**, 1–60.
Olupot, W., Chapman, C. A., and Brown, C. H. (1994). Mangabey (*Cercocebus albigena*) population density, group size and ranging: a 20 year comparison. *American Journal of Primatology*, **32**, 197–205.

Oppenheimer, J. R. (1977). *Presbytis entellus*, the Hanuman langur. In *Primate Conservation*, ed. H. S. H. Prince Ranier & G. H. Bourne. New York: Academic Press, pp. 469–512.

Palacios, E. & Rodriguez, A. (2001). Ranging pattern and use of space in a group of red howler monkeys (*Alouatta seniculus*) in a southeastern Colombian rainforest. *American Journal of Primatology*, **55**, 233–51.

Parnell, R. J. (2002). Group size and structure in western lowland gorillas (*Gorilla gorilla gorilla*) at Mbeli Bai, Republic of Congo. *American Journal of Primatology*, **56**, 193–206.

Pereira, M. A., Clutton-Brock, T. H., & Kappeler, P. M. (2000). Understanding male primates. In *Primate Males*, ed. P. M. Kappeler. Cambridge: Cambridge University Press, pp. 271–7.

Plavcan, J. M. (1990). Sexual dimorphism in the dentition of extant anthropoid primates. Ph. D. thesis, Duke University, Ann Arbor.

— (1999). Mating systems, intrasexual competition and sexual dimorphism in primates. In *Comparative Primate Socioecology*, ed. P. C. Lee. Cambridge: Cambridge University Press, pp. 241–69.

— (2000). Inferring social behavior from sexual dimorphism in the fossil record. *Journal of Human Evolution*, **39**, 327–44.

— (2001). Sexual dimorphism in primate evolution. *Yearbook of Physical Anthropology*, **44**, 25–53.

— (2003). Scaling relationships between craniofacial sexual dimorphism and body mass dimorphism in primates: implications for the fossil record. *American Journal of Physical Anthropology*, **120**, 38–60.

— (2004). Sexual selection, measures of sexual selection, and sexual dimorphism in primates. In *Sexual Selection in Primates: A Comparative Perspective*, ed. P. M. Kappeler & C. P. van Schaik. Cambridge: Cambridge University Press, pp. 230–52.

Plavcan, J. M. & van Schaik, C. P. (1992) Intrasexual competition and canine dimorphism in anthropoid primates. *American Journal of Physical Anthropology*, **87**, 461–77.

— (1997). Intrasexual competition and body weight dimorphism in anthropoid primates. *American Journal of Physical Anthropology*, **103**, 37–68.

Plavcan, J. M., van Schaik, C. P., & Kappeler, P. M. (1995). Competition, coalitions and canine size in primates. *Journal of Human Evolution*, **28**, 245–76.

Pope, B. (1966). Population characteristics of howler monkeys (*Alouatta caraya*) in northern Argentina. *American Journal of Physical Anthropology*, **24**, 361–70.

— (1968). Population characteristics. *Biblio Primatology*, **7**, 13–70.

Poulsen, J. R., Clarke, C. J., & Smith, T. B. (2001). Seasonal variation in the feeding ecology of the gray-cheeked mangabey (*Lophocebus albigena*) in Cameroon. *American Journal of Primatology*, **54**, 91–105.

Purvis, A. & Webster, A. J. (1999). Phylogenetically independent comparisons and primate phylogeny. In *Comparative Primate Socioecology*, ed. P. C. Lee. Cambridge: University of Cambridge Press, pp. 44–70.

Ravosa, M. J. (1998). Cranial allometry and geographic variation in slow lorises (*Nycticebus*). *American Journal of Primatology*, **45**, 225–43.

Reed, C., O'Brein, T. G., & Kinnaird, M. E. (1997). Male social behavior and dominance hierarchy in the Sulawesi crested black macaque (*Macaca nigra*). *International Journal of Primatology*, **18**, 247–60.
Remis, M. (1997). Western lowland gorillas (*Gorilla gorilla gorilla*) as seasonal frugivores: use of variable resources. *American Journal of Primatology*, **43**, 87–109.
Rodriguez, M. A. R. (1985). Algunos aspectos sobre comportamiento, alimentacion y nivel de poblacion de los monos (Primates: Cebidae) en el Refugio De Fauna Silvestre Palo Verde (Guanacaste, Costa Rica). In *Investigaciones Sobre Fauna Silvestr de Costa Rica*. San Jose, Costa Rica: Editorial Universidad Estatal a Distancia, pp. 53–71.
Rose, M. D. (1977). Interspecific play between free ranging guerezas (*Colobus guereza*) and vervet monkeys (*Cercopithecus aeithiops*). *Primates*, **18**, 957–64.
Rudran, R. (1979). The demography and social mobility of a red howler (*Alouatta seniculus*) population in Venezuela. In *Vertebrate Ecology in the Northern Neotropics*, ed. J. F. Eisenberg. Washington, DC: Smithsonian Institution Press, pp. 107–26.
Schlicte, H. (1978). A preliminary report on the habitat utilization of a group of howler monkeys (*Alouatta villosa pigra*) in the National Park of Tikal, Guatemala. In *The Ecology of Arboreal Folivores*, ed. G. G. Montgomery. Washington, DC: Smithsonian Institution Press, pp. 551–9.
Siex, K. S. & Struhsaker, T. T. (1999). Ecology of the Zanzibar red colobus monkey: demographic variability and habitat stability. *International Journal of Primatology*, **20**, 163–92.
Silkiluwasha, F. (1981). The distribution and conservation status of the Zanzibar red colobus. *African Journal of Ecology*, **19**, 187–94.
Silver, S. C., Ostro, L. E. T., Yeager, C. P., & Horwich, R. (1998). Feeding ecology of the black howler monkey (*Alouatta pigra*) in northern Belize. *American Journal of Primatology*, **45**, 263–79.
Simmen, B. & Sabatier, D. (1996). Diets of some French Guianan primates: food composition and food choices. *International Journal of Primatology*, **17**, 661–94.
Smith, C. C. (1977). Feeding behavior of social organization in howling monkeys. In *Primate Ecology*, ed. T. T. Clutton-Brock. London: Academic Press, pp. 97–126.
Smith, R. J. (1999). Statistics of sexual size dimorphism. *Journal of Human Evolution*, **36**, 423–59.
Smith, R. J. & Cheverud, J. M. (2002). Scaling of sexual dimorphism in body mass: a phylogenetic analysis of Rensch's rule in primates. *International Journal of Primatology*, **23**, 1095–135.
Smith, R. J. & Jungers, W. L. (1997). Body mass in comparative primatology. *Journal of Human Evolution*, **32**, 523–59.
Southwick, C. H., Zhang, Y., Jiang, H., Liu, Z., & Qu, W. (1996). Population ecology of rhesus macaques in tropical and temperate habitats. In *Evolution*

and Ecology of Macaque Societies, ed. J. E. Fa & D. G. Lindburg. Cambridge: Cambridge University Press, pp. 95–105.

Srivastava, A. & Dunbar, R. I. M. (1996). The mating system of hanuman langurs: a problem in optimal foraging. *Behavioral Ecology and Sociobiology*, **39**, 219–26.

Stearns, S. C. & J. C. Koella (1986). The evolution of phenotypic plasticity in life-history traits: predictions of reaction norms for age and size at maturity. *Evolution*, **40**, 893–915.

Stevenson, P. (1998). Proximal spacing between individuals in a group of woolly monkeys (*Lagothrix lagothricha*) in Tinigua National Park, Columbia. *International Journal of Primatology*, **19**, 299–312.

Stevenson, P. R. (2000). Seed dispersal by woolly monkeys (*Lagothrix lagothricha*) at Tinigua National Park, Colombia: dispersal distance, germination rates, and dispersal quantity. *American Journal of Primatology*, **50**, 275–89.

Stoner, K. E. (1996). Habitat selection and seasonal patterns of mantled howling monkeys (*Alonatta palliata*) in northeastern Costa Rica. *International Journal of Primatology*, **17**, 1–30.

Strier, K. B. (2003). Demography and temporal scale of sexual selection. In *Sexual Selection and Reproductive Competition in Primates: New Perspectives and Directions*, Vol. 3, ed. C. B. Jones. Norman, OK: The American Society of Primatologists, pp. 45–63.

Struhsaker, T. T. (1975). *The Red Colobus Monkey*. Chicago: University of Chicago Press.

Struhsaker, T. T. & Leland, L. (1979). Socioecology of five sympatric monkey species in the Kibale Forest, Uganda. In *Advances in the Study of Behavior*, Vol. 9, ed. J.Roseblatt, R. A.Hinde, C.Beer, & M. C.Busnel. New York: Academic, pp. 158–228.

— (1987). Colobines: infanticide by adult males. In *Primate Societies*, ed. B. B. Smuts, D. L. Cheney, R. M. Seyfarth, *et al.* Chicago: University of Chicago Press, pp. 83–97.

Suzuki, A. (1979). The variation and adaptation of social groups of chimpanzees and black and white colobus monkeys. In *Primate Ecology and Human Origins*, ed. I. S. Bernstein & E. O. Smith. New York: Garland STPM Press, pp. 153–73.

Thorington, R. W., Ruiz, J. C., & Eisenberg, J. F. (1984). A study of black howler monkey (*Alouatta caraya*) populations in northern Argentina. *American Journal of Primatology*, **6**, 357–66.

Tilson, R. (1977). Social organization of Simakobu monkeys (*Nasalis concolor*) in Siberut Island, Indonesia. *Journal of Mammalogy*, **58**, 202–12.

Turner, T. R., Anapol, F., & Jolly, C. J. (1997). Growth, development, and sexual dimorphism in vervet monkeys (*Cercopithecus aethiops*) at four sites in Kenya. *American Journal of Physical Anthropology*, **103**, 19–35.

Tutin, C. E. G., Ham, R. M., White, L. J. T., & Harrison, M. J. S. (1997). The primate community of the Lope Reserve, Gabon: diets, responses to fruit scarcity, and effects on biomass. *American Journal of Primatology*, **42**, 1–24.

Van Schaik, C. P. (2000). Social counterstrategies against male infanticide in primates and other mammals. In *Primate Males*, ed. P. M. Kappeler. Cambridge: Cambridge University Press, pp. 34–52.

Vogel, C. (1971). Behavioral differences of *Presbytis entellus* in two different habitats. In *Proceedings of the Third International Congress of Primatology, Zurich 1970*, ed. H. Kummer. Basel: S. Karger, pp. 41–7.

Wallace, R. B., Painter, R. L. E., & Taber, A. B. (1998). Primate diversity, habitat preferences, and population density estimates in Noel Kempff Mercado National Park, Santa Cruz Department, Bolivia. *American Journal of Primatology*, **46**, 197–211.

Watanabe, K. (1981). Variation in group composition and population density of the two sympatric Mentawaian leaf monkeys. *Primates*, **22**, 145–60.

Whitten, P. L. (1983). Diet and dominance among female vervet monkeys (*Cercopithecus aethiops*). *American Journal of Primatology*, **5**, 139–59.

Wolf, K. (1978). Preliminary report on the completion of the field phase of a study of the social behavior of the silvered leaf monkeys (*Presbytis cristata*) at Kuala Selangor, Peninsular Malaysia. Kuala Lumpur: Department of Wildlife and National Parks.

Yamakoshi, G. (1998). Dietary responses to fruit scarcity of wild chimpanzees at Bossou, Guinea: possible importance of tool use. *American Journal of Physical Anthropology*, **106**, 283–95.

Yeager, C. P. (1989). Feeding ecology of the proboscis monkey (*Nasalis larvatus*). *International Journal of Primatology*, **10**, 497–530.

— (1991). Proboscis monkey (*Nasalis larvatus*) social organization: intergroup patterns of association. *American Journal of Primatology*, **23**, 73–86.

Ziegler, T., Hodges, K., Winkler, P., & Hiestermann, M. (2000). Hormonal correlates of reproductive seasonality in wild female Hanuman langurs (*Presbytis entellus*). *American Journal of Primatology*, **51**, 119–34.

Part V *Seasonality and community ecology*

15 *Seasonality and primate communities*

CAREL P. VAN SCHAIK
Anthropologisches Institut, University of Zurich, Winterthurerstrasse 190, CH-8057, Zurich, Switzerland

RICHARD MADDEN
Biological Anthropology and Anatomy, Duke University, Box 90383, Durham NC 27708, USA

JÖRG U. GANZHORN
Department of Animal Ecology and Conservation, Hamburg University, Martin-Luther-King Platz 3, 20146 Hamburg, Germany

Introduction

In this chapter, we examine the extent to which environmental seasonality and, hence, phenological seasonality affect aspects of primate communities. Two aspects of community structure are especially interesting in this context: (i) community composition and, hence, species richness, trophic distribution, and size distribution, and (ii) community biomass. Over the past few decades, there have been several major efforts to identify the factors responsible for the variation in these features of primate communities (Bourlière 1985; Terborgh & van Schaik 1987; Fleagle *et al.* 1999; Stevenson 2001). Surprisingly, however, few general conclusions have emerged. There may be several reasons for this lack of success, although at this stage we will have to take an empirical approach.

The general question of how seasonality affects communities can be broken down into two parts. First, we can ask what factors affect the presence of a species at any given site and, hence, more broadly, what factors affect the distributional range of a species. This question about the determinants of species richness and geographical patterns in diversity has spawned a large literature (Huston 1994; Rosenzweig 1995). If biogeographic, i.e. historical, factors are kept constant, then there is an approximately linear relationship between productivity, from the perspective of the taxon under consideration, and local species richness or alpha-diversity, at least for animal consumers such as primates (Ganzhorn *et al.* 1997; Kay *et al.* 1997). The logic is compelling: the continued presence of a species at a site is likely

Seasonality in Primates: Studies of Living and Extinct Human and Non-Human Primates, ed. Diane K. Brockman and Carel P. van Schaik. Published by Cambridge University Press.

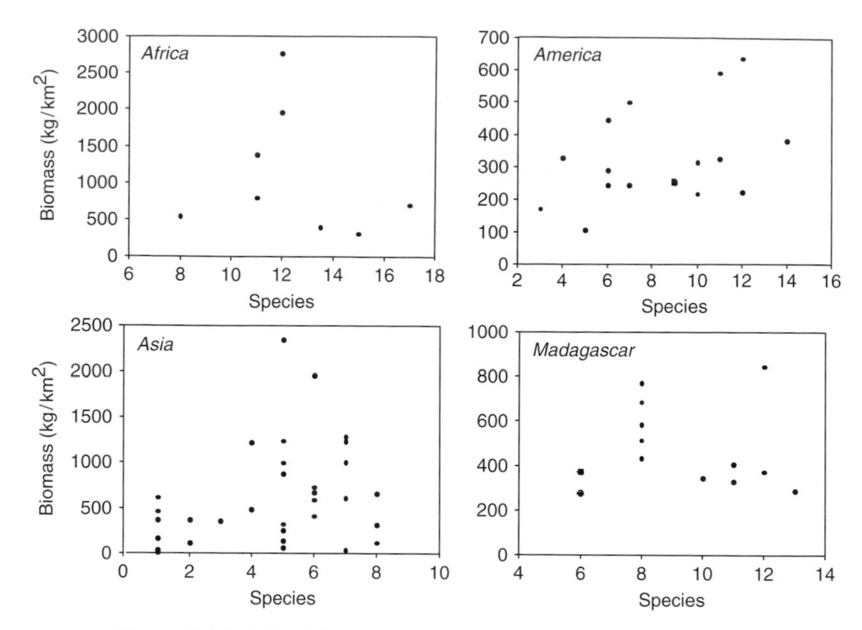

Figure 15.1 Relationship between the number of species and the biomass in primate communities. Data from contributions to Fleagle *et al.* (1999), supplemented with data from evergreen rainforests in Madagascar for altitudes between 700 and 1300 m (from Sterling & Ramaroson [1996], Schmid & Smolker [1998], Feistner & Schmid [1999], Goodman & Rasolonandrasana [1999], Gould *et al.* [1999], Richard *et al.* [2002], Schmid & Rasoloarison [2002], and Sterling & McFadden [2000]).

to depend on a variety of bottom-up forces, such as food abundance, and top-down forces, such as predation and disease. Because it is generally assumed that the top-down forces do not show much spatial variation in intact ecological landscapes, this leaves the bottom-up forces, in particular food, as determinants of species presence. Thus, one possible reason for the lack of success of earlier efforts is that productivity does not follow the same geographic pattern for different trophic groups.

Second, we can ask what factors determine the total biomass of a taxon at a particular site. It seems reasonable to assume that the same factors that determine the presence or absence of a species also play a role in its density. We would therefore expect a positive correlation between species richness and community biomass. However, competition with other species, both primates and non-primates, and other possible ecological interactions should also be considered, and one important empirical result that did emerge from the earlier work on primates is that there is no correlation between the number of species and the biomass of the primate community on any continent or for any primate radiation. Fig. 15.1 illustrates this finding for different continents. Analysis of covariance indicates that biomass is unrelated to the number of species but differs significantly between regions ($F = 4.56$, $P = 0.006$). This suggests strongly that different processes affect species richness and biomass.

The lack of relationship between species richness and biomass may have a variety of explanations. First, it is possible that we must decompose the community into homogeneous groups. If food abundance predicts both presence and densities of all primates, but the abundance of food responds in different ways to rainfall and seasonality for the different trophic groups, then this would explain the heterogeneity of earlier findings. Second, it is possible that the number of species at a site (and perhaps their size and trophic distribution) is governed by a more complex set of processes than is the community's biomass. Either way, it will be useful to examine species richness and biomass correlates separately.

We will now examine the impact of seasonality on the composition (species richness, trophic and size distributions) and biomass of primate communities. The analysis is restricted largely to tropical regions.

Seasonality and community composition

As always, local species richness may have a regional component, being the product of large-scale biogeographical processes (Ricklefs & Schluter 1993). Thus, sites in Amazonia have far higher species numbers than those in Central America, even under similar climate and phenology. If such obvious biogeographical influences are controlled for, then the number of primate species increases with increasing rainfall in America, Africa, and Madagascar, but not in Asia (Reed & Fleagle 1995). This correlation, however, could be due to a variety of more immediately causal factors, including seasonality, and empirical studies to date generally have not looked for an independent effect of seasonality.

A comparison of the relationships between rainfall and habitat productivity, floristic diversity, or seasonality for South American primates by Kay *et al.* (1997) suggested that productivity (as estimated by litter fall) was the best predictor: species richness rises to a peak at around 2500 mm annual rainfall and then declines. Productivity shows the same relationship with rainfall, whereas floristic diversity continues to rise until *c.* 4000 mm annual rainfall, after which it stabilizes, and seasonality declines until *c.* 3000 mm and then remains constant.

In Madagascar, the number of lemur species also seems to be related to rainfall (Fig. 15.2). This correlation is highly significant ($r = 0.63$,

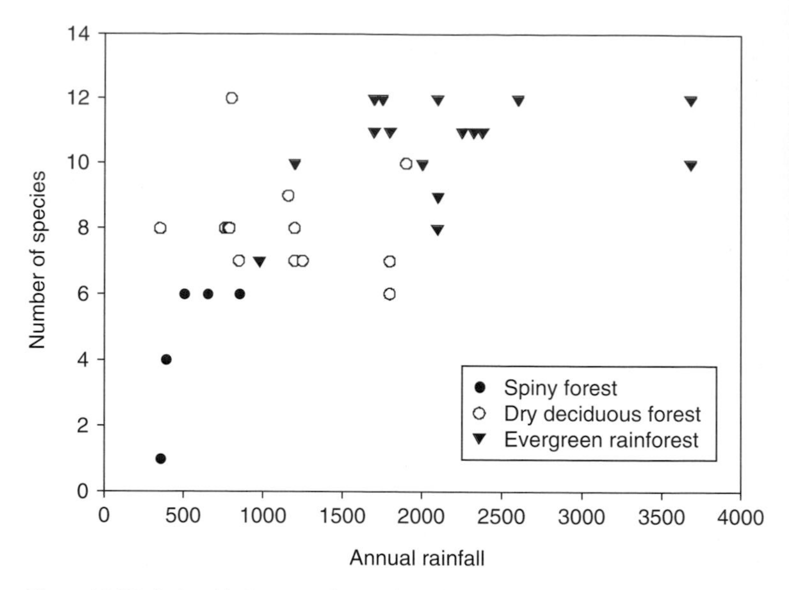

Figure 15.2 Relationship between the number of lemur species and annual rainfall in the three major vegetation types of Madagascar (spiny forest, deciduous forest, evergreen forest). Note that rainfall is insignificant once the effects of the different vegetation types have been accounted for. Data from Ganzhorn *et al.* (1996,1997), and supplemented with new survey data (list of references available from Ganzhorn upon request).

$P < 0.001$, $n = 35$). However, once the three different types of forest (spiny forest, deciduous forest, evergreen forest) are added as another factor in an analysis of covariance, the type of forest is highly significant ($F = 10.13$, $P < 0.001$) but the effect of annual rainfall becomes insignificant ($F = 0.84$, $P = 0.4$) (see Fig. 15.2). The effect of rainfall may operate through its effect on the physiognomy and productivity of the vegetation.

The limited data from Madagascar suggest that species richness reaches a plateau and then remains constant in the highest rainfall regions (Fig. 15.2). In a subset of these data for which more vegetation data are available (Table 15.1), the number of lemur species per site is correlated most strongly with the number of tree species, unlike the pattern for Neotropical primates (Kay *et al.* 1997). The more extensive data from South America suggest that both productivity and primate species richness decline at the highest rainfall, whereas number of tree species continues to remain very high. Thus, although the two detailed data sets show slightly divergent patterns, productivity is a major determinant of primate species richness, although an additional causal role for plant diversity cannot be excluded.

Table 15.1 *Relationships between the number of lemur species, lemur biomass, and the number of tree species (with a diameter at breast height ≥10 cm) and their basal area (in m²) per 0.1 ha across a range of sites. Presented are Pearson's correlation coefficients and sample sizes (in parentheses)*

	Lemur species	Lemur biomass	Tree species	Basal area
Annual rainfall	0.63[a] (35)	– 0.18 (11)	0.78[b] (12)	0.78[b] (13)
Lemur species		– 0.01 (13)	0.83[b] (11)	0.49 (13)
Lemur biomass			– 0.623 (7)	– 0.08 (8)
Tree species				0.57 (9)

[a] $P \leq 0.001$; [b] $P \leq 0.01$.

There may be straightforward explanations for this pattern. In areas with very high rainfall, light becomes a limiting factor due to high cloudiness (e.g. Wright & van Schaik [1994]) and soils tend to get leached. However, primate species numbers decline faster than productivity, and sites with similar productivity have fewer primate species at high rainfall than at low rainfall (Kay *et al.* 1997). This rapid decline could be a consequence of an interaction with the continuing rise in floristic diversity, creating a rarefaction or dilution effect. At sites of high floristic diversity the animals encounter fewer suitable food trees (Ganzhorn *et al.* 1997). Animals may simply be unable to expand their diet, probably because of an increasing proportion of food items with secondary plant chemicals in high rainfall areas to which the animals are not adapted.

This broad explanation so far does not attribute any role to seasonality. If seasonality plays a role in species richness, then it is especially likely in the range of rainfall less than *c.* 2500 mm. A seasonality effect is not inconceivable, because similar annual rainfall totals can be associated with very different phenological patterns (Johnson 2002) (See also Chapter 2). Figure 15.2 supports such an interpretation: species number in the dry deciduous forest of Madagascar is clearly higher than in the spiny forest, even at similar annual rainfall; spiny forest occurs in areas with longer dry seasons.

Especially in highly seasonal environments, we may expect an influence of variance, or extremes, as well as mean values. In fact, two distinct effects of seasonality can be envisaged. First, severe seasonal food shortages may lead to reduced reproduction (e.g. Gould *et al.* [1999]; Richard *et al.* [2002]) or increased mortality (Foster 1982). Predictable serious seasonality has led to the development of the concept of keystone resource: the kind of food that supports a consumer population during the lean season and the density of this keystone resource, rather than the annual mean food

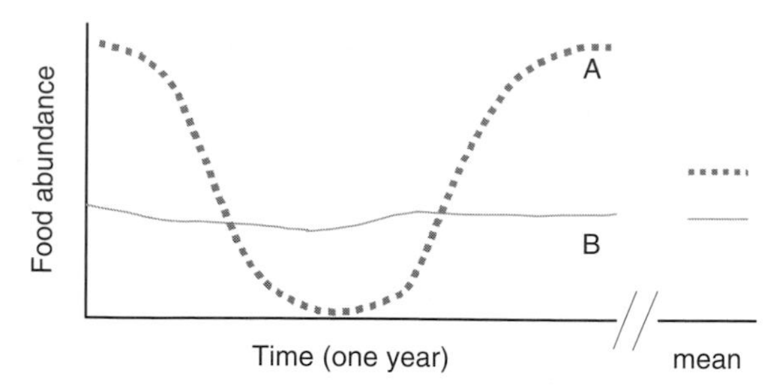

Figure 15.3 Illustration of the question of whether the annual mean or the annual minimum in food abundance is the best predictor for a consumer's local density: mean food abundance over the year is higher in case A, but the minimum is lower than in the much less variable case B, which has the higher minimum food abundance.

abundance, is likely to determine the density of the consumer (Terborgh 1983, 1986). Figure 15.3 illustrates the general point: species A experiences a higher mean food abundance than species B but a lower minimum, set by the abundance of a keystone resource, and therefore ends up having a lower density than species B. Hence, a clear effect of seasonality on community biomass of primates should not be surprising.

Alternatively, if animals can overcome the lean periods through special adaptations and make use of periods of unusually high food abundance for reproduction (see Chapters 1 and 11), then peak fruit abundance rather than mean or minimum abundance may determine the presence and numbers of frugivores. More seasonal habitats may have more predictable peak abundance. In Madagascar, individual trees seem to fruit more predictably in more seasonal habitats. In the dry deciduous forest of western Madagascar, trees fruit with a probability of 72% per year, while lemur food trees fruit with only 40% probability in the evergreen rainforest (Ganzhorn *et al.* 1999: data for Kirindy/Centre de Formation Professionnelle Forestiere (CFPF) and Ranomafana, respectively) and with only 14% probability when non-primate fruit trees are included (Gachet 1969: data for Anamalazaotra). At the same time, however, the evergreen forests have a higher percentage than the deciduous forests of mean monthly fruiting trees. The decrease in the probability of fruiting is paralleled by a decrease in the biomass of frugivorous lemurs. This parallels the relationships found in other continents (Chapman *et al.* 1999; Janson & Chapman 1999).

To sum up, the impact of seasonality on species richness can in principle be negligible (if mean abundance determines primate presence and abundance), negative (if rare keystone resources during the lean fruit season set presence and abundance), or even positive (if peak abundance of fruit determines presence and abundance).

Seasonality and the number of primate species

We examined the possible role of seasonality in a set of sites for South American and Malagasy primates. In order to control for biogeographic effects, we limited the analysis to the South American mainland, excluding all areas west of the Andes as well as the Atlantic region of Brazil. For each site, we determined the number of species from the literature (full data set available upon request from R. M.). Because it is conceivable that different trophic groups or size classes respond differently to seasonality, we examined these relationships for all species combined and for the various trophic and size groups separately. Platyrrhine species therefore were assigned to a diet class (frugivore, folivore, insectivore, seed-eater) based on a detailed assessment of published data (from published naturalistic studies, as we could locate them, supplemented with summary descriptions in secondary sources such as Emmons & Feer [1997]). Lemurs were subdivided into frugivores, folivores and omnivores, according to Ganzhorn (1997).

Species were also classified into three broad categories of body mass: less than 1 kg, between 1 and 5 kg, and more than 5 kg. The weight data for plathyrrhines came primarily from Ford and Davis (1992) and secondarily from Hershkovitz (1987). When body mass was unknown, species were assigned to the same category as all other species of the same genus for which body mass is known. With the exception of *Callicebus brunneus* (835 g), all other species of genus *Callicebus* were assigned to category between 1 and 5 kg. Body mass data for lemurs were taken from Goodman *et al.* (2003).

We added climate data, where necessary, from the nearest climate station. Seasonality was estimated as the number of dry months (a good index for phenological seasonality; see Chapter 2). We included only sites with less than 2500 mm annual rainfall to ensure that we only have the rainfall range in which it shows a linear relationship with productivity.

In the Neotropics (n = 72 sites), there is only a very weak decline in the number of species with increasing seasonality (proxy: number of dry months with <100 mm rainfall). In fact, only the correlation with folivores is significant (but the number of folivores varies between 0 and 2, and this may simply indicate that the genus *Callicebus* can only survive in closed forests). If we examine the effect of productivity (proxy: mean annual rainfall) and seasonality separately in a multiple regression, we find a consistent significant (positive) independent effect of rainfall (Table 15.2a). However, the independent effect of seasonality is now in the opposite direction: with more dry months, we actually find more species at a site. Figure 15.4 illustrates this result for frugivores. The exception is again for folivores, but among them there is no independent effect of rainfall.

Table 15.2a *Number of species in different trophic guilds at 72 Neotropical sites with <2500 mm annual rainfall, as a function of annual rainfall and number of dry months (as proxies for productivity and seasonality)*

	Total: F [2, 69]	Rainfall: t [69]	N dry months: t [69]
N all species	14.39[a]	5.07[a]	2.32[b]
N frugivores	11.44[a]	4.58[a]	2.25[b]
N folivores	6.92[c]	0.56	−2.22[b]
N insectivores	5.13[c]	3.14[c]	1.76
N seed-eaters	4.76[b]	3.03[c]	1.69

[a] $P \leq 0.001$; [b] $P < 0.05$; [c] $P \leq 0.01$.

Table 15.2b *Number of species in different trophic guilds at 33 sites in Madagascar with <2500 mm annual rainfall, as a function of annual rainfall and number of dry months (as proxies for productivity and seasonality) (one variable was excluded from the multiple regression according to the criterium for exclusion at $P > 0.1$)*

	Total: F [1, 32]	Rainfall: t [32]	N dry months: t [32]
N all species	22.26[a]	−0.02	−4.72[a]
N frugivores	5.02[b]	−0.13	−2.24[b]
N folivores	23.89[a]	1.06	−4.85[a]
N omnivores	0.37	−1.57	−0.81
N frugi- and omnivores	4.49[b]	−1.60	−2.12[b]

[a] $P \leq 0.001$; [b] $P < 0.05$.

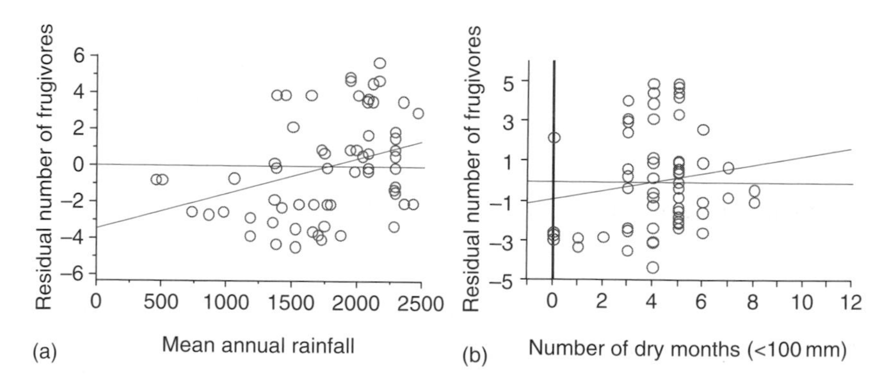

Figure 15.4 Number of frugivorous primate species at a site as a function of (a) annual rainfall (b) and number of dry months per year, while holding the effect of the other factor constant (see also Table 15.2).

Table 15.3a *Number of different sized species at 72 Neotropical sites with <2500 mm annual rainfall, as a function of annual rainfall and number of dry months (as proxies for productivity and seasonality)*

	Total: F [2, 69]	Rainfall: t [69]	N dry months: t [69]
N all species	14.39[a]	5.07[a]	2.32[b]
N small species	11.09[a]	4.65[a]	2.72[c]
N mid-sized species	12.87[a]	4.87[a]	2.42[b]
N large species	10.23[a]	3.16[c]	−0.08

[a]$P \leq 0.001$; [b]$P < 0.05$; [c]$P \leq 0.01$.

Table 15.3b *Number of different sized species at 33 sites in Madagascar with <2500 mm annual rainfall, as a function of annual rainfall and number of dry months (as proxies for productivity and seasonality) (one variable was excluded from the multiple regression according to the criterium for exclusion at P > 0.1)*

	Total: F [1, 32]	Rainfall: t [32]	N dry months: t [32]
N all species	22.26[a]	−0.02	−4.72[a]
N small species	1.63	−1.11	−1.56
N mid-sized species	16.76[a]	4.09[a]	−0.56
N large species	18.14[a]	0.15	−4.26[a]

[a]$P \leq 0.001$.

Thus, if there is an independent effect of seasonality, then it is to increase the number of species. This data set therefore shows no support for the impact of keystone resources and possibly some effect of the effect of peak abundance on the presence of primate species in these Neotropical sites.

Independent of diet, body size may affect a species' response to seasonality. There are suggestions that the smallest species are most sensitive to severe seasonality and therefore should be increasingly rare at sites as seasonality increases (Cowlishaw & Dunbar 2000: 176). To examine this, we divided the species into three size classes. In contrast to this suggestion, small and mid-sized species follow the trend established for species generally, showing independent positive correlations with both rainfall and the number of dry months (Table 15.3a). The largest species show no correlation with seasonality at all.

Overall, then, this analysis finds no independent effect of seasonality on the species richness in Neotropical primate communities. Increased numbers of species with reduced seasonality are explained better by the increase in productivity. This result implies that primates, even in rather seasonal deciduous forests and woodlands, have found ways to cope with seasonal variation in food abundance to the point that seasonality does not affect their ability to survive at these more seasonal sites.

Similar analyses for the lemurs of Madagascar were inconclusive (Tables 15.2b and 15.3b). Multiple regression models using annual rainfall and the number of months with less than 100 mm of rain as independent variables result in significant models for all guilds, except for the small and omnivorous lemurs. As in the Neotropics, annual rainfall and the number of dry months might have independent effects on the number of species in the guilds of frugivores and omnivores. However, the pattern in the results seems to be different. The number of frugivores and omnivores increases with decreasing seasonality (decreasing number of dry months), whereas once seasonality has been accounted for, species numbers seem to decline with increasing rainfall. However, the lack of statistical significance prohibits further speculation. For the time being, we simply have too few reliable data on local scales. Additional data are required for Madagascar to account for possible effects of very fine-grained habitat and meteorological variability on primate community composition in Madagascar.

Community biomass

If one can ignore species interactions, then the determinants of the biomass of the community are simply a linear combination of the determinants of the densities of individual species. However, even if we can make this simplifying assumption, we must reckon with substantial heterogeneity in how species respond to environmental factors. Indeed, the results of empirical studies for primates have been mixed, with some finding a strong relationship (e.g. Stevenson [2001]) but others finding no relationship at all (Chapman *et al.* 1999). These relationships with mean food abundance may cover up an independent effect of seasonality.

Perhaps one major reason for the confusing results is that productivity for different trophic groups shows different relationships with rainfall and/or shows different responses to seasonality. Indeed, Janson and Chapman (1999) showed that the ratio of folivore to frugivore biomass shows a strong negative correlation with rainfall, whereas folivore and frugivore biomass alone show no clear relationship with rainfall. Here, therefore, we analyze the patterns separately for folivores and frugivores, with the latter including also omnivores. Table 15.4 shows the sample used.

Table 15.4 *Estimates of primate biomass (kg/km²) for the total community and for two broad trophic groups, at selected sites with reliable data (largely from Janson & Chapman [1999], with additional data from Feistner & Schmid [1999], Goodman & Rasolonandrasana [1999], Gould* et al. *[1999], Richard* et al. *[2002], Schmid & Rasoloarison [2002], Schmid & Smolker [1998], Sterling & McFadden [2000], Sterling & Ramaroson [1996], Stevenson [2001], and van Schaik, unpublished data)*

Continent	Site	Total biomass	Frugivore biomass	Folivore biomass	Annual rainfall (mm)	N dry months (<100 mm rainfall)
Africa	Budongo	546	262	284	1495	4
	Irangi	710	402	308	2646	0
	Kibale	2759	682	2077	1595	5
	Lope	319	228	91	1506	3
	Taï	802	244	558	2144	5
	Tiwai	1386	786	600	2747	4
America	BCI	445	29	416	2663	4
	C. Cashu	636	456	180	2248	4
	Caparu	497	411	86	3833	0
	Hato Masaguaral	145	48	98	1685	6
	La Pacifica	327	181	146	1539	5
	Los Tuxtlas	171	10	161	4725	5
	Manaus	77	26	51	1897	4
	Nouragues	215	153	62	3175	3
	Raleighv.	251	157	94	2200	3
	Tinigua	471	381	90	2400	4
	Urucu River	381	344	37	3256	0
Asia	Kuala Lompat	933	596	337	1982	1
	Ketambe	837	702	135	3229	0
	Kutai	355	273	82	2177	0
	Polonnaruwa	2480	300	2180	1770	5
	Sepilok	267	184	83	2977	1
	Suaq	975	660	315	3271	0
Madagascar	Ampijoroa	771	444	327	1200	7
	Ranomafana	290	249	41	2200	4
	Kirindy Centre de Formation Professionnelle Forestiere	634	302	332	780	8
	Analamazaotra	375	208	167	1700	7
	Ankarana	346	165	181	1900	6
	Andohahela Parcel 2	311	286	26	500	11
	Andringitra	829	634	195	2097	4
	Anjanaharibe Sud	271	142	129	2300	4
	Marojejy	452	393	59	2300	1
	Beza Mahafaly	485	326	159	522	10
	Ivohibe	483	445	37	2100	8

Folivore biomass

Janson and Chapman (1999) show that the total folivore biomass at sites on continents with forestomach fermenters (Africa, Asia) significantly exceeds that in regions without them (Madagascar, America). They also show a negative correlation with total annual rainfall. Since leaf production increases up to *c.* 2500 mm (Kay *et al.* 1997), this result may seem puzzling at first. However, the relationship is actually an artifact of the correlation between annual rainfall and seasonality. In our data set, if we use presence or absence of forestomach fermenters (Colobinae) as the factor and either mean annual rainfall or mean number of dry months (<100 mm rainfall on average) as covariate, then folivore biomass is far better explained by number of dry months ($F[1,31] = 4.43$, $P = 0.043$) than by annual rainfall ($F[1,31] = 1.26$, $P = 0.27$), indicating that it is increased seasonality rather than reduced rainfall that somehow increases folivore biomass (Fig. 15.5).

Current ecological knowledge is consistent with this finding. Average leaf quality (expressed as protein/fiber ratios) seems to determine the biomass of folivores in all primate radiations (Milton 1979; Waterman *et al.* 1988; Oates *et al.* 1990; Ganzhorn 1992; Peres 1997; Chapman *et al.* 2002). In all of these examples, the biomass increases with increasing protein and decreasing fiber concentrations in representative samples of mature leaves in various forests (i.e. not necessarily of species eaten by the folivores). Hence, for leaves, the relationships with rainfall and seasonality seem straightforward. The metabolic rate of leaves decreases with increasing lifespan of leaves (Reich 2001). Short-lived leaves from more seasonal habitats therefore have higher metabolic rates and higher protein contents than long-lived leaves from evergreen habitats (Coley 1983; Coley *et al.* 1985; Cunningham *et al.* 1999). Consequently, seasonal forests, being more deciduous, have on average higher-quality leaves than evergreen forests, and thus have higher biomass of folivorous primates (Ganzhorn 1992; Aerts & Chapin 2000; Ganzhorn *et al.* 2003). Higher folivore biomass in more seasonal habitats therefore is a result of deciduousness, i.e. food quality.

Frugivore biomass

Previous attempts to predict frugivore biomass as a function of fruit production have met with variable success. Stevenson (2001) found a strong positive correlation between frugivore biomass and different

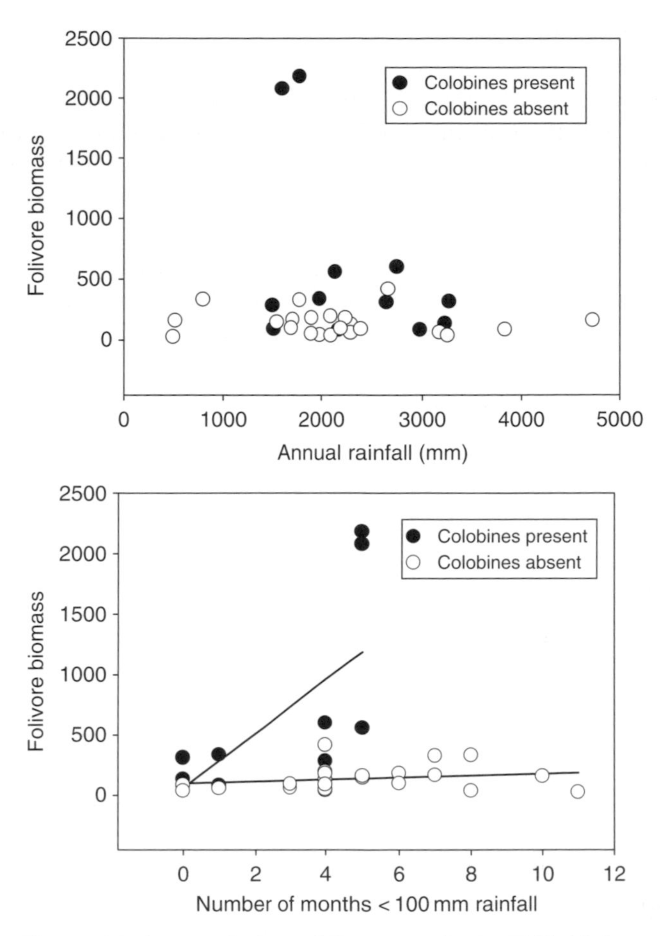

Figure 15.5 Biomass of primate folivores at a site (see Table 15.4) as a function of rainfall and the number of dry months. Sites are separated depending on the presence or absence of foregut fermenters (which dominate where they occur).

measures of mean annual fruit production among platyrrhines, whereas Chapman *et al.* (1999) and Janson and Chapman (1999), using rainfall as a predictor of fruit production, found no significant relationships.

While the idea that keystone resources are the sole determinants of densities of primates in highly seasonal habitats has not been confirmed (e.g. Peres [2000]), diet switching remains one of the most important responses to seasonality (See Chapter 3). The most common switch for frugivores is toward young leaves, but switching to flush is easier where the seasons of peak flush and fruit production are offset by several months (Terborgh & van Schaik 1987). Switching to other fallback resources is, of course, possible, but flush is by far the most abundant fallback resource; thus, species able to switch to flush during times of fruit scarcity should maintain higher densities than those switching to other fallback resources. Van Schaik and Pfennes, in Chapter 2, noted that the most straightforward, albeit somewhat unreliable, estimate of the latter is the interval between the peak month of flushing and that of fruiting. We therefore predict a positive correlation between frugivore biomass and peak interval (negative peak intervals are where fruiting peaks precede flushing peaks, and are found in the most seasonal habitats; see Chapter 2). This leads to a simple prediction: frugivore biomass should show a concave-down relationship with rainfall (as a proxy for mean productivity), but it should also show an independent effect of flush–fruit interval (as a proxy for opportunities for diet switching).

The compilation of primate communities available to us shows no support for the effect of mean productivity: none of rainfall, rainfall squared, or the combination of the two show anything close to a significant correlation with frugivore biomass (the combined variables do best, explaining 7.2 percent of the variation in frugivore biomass). Because studies using direct productivity estimates did find a relationship (Stevenson 2001), it is possible that this lack of relationship is due to the use of the proxy measure of rainfall.

However, in the same data set, the interval between the flushing and fruiting peak shows a strong relationship with frugivore biomass ($r = 0.62$, $n = 19$, $P = 0.005$) (Fig. 15.6). There is no evidence for a continent or radiation effect in this relationship. This correlation is not an artifact of the number of dry months, because that variable shows a far lower correlation with frugivore biomass ($r = -0.295$, NS), although it is still far better than rainfall or rainfall squared.

This result suggests strongly that the abundance of flush at the time of fruit scarcity has a major impact on the density of frugivores and, hence, their combined biomass.

Discussion

Broadly speaking, food abundance predicts both the presence and densities of all primates. Perhaps surprisingly, however, we found very little evidence for a constraining effect of seasonality on the presence of primates. This result suggests that primate consumers may have evolved an effective means of coping with seasonal scarcity of their easily harvested energy-rich resources. In some cases this can be achieved by switching to

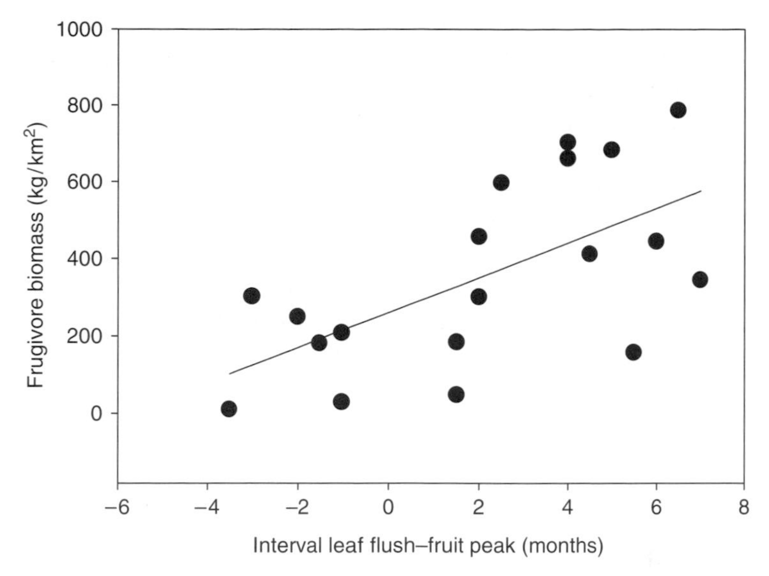

Figure 15.6 Biomass of frugivores in relation to the interval between the peak of leaf flush and peak fruit abundance.

alternative resources, but in others it may also involve metabolic adjustments (Ganzhorn *et al.* 2003). In general, however, our finding casts some doubt on the general importance of keystone resources, at least in the strict sense that they determine the density of the consumers dependent on them. Indeed, recent work suggests that few plant species actually act as keystone resources (Gautier-Hion & Michaloud 1989; Peres 2000).

In fact, it is possible that seasonality may lead to increased primate densities where they can make use of peaks in food abundance to reproduce or replenish body reserves. Thus, individual *Lepilemur ruficaudatus*, a folivorous lemur species from the seasonal dry deciduous forests of western Madagascar, adjust their ranging pattern to optimize food availability and food quality during the rich wet season (which coincides with lactation and weaning) and not to the time of food shortage, which had been assumed to represent the presumed bottleneck period, the dry season (Ganzhorn 2002). In contrast, other lemurs, such as *Lemur catta* and *Propithecus verreauxi*, suffer during times of fruit shortage (Gould *et al.* 1999; Jolly *et al.* 2002; Richard *et al.* 2002; Mertl-Millhollen *et al.* 2003). Thus, individual species respond differently to temporal variation of food quality and abundance, thereby explaining the absence of an effect of seasonality on species richness.

An unambiguous impact of seasonality was found, however, on features of the primate community. We have attributed high folivore biomass to deciduousness, and there is solid evidence in support of this argument. However, evergreen vegetation also becomes increasingly impalatable with increasing rainfall. The causal variable is deciduousness rather than the duration of dry seasons per se.

A more immediate impact of seasonality was found for frugivores. Seasonality in fruit abundance affects frugivore biomass indirectly by bringing the months of peak flushing and peak fruiting closer together and thus constraining the opportunities to switch to the abundant fallback of flush. This result can be reconciled with the findings on species richness: seasonality may not affect the presence of a species, because the species finds ways to cope with the seasonal scarcity of its preferred foods, but its abundance may well depend on how easy it is to cope. Where switching to abundant alternative resources such as young leaves or the insects depending on them is easy, the species numbers may well be higher.

References

Aerts, R. & Chapin, F. S., III (2000). The mineral nutrition of wild plants revisited: a reevaluation of process and patterns. *Advances in Ecological Research*, **30**, 1–61.

Bourlière, F. (1985). Primate communities: their structure and role in tropical ecosystems. *International Journal of Primatology*, **6**, 1–26.

Chapman, C. A., Gautier-Hion, A., Oates, J. F., & Onderdonk, D. A. (1999). African primate communities: determinants of structure and threats to survival. In *Primate Communities*, ed. J. F. Fleagle, C. Janson, & K. E. Reed. Cambridge: Cambridge University Press, pp. 1–37.

Chapman, C. A., Chapman, L. J., Bjorndal, K. A., & Onderdonk, D. A. (2002). Application of protein-to-fiber ratios to predict colobine abundance on different spatial scales. *International Journal of Primatology*, **23**, 283–310.

Coley, P. D. (1983). Herbivory and defensive characteristics of tree species in a lowland tropical forest. *Ecological Monographs*, **53**, 209–33.

Coley, P. D., Bryant, J. P., & Chapin, F. S., III (1985). Resource availability and plant antiherbivore defense. *Science*, **230**, 895–9.

Cowlishaw, G. & Dunbar, R. (2000). *Primate Conservation Biology*. Chicago: University of Chicago Press.

Cunningham, S. A., Summerhayes, B., & Westoby, M. (1999). Evolutionary divergences in leaf structure and chemistry, comparing rainfall and soil nutrient gradients. *Ecological Monographs*, **69**, 569–88.

Emmons, L. H. & Feer, F. (1997). *Guide to the Mammals of Amazonia*. Chicago: University of Chicago Press.

Feistner, A. T. C. & Schmid, J. (1999). Lemurs of the Réserve Naturelle Intégrale d'Andohahela, Madagascar. In *A Floral and Faunal Inventory of the Réserve Naturelle Intégrale d'Andohahela, Madagascar: With Reference to Elevational Variation*, ed. S. M. Goodman. Chicago: Field Museum of Natural History, pp. 269–83.

Fleagle, J. G., Janson, C., & Reed, K. E. (1999). *Primate Communities.* Cambridge: Cambridge University Press.

Ford, S. M. & Davis, I. C. (1992). Systematics and body size: implications for feeding adaptations in New World monkeys. *American Journal of Physical Anthropology*, **88**, 415–68.

Foster, R. B. (1982). Famine on Barro Colorado Island. In *The Ecology of a Tropical Forest*, ed. E. G. Leigh, Jr, A. S. Stanley, & D. M. Windsor. Washington, DC: Smithsonian Institution Press, pp. 201–12.

Gachet, C. (1969). Résultats de deux périodes d'observations phénologiques. Rapport technique du Centre technique forestier tropical: *Sol et Foret*, **47**, Centre de Madagascar.

Ganzhorn, J. U. (1992). Leaf chemistry and the biomass of folivorous primates in tropical forests. *Oecologia (Berlin)*, **91**, 540–47.

— (1997). Test of Fox's assembly rule for functional groups in lemur communities in Madagascar. *Journal of Zoology, London*, 241, 533–42.

— (2002). Distribution of a folivorous lemur in relation to seasonally varying food resources: integrating quantitative and qualitative aspects of food characteristics. *Oecologia (Berlin)*, **131**, 427–35.

Ganzhorn, J. U., Langrand, O., Wright, P. C., *et al.* (1996). The state of lemur conservation in Madagascar. *Primate Conservation*, **17**, 70–86.

Ganzhorn, J. U., Malcomber, S., Andrianantoanina, O., & Goodman, S. M. (1997). Habitat characteristics and lemur species richness in Madagascar. *Biotropica*, **29**, 331–43.

Ganzhorn, J. U., Wright, P. C., & Ratsimbazafy, J. (1999). Primate communities: Madagascar. In *Primate Communities*, ed. J. G. Fleagle, C. Janson, & K. E. Reed. Cambridge: Cambridge University Press, pp. 75–89.

Ganzhorn, J. U., Klaus, S., Ortmann, S., & Schmid, J. (2003). Adaptations to seasonality: some primate and non-primate examples. In *Primate Life Histories and Socioecology*, ed. P. M. Kappeler & M. E. Pereira. Chicago: University of Chicago Press, pp. 132–48.

Gautier-Hion, A. & Michaloud, G.(1989). Are figs always keystone resources for tropical frugivorous vertebrates? A test in Gabon. *Ecology*, **70**, 1826–33.

Goodman, S. M. & Rasolonandrasana, B. P. N. (1999). *Inventaire Biologique de la Réserve Spéciale du Pic d'Ivohibe et du Couloir Forestier qui la relie au Parc National d'Andringitra. Antananarivo: Centre d'Information et de Documentation Scientifique et Technique.*

Goodman, S. M., Ganzhorn, J. U., & Rakotondravony, D. (2003). Introduction to the mammals. In *The Natural History of Madagascar*, ed. S. M. Goodman & J. P. Benstead. Chicago: University of Chicago Press, pp. 1159–86.

Gould, L., Sussman, R. W., & Sauther, M. L. (1999). Natural disasters and primate populations: the effects of a 2-year drought on a naturally occuring population of ring-tailed lemurs (*Lemur catta*) in Southwestern Madagascar. *International Journal of Primatology*, **20**, 69–84.

Hershkovitz, P. (1987). The taxonomy of South American sakis, genus *Pithecia* (Cebidae, Platyrrhini): a preliminary report and critical review with the description of a new species and subspecies. *American Journal of Primatology*, **12**, 387–468.

Huston, M. A. (1994). *Biological Diversity: The Coexistence of Species on Changing Landscapes*. Cambridge: Cambridge University Press.

Janson, C. H. & Chapman, C. (1999). Resources and primate community structure. In *Primate Communities*, ed. J. G. Fleagle, C. H. Janson, & K. Reed. Cambridge: Cambridge University Press, pp. 237–67.

Johnson, S. E. (2002). Ecology and speciation in brown lemurs: white-collared lemurs (*Eulemur albocollaris*) and hybrids (*Eulemur albocollaris* X *Eulemur fulvus rufus*) in southeastern Madagascar. Ph. D. thesis, University of Texas.

Jolly, A., Dobson, A., Rasamimanana, H. M., *et al.* (2002). Demography of *Lemur catta* at Berenty Reserve, Madagascar: effects of troop size, habitat and rainfall. *International Journal of Primatology*, **23**, 327–53.

Kay, R. F., Madden, R. H., van Schaik, C., & Higdon, D. (1997). Primate species richness is determined by plant productivity: implications for conservation. *Proceedings of the National Academy of Science, USA*, **94**, 13023–7.

Mertl-Millhollen, A. S., Moret, E. S., Felantsoa, D., *et al.* (2003). Ring-tailed lemur home ranges correlate with food abundance and nutritional content at a time of environmental stress. *International Journal of Primatology*, **24**, 969–85.

Milton, K. (1979). Factors influencing leaf choice by howler monkeys: a test of some hypotheses of food selection by generalist herbivores. *American Naturalist*, **114**, 362–78.

Oates, J. F., Whitesides, G. H., Davies, A. G., *et al.* (1990). Determinants of variation in tropical forest primate biomass: new evidence from West Africa. *Ecology*, **71**, 328–43.

Peres, C. A. (1997). Effects of habitat quality and hunting pressure on arboreal folivore densities in Neotropical forests: a case study of howler monkeys (*Alouatta* spp.). *Folia Primatologica*, **68**, 199–222.

— (2000). Identifying keystone plant resources in tropical forests: the case of gums from *Parkia* pods. *Journal of Tropical Ecology*, **16**, 287–317.

Reed, K. E. & Fleagle, J. G. (1995). Geographic and climatic control of primate diversity. *Proceedings of the National Academy of Science, USA*, **92**, 7874–6.

Reich, P. B. (2001). Body size, geometry, longevity and metabolism: do plant leaves behave like animal bodies? *Trends in Ecology and Evolution*, **16**, 674–80.

Richard, A. F., Dewar, R. E., Schwartz, M., & Ratsirarson, J. (2002). Life in the slow lane? Demography and life histories of male and female sifaka. *Journal of Zoology, London*, **256**, 421–36.

Ricklefs, R. E. & Schluter, D. (1993). *Species Diversity in Ecological Communities*. Chicago: University of Chicago Press.

Rosenzweig, M. L. (1995). *Species Diversity in Space and Time*. Cambridge: Cambridge University Press.

Schmid, J. & Rasoloarison, R. (2002). Lemurs of the Réserve Naturelle d'Ankarafantsika, Madagascar. In *A Biological Assessment of the Réserve Naturelle Intégrale d'Ankarafantsika, Madagascar*, ed. L. E. Alonso, T. S. Schulenberg, S. Radilofe, & O. Missa. Washington, DC: Conservation International, pp. 73–82.

Schmid, J. & Smolker, R. (1998). Lemurs of the Réserve Spéciale d'Anjanaharibe-Sud, Madagascar. In *A Floral and Faunal Inventory of the Réserve Spéciale d'Anjanaharibe-Sud, Madagascar: With Special Reference to Elevational Variation*, ed. S. M. Goodman. Chicago: Field Museum Natural History, pp. 227–38.

Smith, R. J. & Jungers, W. L. (1997). Body mass in comparative primatology. *Journal of Human Evolution*, **32**, 523–59.

Sterling, E. J. & McFadden, K. (2000). Rapid census of lemur populations in the Parc National de Marojejy, Madagascar. In *A Floral and Faunal Inventory of the Parc National de Marojejy, Madagascar: With Special Reference to Elevational Variation*, ed. S. M. Goodman. Chicago: Field Museum of Natural History, pp. 265–74.

Sterling, E. J. & Ramaroson, M. G. (1996). Rapid assessment of primate fauna of the eastern slopes of the RNI d'Andringitra, Madagascar. In *A Floral and Faunal Inventory of the Eastern Side of the Réserve Naturelle Intégrale d'Andringitra, Madagascar: With Reference to Elevational Variation*, ed. S. M. Goodman. Chicago: Field Museum Natural History, pp. 293–305.

Stevenson, P. R. (2001). The relationship between fruit production and primate abundance in Neotropical communities. *Biological Journal Linnean Society*, **72**, 161–78.

Terborgh, J. (1983). *Five New World Primates*. Princeton, NJ: Princeton University Press.

— (1986). Keystone plant resources in the tropical forest. In *Conservation Biology*, ed. M. E. Soulé. Sunderland: Sinauer, pp. 330–44.

Terborgh, J. & van Schaik, C. P. (1987). Convergence vs. nonconvergence in primate communities. In *Organization of Communities*, ed. J. H. R. Gee & P. S. Giller. Oxford: Blackwell, pp. 205 26.

Waterman, P. G., Ross, J. A. M., Bennett, E. L., & Davies, A. G. (1988). A comparison of the floristics and leaf chemistry of the tree flora in two Malaysian rain forests and the influence of leaf chemistry on populations of colobine monkeys in the Old World. *Biological Journal of the Linnean Society*, **34**, 1–32.

Wright, S. J. & van Schaik, C. P. (1994). Light and the phenology of tropical trees. *American Naturalist*, **143**, 192–9.

16 *Primate diversity and environmental seasonality in historical perspective*

NINA G. JABLONSKI
Department of Anthropology, California Academy of Sciences, 875 Howard Street, San Francisco CA 94103–3009, USA

Introduction

The nature of the relationship between organisms and their environment has figured prominently in studies of evolution and ecology for over a century. Of particular interest has been the investigation of how climatic and environmental changes influence the course of animal evolution, and specifically how important environmental change may be relative to other factors such as migration, predation, and competition, if, in fact, its influence can be isolated sufficiently from these others to be tested. Some of the important questions that have been framed on this topic include: How are animals buffered against their environment? Why do different species appear to respond differently to environmental change? What kinds and degrees of environmental change are sufficient to induce range shifts or the "retuning" of anatomical or physiological tolerances? How directly or immediately do climatic and environmental change influence the origination or extinction of species? These questions have been the subject of many recent and important studies of mammalian evolution and diversification (e.g. Vrba [1985]; McKinney [1998]; Behrensmeyer *et al.* [1997]; Alroy *et al.* [2000]; Hooker [2000]) and have begun to figure prominently in discussions of the more specific context of primate evolution (e.g. Reed [1999]; Jablonski *et al.* [2000]). Different species of mammals have responded differently to changing patterns of environmental seasonality through time, coping by means of combinations of geographical range shifts and evolution of new ecological strategies, or not coping by extinction (Hooker 2000; Jablonski *et al.* 2000). Although these responses may not be immediate, there remains little doubt that long-term climatic changes affecting annual ranges of temperature and thus determining seasonality versus equability have been an important cause of mammalian

Seasonality in Primates: Studies of Living and Extinct Human and Non-Human Primates, ed. Diane K. Brockman and Carel P. van Schaik. Published by Cambridge University Press.

community evolution (Collinson & Hooker 1987; Janis 1989; Hooker 2000).

Studies of the evolutionary dynamics of primate species and communities promise to shed considerable light on general questions of the relationship between environment and organism. This is because the evolutionary and biogeographic history of the order Primates is reasonably well known and because most aspects of the biology of the modern species are well understood. In this chapter, we explore some of the phenomena that have influenced the diversification of primate species through time, in particular the interplay between environmental seasonality and life history evolution. The early evolution of Primates occurred under the most equable and least seasonal environmental conditions known to have existed on Earth (Parrish 1987). This made possible the establishment in the order of a distinctive set of life history parameters typified by a low intrinsic rate of increase of population and characterized by single, relatively long-lived, large-brained offspring and long periods of gestation and infant dependency. As we will see, under conditions of increasing environmental seasonality, these life history characteristics become evolutionary fetters for some primate groups, consigning them to disappearing environments and modes of life. For other groups, however, conditions of increasing seasonality, especially in the latest Tertiary and Quaternary periods, created new ecological opportunities out of which new evolutionary radiations were born.

Changes in environmental seasonality through time created changes in the physical structure of primate habitats and in the availability of particular foods. Both of these aspects of environmental change had profound affects on the evolutionary trajectories of primate lineages, especially those that were adapted to the relatively dense physical structure and the varied and high-quality food resources offered by tropical rainforests.

In order to examine some of the factors that have influenced primate diversity through time and through different periods of environmental seasonality, we will concentrate here on the history of four major groups, the tarsiers, apes, Old World monkeys, and humans. These groups have been chosen because the timing of their radiations span nearly the entire history of the order, and because all groups have living representatives.

Early Tertiary environments and early primates

The first appearance of mammals that can be assigned to the order Primates with some confidence derive from late Paleocene deposits in China (*Decoredon* and *Petrolemur*) and Morocco (*Altiatlasius*) (Hooker

1998). Unfortunately, none of these finds is sufficiently authenticated or well known to be able to say with certainty which one may be the earliest true Primate and which continent was the Primate center of origin. New paleontological and statistical evidence suggests that Asia is the more likely center of origin for primates, perissodactyls, and artiodactyls (Beard 1998a, 2002; Bowen *et al.* 2002), but this interpretation is not accepted universally because of continued disagreements over dating of key paleontological sites and confirmed identifications of putative ancient primate species (Sige *et al.* 1990; Simons & Rasmussen 1994).

The Paleocene/Eocene boundary is characterized by extremely rapid environmental warming, and the first appearances of many mammalian taxa – including several primates, hyaenodontid creodonts, artiodactyls, and perissodactyls – are coupled tightly to this warming (Alroy *et al.* 2000; Wing & Harrington 2001) (Fig. 16.1). This appearance event is thought to reflect rapidly increasing homogeneity of faunas occurring as a result of high-latitude landbridges permitting rapid exchanges of faunas between Holarctic landmasses, which enjoyed extremely salubrious tropical conditions (Alroy *et al.* 2000). The late Paleocene and early Eocene epochs witnessed the expansion of humid, multistratal forests at low and middle paleolatitudes, due to more abundant precipitation, with evergreen rainforests becoming widespread in regions with megathermal temperatures (i.e. with a mean temperature in the coldest month of greater than 18 °C) (Parrish 1987; Upchurch & Wolfe 1987; Morley 2000). The disappearance of many groups of archaic mammals, such as multituberculates, and appearance of all the orders of extant mammals in quick succession in the Early Eocene Epoch carries the important implication that modern trophic specializations in mammals were just developing at this time (Morgan *et al.* 1995). Mammalian faunas of the Early Eocene Epoch were dominated by small mammals (generally weighing less than 1 kg) that were semiterrestrial and scansorial with respect to locomotion and insectivorous or frugivorous in their dietary preferences (Collinson & Hooker 1987; Gunnell 1998; Hooker 1998; Whybrow & Andrews 2000). The mammals inhabiting Early and Middle Eocene forests were broadly analogous to their modern counterparts, but their community structures emphasized terrestrial frugivores and arboreal insectivores to a much greater extent than do the tropical and subtropical forests of the present day (Jablonski 2003).

By Early Eocene times, Primates are represented in the Holarctic by many species of lemur- and tarsier-like primates, generally assigned to the superfamilies Adapoidea and Omomyoidea, respectively. Adapoids and omomyoids were among the most common of Early Eocene mammals,

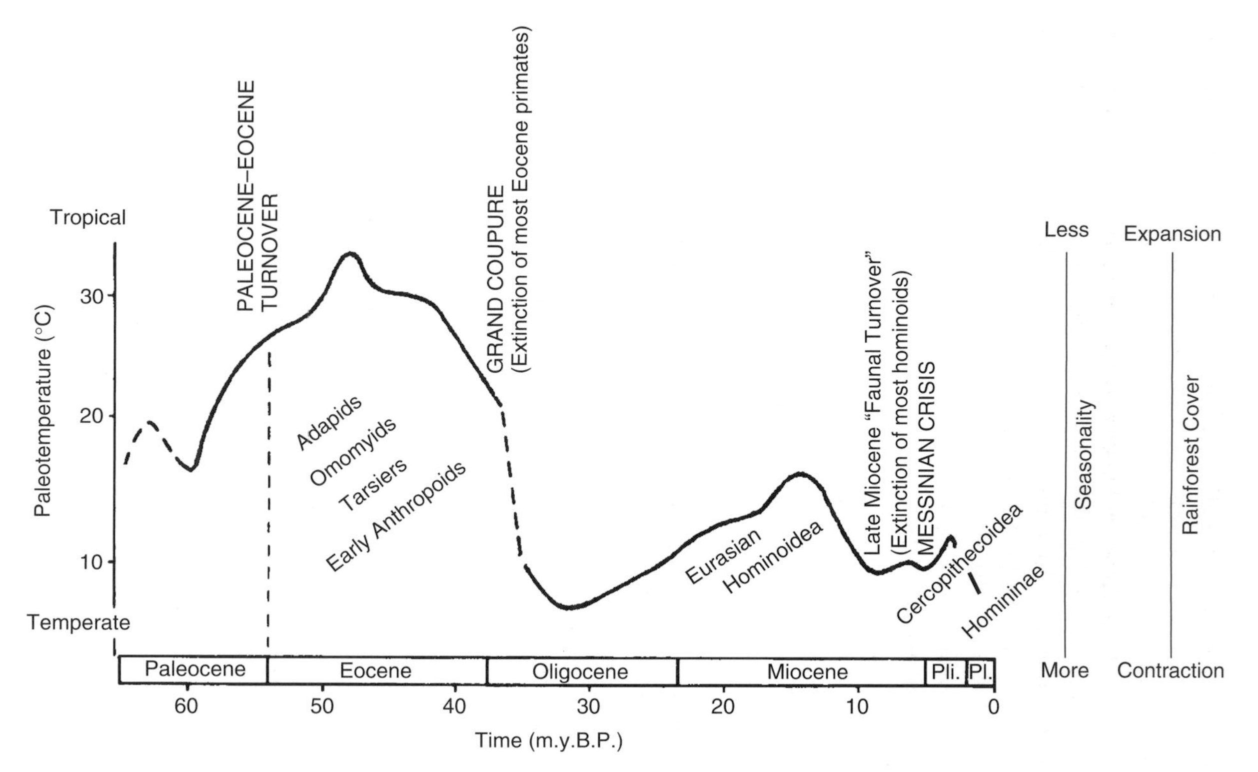

Figure 16.1 The global paleotemperature curve for Tertiary and Quaternary terrestrial environments, illustrating the major trends in environmental seasonality and rainforest cover through time and some of their effects on primate faunas. Modified from Fleagle (1999).

comprising an estimated 34 and 45 species, respectively (based on species counts made from lists in a well-respected primatology textbook [Fleagle 1999]). By the Middle Eocene Epoch, adapoids and omomyoids became less common in North America but were present in substantial numbers in eastern Asia and Africa, with 38 and 43 total species in each superfamily, respectively. Thus, the Early and Middle Eocene epochs were periods of great taxonomic diversity for Primates, but within only the relatively few higher taxa that had emerged by that time.

Among Eocene Primates, leaping was a common locomotor specialization, but it was not necessarily practiced in modes identical to those of modern prosimians and tarsiers. This is especially true of the Omomyoidea, said to have engaged in what is best described as galago-like locomotion (Gunnell & Rose 2002), and the Notharctinae within the Adapoidea, which resembled some living lemurs in their combination of scrambling quadrupedalism and leaping (Gebo 2002). The pervasiveness of leaping in its various forms in Eocene primates almost certainly can be attributed to its efficiency in transporting small-bodied mammals within a tropical forest environment (especially within the understorey) in order to forage for foods such as insects, fruits, and, possibly, tree gums – resources that are not distributed evenly (Crompton 1995; Emmons 1995). The widespread dependence of Eocene prosimians on leaping specializations was certainly one of the phenomena that led to decline and eventual extinction of most species.

The Middle to Late Eocene Epoch witnessed a 15-million-year-long cooling trend, which had disastrous results for many forest-dwelling mammals residing in temperate latitudes. This trend ended with the "Terminal Eocene Event" at 33.5 Ma, involving a coincidence of a sudden climatic shift and a major eustatic sea-level fall (Hooker 2000). The primary consequence of this event for mammals was the dramatic reduction in the extent of tropical forests as a result of the cooling of the climate and increase in the latitudinal temperature gradient (Collinson & Hooker 1987; Janis 1993; Hooker 2000). Here we see, for the first time in Primate history, the devastating effects of the first major episode of increased environmental seasonality outside of tropical latitudes. Under these conditions, the many Holarctic adapoids and omomyoids dependent on tropical-forest angiosperms and their associated insect communities were stuck without food resources and without landbridges or other escape routes that would allow them to track the retreat of the rainforests. Their plight was exacerbated by their leaping specializations, which precluded their safe navigation across the increasingly large discontinuities of non-tropical forests, woodlands, and other more open environments.

In addition to adapoids and omomyoids, true tarsiers are also recognized in the fossil record, beginning in the later part of the Early Eocene Epoch, about 45 million years ago. These animals are of particular interest because, of all known Eocene primates, they have changed least through time. Because of this, they hold the potential to provide greater insight into the nature of the original primate adaptation.

Tarsiers: denizens of Old World rainforests past and present

The presence of lemurs and lorises today in tropical Asia and Africa attests to the fact that the descendants of some Eocene prosimian lineages survived the Terminal Eocene Event, but our knowledge of the evolutionary course of these groups in the intervening period is very poor.

True fossil tarsiers (superfamily Tarsioidea, family Tarsiidae) as opposed to fossils of "tarsier-like" primates have come to light only recently. The fossil record of tarsiers remains poor but clearly records the early origin, persistence, and apparently conservative level of branching evolution in the lineage. The two earliest tarsiid species, *Xanthrorhysis tabrumi* and *Tarsius eocaenus*, are recognized from Middle Eocene deposits of Shanghuang in Jiangsu Province, China (Beard 1998b). These species are known from fragmentary – mostly dental – remains, but they are unequivocally tarsiids. Fossil tarsiers are also known from early Oligocene deposits of Egypt (*Afrotarsius chatrathi*) (Simons & Bown 1985; Rasmussen *et al.* 1998) and early Miocene deposits of Thailand (*Tarsius thailandicus*) (Ginsburg & Mein 1987). In each of these cases, fossil tarsiids are associated with a closed-habitat micromammal fauna characteristic of rainforest ecosystems. The tarsiers themselves occupied the niche of the small-bodied nocturnal insectivore and carnivore of the tropical rainforest understorey, an ecological role that appears to have changed little over the course of over 45 million years (Jablonski 2003). Such forests were of modern aspect, characterized by tall trees with buttressed trunks, lianas, epiphytes, and a distinct understorey, and they would have provided many new or greatly expanded niches for plant and animal adaptation (Upchurch & Wolfe 1987). The forests supported a highly diverse arthropod fauna as well as a suite of mammalian fruit- and seed-eaters. The evolution of large-bodied arthropods was promoted in the forest understorey because of large leaf size and low leaf toxicity (Morley 2000), and it is these animals that appear to have been the main focus of tarsier interest (Jablonski 2003).

Strong similarities in the molar morphology between Eocene and modern tarsiers led to the inference that the diet of ancient and modern tarsiids was essentially the same, consisting mostly of large-bodied arthropods and supplemented by small vertebrates (Jablonski & Crompton 1994; Beard 1998a). The locomotion and foraging behaviors of ancient tarsiids were also comparable to those of modern tarsiers, on the basis of a well-preserved tibiofibula of the early Oligocene African tarsiid, *Afrotarsius chatrathi*, that exhibits a morphology virtually identical to that of modern leaping tarsiers (Rasmussen *et al.* 1998). Large-bodied insects are rare and scattered in their distribution in forest-floor ecosystems, and many are active only at night. Concentration on such widely dispersed prey required the evolution of a mode of locomotion that allowed animals to cross gaps quickly and in the dark.

Like other nocturnal mammals, tarsiers have evolved important adaptations relating to metabolism, energy expenditure, and reproduction. Because small animals facing cool nighttime temperatures could, potentially, exhaust their energy reserves trying to stay warm, prosimians and tarsiers have low basal metabolic rates and low body temperatures (McNab & Wright 1987). Some nocturnal prosimians enter a lengthy period of physiological quiescence (torpor) as an adaptation to seasonal food shortages, but this does not occur in tarsiers (Wright & Martin 1995) (see Chapter 4). However, in tarsiers, a host of reproductive parameters appear to be related to their slower metabolism and reduced body temperatures (McNab & Wright 1987). They have single offspring and long gestation periods (Izard *et al.* 1985), and their rates of fetal and postnatal growth are among the slowest recorded for any mammal (Roberts 1994). This "slow" life history pattern is characteristic of anthropoid primates that have evolved in the stable ecosystems of equable forests (Jablonski *et al.* 2000).

Tarsiids have maintained relictual distributions in Southeast Asia long after their close relatives on other continents became extinct because of the persistence in Southeast Asia of stable humid multistratal rainforest ecosystems from the earliest Tertiary period to the present day. Of all the areas of the Old World once covered by tropical or paratropical rainforest in the Eocene Epoch, it is only small, low-latitude areas of Southeast Asia that have retained such forests through the Neogene Age.

The five recognized species of living tarsiers inhabit parts of Southeast Asia that have been covered with tropical rainforests continuously since the Middle Eocene Epoch (Gursky 1999; Morley 2000). These forests have a dynamic history of their own, having undergone considerable floristic changes as a result of climatic oscillations and migrations resulting from fluctuating land connections. Continuous land connections between Southeast Asia and mid-latitude Asia since the earliest Tertiary period allowed elements of Paleogene Northern Hemisphere rainforests to find refuge in the lower montane forests of Southeast Asia following the mid-Tertiary global climatic deterioration (Morley 2000). This event has no parallel in other regions (Morley 2000) and accounts for the fact the rainforests of Southeast Asia, despite fluctuations in composition and diversity, resemble more closely the rainforests of the Paleogene Age than do rainforests elsewhere. Thus, tarsiers can be said to have, effectively, inhabited the same forests for the past 45 millions of years (myr) and have survived only because their current habitats have retained the relatively non-seasonal character of Early and Middle Eocene forests.

Early catarrhines and hominoids: at home in the rainforests and woodlands

The Terminal Eocene Event brought about a great retraction of tropical rainforests to a narrow belt of continental areas straddling the equator (Janis 1993; Whybrow & Andrews 2000). In many present-day temperate regions, a clear trend toward greater seasonality of rainfall and an increase in the mean annual temperature range can be distinguished from the Late Eocene Epoch onward, on the basis of fossil pollen and leaf assemblages (Leopold *et al.* 1992; Wolfe 1992). These climatic changes led to a great increase, from 34 Ma onward, in areas occupied by "low biomass vegetation" – dry forests, dry woodlands, wooded grasslands, and grasslands (Behrensmeyer *et al.* 1992; Leopold *et al.* 1992; Retallack 1992b). These environments supported a diversity of mammals that could survive in more seasonal environments, characterized by greater annual fluctuations of rainfall, temperature, and food availability. On a global scale, Oligocene mammalian faunas witnessed a great proliferation of Carnivora and Rodentia, as well as Perissodactyla filling most terrestrial herbivore niches (Collinson & Hooker 1987; Collinson 1992; Janis 1993; Whybrow & Andrews 2000).

The primates of the early Neogene Age comprised Eocene survivors and a major radiation of new, larger primates, the Anthropoidea. We surmise on the basis of the distribution of lemurs, lorises, and tarsiers today that the former group was relegated largely to tropical and paratropical forest refuges in Africa and Asia from the Early Oligocene Epoch onward. The latter group, despite probable origins in Asia in the Middle Eocene Epoch (Beard 1998a), became extraordinarily diverse and abundant in the Late Eocene Epoch and Early Oligocene Epoch of northern Africa and came to dominate the arboreal frugivorous and insectivorous niches (Whybrow & Andrews 2000). The earliest fossil catarrhines date to the latest Eocene

Epoch and Early Oligocene Epoch and are derived from deposits representing humid riverine tropical forests and swamp forests close to the southern border of the Tethys Sea (Rasmussen 2002). These catarrhines, such as *Aegyptopithecus zeuxis* and species of *Propliopithecus*, appear to have been generalized quadrupeds that lacked the locomotor specializations of later catarrhines (Fleagle & Simons 1978; Fleagle & Simons 1982; Rasmussen 2002). In their diet and locomotion, the Early Oligocene anthropoids exhibit a range of body sizes and adaptations comparable to those of modern platyrrhines, being mostly frugivorous, insectivorous, and seed-eating, and all arboreal quadrupeds and leapers ranging in approximate size from 5 to 10 kg (Fleagle & Reed 1999). The paucity of Oligocene terrestrial fossils worldwide precludes the making of any conclusive statements about primate diversity during this period, but it appears that – as in earlier epochs – diversity of both prosimians and anthropoids was limited, whether measured by total species numbers or numbers of higher taxa. This is because of the limited distribution of appropriate tropical or non-seasonal forest environments and the apparent absence of anatomical and physiological adaptations that would permit primates to live outside of such habitats.

The end of the Oligocene Epoch and the beginning of the Miocene Epoch, about 23 Ma, marks the first appearance of hominoids (apes *sensu lato*) and Old World monkeys. During the Early Miocene Epoch, climates became warmer and considerably drier, with the establishment of a steeper latitudinal thermal gradient (Kennett 1985; Behrensmeyer *et al.* 1992). This trend was associated with major orogenic events, which resulted in the formation of the Rocky Mountains, the Andes, and the Himalaya (Janis 1993). The Early Miocene Epoch witnessed a return of global climatic equability, although not to Early Eocene levels. This was reflected in an increase in the areal extents of tropical rainforests (Morley 2000), flanking paratropical forests and subtropical woodlands, and the emergence of chapparal or thorn scrub on the western sides of continents (Janis 1993).

The catarrhines of the Early Miocene Epoch are mostly of modern aspect, with most thought to be stem hominoids (e.g. Harrison [2002]). The early apes comprise a mixture of pronograde arboreal quadrupeds such as *Griphopithecus* (unlike most modern apes) and arboreal suspensory feeders such as *Oreopithecus* more reminiscent of modern hominoids (Begun 2002; Pilbeam 2002). Their body sizes were in the approximate range of 5–20 kg (Fleagle & Reed 1999). The early Old World monkeys such as *Victoriapithecus* were terrestrial and smaller, ranging in body size from 3 to 5 kg (Benefit & McCrossin 2002). Of great importance is that the habitats in which most of these catarrhines lived were forested but are best described as subtropical woodlands. These woodlands were probably structurally unlike Equatorial forests (Andrews *et al.* 1997; Harrison 2002; Pilbeam 2002), with larger physical gaps between trees, fewer epiphytes and lianas, and a thinner canopy (Parker 1995). In these habitats, one of the most significant adaptations made by many apes (especially *Pliopithecus* and *Oreopithecus*) involved the evolution of bridging postures, which permitted the animals to cross gaps in the forest canopy and to harvest foods in the terminal branches of trees without recourse to leaping or coming to ground. This adaptation has been interpreted as one promoting the harvesting of widely separated, high-quality food items (Chivers 1991; Andrews *et al.* 1997); it was to prove valuable for many ape taxa but an unexpected liability for others.

Most of the catarrhine fossil record for the Miocene derives from what are thought to have been subtropical woodland habitats that had a moderately seasonal pattern of rainfall and a moderate range of annual mean temperatures (Morley 2000). These parameters were probably associated with a moderate level of seasonality with respect to forest phenology and the productivity of potential catarrhine foods such as fruits, seeds, and young leaves. Most species of Miocene apes appear to have enjoyed mostly frugivorous diets falling more or less within the range of living apes (Ungar & Kay 1995). Using the evidence of dental growth patterns, Kelley (1997, 2002) has shown that, like living apes, Miocene apes exhibited prolonged life histories, which are compatible with stable environments with predictable levels of resource productivity (Jablonski *et al.* 2000). The life history parameters of Miocene Old World monkeys are not known, but they were probably similar to those of modern cercopithecoids, being somewhat faster than those of apes on the fast–slow life history continuum (Jablonski *et al.* 2000). The assumption of continuity or conservatism in life history parameters throughout the history of a lineage is based on consistent evidence for mammals that shows that fundamental patterns of life histories are established early in the history of a lineage and change little, even with changes of body size and habitus (Martin & McLarnon 1990; Read & Harvey 1989; Kelley 2002). This suggests that there was an actual divergence of life histories between apes and Old World monkeys, rather than a slowing in the former or an acceleration in the latter (Kelley 2002).

The most widespread distributions and highest levels of diversity of ape species were attained when Miocene forests (including subtropical woodlands) reached their areal maxima, approximately 12–17 Ma (Bernor 1983; Jablonski *et al.* 2000). For reasons that are still not understood well, Old

World monkey species were uncommon elements of forest faunas at this time (Jablonski & Kelley 1997; Jablonski & Whitfort 1999; Benefit & McCrossin 2002). This evidence indicates that under the relatively equable regimes of the Early Miocene Epoch, the most successful catarrhines were the apes. Despite being selective feeders and obligate tree-dwellers (as a result of their postural and locomotor specializations), and exhibiting slow life history, they not only survived but also diversified. This situation changed dramatically near the close of the Miocene.

The Miocene faunal turnover and the rise of Old World monkeys

The climatic deterioration at the end of the Miocene and its ramifications for mammalian species and communities has been studied in great detail (e.g. Quade *et al.* [1989]; Retallack [1992a]; Cerling *et al.* [1993]; Retallack *et al.* [1995]; Barry *et al.* [2002]). After 13 Ma, a decline in temperature that persisted through the Pliocene is associated with a retreat of tropical floras and the expansion of temperate deciduous trees, grasses, composites, and herbaceous dicotyledons (Behrensmeyer *et al.* 1992). Environment change accelerated from 10 Ma onward, as a result of increasing continental fragmentation due to plate tectonic movements and mountain-building in the Old and New Worlds, which resulted in the elevation of the Himalaya and Tibetan Plateau, the Alps, and the Andes. Changes in the patterns of oceanic and atmospheric circulation resulting from these alterations in geographical conformation were responsible for intensification of the pattern of monsoonal air circulation, which is the primary mechanism of increased environmental seasonality (Quade *et al.* 1989; An *et al*, 2001; Dettman *et al.* 2001).

These Late Miocene climatic events, as well as extensive migrations of fauna between Africa and Eurasia, brought about what has been referred to as the Late Miocene faunal turnover, but what is seen more accurately as a series of high- and low-turnover events occurring in the past five million years of the Miocene (see Barry *et al.* [2002] and references therein). In Africa, the Late Miocene biota was affected greatly by aridification, which took place at the end of the Miocene. Further uplift and rifting in east Africa caused additional restriction of forest and woodland and favored the expansion of savanna–mosaic vegetation (Behrensmeyer *et al.* 1992). As a result, ruminants, elephants, and carnivores diversified extensively (see Behrensmeyer *et al.* [1992] and references therein). In Asia, paleoecological evidence points also to a clear shift to a drier and more seasonal climate, especially after 9.2 Ma, and the appearance of open woodlands or grassy woodlands as early as 7.4 Ma (Barry *et al.* 2002) (Fig. 16.1). Mammalian species richness declined steadily during the Late Miocene Epoch, with the large-mammal fauna being dominated first by equids and then by other grazing ruminants (Barry *et al.* 2002).

For primates, the most important consequence of the gradual but inexorable increase in environmental seasonality beginning in the Late Miocene Epoch was the steady and dramatic decline in the distribution of suitable forest habitats, affecting most especially the belt of subtropical woodland inhabited by the apes of Eurasia (Bernor 1983; Fortelius *et al.* 1996). For most apes, the environmental deterioration of the Late Miocene Epoch meant extinction and the consequent collapse of diversity in a major primate clade. The strong dependence of apes on forest habitats of low seasonality was three-fold – related to their slow life histories, mostly frugivorous "high-quality" diets, and modes of posture and locomotion that precluded easy terrestriality in more open habitats (Jablonski *et al.* 2000). Their slow life histories meant that their reproduction could not be timed within the restrictions of more strongly seasonal cycles of resource availability (Jablonski *et al.* 2000). Apes' dependence on mostly ripe fruits and high-quality keystone foods when fruits were not available (see Jablonski *et al.* [2000] and references therein) meant that they could not subsist in habitats with a strongly seasonal pattern of productivity. This problem would have been exacerbated by large body size and relatively large brain size (Jablonski *et al.* 2000), despite the abilities of apes to store food surpluses as fat (Chivers 1991). Finally, with respect to posture and locomotion, we return to the "curse" of the evolution of bridging postures. Such postures, which are associated with high intermembral and brachial indices (Fleagle 1999), effectively preclude progression on the ground because of the extreme length and weight of the forelimbs. Only in those African ape lineages in which evolved knuckle-walking did the curse fail. The modern orangutan and its predecessors are classic cases of animals shackled to the tropical forests by the very characteristics that created a spectacular evolutionary success for their Miocene ancestors.

The Late Miocene Epoch was not a time of decline for all primate lineages, however, despite the worldwide increase in environmental seasonality and increase in the area covered by habitats receiving highly seasonal rainfall. Notable successes were scored by two groups, the Old World monkeys (superfamily Cercopithecoidea) and a single lineage of apes, the hominins (subfamily Homininae). As is always the case in evolution, the triumph of these groups was accidental, insofar as it was the result of these lineages possessing exaptations that turned into excellent adaptations for the drier and more seasonal environments of the latest Tertiary and Quaternary. The constellations of morphological, physiological, and behavioral characteristics that distinguish species evolved under specific environmental regimes and are being scrutinized constantly by natural selection. These suites of characteristics may or may not work for lineages under environmental conditions different from those under which they originally evolved. In the case of the Old World monkeys, their success can be linked to their possession of life history, dietary, and locomotor characteristics that are distinct from those of apes.

In terms of life history, cercopithecoids exhibit an earlier age for onset of reproduction, shorter gestation times, shorter weaning periods, and short interbirth intervals. Compared with apes, Old World monkeys thus exhibit an overall capacity for a much higher intrinsic rate of increase of population and an ability to time their breeding and birth schedules in highly seasonal habitats (Jablonski *et al.* 2000). With respect to diet, monkeys are more eclectic feeders than apes and can subsist on generally lower-quality foods, when quality is defined in terms of easily realized energy yield (Temerin & Cant 1983; Wrangham *et al.* 1998; Jablonski *et al.* 2000). This applies equally to the two main lineages of Old World monkeys, the cercopithecines (subfamily Cercopithecinae) and colobines (subfamily Colobinae), which appear to have diverged approximately 15 Ma. Cercopithecines have the ability to forage intensively and to "eat everything," from fruits and seeds to leaves, tubers, and insects, without great regard to food quality (Wheatley 1982). Colobines can afford to eat lower-quality vegetation than is generally consumed by cercopithecines and apes because they are equipped with a capacity for fermentation of cellulose and hemicellulose through the action of bacterial symbionts living in the forestomach. Their fermentative digestive tract also is highly effective in neutralizing toxic secondary compounds and antifeedants, which otherwise deter many herbivores (Bauchop 1978; Wrangham *et al.* 1998). This system produces energy for the animals in the form of fatty acids, as well as proteins and other compounds realized from the breakdown of cell walls of the symbiotic bacteria by the enzyme lysozyme (Stewart 1999).

Finally, in connection with posture and locomotion, Old World monkeys have remained relatively generalized quadrupeds that generally can move comfortably in the trees and on the ground through both open and closed environments. This is not to deny the existence of subtly different forms of quadrupedalism or the different degrees of terrestriality and arboreality in the group that reflect adaptations to particular environments (Oxnard 1976). The adaptive flexibility that generalized quadrupedalism provided to Old World monkeys was of increasing importance in the more open environments of the Late Tertiary and Quaternary. This is because generalized quadrupedal postures and modes of progression allowed monkeys to move easily on the ground between forest patches and through discontinuities in the forest in order to forage and maintain contact with other members of their species.

The life history, dietary, and locomotor characteristics that permitted Old World monkeys to undergo tremendous diversification and range expansion in the Late Miocene Epoch continued to work well for the group during the Pliocene and Pleistocene epochs, periods associated with rapidly alternating periods of expansion and contraction of Afromontane, lowland, and subtropical forests (Eeley & Lawes 1999; Jablonski *et al.* 2000). Only under the extreme seasonal conditions and dramatically fluctuating climates of the latest Pleistocene Epoch were monkey species driven to extinction or into more salubrious refugia (Jablonski *et al.* 2000). After this, however, these characteristics improved their evolutionary rebound and permitted monkeys to reclaim habitats significantly faster than apes during the climatic amelioration of the Holocene Epoch (Jablonski *et al.* 2000). It is significant that today's catarrhine fauna comprises 87 Old World monkey species but only 13 apes.

Environmental seasonality and the rise of hominins

Confined to approximately the past seven million years of time, human evolution has occurred entirely within a period of heightened (but fluctuating) environmental seasonality. In this period, humans have established not an evolutionary toehold but hegemony over nearly all terrestrial environments. Given the ape ancestry of humans and its inherent shortcomings (in terms of life history, for instance, as discussed above), it is salutary to consider briefly how this dramatic turn of events may have occurred. If we consider the three major aspects of adaptation recited above in connection with apes and Old World monkeys – life histories, diet, and locomotion – then the reasons for the evolutionary success of humans become immediately clear. As is common practice in modern paleoanthropology, the human or hominin lineage is defined here as that comprising all species of habitually bipedal ape.

Humans are strongly ape-like in their life history parameters (Jablonski *et al.* 2000; Kelley 2002), but they have overcome the constraints of these parameters through social means. Reductions in the length of the period of lactation and the length of the interbirth interval were probably the most critical changes affected and can be attributed to the increasing importance of collective childcare through time (Lancaster 1978; Blurton Jones *et al.*

1999; Hrdy 1999). With respect to diet, modern humans have been described as the ultimate omnivores, and yet they appear to concentrate on hard-to-acquire but highly rewarding foods (Blurton Jones *et al.* 1999). An emphasis on such foods has also been associated with enhanced abilities through time to realize the full nutritional potential of those resources through cooking and food storage (Wrangham *et al.* 1999). Finally, in connection with locomotion, hominins were the only apes to have adopted habitual bipedalism, the best alternative to generalized quadrupedalism for crossing open ground. If, as is now suspected, bipedalism is shown to have evolved in pre-hominins while they were still living in relatively closed habitats (White *et al.* 1994), then this would be another excellent demonstration of an exaptation that led to spectacularly successful adaptation. It is significant that hominins underwent only a modest adaptive radiation in terms of number of species in existence at any time, but tremendous range expansion (Fig. 16.2). This was certainly because the hominin way of life embraced many cultural attributes (e.g. tool use, hunting) that rendered humans more adaptable and effective so that a single species could occupy the ecological space of many (DeVore & Washburn 1963).

Conclusions

For roughly the first third of their known evolutionary history, from about 50–33 Ma, primates lived in equable tropical and paratropical forests. In these warm, humid, and non-seasonal environments, primates evolved their defining characteristics, including their pattern of slow life histories, and diversified into many, mostly small-bodied (<5 kg) taxa. From that time onward, primates have borne the "imprint" of this life-history legacy. Since the Eocene Epoch, increasing fragmentation of continental masses, changes in patterns of atmospheric and oceanic circulation, and major orogenic events have created profound changes in patterns of global climate. Long periods of global climatic equability, such as those witnessed during the Early and Middle Eocene epochs, have been supplanted by less stable regimes, characterized in general by heightened seasonality of temperature and rainfall. The Terminal Eocene Event extirpated most of the world's primate fauna, except for those forms, such as tarsiers, lemurs, and lorises, that survived in the relict rainforests and other lowland forests of Asia and Africa. The next major period of primate diversification occurred in the Early Miocene Epoch, beginning about 20 Ma, when climates again became more equable and tropical forests and subtropical woodlands became more widespread. This radiation involved, primarily, early

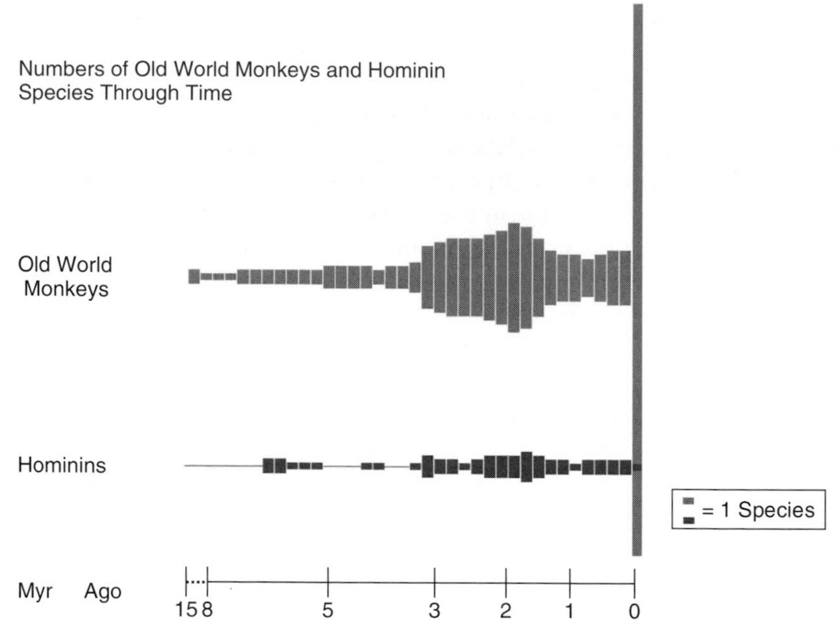

Figure 16.2 Comparison of numbers of Old World monkey and hominin species in the past 15 million years. Both lineages have adapted successfully to seasonal environments, but the number of species that have evolved in the two lineages is remarkably different. During most of the Pliocene and Pleistocene epochs, the Old World monkeys comprised 10–15 species, many of which were sympatric. Modern cercopithecoids number around 87 species. In contrast, the Homininae have been far less speciose, comprising only four to five species at any given time in the Pliocene Epoch, dropping to two to three in the Pleistocene Epoch, and finally dropping to a single extant species. Compared with cercopithecoids, hominins exhibit not only fewer species at any time, but far fewer in any place, with sympatry of multiple hominin species being uncommon. Many of these differences can be attributed to the power of cultural adaptations to the environment in hominins, which have dramatically increased their ecological niche breadth, especially since the Late Pleistocene Epoch. (Species counts based on numbers of species enumerated in Fleagle [1999].)

hominoids (apes), which were primarily frugivorous, larger in body size (5–20 kg), and mostly agile tree climbers. This hugely successful radiation came slowly to an end, however, with the slow deterioration of climates at the end of the Miocene Epoch, especially beginning about 10 Ma. Under the drier conditions of the Late Miocene Epoch, forests shrank in extent and gave way to new, more heterogeneous environments such as wooded grasslands and grassy woodlands, as well as open savannas and grasslands. Most apes became extinct at this time, seemingly because of the constraints of their slow life histories, their dependence on year-round supplies of high-quality foods such as fruits, and modes of locomotion that precluded their safe and efficient progression on the ground in more open environments.

The last major groups of primates to undergo major periods of diversification and range expansion were the Old World monkeys and humans, who did so beginning under the strongly seasonal conditions of the terminal Miocene Epoch, 6–7 Ma. The Old World Monkeys appear to have succeeded where apes failed, by establishing a pattern of somewhat faster life histories, an ability to survive on eclectic, often low-quality diets, and the retention of an unspecialized quadrupedal locomotor habitus. Humans overcame the constraints of the ape ancestry and expanded into a wide variety of non-seasonal and seasonal environments. This was made possible through the evolution of a truly efficient mode of terrestrial locomotion, bipedalism, the evolution of an omnivorous diet characterized by high-quality food items often nutritionally enhanced by cooking, and the evolution of cultural mechanisms, such as collective childcare, that helped to overcome the constraints of long periods of infant dependency and the long interbirth interval.

References

Alroy, J., Koch, P. L., & Zachos, J. C. (2000). Global climate change and North American mammalian evolution. In *Deep Time: Paleobiology's Perspective*, ed. D. H. Erwin & S. L. Wing. Kansas: The Paleontological Society, pp. 259–88.

An, Z.-S., Kutzbach, J. E., Prell, W. L., & Porter, S. C. (2001). Evolution of Asian monsoons and phased uplift of the Himalaya-Tibetan plateau since Late Miocene times. *Nature*, **411**, 62–6.

Andrews, P., Begun, D. R., & Zylstra, M. (1997). Interrelationships between functional morphology and paleoenvironments in Miocene hominoids. In *Function, Phylogeny, and Fossils: Miocene Hominoid Evolution and Adaptations*, ed. D. Begun. New York: Plenum Press, pp. 29–58.

Barry, J. C., Morgan, M. E., Flynn, L. J., *et al.* (2002). *Faunal and Environmental Change in the Late Miocene Siwaliks of Northern Pakistan*. Lawrence, KS: The Paleontological Society.

Bauchop, T. (1978). Digestion of leaves in vertebrate arboreal folivores. In *The Ecology of Arboreal Folivores*, ed. G. G. Montgomery. Washington, DC: Smithsonian Institution Press, pp. 193–204.

Beard, K. C. (1998a). East of Eden: Asia as an important center of taxonomic origination in mammalian evolution. *Bulletin of the Carnegie Museum of Natural History*, **34**, 5–39.

— (1998b). A new genus of Tarsiidae (Mammalia: Primates) from the middle Eocene of Shanxi Province, China, with notes on the historical biogeography of tarsiers. In *Dawn of the Age of Mammals in Asia*, ed. K. C. Beard & M. R. Dawson. Pittsburgh: Carnegie Museum of Natural History, pp. 260–77.

Beard, C. (2002). East of Eden at the Paleocene/Eocene boundary. *Science*, **295**, 2028–9.

Begun, D. (2002). The Pliopithecoidea. In *The Primate Fossil Record*, ed. W. C. Hartwig. Cambridge: Cambridge University Press, pp. 221–40.

Behrensmeyer, A. K., Damuth, J. D., DiMichele, W. A., *et al.* (1992). *Terrestrial Ecosystems Through Time*. Chicago: University of Chicago Press.

Behrensmeyer, A. K., Todd, N. E., Potts, R., & McBrinn, G. E. (1997). Late Pliocene faunal turnover in the Turkana Basin of Kenya and Ethiopia. *Science*, **278**, 1589–94.

Benefit, B. R. & McCrossin, M. L. (2002). The Victoriapithecidae, Cercopithecoidea. In *The Primate Fossil Record*, ed. W. C. Hartwig. Cambridge: Cambridge University Press, pp. 241–53.

Bernor, R. L. (1983). Geochronology and zoogeographic relationships of Miocene Hominoidea. In *New Interpretations of Ape and Human Ancestry*, ed. R. L. Ciochon & R. S. Corruccini. New York: Plenum Press, pp. 21–64.

Blurton Jones, N., Hawkes, K., & O'Connell, J. F. (1999). Some current ideas about the evolution of human life history. In *Comparative Primate Socioecology*, ed. P. C. Lee. Cambridge: Cambridge University Press, pp. 140–66.

Bowen, G. J., Clyde, W. C., Kock, P. L., *et al.* (2002). Mammalian dispersal at the Paleocene/Eocene boundary. *Science*, **295**, 2062–5.

Cerling, T. E., Wang, Y., & Quade, J. (1993). Expansion of C4 ecosystems as an indicator of global ecological change in the late Miocene. *Nature*, **361**, 344–5.

Chivers, D. J. (1991). Species differences in tolerance to environmental change. In *Primate Responses to Environmental Change*, ed. H. O. Box. London: Chapman & Hall, pp. 5–37.

Collinson, M. E. (1992). Vegetational and floristic changes around the Eocene/Oligocene boundary in western and central Europe. In *Eocene-Oligocene Climatic and Biotic Evolution*, ed. D. R. Prothero & W. A. Berggren. Princeton: Princeton University Press, pp. 437–50.

Collinson, M. E. & Hooker, J. J. (1987). Vegetational and mammalian faunal changes in the Early Tertiary of southern England. In *The Origins of Angiosperms and Their Biological Consequences*, ed. E. M. Friis, W. G. Chaloner, & P. R. Crane. Cambridge: Cambridge University Press, pp. 259–304.

Crompton, R. H. (1995). "Visual predation," habitat structure, and the ancestral primate niche. In *Creatures of the Dark: The Nocturnal Prosimians*, ed. L. Alterman, G. A. Doyle, & M. K. Izard. New York: Plenum Press, pp. 11–30.

Dettman, D. L., Kohn, M. J., Quade, J., *et al.* (2001). Seasonal stable isotope evidence for a strong Asian monsoon throughout the past 10.7 m.y. *Geology*, **29**, 31–4.

DeVore, I. & Washburn, S. L. (1963). Baboon ecology and human evolution. In *African Ecology and Human Evolution*, ed. F. C. Howell & F. Bourliere. Chicago: Aldine Publishing Company, pp. 335–67.

Eeley, H. A.C. & Lawes, M. J. (1999). Large-scale patterns of species richness and species range size in anthropoid primates. In *Primate Communities*, ed. J. G. Fleagle, C. Janson, & K. E. Reed. Cambridge: Cambridge University Press, pp. 191–219.

Emmons, L. H. (1995). Mammals of rain forest canopies. In *Forest Canopies*, ed. M. D. Lowman & N. M. Nadkarni. San Diego: Academic Press, pp. 199–223.

Fleagle, J. G. (1999). *Primate Adaptation and Evolution*. San Diego: Academic Press.

Fleagle, J. G. & Reed, K. E. (1999). Phylogenetic and temporal perspectives on primate ecology. In *Primate Communities*, ed. J. G. Fleagle, C. Janson, & K. E. Reed. Cambridge: Cambridge University Press, pp. 92–115.

Fleagle, J. G. & Simons, E. L. (1978). Humeral morphology of the earliest apes. *Nature*, **276**, 705–7.

— (1982). Skeletal remains of *Propliopithecus chirobates* from the Egyptian Oligocene. *Folia Primatologica*, **39**, 161–77.

Fortelius, M., Werdelin, L., Andrews, P., *et al.* (1996). Provinciality, diversity, turnover, and paleoecology in land mammal faunas of the Later Miocene of Western Eurasia. In *The Evolution of Western Eurasian Neogene Mammal Faunas*, ed. R. L. Bernor, V. Fahlbusch, & H.-W. Mittmann. New York: Columbia University Press, pp. 414–48.

Gebo, D. L. (2002). Adapiformes: phylogeny and adaptation. In *The Primate Fossil Record*, ed. W. C. Hartwig. Cambridge: Cambridge University Press, pp. 21–43.

Ginsburg, L. & Mein, P. (1987). *Tarsius thailandica* nov. sp., premier Tarsiidea (Primates, Mammalia) fossile d'Asie. *C. R. Academy of Sciences Paris*, **304**, 1213–15.

Gunnell, G. F. (1998). Mammalian faunal composition and the Paleocene/Eocene Epoch/Series boundary: evidence from northern Bighorn Basin, Wyoming. In *Late Paleocene–Early Eocene Climatic and Biotic Events in the Marine and Terrestrial Records*, ed. M.-P. Aubry, S. Lucas, & W. A. Berggren. New York: Columbia University Press, pp. 409–27.

Gunnell, G. F. & Rose, K. D. (2002). Tarsiiformes: evolutionary history and adaptation. In *The Primate Fossil Record*, ed. W. C. Hartwig. Cambridge: Cambridge University Press, pp. 45–82.

Gursky, S. (1999). The Tarsiidae: taxonomy, behavior, and conservation status. In *The Nonhuman Primates*, ed. P. Dolhinow & A. Fuentes. Mountain View: Mayfield Publishing Company, pp. 140–45.

Harrison, T. (2002). Late Oligocene to middle Miocene catarrhines from Afro-Arabia. In *The Primate Fossil Record*, ed. W. C. Hartwig. Cambridge: Cambridge University Press, pp. 311–38.

Hooker, J. J. (1998). Mammalian faunal change across the Paleocene-Eocene transition in Europe. In *Late Paleocene-Early Eocene Climatic and Biotic Events in the Marine and Terrestrial Records*, ed. M.-P. Aubry, S. Lucas, & W. A. Berggren. New York: Columbia University Press, pp. 428–50.

— (2000). Paleogene mammals: crises and ecological change. In *Biotic Response to Global Change: The Last 145 Million Years*, ed. S. J. Culver & P. F. Rawson. Cambridge: Cambridge University Press, pp. 333–49.

Hrdy, S. B. (1999). *Mother Nature: A History of Mothers, Infants and Natural Selection*. New York: Pantheon.

Izard, M. K., Wright, P. C., & Simons, E. L. (1985). Gestation length in *Tarsius bancanus*. *American Journal of Primatology*, **9**, 327–31.

Jablonski, N. G. (2003). The evolution of the tarsiid niche. In *Tarsiers: Past, Present, and Future*, ed. P. Wright, E. Simons, & S. Gursky. New Brunswick, NJ: Rutgers University Press, pp. 35–49.

Jablonski, N. G. & Crompton, R. H. (1994). Feeding behavior, mastication, and tooth wear in the western tarsier, *Tarsius bancanus*. *International Journal of Primatology*, **15**, 29–59.

Jablonski, N. G. & Kelley, J. (1997). Did a major immunological event shape the evolutionary histories of apes and Old World monkeys? *Journal of Human Evolution*, **33**, 513–20.

Jablonski, N. G. & Whitfort, M. G. (1999). Environmental change during the Quaternary in East Asia and its consequences for mammals. *Records of the Western Australian Museum*, Suppl. **57**, 307–15.

Jablonski, N. G., Whitfort, M. G., Roberts-Smith, N., & Xu, Q.-Q. (2000). The influence of life history and diet on the distribution of catarrhine primates during the Pleistocene in eastern Asia. *Journal of Human Evolution*, **39**, 131–57.

Janis, C. M. (1989). A climatic explanation for patterns of evolutionary diversity in ungulate mammals. *Paleoanthropology*, **32**, 463–81.

— (1993). Tertiary mammal evolution in the context of changing climates, vegetation, and tectonic events. *Annual Review of Ecology and Systematics*, **24**, 467–500.

Kelley, J. (1997). Paleobiological and phylogenetic significance of life history in Miocene hominoids. In *Function, Phylogeny, and Fossils: Miocene Hominoid Evolution and Adaptation*, ed. D. R. Begun, C. V. Ward, & M. D. Rose. New York: Plenum Press, pp. 173–208.

— (2002). Life-history evolution in Miocene and extant apes. In *Human Evolution Through Developmental Change*, ed. N. Minugh-Purvis & K. J. McNamara. Baltimore: Johns Hopkins University Press, pp. 223–48.

Kennett, J. P. (1985). Neogene paleoceanography and plankton evolution. *South African Journal of Science*, **81**, 251–3.

Lancaster, J. (1978). Carrying and sharing in human evolution. *Human Nature*, **1**, 82–9.

Leopold, E. B., Liu, G., & Clay-Poole, S. (1992). Low-biomass vegetation in the Oligocene? In *Eocene-Oligocene Climatic and Biotic Evolution*, ed. D. R. Prothero & W. A. Berggren. Princeton: Princeton University Press, pp. 399–420.

Martin, R. D. & McLarnon, A. M. (1990). Reproductive patterns in primates and other mammals: the dichotomy between altricial and precocial offspring. In *Primate Life History and Evolution*, ed. C. J. De Rousseau. New York: Wiley-Liss, pp. 47–79.

McKinney, M. L. (1998). Biodiversity dynamics: niche preemption and saturation in diversity euilibria. In *Biodiversity Dynamics: Turnover of Populations, Taxa, and Communities*, ed. M. L. McKinney & J. A. Drake. New York: Columbia University Press, pp. 3–16.

McNab, B. K. & Wright, P. C. (1987). Temperature regulation and oxygen consumption in the Philippine tarsier *Tarsius syrichta*. *Physiological Zoology*, **60**, 596–600.
Morgan, M. E., Badgley, C., Gunnell, G. F., *et al.* (1995). Comparative paleoecology of Paleogene and Neogene mammalian faunas: body-size structure. *Palaeogeography, Palaeoclimatology, and Palaeoecology*, **115**, 287–317.
Morley, R. J. (2000). *Origin and Evolution of Tropical Rain Forests*. New York: John Wiley & Sons.
Oxnard, C. E. (1976). Primate quadrupedalism: some subtle structural correlates. *Yearbook of Physical Anthropology* **20**: 538–54.
Parker, G. G. (1995). Structure and microclimate of forest canopies. In *Forest Canopies*, ed. M. D. Lowman & N. M. Nadkarni. New York: Academic Press, pp. 73–106.
Parrish, J. T. (1987). Global palaeogeography and palaeoclimate of the Late Cretaceous and Early Tertiary. In *The Origins of Angiosperms and Their Biological Consequences*, ed. E. M. Friis, W. G. Chaloner, & P. R. Crane. Cambridge: Cambridge University Press, pp. 51–73.
Pilbeam, D. (2002). Perspectives on the Miocene Hominoidea. In *The Primate Fossil Record*, ed. W. Hartwig. Cambridge: Cambridge University Press, pp. 303–10.
Quade, J., Cerling T. E., & Bowman J. R. (1989). Development of Asian monsoon revealed by marked ecological shift during the latest Miocene in northern Pakistan. *Nature*, **342**, 163–6.
Rasmussen, D. T. (2002). Early catarrhines of the African Eocene and Oligocene. In *The Primate Fossil Record*, ed. W. C. Hartwig. Cambridge: Cambridge University Press, pp. 203–20.
Rasmussen, D. T., Conroy G. C., & Simons E. L. (1998). Tarsier-like locomotor specializations in the Oligocene primate *Afrotarsius*. *Proceedings of the National Academy of Sciences, USA*, **95**, 14848–50.
Read, A. F. & Harvey, P. H. (1989). Life history differences among the eutherian radiations. *Journal of Zoology, London*, **219**, 329–53.
Reed, K. E. (1999). Population density of primates in communities: differences in community structure. In *Primate Communities*, ed. J. G. Fleagle, C. Janson, & J. L. Rees. Cambridge: Cambridge University Press, pp. 116–40.
Retallack, G. J. (1992a). Middle Miocene fossil plants from Fort Ternan (Kenya) and evolution of African grasslands. *Paleobiology*, **18**, 383–400.
— (1992b). Paleosols and changes in climate and vegetation across the Eogene/Oligocene boundary. In *Eocene-Oligocene Climatic and Biotic Evolution*, ed. D. R. Prothero & W. A. Berggren. Princeton: Princeton University Press, pp. 382–98.
Retallack, G. J., Bestland, E. A., & Dugas, D. P. (1995). Miocene paleosols and habitats of *Proconsul* on Rusinga Island, Kenya. *Journal of Human Evolution*, **29**, 53–91.
Roberts, M. (1994). Growth, development, and parental care in the western tarsier (*Tarsius bancanus*) in captivity: evidence for a "slow" life-history and nonmonogamous mating system. *International Journal of Primatology*, **15**, 1–28.
Sige, B., Jaeger, J. J., Sudre, J., & Vianey-Liaud, M. (1990). *Altiatlasius koulchii* n. gen. et sp. Primate omomyide du Paleocene Superieur du Moroc, et les origines des euprimates. *Palaeontographica (A)*, **214**, 31–56.

Simons, E. L. & Bown, T. M. (1985). *Afrotarsius chatrathi*, first tarsiiform primate (?Tarsiidae) from Africa. *Nature*, **313**, 475–7.

Simons, E. L. & Rasmussen, D. T. (1994). A remarkable cranium of *Pleisopithecus teras* (Primates, Prosimii) from the Eocene of Egypt. *Proceedings of the National Academy of Sciences, USA*, **91**, 9946–50.

Stewart, C.-B. (1999). The colobine Old World monkeys as a model system for the study of adaptive evolution at the molecular level. In *The Nonhuman Primates*, ed. P. Dolhinow & A. Fuentes. Mountain View: Mayfield Publishing Company, pp. 29–38.

Temerin, L. A. & Cant, J. G. H. (1983). The evolutionary divergence of Old World monkeys and apes. *American Naturalist*, **122**, 335–51.

Ungar, P. S. & Kay, R. F. (1995). The dietary adaptations of European Miocene catarrhines. *Proceedings of the National Academy of Sciences, USA*, **92**, 5479–81.

Upchurch, G. R. & Wolfe, J. A. (1987). Mid-Cretaceous to Early Tertiary vegetation and climate: evidence from fossil leaves and woods. In *The Origins of Angiosperms and their Biological Consequences*, ed. E. M. Friis, W. G. Chaloner, & P. R. Crane. Cambridge: Cambridge University Press, pp. 75–105.

Vrba, E. S. (1985). Environment and evolution: alternative causes of the temporal distribution of evolutionary events. *South African Journal of Sciences*, **81**, 229–36.

Wheatley, B. P. (1982). Energetics of foraging in *Macaca fascicularis* and *Pongo pygmaeus* and a selective advantage of large body size in the orang-utan. *Primates*, **23**, 348–63.

White, T. D., Suwa, G., & Asfaw, B. (1994). *Australopithecus ramidus*, a new species of early hominid from Aramis, Ethiopia. *Nature*, **371**, 306–12.

Whybrow, P. J. & Andrews, P. (2000). Response of Old World terrestrial vertebrate biotas to Neogene climate change. In *Biotic Response to Global Change: The Last 145 Million Years*, ed. S. J. Culver & P. F. Rawson. Cambridge: Cambridge University Press, pp. 350–66.

Wing, S. L. & Harrington, G. J. (2001). Floral response to rapid warming in the earliest Eocene and implications for concurrent faunal change. *Paleobiology*, **27**, 539–63.

Wolfe, J. A. (1992). Climatic, floristic, and vegetational changes near the Eocene/Oligocene boundary in North America. In *Eocene-Oligocene Climatic and Biotic Evolution*, ed. D. R. Prothero & W. A. Berggren. Princeton: Princeton University Press, pp. 421–36.

Wrangham, R. W., Conklin-Brittain, N. L., & Hunt, K. D. (1998). Dietary response of chimpanzees and cercopithecines to seasonal variation in fruit abundance. I. Antifeedants. *International Journal of Primatology*, **19**, 949–70.

Wrangham, R. W., Jones, J. H., Laden, G., Pilbeam, D., & Conklin-Brittain, N. (1999). The raw and the stolen. *Current Anthropolology*, **40**, 567–94.

Wright, P. C. & Martin, L. B. (1995). Predation, pollination and torpor in two nocturnal prosimians: *Cheirogaleus major* and *Microceubs rufus* in the rain forest of Madagascar. In *Creatures of the Dark: The Nocturnal Prosimians*, ed. L. Alterman, G. A. Doyle, & M. K. Izard. New York: Plenum Press, pp. 45–60.

Part VI *Seasonality and human evolution*

17 *Tropical and temperate seasonal influences on human evolution*

KAYE E. REED
Department of Anthropology/Institute of Human Origins, Arizona State University, Tempe AZ 85287, USA

JENNIFER L. FISH
Max Planck Institute of Molecular Cell Biology and Genetics, Pfotenhauerstrasse 108, 01307 Dresden, Germany

Introduction

Climatic and subsequent habitat change has often been invoked as a driving force of evolutionary change in hominins,[1] mammals, and other taxa (Vrba 1988a, 1988b, 1992, 1995; Bromage & Schrenk 1999; Potts 1998; Bobe & Eck 2001; Janis *et al.* 2002). Global climatic change in the mid Pliocene Epoch has been suggested as a cause for hominin speciation events (Vrba 1995) and is correlated with changes in dentition and jaw morphology of two hominin lineages (Teaford & Ungar 2000). Climatic change also influences seasonality, such that drying trends, for example, likely instigate short wet seasons, while the reverse is also true. Although Foley (1987) indicated that seasonal differences were likely important in determining hominin foraging effort, and Blumenschine (1987) posited a dry-season scavenging niche for Pleistocene hominids, little attention has been given to how seasonal changes over time might contribute to differences among hominin behavioral adaptations. Seasonal changes refer to changes in the lengths of regular four-season patterns in temperate climates or wet and dry seasonal differences in the tropics over geological time.

Evolutionary changes in fossil hominins are detected through changes in morphology that represent different behavioral adaptations. Fossil hominin diets are inferred from comparisons with extant primates in features such as tooth size (Hylander 1975; Kay 1984; Ungar & Grine 1991), molar

[1] Hominins are members of the Hominidae, which, based on genetic evidence, include humans and all great apes. Bipedal taxa in this group are placed in the tribe Hominini, and the term "hominin" is thus used throughout this chapter.

Seasonality in Primates: Studies of Living and Extinct Human and Non-Human Primates, ed. Diane K. Brockman and Carel P. van Schaik. Published by Cambridge University Press.

shearing crests (Kay 1984), dental microwear (Grine 1981; Teaford 1988; Ungar 1998), biomechanics (Hylander 1988; Daegling & Grine 1991), and isotopic signatures (Sponheimer & Lee-Thorp 1999; van der Merwe *et al.* 2003). These studies suggest types of food consumed, e.g. hard, grit-covered foods, fruit or leaves, or C3/C4 plant food. It is unknown, however, whether the morphology of early hominins reflects year-round diet or allows survival through lean seasons in which regular foods are scarce (Rosenberger & Kinzey 1976). In this chapter, the importance of the number of dry-season months with which hominins may have coped, and how changes to lengthier dry seasons over time may have contributed to hominin dietary adaptations and behavioral differences, is discussed. We also address foraging issues faced by hominins in environments of wet and dry seasons as well as those that lived in four-season temperate or cold environments. Hypotheses about the effect of seasonal changes through time on hominin behavior and evolution are formulated, based on information from changes of masticatory morphology compared with both habitat and seasonality evidence.

Early hominin taxa (up to the origin of *Homo* species) have been recovered from Ethiopia, Kenya, Tanzania, Chad, and South Africa. These sites (Table 17.1) are located within 33 degrees of the Equator, indicating a tropical to subtropical climate, although most localities are in the tropical range. Later *Homo* species occupied subtropical zones in Africa and Southeast Asia, and the earliest known hominin occupation of temperate climates occurred 1.7 Ma in Eurasia, at Dmanisi, Georgia (Gabunia *et al.* 2000a; Vekua *et al.* 2002). First appearance dates (FAD) and last appearance dates (LAD) for these hominin species as they are currently known (Kimbel 1995; Woldegabriel *et al.* 1994; Haile-Selassie 2001) are given in Table 17.1 so that they can be contrasted with climatic and habitat shifts through time (Table 17.2).

Climates, habitats, and seasonality

Climatic patterns

In the past ten million years, the African continent, like most of the globe, has been influenced by glacial cycles that have caused, overall, drier and more open habitats than existed previously (Crowley & North 1991). Before this time, forests and deciduous woodlands existed in eastern Africa, indicating higher rainfall (Foley 1987; Andrews & Humphrey 1999; Jacobs 2002). Water held in glaciers at both poles led to increasing

Table 17.1 *Early Pliocene to recent hominin taxa. Species are listed with their estimated first appearance date (FAD) and last appearance date (LAD). Latitude refers to the zone that we have assigned to the regions inhabited by each species: tropical is within 20° north and south of the Equator; subtropical is between 20 and 30° latitudes; temperate is between 30 and 60° latitudes; and cold is >60° latitudes. Seasonal differences are those that would likely occur based on latitude*

Species	Estimated FAD	Estimated LAD	Latitude	Reconstructed habitats	Seasonal differences	Localities
Ardipithecus species	5.6	4.4	Tropical	Forests/closed woodlands	Wet/dry	East Africa
Australopithecus anamensis	4.2		Tropical	Open woodlands	Wet/dry	East Africa
A. afarensis	3.6 (?4.2)	2.9	Tropical	Closed/open woodlands, bushlands, edaphic grasslands	Wet/dry (~3 months)	East Africa
A. africanus	3	2.4	Subtropical	Woodlands, bushlands, edpahic grasslands, ?secondary grasslands	Wet/dry	South Africa
A. bahrelgazalia	3		Subtropical	Open woodlands	Wet/dry	Northern central Africa
A. garhi	2.5		Tropical		Wet/dry	East Africa
A. aethiopicus	2.7	?1.9	Tropical	Woodlands, edaphic grasslands	Wet/dry	East Africa
A. boisei	2.3	1.4		Open woodlands, edaphic grasslands	Wet/dry	East Africa
A. robustus	1.9	?1.0		Open woodlands, edaphic grasslands	Wet/dry	South Africa
Kenyanthropus platyops	3.5	3.3	Tropical	Woodlands, edaphic grasslands	Wet/dry	East Africa
Homo habilis	2.33	1.6	Tropical	Woodlands	Wet/dry	East and South Africa
H. rudolfensis	2.4	1.9	Tropical	Woodlands	Wet/dry	East Africa
H. ergaster/erectus	1.8	0.4	Tropical/temperate	Grasslands, riparian regions	Wet/dry and four season	Africa and Eurasia
H. heidelbergensis	0.8	0.15	Tropical/temperate		Four season	Africa and Eurasia
H. neandertalensis	0.12	0.032	Temperate/cold		Four season	Eurasia
H. sapiens	0.12		Tropical/temperate/cold		Wet/dry and four season	Global

Table 17.2 *Climate, habitat, and seasonal differences through time*

Geologic epoch	Approximate dates	Climate evidence	Reconstructed habitats	Possible seasonal patterns
Mid–Late Miocene	*10–5.4 Ma*	*Reduced rainfall and cooler temperatures*	*Overall trend toward habitat aridification from forests to woodlands in Africa*	*Fluctuating*
Middle Miocene	9 Ma	Reduced rainfall and cooler temperatures	Deciduous woodlands in Africa	3.3–6.9 dry-season months
Late Miocene	6.8 Ma	Beginning of warming trend, with less glaciation	Open habitats in northern Africa; more wooded habitats in east Africa	0–3.3 dry-season months
Pliocene	*5.4–1.8 Ma*	*Overall trend to arid conditions, with reduced rainfall and cooler temperatures*	*Closed woodlands/grasslands and shrublands in eastern and southern Africa*	*Fluctuating*
Early Pliocene	5.4 Ma	End of Mediterreanean crisis, and further rainfall and warming in eastern Africa	Closed/open woodlands in eastern Africa	0–3 dry-season months?
Mid Pliocene	3.2–2.2 Ma	Aolian dust levels increase and cooling evidence in SSTs; lower CO_2 levels; terrestrial temperatures and MAR lowered gradually across this time period	Woodland/shrubland in eastern and southern Africa	Increasing dry-season months
Late Pliocene	2.1–1.9 Ma	Rapid reduction of SSTs	Spread of open grasslands in Africa among woodland environments	Increasing dry-season months
Pleistocene	*1.8–10 Ka*	*Development of glacial/interglacial cycles*	*Fluctuating*	*Fluctuating*
Early Pleistocene	1.8–1.6 Ma	Aridification	Open grasslands dominate in Africa; temperate climates warmer than present day	Increasing dry-season months
Early Pleistocene	1.6 Ma	Wetter, warmer temperatures	Woodlands return to eastern and southern Africa	Decreasing dry-season months
Mid Pleistocene	1.0–650 Ka	Extreme reduction in SSTs and intensive aridification	Grasslands and shrublands in Africa; glacial steppe in Eurasia	Increasing dry-season months in Africa; four season with extreme cold in Eurasia

Late Pleistocene	*650–10 Ka*	*100 000 glacial/interglacial cycles*	*Advance of cold habitats in Eurasia; African habitats during interglacial similar to those of today; drier during glacial intervals*	*Fluctuating numbers of wet- and dry-season months; bimodal east African pattern and unimodal South African pattern possible at times during this interval*
	150 Ka	Moderately severe climate in Europe	Polar desert and intermittent herbaceous plant cover in Europe	Four season
	130–117 Ka	Milder climate in Europe	Boreal forests Europe	Four season
	116–74 Ka	Fluctuating cold/warmer temperatures in Europe	Alternating steppe and forest	Four season
	74–59 Ka	Advancing ice sheet; colder and more arid in Europe	Gradual change from forests to tundra	Four season

MAR, mean annual rainfall; SST, sea surface temperature.
Note: Italics indicates the overall climate and habitats of entire epoch or long time span.

aridity from ~10 Ma to ~7.0 Ma, but towards the end of the Miocene Epoch this trend was reversed and only the Antarctic remained glaciated. The early Pliocene Epoch was consequently characterized by a wetter, warmer climate in Africa than in the previous five million years. Warmer temperatures and higher rainfall lasted from 5.4 Ma until ~3.0 Ma (Marlow *et al.* 2000).

Another glacial-driven cooling trend, represented by lower sea surface temperatures (SSTs), began at 3.2 Ma and continued through ~2.2 Ma. Accordingly, terrestrial temperatures and mean annual rainfall (MAR) were lowered gradually during the same time period. By 2.8 Ma, this trend caused aolian dust levels to rise, hence indicating xerification (deMenocal & Bloemendal 1995). From 2.1 to 1.9 Ma, SSTs indicate rapid cooling of sea and land temperatures (Marlow *et al.* 2000). Climates remained extremely arid until ~1.6 Ma, (deMenocal & Bloemendal 1995), when temperatures, rainfall, and aolian dust levels returned to pre-1.8-Ma values. Another climatic transition, lasting from 1.0 Ma through 650 Ka, was a time of extremely decreased SSTs and further aridification in tropical Africa (Schefuss *et al.* 2000).

Hominin occupation of other continents began at ~1.8 Ma, those occupying latitudes beyond the subtropics experiencing markedly different climatic patterns. For fossil hominin sites, we define ancient temperate climates as those similar to climates in latitudes between ~30° and ~60° today, and cold climates as those that are similar to the climate north of 60° latitude today. There are regions in which winter cold is mild within the temperate band due to ocean currents and local factors, but many temperate climates experience traditional four-season shifts, with freezing winter temperatures. We assume temperate to cold climates and four seasons for much of Pleistocene Eurasia, which, after 630 Ka, experienced 100 Ka cycles of glacial and interglacial climate changes (Marlow *et al.* 2000). These changes influenced MAR and temperatures, such that as cycles progressed, the climate became colder and precipitation decreased substantially (Gamble 1986).

Habitats

Hominin species from early to mid-Pliocene African sites existed in a variety of habitats. These environments (Table 17.3) were usually mosaic and consisted of combinations of closed to open woodlands, bushlands, riverine forests, and seasonal flood plains that produced edaphic grasslands and wetlands (Andrews 1989; Bonnefille 1995; Spencer 1997; Reed 1997, 1998, 2002;

Table 17.3 *Categories of habitats and mammalian adaptations. Brief descriptions of each habitat are provided. These are based on broad characteristics, as many localities defined as forests, for example, include other microhabitats (Lind & Morrison 1974; Pratt & Gwynne 1977; White 1983; Cole 1986). Mammalian substrate and trophic adaptations are also explained. Note that the terms "browser," "mixed feeder," and "grazer" refer to terrestrial animals. Strictly speaking, monkeys could be described as browsers as they often eat leaves, but these terms usually are reserved for ungulates*

Habitat/mammal adaptation	Definition	Classification
Habitat		
Tropical forests	Tall trees with interlaced, multi-level canopies; epiphyte ground cover	
Tropical closed woodland	Approximately 60–80% tree density (shorter trees); no interlacing or multiple canopy levels; grass ground cover	
Tropical forest/grassland (mixed)	Combination of forest and grassland	
Tropical bushlands	Bushes <3 m in height; patches of thicket, sparse grass cover	
Tropical open woodlands	Approximately 20–30% tree density; grass ground cover	
Tropical shrublands	Shrubs <1 m in height; arid adapted plants; sparse grass cover	
Tropical grasslands	<2% trees; continuous grass ground cover	
Mammal adaptations		
Trophic		
Carnivore	Meat (flesh)	Meat
	Meat and bone	Meat/bone
	Small mammals and invertebrates	Fauna/insects
Insectivore	Ants, termites, etc.	Insects
Herbivores	Dicotyledon leaves	Browser
	Dicotyledon and monocotyledon leaves	Mixed feeder
	Monocotyledon leaves	Grazer
	Monocotyledon leaves from wetlands	Fresh grass grazer

Table 17.3 (*cont.*)

Habitat/mammal adaptation	Definition	Classification
	Roots and bulbs	Roots and bulbs
Frugivores	Fruit	Fruit
	Fruit and leaves/ leaves and fruit	Fruit
	Fruit and insects	Fruit
Omnivore	No preference	Omnivore
Substrate		
Aquatic	Uses for foraging or necessary for survival	
Arboreal	Forages and rarely descends to terra firma	
Fossorial	Forages by digging or burrows necessary for survival	
Terrestrial	Forages and rarely ascends to trees or swims	
Terrestrial/arboreal	Forages and travels both terrestrially and arboreally, or travels terrestrially and forages arboreally	

Woldegabriel *et al.* 1994; Bobe & Eck 2001; Wynn 2000; Leakey *et al.* 2001). There are also a few early localities in which hominins are suggested to have existed in fairly dry shrubland environments (Reed 1997; Wynn 2000). Different hominin species consequently lived in slightly different environments from one another, and, in the Hadar Formation, *Australopithecus afarensis* habitats fluctuate from closed woodland to arid open shrublands from >3.4 to ~ 2.94 Ma (Reed 1997) (Fig. 17.1).

In general, hominin species from 2.8 to 2.0 Ma lived in African habitats reconstructed as open woodlands, bushlands, and shrublands with wetlands (Reed 1997; Spencer 1997; Bobe & Eck 2001). Based on the proportions of grazing mammals, there was also more grassland area within mosaic environments than found in the early Pliocene Epoch (Reed 1997). The spread of open grasslands is indicated by carbon isotopes in soils and the increased numbers of grazing mammals at ~1.8 Ma (Cerling 1992; Reed 1997; Spencer 1997; Marlow *et al.* 2000). Hominins recovered from African sites dated to 2.0–1.0 Ma lived in habitats that were more open; grasslands dominated at ~1.7 and 1.2 Ma (Cerling 1992).

Although the overall trend in Africa was toward more open arid habitats from 3 to 1.0 Ma, the transition occurred at different rates and with different timing depending on location. For example, Bobe and Eck (2001) observe a species turnover and habitat change in the Omo at 2.8 Ma, while a habitat shift occurred slightly earlier at the Hadar locality, ~3.18–2.95 Ma. Behrensmeyer *et al.* (1997) report gradual species turnover reflecting habitat change across this entire time period at West Turkana. Hominins outside of the subtropics at ~1.7 Ma experienced temperate deciduous woodlands and grasslands or tundra (Van Andel & Tzedakis 1996). Gamble (1986) notes that deciduous forests with various trees providing fruits and nuts would have been available in some interglacial cycles. Hominins living around the Mediterranean Sea during the Pleistocene Epoch experienced milder temperatures at times than those further north, but habitats still fluctuated from evergreen and deciduous woodland to arid cold steppe (Van Andel & Tzedakis 1996).

Seasonality

Climate, as influenced by latitude, partially dictates seasonality, which in turn affects habitat type. Low-latitude forest habitats experience long wet seasons and short dry seasons, while open woodlands at the same latitudes have shorter wet seasons and longer dry seasons. Local and regional factors,

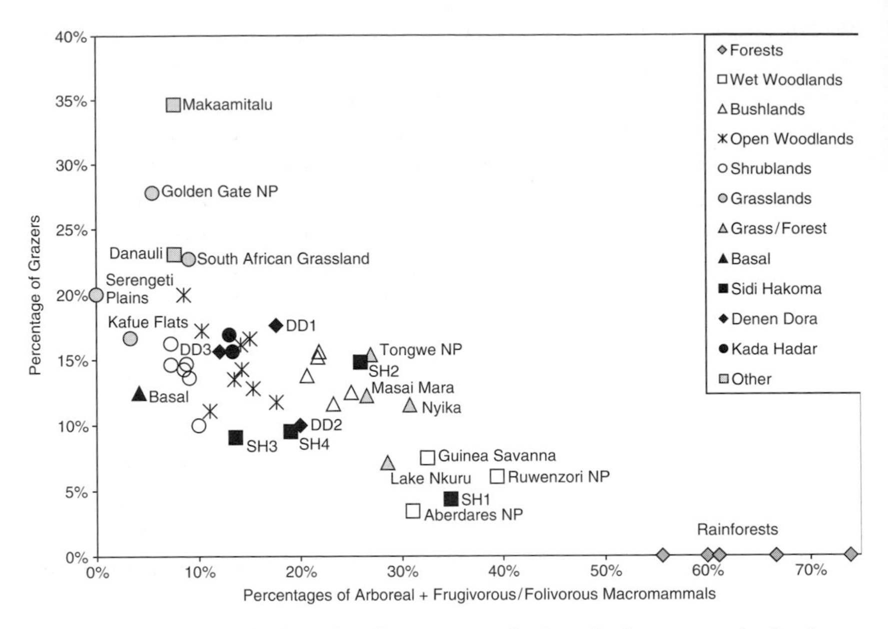

Figure 17.1 Bivariate plot of percentages of arboreal substrate use plus frugivory versus grazing animals. In general, extant sites group together, depending on the habitat from which they were derived. Fossil localities, represented by those from the Hadar Formation, group with extant habitats and, thus, their habitat is estimated from the percentages of these adaptations in the fossil fauna recovered at the site. Reconstructions for other fossil sites can be found in Reed (1997, 2002). Basal = Basal Member ≥3.4 Ma; Sidi Hakoma (SH) = 3.4–3.22 Ma, divided by sand units into four submembers, with SH1 being the oldest; Denen Dora (DD) = 3.22–3.18 Ma, divided by sand units into three submembers; Kada Hadar = (lower KH Member) 3.18–2.92 Ma, divided by volcanic tephras into two submembers; Upper = Kada Hadar (upper) <2.33 Ma and representing two sites, Makaamitalu and Danauli. Extant species lists compiled by Reed (1997) from reports on African national parks, game reserves, and South African and Nigerian biomes. Classifications of habitats from forest through grassland are dependent on tree cover and overall plant type and derived from White (1983), Pratt and Gwynne (1977), Lind and Morrison (1974), and Cole (1986). All habitats except forests are considered part of the African savanna ecosystem. Grass/forest refers to grassland habitats with extensive amounts of riverine forests.

as well as monsoonal patterns, also contribute to differences in the lengths of wet and dry months and to severity of winter in four-season latitudes.

Seasons in the mid Miocene Epoch correspond to what is expected for forested habitats. Jacobs (1999a) has estimated dry-season months for Tugen Hills, Kenya, a Miocene fossil site using fossil plant leaves. At 12.8 Ma, dry-season months are estimated at 0–3.8 months per year (MPY). This is followed by evidence for 3.3–6.9 dry-season MPY at

9.0 Ma and corresponds to the trend of aridification (Jacobs 1999a, 1999b, 2002, Marlow *et al.* 2000). In the late Miocene Epoch, Jacobs (1999a) detects a reduction in dry season months to 0–3.7 MPY, which is correlated with the beginning of the wet warming trend.

Early Pliocene hominin habitats indicate that dry-season months should range from four to six months per annum, with the more open habitats having the lengthier dry seasons. However, Hailemichael (1999) has identified seasonal growth changes in mollusks acquired from the Sidi Hakoma Member of the Hadar Formation, from which *A. afarensis* has also been recovered. They indicate that from 3.4 to 3.2 Ma, at least, there were yearly seasonal changes with dry seasons of approximately three months. Extant closed woodland and bushland habitats have longer dry seasons, raising the question of why there is a difference between extant and ancient habitats. We will address this by examining mammalian community structure.

Mammalian community structure is represented by the trophic and substrate adaptations of resident mammal species in extant localities of particular habitats or recovered from fossil sites (Andrews 1989; Reed 1997, 1998). There are differences between extant and Pliocene mammalian community structures that may relate to shorter dry seasons per year and greater primary productivity. Proportions of arboreal, frugivorous, and grazing adaptations within communities are dependent on habitat (Fig. 17.2 & 17.3) and these percentages have been used to reconstruct ancient habitats (Reed 1997, 1998). The complete array of community trophic and substrate adaptations also groups similar extant habitats together in a principal components analysis (PCA), but major differences are identified between early Pliocene and extant African mammal communities (Fig. 17.4). These discrepancies are related to significant differences in percentages of browsers and mixed feeders and indicate that the mammalian community structure of the African Pliocene Epoch was fundamentally different from extant mammalian communities.

Higher proportions of browsing species are unlikely to reflect taphonomic biases as there are significantly higher proportions of these mammals in early Pliocene deposits (Table 17.4) with different modes of accumulation: fluvial, lacustrine, volcanic, and animal (Reed 2002). One would not expect either mode of accumulation or time-averaging to consistently raise the frequency of the same two adaptations, or decrease all the others. We suggest that greater proportions of browsing species as well as mixed feeders possibly reflect greater primary productivity and, thus, seasonal differences between extant and Pliocene habitats.

Pliocene habitats could support greater proportions of browsers through increased productivity in one of two ways: either plants were

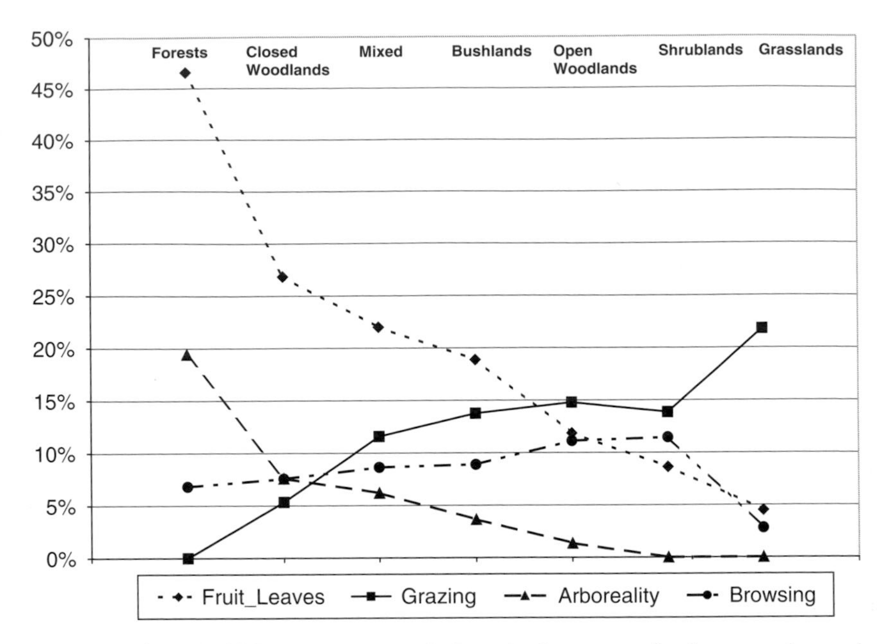

Figure 17.2 Mean percentages of arboreal substrate use, frugivory, grazing, and browsing in African habitats. The mean percentages of frugivory and aboreality decline from forests through grasslands. This corresponds to a gradient of declining annual rainfall. Conversely, the mean percentages of grazing mammals increase across the same gradient. Thus, mean annual rainfall correlates with tree cover, such that as annual rainfall declines, tree species become sparse or bush-like, finally giving way to shrubs and open grasslands. Browsing species have no particular pattern and are not significantly different among extant habitats.

more abundant or plants produced more leaves over the course of the year. Browsers eat C_3 plants, which include trees, bushes, shrubs, and herbs. It is unlikely that these plants were more abundant, as greater numbers of trees or bushes would be reflected in higher percentages of arboreal mammals and, thus, would change the reconstructed habitat, e.g. from closed woodland to forest.

An extended growing season caused by longer wet seasons rather than increased annual rainfall would account for increased primary productivity. In fact, some plants actually increase leafing productivity at the expense of fruits in a situation of prolonged wet seasons (Foley 1987). Greater production can result in increased species diversity (Connell & Orias 1964; Ritchie & Olff 1999). As the major spread of C_4 grasslands did not occur until around 1.8 Ma (Cerling 1992), greater species diversity in

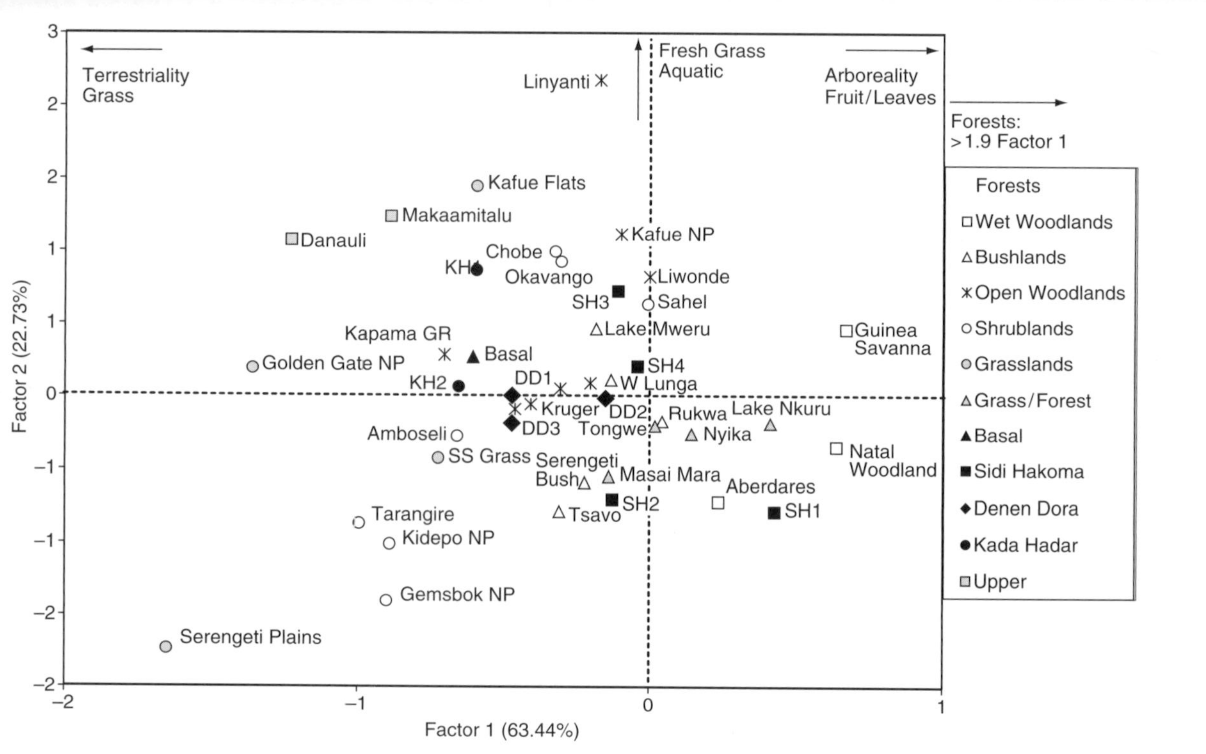

Figure 17.3 Principal components analysis using six substrate and trophic adaptations from mammals in extant and fossil communities. Adaptations found to be significantly different from one another among habitats were used to further explore relationships among fossil localities and extant habitats. These six adaptations account for 86.1% of the variation among extant communities. Fossil submembers from Hadar are again estimated to represent closed woodlands through grassland habitats. (Extant forests are eliminated from the diagram.) Those localities high on the *y*-axis have significant proportions of aquatic substrate use and fresh grass grazing from edaphic grassland. Edaphic grasslands are those that are flooded for a portion of the year. Selected extant localities labeled for reference. See Figure 17.1 for definitions.

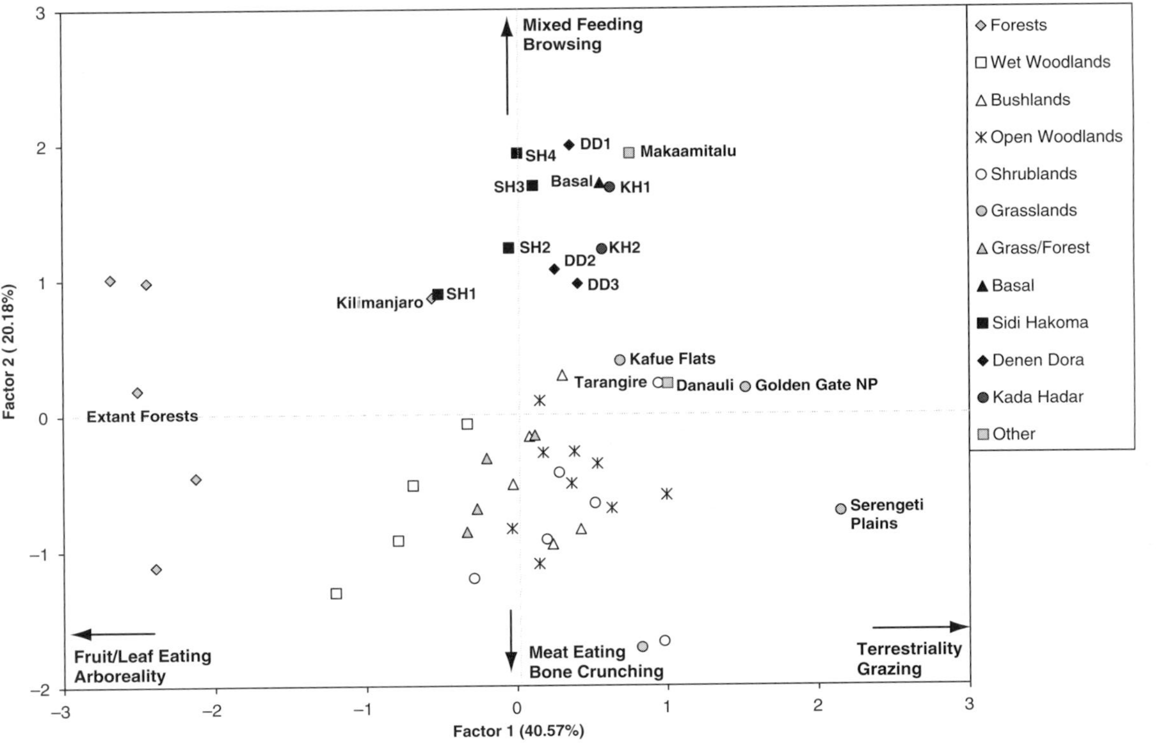

Figure 17.4 Principal components analysis (PCA) using 17 substrate and trophic adaptations from mammals in extant and fossil communities. In this PCA, only 60.8% of the variation is accounted for among extant localities, indicating that community structure differs more than specific adaptations that indicate habitats. Fossil sites plotted within this analysis show that they are still within closed woodland through grassland habitats (*x*-axis), but they have greater proportions of mixed feeding and browsing adaptations than extant localities (*y*-axis). Selected extant localities labeled for reference. See Figure 17.1 for definitions.

Table 17.4 *Analysis of variance (ANOVA) and Tukey's studentized range test (HSD) for unequal sample sizes between extant and fossil proportions of browsing and mixed feeding adaptations. The ANOVA reveals significantly different proportions between the Hadar fossil localities and all extant habitats in both browsing and mixed feeding adaptations. The Tukey HSD further shows that the Hadar fossil localities ($\bar{x} = 15.73\%$) are also significantly different from extant forest localities ($\bar{x} = 6.81\%$) in browsing adaptations. Hadar fossil localities are not significantly different in mixed feeding adaptations from extant forest localities, as extant forests include mixed feeding species that graze on grasses in riverine and swampy areas.*

	SS effect	df effect	MS effect	SS error	df error	MS error	*F*	*P*
Browsing	0.04768189	2	0.0238409	0.101214	47	0.002153	11.07081	**0.0001**
Mixed Feeding	0.02436361	2	0.0121818	0.050254	47	0.001069	11.39297	**0.0001**
Tukey HSD for uneven sample sizes								
	Extant Localities, No Forests	**Fossil Localities**	**Extant Forest Localities**					
Mixed feeding	**Mean =** 5.38%	**Mean =** 10.49%	**Mean =** 8.51%					
Extant Localities		**0.0002**						
Extant Forests	0.0903	0.4542						
Browsing	**Mean =** 9.04%	**Mean =** 15.73%	**Mean =** 6.81%					
Extant localities		**0.0004**						
Extant forests	0.5324	**0.0012**		Significant P values in bold				

df, degrees of freedom; MS, mean squares; SS, sum of squares.

the early Pliocene Epoch would affect browsing and mixed feeding species that focused on C_3 plants. Frugivore diversity would not increase if greater leafing occured at the expense of fruit production. Increased production in C_3 plants would account for the presence of four species of browsing giraffids, a browsing perissodactyl, and higher percentages of browsing bovids in Pliocene sites than extant ecosystems.

Seasonal changes in the subtropics fluctuated from 1.8 Ma to the present as further aridification and subsequent glacial cycles created periods of longer and shorter wet seasons over time. This was likely coupled, as it is today, with monsoonal differences causing both unimodal and bimodal patterns of wet and dry seasonal change (O'Brien 1998). At the same time, glaciers in the Northern Hemisphere caused habitats to shift north and south. Other local conditions determined by immediacy to seas, mountain ranges, etc. may have caused various pockets of milder climates, and thus milder winter conditions, in areas of Eurasia (van Andel & Tzedakis 1996).

Hominin diets

Teaford and Ungar (2000) summarize differences between early hominin species and Miocene hominoids in the masticatory system. Large, low-relief molars, increased enamel thickness, increased mandibular robusticity, and broader incisors typify early hominins. Teaford & Unger (2000) review research that shows vegetation in *Australopithecus* diets probably included soft fruits, hard smaller fruits that did not require incisors to peel or extensively prepare, and possibly underground storage organs (USOs), such as roots and rhizomes (see also Hatley & Kappelman [1980]). Early *Australopithecus* species probably were not able to eat tough foods such as meat, mature or fibrous leaves, or pithy fruits (Teaford & Ungar 2000). Other evidence of early hominin diet is provided from microwear striations and pitting on incisors, suggested by Ryan and Johanson (1989) to indicate savanna vegetation, and striations on molars, suggested by Grine (1981) to indicate abrasive vegetation. Janis (1988) suggests that the height above the ground from which the food comes, rather than the properties of the foods themselves, often causes tooth abrasion, i.e. grazing close to the ground would cause ingestion of more grit compared with browsing from tall trees. The findings of Ryan and Johanson (1989) and Grine (1981) thus indicate that early hominins were eating some grit- or dust-covered foods and, therefore, possibly foraging from bushes or shrubs close to the ground.

Carbon isotope data provide additional information about the foods of early hominins. For example, *Australopithecus africanus* is suggested to have eaten food with some C_4 isotopes (Sponheimer & Lee-Thorp 1999). Grasses and sedges, including their roots, seeds, and leaves, result in C_4 isotope traces in enamel, as does eating animals that eat those plant foods. While Sponheimer and Lee-Thorp (1999) favor *A. africanus* eating meat from grazing animals, others suggest that this hominin may have focused on grasses (seeds or rhizomes), sedges, or animals (van der Merwe *et al.* 2003). Although C_4 grasslands expanded throughout Africa at 1.8 Ma, these plants have existed since the Miocene Epoch. We think that meat-eating is unlikely for the following reasons: first, even if *A. africanus* ate meat occasionally, there is no particular reason for any early hominin to focus on ungulates that ate grasses in areas with abundant browsing animals. Second, early hominins would not have been able to process meat or marrow by relying on their dentition. They would have needed stone tools to expose marrow cavities and to remove meat from bones, but there are no stone tools associated with this species. Third, hominin masticatory morphology indicates adaptations to harder, not tougher (such as meat), food items (Teaford & Ungar 2000), and microwear studies show a certain amount of grit in the diet (Grine 1981). These data suggest that *A. africanus* and other early hominins ate plant products that were hard and gritty fairly often. USOs (both C_3 and C_4 bulbs and rhizomes) would fit these data, but species exhibiting higher C_4 isotope values may have selected USOs from grasses. Teaford and Ungar (2000) note that early hominins did not specialize on hard resources but were able to eat these foods as a critical fallback. While it is possible that these early species also scavenged in a limited way, either USOs or termites are more likely resources causing the C_4 values.

Paranthropus species' craniodental morphology includes very large cheek teeth, extremely thick enamel, small incisors, and inflated malar regions (Fleagle 1999). It has been hypothesized that *Paranthropus* species were eating grittier food items than were earlier hominins, as evinced by dental microwear that shows heavy pitting on the cheek teeth (Grine 1981). Based on skull morphology, it has also been suggested that these species were eating harder foods than were previous hominins (see review by Teaford & Ungar [2000]), or they were eating hard foods more consistently (Foley 1987). In addition, tools must now be considered as part of the food acquisition process. *P. robustus*, at least, is associated with polished bone tools that are indicative of digging (Brain & Shipman 1993; d'Errico *et al.* 2001), while stone tools are allocated to *Homo* species. The masticatory morphology associated with *Paranthropus* species is likely the result of further, intensive focus on USOs (Hatley & Kappelman 1980; Teaford & Ungar 2000).

Early *Homo* species such as *H. rudolfensis* and *H. habilis* have different morphology from *Paranthropus* and earlier hominin species, such that their cheek teeth are smaller, their incisors are larger, and their faces are reduced (Fleagle 1999). They are also usually associated with Oldowan stone tools, which although appearing around 2.6 Ma (Semaw *et al.* 1997), become ubiquitous in Africa after 2.3 Ma. These features suggest that early *Homo* species were eating foods that were not as gritty or hard as those consumed consistently by the *Paranthropus* lineage, and they were also processing food with stone tools. *Homo* species probably focused on softer fruits and vegetable products as well as incorporating meat into their diets. Later *Homo* species added more meat into their diets and likely began processing vegetable products beyond simple collection (Wrangham *et al.* 1999; Aiello & Wheeler 1995). We discuss the evolutionary significance of changes in seasonality to hominin diets and behavior in the next section.

Seasonal changes and hominin evolution

Early hominins

Ecological evidence from early hominin sites shows that they were able to exist in various habitats and through changes in climate. Extended rainy seasons probably supplied preferred food items for long periods of each year. There were dry seasons, however, in which production of fruits, flowers, and flush leaves were diminished.

Due to global climate changes that began at 3.2 Ma (Marlow *et al.* 2000), primary productivity likely declined and resources were difficult to find in the dry season. As a result, there was a progressive reduction in the amount of above-ground plant material available for hominin exploitation. Plants are affected by longer periods of drought, such that, if deciduous, they store energy in USOs instead of producing leaves, shoots, or fruits (Archibold 1995). Bulbs and rhizomes therefore would be available throughout much of the dry season and provide needed nutrition to hominins.

This has implications for *A. afarensis* and other early hominins, such that a change to longer dry seasons instigated by increased aridification (Vrba, 1995; Behrensmeyer *et al.* 1997; Bobe & Eck 2001) could have pushed the species to extinction. Some populations of early *Australopithecus* species may have switched their foraging effort to USOs during long dry seasons, and over time *Paranthropus* species evolved. Other early hominin populations switched to critical resources in response to lengthening dry seasons by incorporating meat or marrow into their diets, obtained through the use of stone tools. There may be other species, such as *A. garhi*, that focused on other, yet unknown, resources. In any event, we conclude that longer dry seasons initiated by climatic changes around 3.0–2.8 Ma contributed to hominin extinction and speciation events.

Robust **Australopithecine** *(***Paranthropus***) lineage*

Aridification caused a change in resource availability of above-ground plant material after ~2.8 Ma. USOs would have been a reliable food source, possibly requiring increased foraging effort as dry seasons lengthened. *Paranthropus* existed into the Pleistocene Epoch and is almost always associated with edaphic grasslands in both eastern and southern Africa (Reed 1997). Edaphic grasslands and wetlands are often present in dry seasons because of floodwaters caused by rain in areas far away from the immediate climate (e.g. Okavango Delta). For this reason, it is possible that food resources used by *Paranthropus* species survived the aridification at the beginning of the Pleistocene Epoch (1.8 Ma) but disappeared with more extreme climate change at ~1.0 Ma. The sudden drop in SSTs and subsequent glaciation of both poles would have caused serious droughts across Equatorial areas during this time period. *Paranthropus* disappears from the fossil record at some stage in this period of increased aridity.

Early **Homo**

Although *Homo* species no doubt were predominantly vegetarians, this lineage most likely began consuming animal resources sometime after ~2.7 Ma, as indicated by the presence of stone tools (Semaw *et al.* 1997). Seasonal stresses, i.e. reduced availability of edible above-ground plant material in dry-season months, probably contributed to the change in diet. *Homo* species were able to scavenge carcasses for bone marrow and any remaining meat during these lean periods (Blumenschine 1987). The masticatory morphology of *Homo* suggests that they utilized different food resources from *Paranthropus*, and stone tools would have allowed the initial processing of marrow or meat. Scavenging opportunities probably occurred in riparian woodlands, and many early *Homo* sites have been located in these habitats (Blumenschine 1987; Marean 1989). In addition, scavenging was probably dangerous for *Homo*, even if they carried stone tools, so the behavior was likely opportunistic rather than consistent or confrontational. Opportunistic scavenging could have been an option for *Australopithecus* species as well, but it probably became critical for early *Homo*. Within the changing climate and longer dry seasons, meat would have been available relatively consistently, in contrast with usual vegetation sources utilized by *Homo*, which may have failed seasonally. Animals were likely a critical resource by the time the Oldowan industry appears. Animal resources allowed this lineage of Pliocene hominins to survive seasonal fluctuations in plant resources, whereas *Paranthropus* species survived seasonal fluctuations by turning to underground vegetable resources.

Homo erectus

The Dmanisi locality demonstrates that hominins had colonized temperate climates by the Early Pleistocene Epoch. The fauna associated with the Dmanisi hominins includes both palearctic and paleotropical species, indicating a relatively mild temperate climate (Gabunia *et al.* 2000b). Gabunia *et al.* (2000b) also report a high ratio of felids to hyaenids at Dmanisi, and they suggest that hominins would have had minimal competition from other scavengers and that these earliest immigrants were most likely opportunistic scavengers. Given the limited evidence for hominins in temperate climates at this early date, long-term survival via opportunistic scavenging may have been limited, particularly as the climate cooled after 1.6 Ma (de Menocal & Bloemendal 1995).

At the same time in Africa, Early Pleistocene deposits yield evidence for the first appearance of *Homo erectus sensu lato* at Koobi Fora 1.8 Ma. It has recently been argued that *H. erectus*' anatomy, social structure, and life history traits are best explained by a subsistence strategy heavily dependent upon foraging for tubers (O'Connell *et al.* 1999). We consider this scenario unlikely for *H. erectus*, especially in Africa. Given that *H. erectus* and *Paranthropus* were synchronic and sympatric in Africa, it is unlikely that they shared a dietary specialization. Further evidence may show that *H. erectus* focused on tubers as well, but we argue that *H. erectus* moved from opportunistic to confrontational scavenging, relying on animals to buffer seasonal stress for the following reasons. *H. erectus* displays an increase in stature and is the first hominin species with modern human limb proportions (McHenry 1992). Additionally, the thoracic cavity of *H. erectus* is barrel-shaped, as in modern humans and in contrast to australopithecines, suggesting both a reduction in gut size and a more efficient cardiovascular system (Aiello & Wheeler 1995). These anatomical features represent an adaptation to open arid habitats with high levels of activity and an extended foraging range (Cachel & Harris 1998).

In Africa, *H. erectus* is associated consistently with stone tools and large animal remains. In particular, the Acheulean industry first appears 1.5 Ma with *Homo erectus* and includes heavy-duty butchering tools such as hand axes and cleavers (Klein 1999). Recent analyses at Olduvai indicate that *H. erectus* had early access to vertebrate carcasses and focused on long bones of larger mammals for meat rather than marrow (Monahan 1996). Together, these data are consistent with the hypothesis that *H. erectus* practiced confrontational scavenging. Despite the long-term success that this strategy provided *H. erectus*, colonization of cold environments was not achieved by this species (Roebroeks *et al.* 1992; Roebroeks 2001). Survival in cold environments required primary access to animals, which can be achieved only through hunting.

Later Homo *species and the evolution of hunting*

The first hominin species to occupy cold climates was *Homo heidelbergensis*, which first appeared in Africa ~800 ka, and then in Europe ~500 ka (Rightmire 1996). Although the archaeological record for *H. heidelbergensis* is sparse, there is reason to argue that this species was a capable hunter (Thieme 1997). The relative brain size of *H. heidelbergensis* is significantly greater than in *H. erectus* (Ruff *et al.* 1997), suggesting an increase in diet quality (Aiello & Wheeler 1995; Fish & Lockwood 2003). Additionally, the continuous record of occupation in the northern latitudes of Europe after 500 Ka indicates an ability to consistently acquire high-quality animal resources to survive the long winters (Roebroeks 2001).

The importance of hunting for survival in cold climates is best illustrated by *Homo neanderthalensis*, the European descendant of *H. heidelbergensis*. Neandertals date from ~150–28 ka and appear in Europe during the height of glacial activity (Isotope Stage 6). Neandertals have also been found in the temperate localities of Spain, Italy, and the Levant, indicating that they were not restricted to cold habitats.

The Neandertal postcranial skeleton is exceptionally robust (Pearson 2000) and has a high incidence of trauma (Trinkaus & Zimmerman 1982), which has been interpreted as resulting from intensive hunting of large animals. Neandertal faunal accumulations from sites such as Salzgitter Levenstedt, Germany (Gaudzinski & Roebroeks 2000), Marillac, France

(Fizet *et al.* 1995), Mauran, France (Gaudzinski 1996), Zafarraya, Spain (Geraads 1997), Mezmaiskaya, Caucasus (Golovanova *et al.* 1999), and the Levantine sites of Kebara (Speth & Tchernov 1998), Amud (Suzuki & Takai 1970), and Kobeh (Marean & Kim 1998) are regularly dominated by a single ungulate species with cut marks, element representation, and age profiles indicative of specialized hunting. Additional evidence for a diet high in animal protein comes from analyses of buccal microwear patterns on the teeth of Neandertals (Lalueza Fox & Perez-Perez 1993; Lalueza *et al.* 1996) and the isotopic composition of their bones (Bocherens *et al.* 1999; Richards *et al.* 2000).

This specialized strategy is in sharp contrast with the hunting pattern associated with contemporary anatomically modern *Homo sapiens* (AMHS). The earliest AMHS appear between 190 and 160 Ka in Ethiopia (White *et al.* 2003; McDougall *et al.* 2005). AMHS localities of Klasies River Mouth and Die Kelders, South Africa (125–110 Ka), contain numerous species and include marine and avian fauna (Klein 1976; Grine *et al.* 1991). The abundant resources in the habitats surrounding these sites allowed AMHS to hunt a variety of mammals and gather other animal resources with no single species favored.

The different hunting strategies of Neandertals and AMHS are demonstrated most clearly by contrasting cave sites in Israel where fossil remains from both species have been identified. Both Neandertal fossils from Amud and Kebara and AMHS remains at Qafzeh and Skhul were found associated with similar Levaillois-Mousterian stone tool technology (Bar-Yosef & Meignen 1992), and many of the same ungulate species are present. However, Neandertal accumulations are significantly less rich and diverse than those of AMHS. Gazelles represent 58 and 66% of the ungulates out of six and seven total species at Amud and Kebara, respectively (Suzuki & Takai 1970; Speth & Tchernov 1998). In contrast, among the ten species found at Qafzeh, the most prevalent accounts for only 35% of the accumulation (Rabinovich & Tchernov 1995). It is also probable that Neandertals occupied caves in the Levant multiseasonally, while AMHS occupied the region in the winter and spring seasons (Leiberman 1993; Shea 2003). Despite occupying broadly similar habitats within the same geographic region, Neandertals utilized a narrower range of animal resources and hunted more (Shea 2003) compared with the generalized strategy employed by AMHS. That is, Neandertals focused on hunting irrespective of the season in the Levant (Leiberman & Shea 1994).

The specialized nature of Neandertal subsistence was certainly related to seasonal limitations on resources. Marean (1997) has suggested that the most optimal strategy for cold grassland inhabitants is tactical hunting, because it is the best way to create an abundance to survive through the long winter. Given the evidence, it is reasonable to propose that Neandertals had adapted to millennia of living in cold environments by developing a specialized hunting strategy. The importance of this strategy for Neandertal survival in cold climates can be measured in the fact that it seems to have become part of their lifestyle and was practiced in warmer habitats such as the Levant. In contrast AMHS evolved in a tropical climate with only wet and dry seasonal changes to face during foraging. Further improvements and innovations related to animal resource utilization first appear in subtropical Africa, and this suggests that resource excess was required for these developments. AMHS developed behaviors and adaptations that allowed more flexibility in foraging behavior over the long term.

Conclusions

Climatic changes caused by glacial cycles have influenced human evolution by changing habitats and ultimately altering resource availability. Climate change also influenced long-term seasonal differences by shifting wet and dry yearly seasonal patterns. Alterations in the number of dry season months in particular caused greater levels of foraging stress for early hominins.

Increases in dry season months between ~3.0 and 2.5 Ma probably forced important modifications in resource use. Some populations of early *Australopithecus* or other hominins turned to reliance on USOs, while other populations focused on animal resources. This assumes that the earliest hominins previously had incorporated limited quantities of both underground tuber and animal resources into their diets. Overall, increased seasonality prompted by global aridification influenced hominin diversification into two separate lineages: *Paranthropus* and *Homo*.

Paranthropus species most likely focused on USOs, at least for possibly lengthy dry seasons, while the critical resource in dry seasons for early *Homo* was marrow and/or meat scavenged from available carcasses (Blumenschine 1987). *Homo* later moved to more confrontational scavenging, possibly due to grassland expansion and longer dry seasons potentially providing more carcasses. The habitat and ecological tolerance of tropical hominins improved with the ability to acquire animals as they became more independent of plant resources. However, the lack of numerous hominin sites in temperate climates until the advent of *Homo heidelbergensis* implies that hunting in these climates was not a major foraging technique of earlier hominins.

As hominins began to occupy cold habitats with greater seasonal stress, meat became more important to the diet and Neandertals focused on acquiring meat. Extreme seasonal stress, even though tolerated well by Neandertals, was apparently not conducive to innovation beyond tactical hunting patterns because of possible lack of overall resources during extreme cold. AMHS, originating in subtropical Africa, had different foraging and hunting practices that better suited a variety of habitats. Abundances of resources led them to various seasonal practices that did not rely on any one resource or acquisition method. These subsistence strategies likely enabled the species to expand worldwide.

Acknowledgments

We would like to thank Diane Brockman and Carel van Schaik for asking us to contribute this chapter. We also thank Diane Brockman, Steve Churchill, Geoff Clark, John Fleagle, Charles Lockwood, Curtis Marean, Lillian Spencer, and Carel van Schaik for helpful comments on earlier versions of this manuscript. Research on the Hadar locality fauna was funded by a field grant from the National Science Foundation (NSF) and analysis grants from the Leakey Foundation and the Institute of Human Origins.

References

Aiello, L. & Wheeler, P. (1995). The expensive-tissue hypothesis. *Current Anthropology*, **36**, 199–221.

Andrews, P. (1989). Palaeoecology of Laetoli. *Journal of Human Evolution*, **18**, 173–81.

Andrews, P. & Humphrey, L. (1999). African Miocene environments and the transition to early hominins. In *Paleoclimate and Evolution with Emphasis on Human Origins*, ed. E. S. Vrba, G. H. Denton, T. C. Partridge, & L. C. Burckle. New Haven, CT: Yale University Press, pp. 282–300.

Archibold, O. W. (1995). *Ecology of Wild Vegetation*. London: Chapman & Hall.

Bar-Yosef, O. & Meignen, L. (1992). Insights into Middle Paleolithic cultural variability. In *The Middle Paleolithic: Adaptation, Behavior, and Variability*, ed. H. Dibble & P. Mellars. Philadelphia: The University Museum, University of Pennsylvania, pp. 163–82.

Behrensmeyer, A. K., Todd, N., Potts, R., & McBrinn, G. E. (1997). Late Pliocene faunal turnover in the Turkana Basin, Kenya and Ethiopia. *Science*, **278**, 1589–94.

Blumenschine, R. J. (1987). Characteristics of an early hominid scavenging niche. *Current Anthropology*, **28**, 383–407.

Bobe, R. & Eck, G. G. (2001). Responses of African bovids to Pliocene climatic change. *Paleobiology*, **27** (Suppl), 1–47.

Bocherens, H., Billiou, D., Mariotti, A., *et al.* (1999). Palae-environmental and Palaeodietary implications of isotopic biogeochemistry of last interglacial Neanderthal and mammal bones in Scladina Cave (Belgium). *Journal of Archaeological Science*, **26**, 599–607.

Bonnefille, R. (1995). A reassessment of the Plio-Pleistocene pollen record of East Africa. In *Paleoclimate and Evolution with Emphasis on Human Origins*, ed. E. S. Vrba, G. H. Denton, T. C. Partridge, & L. C. Burckle. New Haven, CT: Yale University Press, pp. 299–310.

Brain, C. K. & Shipman, P. (1993). The Swartkrans bone tools. In *Swartkrans: A Cave's Chronicle of Early Man*, ed. C. K. Brain. Pretoria, South Africa: Transvaal Museum Monograph, pp. 195–215.

Bromage, T. G. & Schrenk, F. (1999). *African Biogeography, Climate Change & Human Evolution*. New York: Oxford University Press.

Cachel, S. & Harris, J. W. K. (1998). The lifeways of *Homo erectus* inferred from archaeology and evolutionary ecology: a perspective from east Africa. In *Early Human Behaviour in Global Context: The Rise and Diversity of the Lower Palaeolithic Record*, ed. M. D. Petraglia & R. Korisettar. London: Routledge, pp. 280–303.

Cerling, T. (1992). Development of grasslands and savannas in East Africa during the Neogene. *Palaeogeography, Palaeoclimatology, Palaeoecology*, **97**, 241–7.

Cole, M. M. (1986). *The Savannas: Biogeography and Geobotany*. London: Academic Press.

Connell, J. H. & Orias, E. (1964). The ecological regulation of species diversity. *American Naturalist*, **98**, 399–414.

Crowley & North (1991) *Paleoclimatology*. Oxford: Oxford University Press.

Daegling, D. & Grine, F. E. (1991). Compact-bone distribution and biomechanics of early hominid mandibles. *American Journal of Physical Anthropology*, **86**, 321–39.

DeMenocal, P. B., & Bloemendal, J. (1995). Plio-Pleistocene subtropical African climate variability and the paleoenvironment of hominid evolution: a combined data-model approach, In *Paleoclimate and Evolution with Emphasis on Human Origins*, ed. E. S. Vrba, G. H. Denton, T. C. Partridge, & L. C. Burckle. New Haven, CT: Yale University Press, pp. 262–88.

D'Errico F., Backwell L. R., & Berger L. R. (2001). Bone tool use in termite foraging by early hominids and its impact on our understanding of early hominid behaviour. *South African Journal of Science*, **3–4**: 71–5.

Fish, J. L. & Lockwood, C. A. (2003). Dietary constraints on encephalization in Primates. *American Journal of Physical Anthropology*, **120**, 171–81.

Fizet, M., Mariotti, A., & Bocherens, H. (1995) Effect of diet, physiology, and climate on carbon and nitrogen stable isotopes of collagen in a Late Pleistocene anthropic palaeoecosystem: Marillac, Charente, France. *Journal of Archaeological Science*, **22**, 67–79.

Fleagle, J. G. (1999). *Primate Adaptation and Evolution*. New York: Academic Press.

Foley, R. (1987). *Another Unique Species*. Harlow, UK: Longman Scientific & Technical.

Gabunia, L., Vekua, A., Lordkipanidze, D., *et al.* (2000a). Earliest Pleistocene hominid cranial remains from Dmanisi, Republic of Georgia: taxonomy, geological setting, and age. *Science*, **288**, 1019–25.

— (2000b). The environmental context of early human occupation in Georgia (Transcaucasia). *Journal of Human Evolution*, **34**, 785–802.

Gamble, C. S. (1986). *The Palaeolithic Settlement of Europe*. Cambridge: Cambridge University Press.

Gaudzinski, S. (1996). On bovid assemblages and their consequences for the knowledge of subsistence patterns in the Middle Paleolithic. *Proceedings of the Prehistoric Society*, **62**, 19–39.

Gaudzinski, S. and Roebroeks, W. (2000). Adults only: reindeer hunting at the Middle Paleolithic site Salzgitter Lebenstedt, Northern Germany. *Journal of Human Evolution*, **38**, 497–521.

Geraads, D. (1997). Le grande faune associee aux derniers Neandertaliens de Zafarraya (Andalousie, Espagne): systematique et essai d'interpretation. *CRAS, Paris, Sciences de la terre et des planetes*, **325**, 725–31.

Golovanova, L., Hoffecker, J., Kharitonov, V., & Romanova, G. (1999). Mezmaiskaya Cave: a Neanderthal occupation in the Northern Caucasus. *Current Anthropology*, **40**, 77–86.

Grine, F. E. (1981). Trophic differences between gracile and robust australopithecines: a scanning electron-microscope analysis of occlusal events. *South African Journal of Science* **77**, 203–30.

Grine, F., Klein, R., and Volman, T. (1991). Dating, archaeology and human fossils from the Middle Stone Age levels of Die Kelders, South Africa. *Journal of Human Evolution*, **21**, 363–95.

Hailemichael, M. (1999). The Pliocene environment of Hadar, Ethiopia: a comparative isotopic study of paleosol carbonates and lacustrine mollusk shells of the Hadar Formation and of modern analog. Ph. D. thesis, Case Western Reserve University.

Haile-Selassie, Y. (2001). Late Miocene hominids from the Middle Awash, Ethiopia. *Nature*, **412**, 178–81.

Hatley, T. & Kappelman, J. (1980). Bears, pigs, and Plio-Pleistocene hominids: a case for the exploitation of below ground food resources. *Human Ecology*, **8**, 371–87.

Hylander, W. (1975). Human mandible: lever or link. *American Journal of Physical Anthropology*, **43**, 227–42.

— (1988). Implications of in vivo experiments for interpreting the functional significance of "robust" Australopithecine jaws. In *Evolutionary History of the "Robust" Australopithecines*, ed. F. E. Grine. New York: Aldine de Gruyter, pp. 55–8.

Jacobs, B. F. (1999a). The use of leaf form to estimate Miocene rainfall variables in tropical Africa. *XVI International Botanical Congress*, Abstract #4570.

— (1999b). Estimation of rainfall variables from leaf characters in tropical Africa. *Palaeogeography, Palaeoclimatology, Palaeoecology*, **145**, 231–50.

— (2002). Estimation of low latitude paleoclimates using fossil angiosperm leaves: examples from the Miocene Tugen Hills, Kenya. *Paleobiology*, **28**, 399–421.

Janis, C. M. (1988). An estimation of tooth volume and hypsodonty indices in ungulate mammals, and the correlation of these factors with dietary preference. In *Teeth Revisited: Proceedings of the VIIth International Symposium on Dental Morphology, Paris, 1986*, ed. D.E. Russell, J.P. Santoro, & D. Sigogneau-Russell. *Memoires du Museum Nationale d' Histoire Naturelle, Paris, Serie C*, **53**, 367–87.

Janis, C. M., Damuth, J. M., and Theodor, J. (2002). The origins and evolution of the North American grassland biome: the story from hoofed mammals. *Palaeogeography, Palaeclimatology, Palaeoecology*, **177**, 183–98.

Kay, R. F. (1984). On the use of anatomical features to infer foraging behavior in extinct primates. In *Adaptations for Foraging in Nonhuman Primates: Contributions to an Organismal Biology of Prosimians, Monkeys and Apes*, ed. P.S. Rodman & J.G.H. Cant. New York: Columbia University Press, pp. 21–53.

Kimbel, W.H. (1995). Hominid speciation and Pliocene climate change. In *Paleoclimate and Evolution with Emphasis on Human Origins*, ed. E. S. Vrba, G.H. Denton, T.C. Partridge, & L.C. Burckle. New Haven, CT: Yale University Press, pp. 425–37.

Klein, R. (1976). The mammalian fauna of the Klasies River Mouth sites, Southern Cape Province, South Africa. *South African Archaeological Bulletin*, **31**, 75–98.

— (1999) *The Human Career: Human Biological and Cultural Origins*. Chicago: Chicago University Press.

Lalueza Fox, C. & Perez-Perez, A. (1993). The diet of the Neanderthal child Gibraltar 2 (Devil's Tower) through the study of the vestibular striation pattern. *Journal of Human Evolution*, **24**, 29–41.

Lalueza, C., Perez-Perez, A., & Turbon, D. (1996) Dietary inferences through buccal microwear analysis of Middle and Upper Pleistocene human fossils. *American Journal of Physical Anthropology*, **100**, 367–87.

Leakey, M. G., Spoor, F., Brown, F. H., *et al.* (2001). New hominin genus from eastern Africa shows diverse middle Pliocene lineages. *Nature*, **410**, 433–40.

Leiberman, D. (1993). The rise and fall of seasonal mobility among hunter-gatherers. *Current Anthropology*, **35**, 569–98.

Leiberman, D., & Shea, J. (1994). Behavioral differences between archaic and modern humans in the Levantine Mousterian. *American Anthropologist*, **96**, 300–332.

Lind, E. M. and Morrison, M. E. S. (1974). *East African Vegetation*. Bristol, UK: Longman.

Marean, C. W. (1989). Sabertooth cats and their relevance for early hominid diet and evolution. *Journal of Human Evolution*, **18**, 559–82.

— (1997). Hunter-gatherer foraging strategies in tropical grasslands: model building and testing in the East African Middle and Later Stone Age. *Journal of Anthropological Archaeology*, **16**, 189–225.

Marean, C. W. and Kim, S. Y. (1998). Mousterian large mammal remains from Kobeh Cave (Zagros Mountains, Iran): behavioral implications for Neandertals and early modern humans. *Current Anthropology*, **39**, S79–114.

Marlow, J. R., Lange, C. B., Wefer, G., & Rosell-Mele, A. (2000). Upwelling intensification as part of the Pliocene–Pleistocene climate transition. *Science*, **290**, 288–91.

McDougal, I., Braoen, F. H., & Fleagle, J. G. (2005). Stratigraphic placement and age of modern humans from Kibish, Ethiopia. *Nature*, **433**, 733–6.

McHenry, H. (1992). How big were early hominids? *Evolutionary Anthropology*, **1**, 15–20.

Monahan, C. M. (1996). New zooarchaeological data from Bed II, Olduvai Gorge, Tanzania: implications for hominid behavior in the Early Pleistocene. *Journal of Human Evolution*, **31**, 93–128.

O'Brien, E. (1998). Water-energy dynamics, climate, and prediction of woody plant species richness: an interim general model. *Journal of Biogeography*, **25**, 379–98.

O'Connell, J., Hawkes, K., & Blurton Jones, N. (1999). Grandmothering and the evolution of *Homo erectus*. *Journal of Human Evolution*, **36**, 461–85.

Pearson, O. M. (2000). Activity, climate, and postcranial robusticity. *Current Anthropology*, **41**, 569–607.

Potts, R. (1998). Environmental hypotheses of hominin evolution. *Yearbook of Physical Anthropology*, **41**, 93–136.

Pratt, D. J. and Gwynne, M. D. (1977). *Rangeland Management and Ecology in East Africa*. London: Hodder and Stoughton.

Rabinovich, R. & Tchernov, E. (1995). Chronological, paleoecological and taphonomical aspects of the Middle Paleolithic Site of Qafzeh, Israel. In *Archaeozoology of the Near East II: Proceedings of the Second International Symposium on the Archaeozoology of Southwestern Asia and Adjacent Areas*, ed. H. Buitenhuis & H.-P. Uerpman Leiden: Backhuys Publishers, pp. 5–44.

Reed, K. E. (1997). Early hominid evolution and ecological change through the African Plio-Pleistocene. *Journal of Human Evolution*, **32**, 289–322.

— (1998). Using large mammal communities to examine ecological and taxonomic organization and predict vegetation in extant and extinct assemblages. *Paleobiology*, **24**, 384–408.

— (2002). The use of paleocommunity and taphonomic studies in reconstructing primate behavior. In *Reconstructing Primate Behavior in the Fossil Record*, ed. M. J. Plavcan, R. Kay, C. van Schaik, & W. L. Jungers. New York: Kluwer Academic/Plenum Press, pp. 217–59.

— (in preparation) Paleoecology of the Hadar Formation, Ethiopia.

Richards, M., Pettitt, P., Trinkaus, E., Smith, F., Paunovic, M., & Karavanic, I. (2000). Neanderthal diet at Vindija and Neanderthal predation: the evidence from stable isotopes. *Proceedings of the National Academy of Sciences, USA*, **97**, 7663–66.

Rightmire, P. (1996). The human cranium from Bodo, Ethiopia: evidence for speciation in the Middle Pleistocene? *Journal of Human Evolution*, **31**, 21–39.

Ritchie, M. E. & Olff, H. (1999). Spatial scaling laws yield a synthetic theory of biodiversity. *Nature*, **400**, 557–60.

Roebroeks, W. (2001). Hominid behaviour and the earliest occupation of Europe: an exploration. *Journal of Human Evolution*, **41**, 437–61.
Roebroeks, W., Conrad, N., & van Kolfschoten, T. (1992). Dense forests, cold steppes, and the Paleolithic settlement of Northern Europe. *Current Anthropology*, **33**, 551–86.
Rosenberger, A. L. & Kinzey, W. G. (1976). Functional patterns of molar occlusion in platyrrhine primates. *American Journal of Physical Anthropology*, **45**, 281–98.
Ruff, C., Trinkaus, E., & Holliday, T. (1997). Body mass and encephalization in Pleistocene *Homo*. *Nature*, **387**, 173–6.
Ryan A. C. & Johanson, D. C. (1989). Anterior dental microwear in *Australopithecus afarensis*: comparisons with human and nonhuman primates. *Journal of Human Evolution*, **18**, 235–68.
Schefuss, E., Pancost, R. D., Jansen, J. H. F., & Damste, J. S. S. (2000). The mid-Pleistocene climate transition: insight from organic geochemical records from the tropical Atlantic. *Journal of Conference Abstracts*, **5**, 886.
Semaw, S., Renne, P., Harris, J. W. K., *et al.* (1997). 2.5-million-year-old stone tools from Gona, Ethiopia. *Nature*, **385**, 333–6.
Shea, J. (2003). Neandertals, competition, and the origin of modern human behavior in the Levant. *Evolutionary Anthropology*, **12**, 173–87.
Spencer, L. M. (1997). Dietary adaptations of Plio-Pleistocene Bovidae: implications for hominid habitat use. *Journal of Human Evolution*, **32**, 201–28.
Speth, J. & Tchernov, E. (1998). The role of hunting and scavenging in Neandertal procurement strategies: new evidence from Kebara cave (Israel). In *Neandertals and Modern Humans in Western Asia*, ed. T. Akazawa, K.Aoki, & B. Bar-Yosef. New York: Plenum Press, pp. 223–39.
Sponheimer, M. & Lee-Thorp, J. (1999). Isotopic evidence for the diet of an early hominid, *Australopithecus africanus*. *Science*, **283**, 368–70.
Suzuki, H. & Takai, F. (1970). *The Amud Man and His Cave Site*. Tokyo: Academic Press of Japan.
Teaford, M. (1988). Scanning electron-microscope diagnosis of wear patterns versus artifacts on fossil teeth. *Scanning Microscopy*, **2**, 1149–66.
Teaford, M. & Ungar, P. (2000). Diet and the evolution of the earliest human ancestors. *Proceedings of the National Academy of Sciences*, USA, **97**, 13506–11.
Thieme, H. (1997). Lower Palaeolithic hunting spears from Germany. *Nature*, **385**, 807–10.
Trinkaus, E. & Zimmerman, M. R. (1982). Trauma among the Shanidar Neanderthals. *American Journal of Physical Anthropology*, **57**, 61–76.
Ungar P. (1998). Dental allometry, morphology and wear as evidence for diet in fossil primates. *Evolutionary Anthropology*, **6**, 205–17.
Ungar, P. & Grine, F. (1991). Incisor size and wear in *Australopithecus africanus* and *Paranthropus robustus*. *Journal of Human Evolution*, **20**, 313–40.
Van Andel, T. H. & Tzedakis, P.C. (1996). Paleolithic landscapes of Europe and environs, 150,000–25,000 years ago: an overview. *Quaternary Science Review*, **15**, 481–500.
Van der Merwe, N. J., Thackeray, J.F., Lee-Thorp, J. A., & Luyt, J. (2003). The carbon isotope ecology and diet of *Australopithecus africanus* at Sterkfontein, South Africa. *Journal of Human Evolution*, **44**, 581–97.

Vekua, A., Lordkipanidze, D., Rightmire, G. P., *et al.* (2002). A new skull of early *Homo* from Dmanisi, Georgia. *Science*, **297**, 85–9.

Vrba, E. S. (1988a). Late Pliocene climatic events and hominid evolution. In *Evolutionary History of the "Robust" Australopithecines*, ed. F. E. Grine. New York: Aldine de Gruyter, pp. 405–26.

— (1988b). Evolution, species and fossils: how does life evolve? *South African Journal of Science*, **76**, 61–84.

— (1992). Mammals as a key to evolutionary theory. *Journal of Mammalogy*, **73**, 1–28.

— (1995). On the connections between paleoclimate and evolution. In *Paleoclimate and Evolution with Emphasis on Human Origins*, ed. E. S. Vrba, G. H. Denton, T. C. Partridge, & L. C. Burckle. New Haven, CT: Yale University Press, pp. 385–424.

White, F (1983).*The Vegetation of Africa: A Descriptive Memoir to Accompany UNESCO/AETFAT/UNSO Vegetation Maps of Africa*. Paris: United Nations Educational, Scientific and Cultural Organisation.

White, T. D., Asfar, B., Dectusta, D., *et al.* (2003). Pleistocene *Homo sapiens* from Awash, Ethiopia. *Nature*, **423**, 742–7.

Woldegabriel, G. White, T. D., Suwa, G., *et al.* (1994). Ecological and temporal placement of early Pliocene hominids at Aramis, Ethiopia. *Nature*, **371**, 330–33.

Wrangham, R. W., Jones, J. H., Laden, G., Pilbeam, D., & Conklin-Brittain, N. (1999). The raw and the stolen: cooking and the ecology of human origins. *Current Anthropology*, **40**, 567–94.

Wynn, J. (2000). Paleosols, stable carbon isotopes, and paleoenvironmental interpretation of Kanapoi, Northern Kenya. *Journal of Human Evolution*, **39**, 411–32.

18 *Orbital controls on seasonality*

JOHN D. KINGSTON
Department of Anthropology, Emory University, Atlanta GA 30322, USA

Introduction

Given the significant influence of seasonality patterns on many aspects of modern human and non-human primate tropical ecology (Foley 1993; Jablonski *et al.* 2000), it is reasonable to assume that factors associated with seasonality provided key selective forces in the evolution of the human lineage in Equatorial Africa. Reconstructing the climatic and ecological context of early hominin innovations ultimately is critical for interpreting their adaptive significance, and much research has focused on establishing links between hominin evolutionary events and global, regional, and local environmental perturbations (Brain 1981; Grine 1986; Vrba *et al.* 1989, 1995; Bromage & Schrenk 1999). Attempts to correlate hominin evolution with climatic trends have typically invoked models of progressively more arid and seasonal terrestrial conditions in Africa, ultimately resulting in the expansion of grassland ecosystems. Alternative interpretations of the Pliocene fossil record of east Africa suggest pulses (Vrba 1985) or multiple episodes (Bobe & Eck 2001; Bobe *et al.* 2002; Bobe & Behrensmeyer 2004) of high faunal turnover correlated with major global climatic change, set within a gradual shift from forest dominance to more open habitats.

While long-term trends or abrupt turnover events may have influenced human evolution, it has become evident that climatic control of mammalian evolution, including hominins, in Equatorial Africa is much more complex than supposed previously and that this region is characterized by almost continuous flux and oscillation of climatic patterns driven primarily by orbital forcing (e.g. Rossignol-Strick *et al.* 1982; Pokras & Mix 1987; deMenocal 1995; Kutzbach *et al.* 1996; Thompson *et al.* 2002; Hughen *et al.* 2004). These short-term changes in the Earth's orbit occur at varying cycles (Milankovitch cycles) ranging in duration from 20 to 400 ka and affect climate by altering the amount of solar radiation by season and latitude. While establishing links between orbital shifts and

Seasonality in Primates: Studies of Living and Extinct Human and Non-Human Primates, ed. Diane K. Brockman and Carel P. van Schaik. Published by Cambridge University Press.

climate focused initially on the extent of ice sheets in high latitudes, changes in solar radiation also have had significant, independent effects on climatic systems in middle to lower latitudes. Regional Equatorial African climatic patterns have been shown to be highly responsive to orbitally forced shifts in radiation, which have dramatic effects on seasonal variation in precipitation and temperature. The magnitude of changes in African seasonality patterns associated with short-term orbital forcing matches those previously hypothesized for much longer intervals spanning millions of years. Successfully correlating and developing causal links between specific hominin evolutionary events and variations in climate, such as seasonality, therefore, ultimately requires the development of high-resolution records of the nature and timing of climatic, ecological, and phenotypic changes in the fossil record within the temporal and dynamic framework of Milankovitch cycling.

Principal controls on the distribution of vegetation, and, by extension, dietary resources and community ecology of animal consumers, in tropical and subtropical regions of Africa are total annual rainfall and the timing, duration, and intensity of the dry season(s). The complexity and difficulty in unraveling and isolating subtle and interrelated effects of these seasonality parameters in modern African ecosystems is amplified when attempting to untangle these dynamic patterns and associations in the past. The integrity and quantity of data on general aspects of plant physiognomy and community structure in early hominin paleohabitats are sufficient at this point to initiate interpretations of hominin evolutionary innovations in a very broad ecological context. However, the lack of sufficiently developed empirical proxies for paleoseasonality makes it difficult to assess human evolution rigorously within the context of specific aspects of seasonality. Seasonality has proven to be a highly elusive feature to tease from the fossil record, due primarily to a lack of necessary temporal resolution in seasonal proxies, limited relevant data sets, and significant taphonomic biases inherent in the terrestrial fossil archive.

Given these caveats and concerns, the goal of this chapter is to explore a global and regional climatic framework based on orbital forcing (Milankovitch cycling) that highlights major factors controlling paleoseasonality in Equatorial Africa. Examining modern African climatic systems controlling seasonality provides a starting point for interpreting the evolution of regional climatic systems. These patterns are projected into the past to evaluate regional environmental oscillations linked to orbital forcing of climate and seasonality in Africa during the Pliocene and Pleistocene epochs. A key component involved in scaffolding hominin evolution within astronomically driven environmental change is establishing "ground truth," documenting specific aspects of paleoseasonality in the terrestrial fossil record that can be correlated with Milankovitch cycling. The limited number of approaches for developing these high-resolution proxies for paleoseasonality in tropical Africa is reviewed and, finally, the implications of environmental change for the evolution of African ecosystems are discussed.

Astronomical control of solar radiation

Evaluating human ecology and evolution within the framework of orbital forced seasonality shifts in Equatorial Africa ultimately requires establishing which and how specific seasonality parameters vary at the intersection of insolation oscillations with climatic systems at local, regional, and global levels. This requires a rudimentary understanding of Milankovitch cycles as well as climatic systems that have prevailed over tropical and subtropical Africa during the course of human evolution. Following is an overview of features of the Earth's orbit that alter cyclically in response to gravitational forces in the Solar System and an explanation of how these orbital shifts effect climate.

Seasonality in tropical and subtropical Africa (and globally in general) results ultimately from variation in solar radiational heating on the Earth, which primarily is a function of the geometry of the Earth's solar orbit. Two fundamental motions describe this orbit – the Earth's rotation around an axis that passes through its poles and the Earth's once-a-year revolution around the Sun (Bradley 1999; Ruddiman 2001). The tilt, or obliquity, of the Earth's axis of rotation (currently 23.5°) relative to the plane of its orbit around the Sun causes seasonal changes in solar radiation received in each hemisphere as the Earth revolves about the Sun (Fig. 18.1a). During each annual revolution around the Sun, the Earth maintains a constant tilt and a constant direction of tilt in space, such that the northern and southern hemispheres alternatively receive more direct radiation in their respective summers. In addition, the Earth's orbit is slightly eccentric or elliptical, and the distance to the Sun changes with Earth's position in its orbit, which directly affects the amount of solar radiation that the Earth receives. At present, Earth is in the perihelion position (closest to the Sun) on January 3 and in aphelion (furthest from the Sun) on July 4 (Fig. 18.1a). These dates correspond roughly to the winter solstice (shortest day) and summer solstice (longest day), respectively, in the northern hemisphere, resulting in higher winter and lower summer radiation in the northern hemisphere (the opposite is true for the southern hemisphere) than would occur in a perfectly

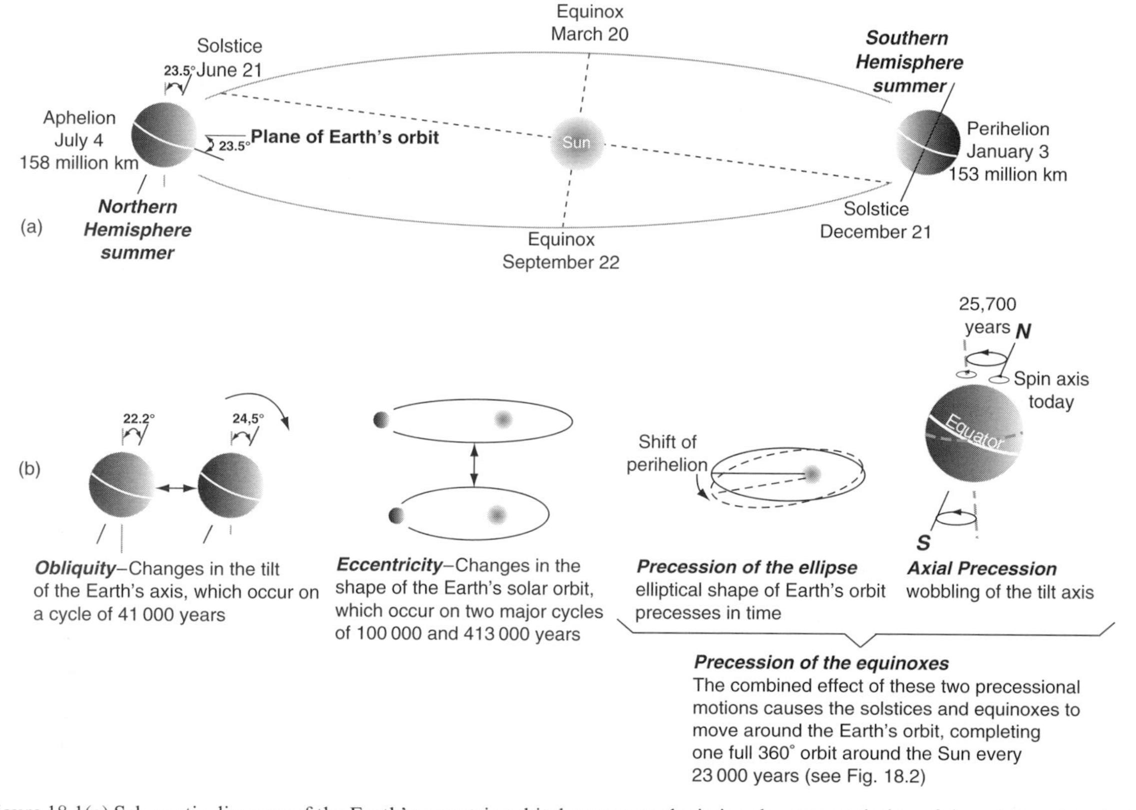

Figure 18.1(a) Schematic diagram of the Earth's eccentric orbital geometry, depicting the current timing of the solstices (shortest and longest days) and equinoxes (equal days and nights) and the tilt of the Earth's rotational spin axis relative to the plane of the Earth's orbit (23.5 °). Earth is closest to the Sun at perihelion on January 3, just after the December 21 solstice, and most distant at aphelion on July 4, shortly after the June 21 solstice. (b) Cyclical variations in the Earth's orbit (obliquity, eccentricity of orbit, relative position of the solstices and equinoxes around the elliptical orbit) due to the mass gravitational attractions among Earth, it's moon, the Sun, and other planets and their moons.

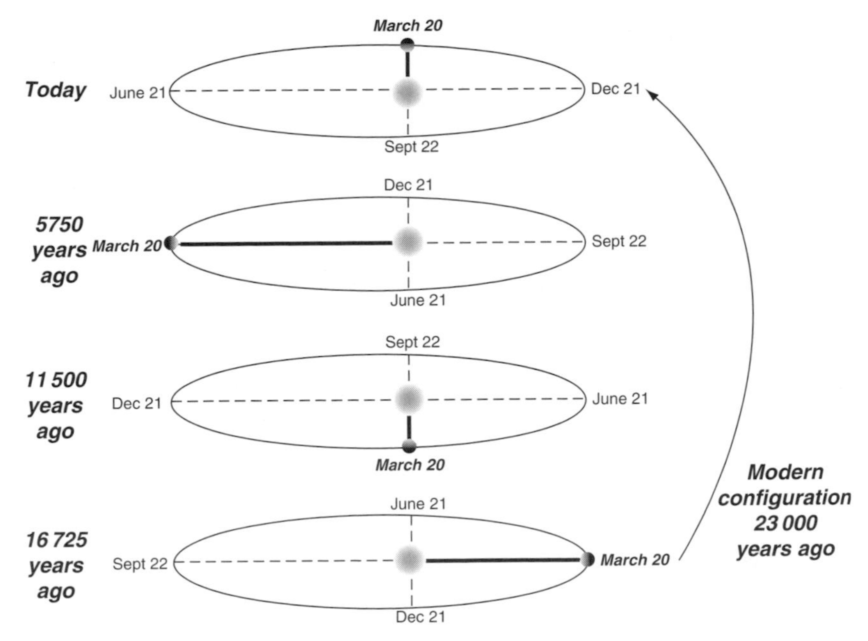

Figure 18.2 As a result of the wobble in the Earth's axis (axial precession) and precession of the Earth's elliptical orbit, the positions of the equinoxes and solstices in the Earth's orbit change slowly (precession of the equinoxes). The diagram shows how the relative position of the March 20 equinox shifts around the Earth's orbit in a 23 000-year cycle. The effects of precession of the equinoxes on radiation receipts will be modulated by the variations in eccentricity. When the orbit is near-circular, the seasonal timing of perihelion is inconsequential, but at maximum eccentricity, when differences in solar radiation may amount to 30%, seasonal timing is crucial (adapted from Imbrie & Imbrie [1979]).

circular orbit. Relative to tilt, the effect of the Earth's elliptical orbit on seasonality is small, enhancing or reducing the intensity of radiation received by only about 3–4%.

The position and orientation of the Earth relative to the Sun, however, is not fixed over long intervals of time. Due to gravitational effects on the Earth of the Sun, the Moon, and other planets, variations occur in the degree of orbital eccentricity around the Sun, the obliquity of the Earth, and the seasonal timing of perihelion and aphelion (precession of the equinoxes) (Figs. 18.1b and 18.2). These longer-term variations in Earth's orbit result in cycles ranging from about 20 ka to 400 ka and cause cyclic variations in the amount of insolation (solar radiation received at the top of the atmosphere) by latitude and season (Fig. 18.3). Changes in solar heating driven by variations in Earth's orbit are the major cause of cyclic climatic change over the ten thousand- to 100 thousand-years

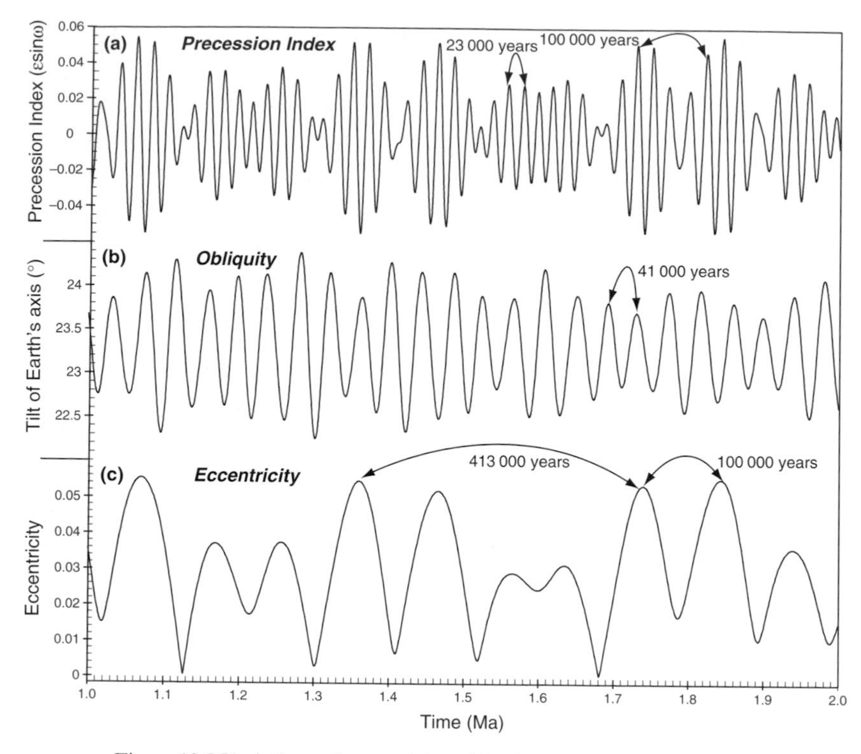

Figure 18.3 Variations of eccentricity, obliquity, and precession (precession index) between one and two million years ago (generated using AnalySeries v.1.2 [Paillard *et al.* 1996]). Superimposing these orbital variations creates complex patterns of insolation through geologic time. (a) The precessional index changes mainly at a cycle of 23 000 years, with the amplitude of the cycle modulated at eccentricity periods of 100 000 and 413 000 years. (b) Changes in axial tilt are periodic, with a mean period of 41 000 years. (c) The eccentricity (ε) of the Earth's orbit varies at periods of 100 000 and 413 000 years. Eccentricity of an ellipse is related to the lengths of its longer axis (a) and shorter axis (b), such that $\varepsilon = (a^2 - b^2)^{1/2}/a$. Eccentricity has varied between values of 0.005 and nearly 0.0607, and the current value of 0.0167 lies toward the almost circular end of the range.

timescale, and explicit theoretical links between orbital parameters and climate have been recognized and developed over the past 150 years (Croll 1875; Milankovitch 1941; Berger 1978; Imbrie & Imbrie 1979). These variations in insolation underpin the overall basis for seasonality and provide a framework in which to assess changes in seasonality in Equatorial Africa relevant to early hominin evolution.

The basics of the three key orbital variables that mediate climate change at different frequencies are summarized below (see also Figs. 18.1–18.3).

1. *Obliquity (tilt)*. Over time, the axial tilt of the Earth varies periodically in a narrow range, cycling back and forth between 22.2 and 24.5 °, with a mean period of 41 ka. Changes in tilt cause long-term variations in seasonal solar insolation received on Earth, and the main effect is to amplify or suppress the seasons. Variation in obliquity has relatively little effect on radiation receipts at low latitudes, but the effect increases towards the poles. Increased tilt amplifies seasonal differences, whereas decreases in tilt diminish the amplitude of the seasonal differences. As the tilt is the same in both hemispheres, changes in obliquity affect radiation receipts in the southern and northern hemispheres equally and cycles are fairly regular, both in period and amplitude (Figs. 18.1b and 18.3b).
2. *Eccentricity*. Variations in orbital eccentricity are quasi-periodic, with two main periods, both of which are far more irregular than the 41 000-year obliquity cycle (Figs. 18.1b and 18.3c). One eccentricity cycle consists of four cycles of nearly equal strength and with periods ranging between 95 and 131 ka, which blend to form a single cycle near 100 000 years. This cycle controls insolation variability associated with glacial/interglacial cycles. The second major eccentricity cycle has a wavelength of 413 000 years. The orbit has varied from almost circular – essentially no difference between perihelion and aphelion – to maximum eccentricity, when insolation varied by about 30% between perihelion and aphelion. Eccentricity variations affect the relative intensities of the seasons, with an opposite effect in each hemisphere.
3. *Precession*. Changes in the seasonal timing of perihelion and aphelion result from a slight wobble in the Earth's axis of rotation as it moves around the Sun. The Earth's wobbling motion (axial precession) is caused by the gravitational pull of the Sun and the Moon on the slight bulge in Earth's diameter at the Equator and effectively results in the slow turning of Earth's axis of rotation through a circular path, with one full turn every 25 700 years (Figs 18.1b and 18.2). Axial precession, when combined with the slow rotation of the elliptical shape of Earth's orbit in space in time (precession of the ellipse), causes the solstices and equinoxes to move around the Earth's orbit, completing one full 360-degree orbit around the sun every 23 000 years. This combined movement, known as the precession of the equinoxes, describes the absolute motion of the equinoxes and solstices in the larger reference frame of the universe and consists of a strong cycle near 23 000 years and a weaker one near 19 000 years, which together average one cycle every 21 700 years. Thus, roughly 11 500 years in the future, perihelion will occur when the northern hemisphere is tilted towards the Sun (mid June) rather than in the northern hemisphere's midwinter, as is the current configuration

(Fig. 18.2). Ultimately, the effects of precession of the equinoxes on insolation are modulated by changes in the eccentricity of Earth's orbit (discussed above). When the orbit is near-circular, the seasonal timing of perihelion is inconsequential, whereas at maximum eccentricity, when differences in solar radiation are up to 30%, seasonal timing is crucial. The combined effects of the precession of equinoxes and changing eccentricity are termed the precessional index or eccentricity-modulated precession. Long-term variations in the precessional index occur in cycles with periods near 23 000 years and exhibit large variations in amplitude due to the modulation effect caused by eccentricity (Fig. 18.3a).

The variability of these orbital parameters and the resulting amount of insolation arriving on Earth at any latitude and season can be calculated for any point or range of time in the past (Fig. 18.4a). In general, monthly seasonal insolation changes are dominated by precession at low and middle latitudes, whereas high latitudes are affected mostly by variations in obliquity. Eccentricity is not significant as a direct cycle of seasonal insolation and is important primarily in modulating the amplitude of the precession cycle. Considered together, the superimposition of variations in eccentricity, obliquity, and precession produces a complex, ever-varying pattern of insolation through evolutionary time. The cumulative effects of these cycles through time are complicated further by the fact that the relative strength of the various cycles can shift at any given location, there can be varying modulation of amplitudes, and there are a number of additional astronomical and Earth-intrinsic climatic feedback mechanisms that mediate external forcing. Despite this complexity, the periodicities associated with orbital variations have been identified in many paleoclimatic records and it is clear that orbital forcing is an important factor in climatic fluctuations on the timescale of 10 Ka to 1 Ma and perhaps much longer throughout most of the Earth's history. The precise mechanism of how orbital forcing is translated into a climate response remains unclear, and much research currently is focusing on exploring links between cyclic variations in insolation and climate change, including changes in seasonality through time.

Modern climate and seasonality in Equatorial Africa

Climatic and atmospheric circulation patterns of Africa today provide an appropriate construct for interpreting climatic parameters in the Pliocene and Pleistocene, given that regional paleogeography (Guiraud & Bosworth 1999) and general climatic variables have remained relatively

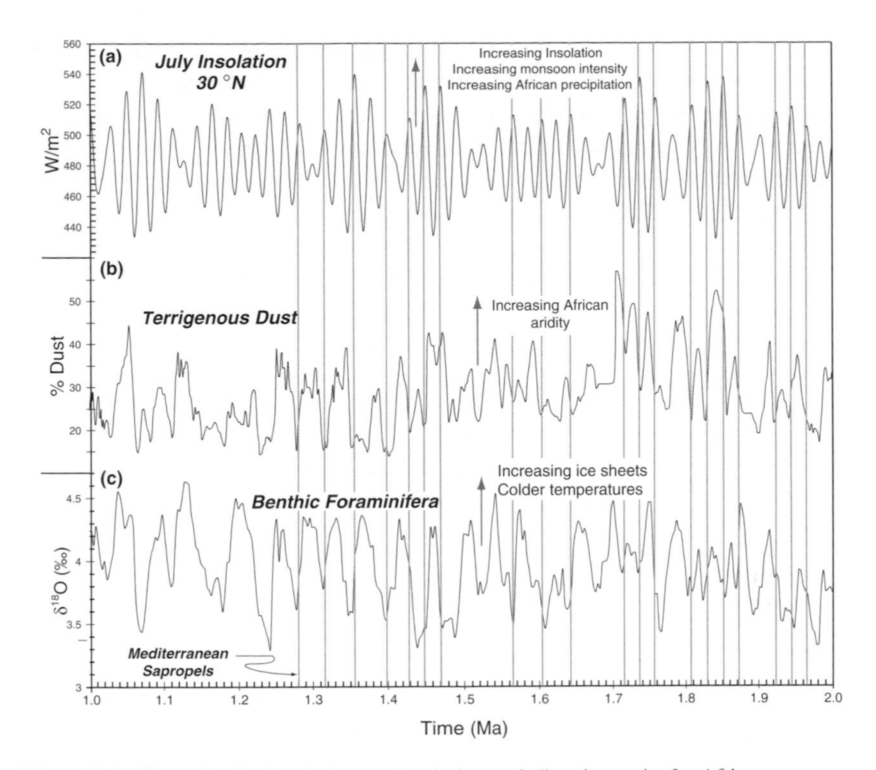

Figure 18.4 Effects of orbital variations on insolation and climatic proxies for African environments between one and two million years ago. (a) Insolation for July at a latitude of 30 °N plotted using Analyseries (Paillard *et al.* 1996). Increases in the amount of summer (e.g. July) insolation at approximately this latitude have been linked to stronger monsoonal circulation (Kutzbach 1981), resulting in increased seasonal precipitation in tropical and subtropical Africa. The amplitude of the monsoon response should be related to the amount of increase in summer insolation forcing and follow the tempo of the dominant cycle, in this case orbital precession. (b) Increasing terrigenous dust flux into marine sediments has been utilized to assess source-area aridity. Variability in the aeolian dust contribution to Ocean Drilling Program (ODP) core sites 721/722 in the Arabian Sea correlates with periodicities associated with orbital forcing. Aeolian variability before 2.8 Ma occurs at the 23–19-ka precession cycle but then shifts to increased 41-Ka variance, with further increases in 100-ka variance after 0.9 Ma, suggesting that the African climate, including seasonality, became more arid and dependent on high-latitude climatic forcing (deMenocal 1995). (c) The oxygen isotope record ($\delta^{18}O$) of foraminifera shells provides a proxy for past global climate states. These data record combined effects of changing ice sheet size, changing sea-water temperature, and changing salinity. Shown here is a compilation of isotopic values from sites ODP 677 and ODP 846 in the eastern Pacific (Shackleton 1995) and indicate variable global temperatures associated with orbital forcing. Global cooling, as reflected by this record, typically is associated with increasing aridity and seasonality in low latitudes, which generally is supported by comparison of the oxygen isotope and dust flux profiles depicted here. Sapropels (organic-rich mud layers) identified in eastern Mediterranean Sea sediment cores have been linked to high freshwater and organic influx from the Nile. As the Nile headwaters drain the east African highlands, this record reflects east African precipitation patterns (Rossignol-Strick *et al.* 1982). Plotted on the figure are sapropel occurrences between 1 and 2 Ma as vertical lines based on the timescale of Lourens *et al.* (1996). Sapropels occur at regular 23 000-year intervals that coincide with summer insolation maxima as well as high African lake levels. Also apparent is that sapropels correlate roughly with low aeolian dust influx, both indicative of moist conditions.

conservative during this interval. In general, the distribution of modern climates is more or less symmetric about the Equator (Fig. 18.5a). The northern and southern extremities of the continent project into temperate areas and experience Mediterranean summer-dry climates. These temperate areas are bordered on the Equator-ward side by subtropical deserts, the Sahara Desert to the north, and the Namib coastal desert to the southwest. A wide belt of tropical climates separates the two arid subtropical zones, forming the climatic region in which much of early hominin evolution likely occurred. Almost all hominin material recovered outside of South Africa is from localities associated with the eastern branch of the East African Rift System, and most interpretations of early human evolution have focused on reconstructions of east African ecosystems. However, recent hominin discoveries in Chad and a complete lack of fossils that can be linked unequivocally to *Pan*, *Gorilla*, or a last common ancestor of the African ape and human lineages open the possibility that significant events in human evolution may have occurred in other portions of tropical and subtropical Africa where fossil recovery is limited by preservational and recovery biases.

Equatorial African climate currently is controlled by the intersection of three major air streams and three convergence zones, superimposed on and influenced by regional factors associated with topography, lakes, coastal currents and upwelling, and sea-surface temperature fluctuations in the Indian and Atlantic oceans (Nicholson 1996a). As a result, climatic patterns are markedly complex and highly variable. In general, most of Africa is characterized by a strong seasonal cycle in rainfall regime and a pronounced dry season. The pattern of wet and dry seasons is produced by a shift of all general circulation features toward the summer hemisphere (Fig. 18.5b,c). In tropical Africa, precipitation patterns are controlled mainly by the strength of the African-Asian monsoonal circulation and the seasonal migration of the Intertropical Convergence Zone (ITCZ), produced when the northeast and southeast trade winds meet. The zone of maximum rainfall follows the latitudinal position of the overhead Sun, resulting in a general bimodal seasonal distribution of rainfall, with maxima occurring in the two transitional seasons (April–May and October–November) associated with the passage of the ITCZ. Seasonal distribution of rainfall varies, in general, along a west–east gradient across Equatorial Africa due to different circulation patterns, rift-related topography, and large inland lakes that may modify African monsoon flows and the influence of the Indian monsoon. In addition, aspects of seasonality change dramatically within distances on the order of tens of kilometers, and there are regions in Equatorial Africa with one, two, and

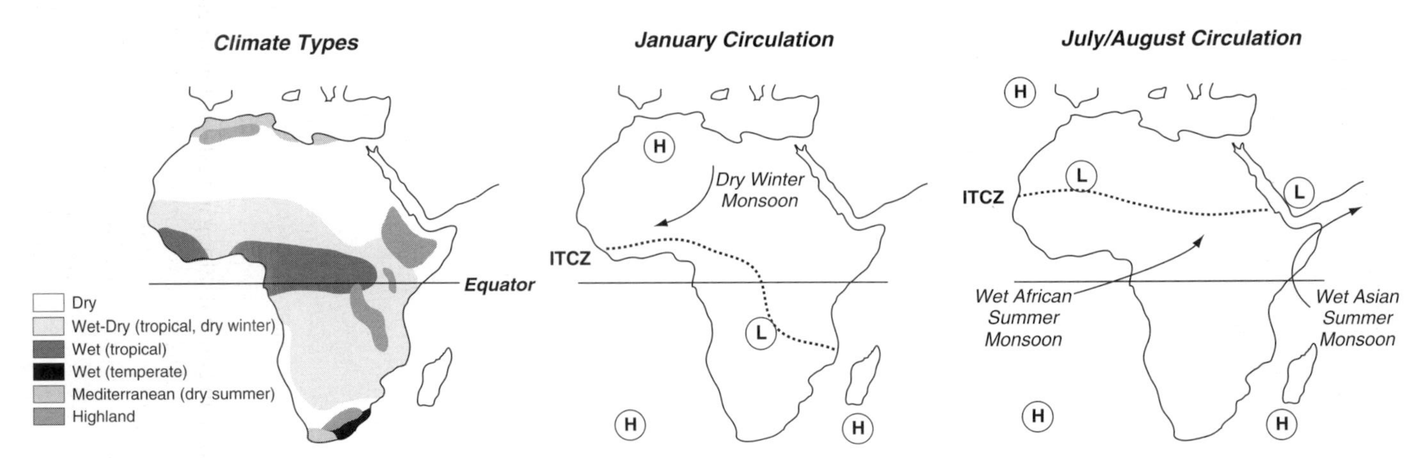

Figure 18.5 (a) Generalized climatic types in Africa (Nicholson 1996b). (b,c) Monsoonal circulation patterns over Africa, showing a moist inflow of monsoonal air toward the low-pressure center over north Africa in boreal (northern hemisphere) summer and a dry monsoonal outflow from the high-pressure center over the land in boreal winter. Precipitation follows the northward shift of the Intertropical Convergence Zone (ITCZ) in boreal summer and Equatorial regions; some areas experience two rainfall maxima during the course of the year as the ITCZ traverses the Equator as it shifts toward the summer hemisphere. Superimposed on this basic pattern are numerous localized patterns of rainfall dictated by topographic relief, shoreline and maritime effects, and local wind systems (Nicholson 1996a).

even three rainfall maxima during the seasonal cycle (Nicholson 1996a). The more semi-arid regions of Equatorial Africa are notable in their extreme year-to-year rainfall variability, and drought is a common occurrence. These localized variations in precipitation probably relate to factors such as topographic relief, shoreline and maritime effects, and local wind systems.

Evidence of orbital forcing and seasonality in Equatorial Africa

Solar radiation is significant in sustaining tropical convergence, and studies of shifts in Equatorial African paleoenvironments and paleoclimate have implicated orbitally forced changes in insolation. Although obliquity (41 ka), eccentricity (100 ka), and precession (23 ka) signals have all been identified in paleoclimatic proxies, Pliocene and Pleistocene circulation in the tropics appears to have been paced primarily by precessional variations in insolation (Fig. 18.4) (Pokras & Mix 1985, 1987; Bloemendal & deMenocal 1989; Molfino & McIntyre 1990; Tiedemann *et al.* 1994).

Equatorial African climate is affected not simply by insolation changes at low-latitudes but also by mid- to upper-latitude insolation, which controls systems such as the Asian monsoon and Atlantic sea-surface temperatures that are central ingredients in African climate. Climatic flux in Equatorial Africa has been linked to high-latitude glacial–interglacial amplitudes, low-latitude precessionally controlled insolation changes driving monsoonal circulation intensity, and direct insolation at the Equator at half-precession cycles. Climatic proxies including eolian and biogenic dust transport (Pokras & Mix 1987; Clemens *et al.* 1991; deMenocal *et al.* 1993; deMenocal 1995; Moreno *et al.* 2001), upwelling patterns adjacent to the African coast (Gorgas & Wilkens 2002), biochemical and biotic bedding patterns in coastal sediments (D'Argenio *et al.* 1998), and rhythmic stratification patterns in subtropical African lakes (Scholz 2000; Johnson *et al.* 2002; Ficken *et al.* 2002) indicate eccentricity and obliquity amplitudes or timing that suggest that sub-Saharan moisture patterns are linked to glacial/interglacial cycling. Numerous studies have also shown that low-latitude precessionally controlled insolation changes driving monsoonal circulation intensity exert control over hydrological patterns in Equatorial Africa (Street-Perrott & Harrison 1984; Kutzbach & Street-Perrott 1985; Pokras & Mix 1985, 1987; Gasse *et al.* 1989; Lamb *et al.* 1995). Other studies (Trauth *et al.* 2003) suggest that precipitation patterns are, in part, also triggered by increased March or September insolation on the Equator, which intensifies the intertropical convergence and convective rainfall in the region.

The coexistence of all main Milankovitch cycles in spectral data may provide evidence for a complex interdependence and influence of high- and low-latitude orbital parameters influencing climatic regimes in Equatorial Africa. Fluctuations in precipitation linked to the El Niño Southern Oscillations (Verschuren *et al.* 2000), threshold effects resulting in abrupt transitions (deMenocal *et al.* 2000; Thompson *et al.* 2002), tentative correlations with millennial-scale phenomena such as Heinrich events and Dansgaard–Oeschger cycles (Stager *et al.* 2002), and linear and non-linear Earth-intrinsic feedback mechanisms of the climatic system (Kutzbach *et al.* 1996; Gorgas & Wilkens 2002) further complicate simple interpretations of external forcing of tropical African paleoclimates.

Pliocene and Pleistocene evidence of short-term climatic change in the East African Rift has been documented at a number of early hominin sites, including Olduvai (Hay 1976; Liutkus *et al.* 2000; Ashley & Driese 2003), Olorgesailie (Potts *et al.* 1999), and Turkana (Feibel *et al.* 1989; Bobe & Behrensmeyer 2004), but it has been difficult to link these perturbations to astronomically mediated insolation patterns and seasonality patterns unequivocally. Studies in the Baringo Basin of the Central Kenyan Rift Valley have identified lake cycling 2.5 and 2.7 Ma at hominin localities consistent with 23-ka Milankovitch precessional periodicity (Kingston *et al.* 2000). These shifts in paleohydrological patterns in the Baringo Basin represent a window on a much more pervasive pattern of environmental change that is generally obscured locally and regionally by rifting activity. DeMenocal (2004) notes that the periodicity of lacustrine intervals at Lokochot in the Turkana Basin (Kenya) and the Middle Awash sequences in Ethiopia compares favorably with precessional cycles documented in the marine paleoclimate record during this time. Research at Koobi Fora (Turkana Basin) indicates lacustrine transgressions occurring at the precessional cycle, while a regressive tendency may occur at the 100 ka eccentricity periodicity (Lepre & Quinn 2005). Environmental variability, reconstructed from pollen data at Hadar, corresponds with global climatic events documented in the marine record and may track a precessional cycle (Bonnefille *et al.* 2004).

Determining the specific effects of astronomical cycling on climatic patterns is complex and remains poorly understood. Pleistocene and Holocene lake-level fluctuations in Equatorial Africa document important changes in the precipitation–evaporation balance and indicate that the timing and duration of humid and dry periods are controlled by a number of factors, including Equatorial insolation, northern hemisphere insolation, and monsoonal intensity (Fig. 18.4) (Street-Perrott & Harrison 1984; Pokras & Mix 1985, 1987; Trauth *et al.* 2003). Interannual variation in rainfall amounts indicated by high lake stands is suspected to result mainly from prolonged rainy seasons rather than more intense rainfall. Precession-modulated cycling of precipitation and seasonality patterns in the rift valley presumably resulted in dynamic fluctuations in the physiognomy and taxonomic representation of flora in the region, with more arid and seasonal climatic regimes associated with more widespread open woodland and grassland habitats. The main response of vegetation to short-term climatic variability would likely have involved a reshuffling of the relative coverage of different plant communities rather than whole-scale extinction or speciation events. Studies of Quaternary paleovegetation records indicate that Equatorial African ecosystems are highly sensitive to orbitally forced climate change, associated atmospheric CO_2 shifts, and vegetation–soil feedbacks, resulting in rapid shifts in pollen assemblage (Lezine 1991; Bonnefille & Mohammed 1994; Elenga *et al.* 1994), charcoal fluxes (Verardo & Ruddiman 1996), and relative proportions of C_3 (trees, shrubs, cold-season grasses and sedges) and C_4 (warm-season grasses and sedges) biomarkers (Huang *et al.* 1999; Ficken *et al.* 2002; Schefuss *et al.* 2003).

Cumulatively, these data suggest that Milankovitch cycles are highly relevant to understanding variation in climate and seasonality in equatorial Africa and that hominin evolution in the rift valley, and tropical Africa in general, should be evaluated in the context of continuous orbitally forced environmental flux as well as long-term ($>10^5$ years) trends.

Proxies for paleoseasonality

While evidence suggests that orbital forced climate change was significant throughout hominin evolution in Equatorial Africa, it remains essential to document the specific effects of these oscillations on paleoseasonality at discrete times and locations relevant to human evolution. The fundamental constraint in characterizing seasonality in the past, especially in the terrestrial realm, is accessing climatic archives that resolve at the level of weeks or at least less than one month. Proxies for intra-annual variation in precipitation, temperature, humidity, and precipitation/evaporation ratios and resultant shifts in vegetation and the diet, ranging behavior, life history patterns, and stress indicators of animals provide means of assessing the timing, intensity, and relative lengths of wet/dry portions of seasonality. A number of approaches to documenting seasonality in the terrestrial fossil record are being developed, including: (i) isotope analyses of monthly growth increments in invertebrate shells that correlate with rainfall patterns (Rye & Sommer 1980; Abell *et al.* 1995; Tojo & Ohno 1999; Rodrigues *et al.* 2000); (ii) periodicity of repetitive linear enamel hypoplasia in primate teeth that have been linked to seasonal fluctuations in environmental parameters (Macho *et al.* 1996, 2003; Dirks *et al.* 2002; Skinner & Hopwood 2004); (iii) serial microsampling of vertebrate enamel to examine intra- and intertooth isotopic variability reflecting seasonal shifts in diet, migratory patterns, and precipitation sources during the lifetime of the animal (Longinelli 1984; Luz & Kolodny 1985; Koch *et al.* 1989; Kohn 1996; Bryant & Froelich 1996; Kohn *et al.* 1996, Fricke *et al.* 1998; Sharp & Cerling 1998; Sponheimer & Lee-Thorp 1999; Gadbury *et al.* 2000; Kingston 2003); and (iv) fluctuations in paleolimnological proxies such as sedimentary accumulation rates, lithologic patterns, molecular and compound-specific analyses of biomarkers in sediments, pollen assemblages, diatom taxonomy, and paleolakes levels (Pilskahn & Johnson 1991; Owen & Crossley 1992; Huang *et al.* 1999; Verschuren *et al.* 2000; Ficken *et al.* 2002; Johnson *et al.* 2002; Hughen *et al.* 2004).

These techniques for characterizing seasonality in the past hold great potential, despite the complications and confounding variables inherent in the approaches. Currently, proxies for paleoseasonality in Equatorial Africa are poorly developed and are too scarce in space or time to construct a comprehensive profile of seasonality parameters or patterns during the Late Miocene and Pliocene epochs. What is suggested by existing data sets is varying patterns and intensity of seasonality at various localities through time.

Climate change and evolution of Equatorial African ecosystems

If short-term climatic fluctuations result in oscillating seasonality patterns, then what specifically are the evolutionary consequences for mammalian paleocommunities in general and hominins in particular? Mean duration of fossil mammalian species typically ranges from 1 to 3 Ma (Van Valen 1985; Sepkoski 1992), and most species have experienced many orbitally forced climatic oscillations on the order of 10^5 to 10^7 years. Various adaptive strategies have been proposed as options for terrestrial taxa faced with environmental instability, including habitat tracking, phenotypic plasticity, developmental plasticity, and behavioral flexibility (generalism). It has been argued that recurrent and rapid climatic oscillations favor speciation, resulting in higher taxonomic diversity and phenotypic disparity (Valentine 1984; Vrba 1992; Moritz *et al.* 2000). Shifting periodicities can affect the duration of phases during which species' geographic distributions remain continuously fragmented and organisms must exist beyond vicariance thresholds (outside optimal range), increasing the incidence of speciation and extinction (Vrba 1995). Others have suggested that orbitally forced climatic shifts generally increase gene flow directly by reshuffling gene pools and that any microevolutionary change that accumulates on a timescale of thousands of years consequently is lost as communities reorganize following climate change (Bennett 1990). Fluctuations select indirectly for vagility (dispersal ability and propensity) and generalism, both of which reduce extinction in the long run and also slow speciation rates (Dynesius & Jansson 2000; Jansson & Dynesius 2002). According to the variability selection hypothesis (Potts 1998), the complex intersection of orbitally forced changes in insolation and Earth-intrinsic feedback mechanisms results in extreme, inconsistent environmental variability selecting for behavioral and morphological mechanisms that enhance adaptive variability. Potts (1998) notes specifically that orbital-scale variations alter landscapes and resources to a far greater extent than do seasonal or annual variations, with the former resulting in overall revision of regional and local hydrology, vegetation, and animal communities rather than simply shifts in temperature and moisture. Evidence from other research suggests that specific taxa appear to respond individually to climate change and that communities are rearranged massively and repeatedly (Graham 1986; Bennett 1990). Strong selective pressures may develop in response to cumulative effects or shifts in mode of dominant oscillations, resulting in differential response of various lineages. Assessing the relative validity of these evolutionary models ultimately hinges on developing a level of temporal and spatial resolution in the faunal record to anchor speciation and extinction events in the context of short-term environment shifts that specifically include seasonality patterns.

Summary

Climatic and geological instability have characterized much of Equatorial Africa during the course of hominin evolution over the past 8–7 Ma. Combined with regional and local habitat heterogeneity, environmental flux has resulted in highly dynamic selective forces on early hominins and their communities. An emerging consensus is that although long-term global and regional climatic trends and specific events have influenced hominin evolution, short-term astronomically driven climatic change coupled with intrinsic feedback loops, threshold effects, and local or regional factors may represent a more significant factor in forcing evolutionary change. This climatic flux can be linked to cyclical shifts in the geometry of the Earth's solar orbit, which alter the amount of solar radiation reaching different latitudes and hemispheres at different times of the year. These changes ultimately affect systems that control Equatorial African climate, such as the intensity and nature of African-Asian monsoonal circulation and the seasonal migration of the ITCZ. These regional changes are experienced by ecosystems in tropical and subtropical Africa primarily by dynamic shifts in seasonality patterns, specifically the length, intensity, and duration of wet–dry seasons. Consequently, paleoseasonality in Equatorial Africa during hominin evolution must be assessed at the temporal scale of orbital cycles or even millennial events. Given the intimate connections and significant effect of seasonality patterns on primate foraging, ranging, reproduction, and social strategies, even subtle changes in seasonality patterns could significantly alter selective pressures. Invoking notions of increasing seasonality over timescales of a million years or more, while perhaps true, does not necessarily inform attempts to provide adaptive scenarios for hominin evolutionary innovations, given the occurrence of short-term changes in seasonality of the same or even greater magnitude. Within the framework of climatic flux in Equatorial Africa, manifested to a large extent by oscillations in seasonality, the challenge remains to identify and link specific shifts in early hominin ecological paleocommunities, intraspecific diversity, and biogeographic patterns to specific varying seasonality parameters. A lack of developed paleoseasonality records from low-latitude regions such as Africa currently limits our understanding of specific effects of global climatic variability, including Milankovitch cycling, on precipitation patterns in Equatorial Africa and ultimately its effect on early hominin evolution.

References

Abell, P. I., Amegashitsi, L., & Ochumba, P. B. O. (1995). The shells of *Etheria elliptica* as recorders of seasonality at Lake Victoria. *Palaeogeography, Palaeoclimatology, Palaeoecology*, **119**, 215–19.

Ashley, G. M. & Driese, S. G. (2003). Paleopedology and paleohydrology of a volcaniclastic paleosol interval; implications for early Pleistocene stratigraphy and paleoclimate record, Olduvai Gorge, Tanzania. *Journal of Sedimentary Research*, **70**, 1065–80.

Bennett, K. D. (1990). Milankovitch cycles and their effects on species in ecological and evolutionary time. *Paleobiology*, **16**, 11–21.

Berger, A. L. (1978). Long-term variations of caloric insolation resulting from the Earth's orbital elements. *Quaternary Research*, **9**, 139–67.

Bloemendal, J. & deMenocal, P. (1989). Evidence for a change in the periodicity of tropical climate cycles at 2.4 Myr from whole-core magnetic susceptibiliby measurements. *Nature*, **342**, 897–9.

Bobe, R. & Behrensmeyer, A. K. (2004). The expansion of grassland ecosystems in Africa in relation to mammalian evolution and the origin of the genus *Homo*. *Palaeogeography, Palaeoclimatology, Palaeoecology*, **207**, 399–420.

Bobe, R. & Eck, G. G. (2001). Responses of African bovids to Pliocene climatic change. *Paleobiology Memoirs*, **27** (suppl. 2), 1–47.

Bobe, R., Behrensmeyer, A. K., & Chapman, R. (2002). Faunal change, environmental variability and late Pliocene hominin evolution. *Journal of Human Evolution*, **42**, 475–97.

Bonnefille, R. & Mohammed, U. (1994). Pollen-inferred climatic fluctuations in Ethiopia during the last 3000 years. *Palaeogeography, Palaeoclimatology, Palaeoecology*, **109**, 331–43.

Bonnefille, R., Potts, R., Chalie, F., Jolly, D., & Peyron, O. (2004). High-resolution vegetation and climate change associated with Pliocene *Australopithecus afarensis*. *Proceedings of the National Academy of Sciences, USA*, **101**, 12125–9.

Bradley, R. S. (1999). *Paleoclimatology: Reconstructing Climates of the Quaternary*. San Diego: Harcourt/Academic Press.

Brain, C. K. (1981). The evolution of man in Africa: was it a consequence of Cainozoic cooling? *Transactions of the Geological Society of South Africa*, annex 84, 1–19.

Bromage, T. G. & Schrenk, F. (1999). *African Biogeography, Climate Change, and Human Evolution*. New York: Oxford University Press.

Bryant, J. D. & Froelich, P. N. (1996). Oxygen isotope composition of human tooth enamel from medieval Greenland: linking climate and society: comment and reply. *Geology*, 477–9.

Clemens, S. C., Murray, D. W., & Prell, W. L. (1991). Forcing mechanisms of the Indian Ocean monsoon. *Nature*, **353**, 720–5.

Croll, J. (1875). *Climate and Time*. New York: Appleton & Co.

D'Argenio, B., Fischer, A. G., Richter, G. M., *et al.* (1998). Orbital cyclicity in the Eocene of Angola: visual and image-time-series analysis compared. *Earth and Planetary Science Letters*, **160**, 147–61.

DeMenocal, P. B. (1995). Plio-Pleistocene African climate. *Science*, **270**, 53–9.

— (2004). African climate change and faunal evolution during the Pliocene–Pleistocene. *Earth and Planetary Science Letters*, **220**, 3–24.

DeMenocal, P. B., Ruddiman, W. F., & Pokras, E. M. (1993). Influences of high- and low-latitude processes on African terrestrial climate: Pleistocene eolian records from equatorial Atlantic Ocean Drilling Program Site 663. *Paleoceanography*, **8**, 209–42.

DeMenocal, P., Ortiz, J., Guilderson, T., *et al.* (2000). Abrupt onset and termination of the African Humid Period: rapid climate responses to gradual insolation forcing. *Quaternary Science Reviews*, **19**, 347–61.

Dirks, W., Reid, D. J., Jolly, C. J., Phillips-Conroy, J. E., & Brett, F. L. (2002). Out of the mouths of baboons: stress, life history, and dental development in the Awash National Park Hybrid Zone, Ethiopia. *American Journal of Physical Anthropology*, **118**, 239–52.

Dynesius, M. & Jansson, R. (2000). Evolutionary consequences of changes in species' geographical distributions driven by Milankovitch climate oscillations. *Proceedings of the National Academy of Sciences, USA*, **97**, 9115–20.

Elenga, H., Schwartz, D., & Vincens, A. (1994). Pollen evidence of late Quaternary vegetation and inferred climate changes in Congo. *Palaeogeography, Palaeoclimatology, Palaeoecology*, **109**, 345–56.

Feibel, C. S., Brown, F. H., & McDougall, I. (1989). Stratigraphic context of fossil hominids from the Omo Group deposits: northern Turkana Basin, Kenya and Ethiopia. *American Journal of Physical Anthropology*, **78**, 595–622.

Ficken, K. J., Woodler, M. J., Swain, D. L., Street-Perrott, F. A., & Eglinton, G. (2002). Reconstruction of a subalpine grass-dominated ecosystem, Lake Rutundu, Mount Kenya: a novel. *Palaeogeography, Palaeoclimatology, Palaeoecology*, **177**, 137–49.

Foley, R. (1993). The influence of seasonality on hominid evolution. In *Seasonality and Human Ecology*, ed. S. J. Ulijaszek & S. Strickland. Cambridge: Cambridge University Press, pp. 17–37.

Fricke, H. C., Clyde, W. C., & O'Neil, J. R. (1998). Intra-tooth variations in 18O (PO4) of mammalian tooth enamel as a record of seasonal variations in continental climate variables. *Geochimica et Cosmochimica Acta*, **62**, 1839–51.

Gadbury, C., Todd, L., Jahre, A. H., & Amundson, R. (2000). Spatial and temporal variations in the istopic composition of bison tooth enamel from the Early Holocene Hudson-Meng Bone Bed, Nebraska. *Palaeogeography, Palaeoclimatology, Palaeoecology*, **157**, 79–93.

Gasse, F., Ledee, V., Massault, M., & Fontes, J.-C. (1989). Water-level fluctuations of Lake Tanganyika in phase with oceanic changes durig the last glaciation and deglaciation. *Nature*, **342**, 57–9.

Gorgas, T. J. & Wilkens, R. H. (2002). Sedimentation rates off SW Africa since the late Miocene deciphered from spectral analyses of borehole and GRA bulk density profiles: ODP sites 1081–1084. *Marine Geology*, **180**, 29–47.

Graham, R. W. (1986). Response of mammalian communities to environmental changes during the late Quaternary. In *Community Ecology*, ed. J. Diamond, & T. J. Case. New York: Harper and Row, pp. 300–13.

Grine, F. E. (1986). Ecological causality and the pattern of Plio-Pleistocene hominid evolution in Africa. *South Africa Journal of Science*, **82**, 87–9.

Guiraud, R. & Bosworth, W. (1999). Phanerozoic geodynamic evolution of northeastern Africa and the northwestern Arabian platform. *Tectonophysics*, **315**, 73–108.

Hay, R. L. (1976). *Geology of Olduvai Gorge*. Berkeley: University of Berkeley Press.

Huang, Y., Street-Perrott, F. A., Perrott, R. A., Metzger, P., & Eglinton, G. (1999). Glacial-interglacial environmental changes inferred from molecular and compound-specific $d^{13}C$ analyses of sediments from Sacred Lake, Mt. Kenya. *Geochimica et Cosmochimica Acta*, **63**, 1383–404.

Hughen, K. A., Eglinton, T. I., Xu, L., & Makou, M. (2004). Abrupt tropical vegetation response to rapid climate changes. *Science*, **304**, 1955–9.

Imbrie, J. & Imbrie, K. P. (1979). *Ice Ages: Solving the Mystery*. Short Hills, NJ: Enslow.

Jablonski, N. G., Whitford, M. J., Roberts-Smith, N., & Qinqi, X. (2000). The influence of life history and diet on the distribution of catarrhine primates during the Pleistocene in eastern Asia. *Journal of Human Evolution*, **39**, 131–57.

Jansson, R. & Dynesius, M. (2002). The fate of clades in a world of recurrent climate change: Milankovitch oscillations and evolution. *Annual Review of Ecology and Systems*, **33**, 741–77.

Johnson, T. C., Brown, E. T., McManus, J., *et al.* (2002). A high-resolution paleoclimate record spanning the past 25,000 years in southern East Africa. *Science*, **296**, 113–32.

Kingston, J. D. (2003). Sources of variability in modern East African herbivore enamel: implications for paleodietary and palaeoecological reconstructions. *American Journal of Physical Anthropology, Supplement*, **36**, 130.

Kingston, J. D., Hill, A., & Deino, A. L. (2000). Pliocene hominid evolution and astronomically forced climate change: evidence from the rift valley of Kenya. *Journal of Human Evolution*, **38**, A15.

Koch, P. L., Fisher, D. C., & Dettman, D. (1989). Oxygen isotope variation in the tusks of extinct proboscideans: a measure of season of death and seasonality. *Geology*, **17**, 515–19.

Kohn, M. J. (1996). Predicting animal 18O: accounting for diet and physiological adaptation. *Geochimica et Cosmochimica Acta*, **60**, 4811–29.

Kohn, M. J., Schoeninger, M. J., & Valley, J. W. (1996). Herbivore tooth oxygen isotope compositions: effects of diet and physiology. *Geochimica et Cosmochimica Acta*, **60**, 3889–96.

— (1998). Variability in oxygen isotope compositions of herbivore teeth: reflections of seasonality or developmental physiology? *Chemical Geology*, **152**, 97–112.

Kutzbach, J. E. (1981). Monsoon climate of the Early Holocene: climate experiment with Earth's orbital parameters for 9000 years ago. *Science*, **214**, 59–61.

Kutzbach, J. E. & Street-Perrott, R. A. (1985). Milankovitch forcing of fluctuations in the level of tropical lakes from 19 to 0 kyr BP. *Nature*, **317**, 130–34.

Kutzbach, J. E., Bonan, G., Foley, J., & Harrison, S. P. (1996). Vegetation and soil feedbacks on the response of the African monsoon to orbital forcing in the early to middle Holocene. *Nature*, **384**, 623–6.

Lamb, H., Gasse, G., Benkaddour, A., *et al.* (1995). Relationship between century-scale Holocene arid intervals in tropical and temperate zones. *Nature*, **373**, 134–7.

Lepre, C. J. & Quinn, R. L. (2005). Precessional forcing of Plio-Pleistocene lake levels from Koobi Fora, Kenya. *Paleoanthropology*, **5**, A38.

Lezine, A.-M. (1991). West African Paleoclimates during the last climatic cycle inferred from an Atlantic deep-sea pollen record. *Quaternary Research*, **35**, 456–63.

Liutkus, C. M., Ashley, G. M., & Wright, J. D. (2000). Short-term changes and Milankovitch cyclicity at Olduvai Gorge, Tanzania: evidence from sedimentology and stable isotopes. *Abstracts with Programs – Geological Society of America*, **32**, 21.

Longinelli, A. (1984). Oxygen isotopes in mammal bone phosphate: a new tool for paleohydrological and paleoclimatological research? *Geochimica et Cosmochimica Acta*, **48**, 385–90.

Lourens, L. J., Antonarakaou, A., Hilgen, F. J., *et al.* (1996). Evaluation of the Plio-Pleistocene astronomical timescale. *Paleoceanography*, **11**, 391–413.

Luz, B. & Kolodny, Y. (1985). Oxygen isotope variations in phosphate of biogenic apatites. IV: mammal teeth and bones. *Earth and Planetary Science Letters*, **75**, 29–36.

Macho, G. A., Reid, D. J., Leakey, M. G., Jablonski, N., & Beynon, A. D. (1996). Climatic effects on dental development of *Theropithicus oswaldi* from Koobi Fora and Olorgesailie. *Journal of Human Evolution*, **30**, 57–70.

Macho, G. A., Leakey, M. G., Williamson, D. K., & Jiang, Y. (2003). Palaeoenvironmental reconstruction: evidence for seasonality at Allia Bay, Kenya, at 3.9 million years. *Palaeogeography, Palaeoclimatology, Palaeoecology*, **199**, 17–30.

Milankovitch, M. M. (1941). *Canon of Insolation and the Ice-Age Problem.* Beograd: Koniglich Serbische Akademie.

Molfino, B. & McIntyre, A. (1990). Nutricline variation in the equatorial Atlantic coincident with the Younger Dryas. *Paleoceanography*, **5**, 997–1008.

Moreno, A., Targarona, J., Henderiks, J., *et al.* (2001). Orbital forcing of dust supply to the North Canary Basin over the last 250 kyr. *Quaternary Science Reviews*, **20**, 1327–39.

Moritz, C., Patton, J. L., Schneider, C. J., & Smith, T. B. (2000). Diversification of rainforest faunas: an integrated molecular approach. *Annual Review of Ecology and Systems*, **31**, 533–63.

Nicholson, S. E. (1996a). A review of climate dynamics and climate variability in eastern Africa. In *The Limnology, Climatology and Paleoclimatology of the East African Lakes*, ed. T. C. Johnson & E. O. Odada. Amsterdam: Gordon and Breach Publishers, pp. 25–56.

— (1996b). Africa. In *Encyclopedia of Climate and Weather*, ed. S. H. Schneider. New York: Oxford University Press, pp. 13–19.

Owen, R. B. & Crossley, R. (1992). Spatial and temporal distribution of diatoms in sediments of Lake Malawi, Central Africa, and ecological implications. *Journal of Paleolimnology*, **7**, 55–71.

Paillard, D., Labeyrie, L., & Yiou, P. (1996). Macintosh program performs time-series analysis. *Eos*, **77**, 379.

Pilskahn, C. H. & Johnson, T. C. (1991). Seasonal signals in Lake Malawi. *Limnology and Oceanography*, **36**, 544–57.

Pokras, E. M. & Mix, A. C.(1985). Eolian evidence for spacial variability of late Quaternary climates in tropical Africa. *Quaternary Research*, **24**, 137–49.

— (1987). Earth's precession cycle and Quaternary climatic change in tropical Africa. *Nature*, **326**, 486–7.

Potts, R. (1998). Environmental hypotheses of hominin evolution. *Yearbook of Physical Anthropology*, **41**, 93–136.

Potts, R., Behrensmeyer, A. K., & Ditchfield, P. (1999). Paleolandscape variation and early Pleistocene hominid activities: Members 1 and 7, Olorgesailie Formation, Kenya. *Journal of Human Evolution*, **37**, 747–88.

Rodrigues, D., Abell, P. I., & Kropelin, S. (2000). Seasonality in the early Holocene climate of Northwest Sudan: interpretation of *Etheria elliptica* shell isotopic data. *Global and Planetary Change*, **26**, 181–7.

Rossignol-Strick, M., Nesteroff, W., Olice, P., & Vergnaud-Grazzini, C. (1982). After the deluge: Mediterranean stagnation and sapropel formation. *Nature*, **295**, 105–10.

Ruddiman, W. F. (2001). *Earth's Climate: Past and Future*. New York: W. H. Freeman.

Rye, D. M. & Sommer, M. A. (1980). Reconstructing paleotemperature and paleosalinity regimes with oxygen isotopes. In *Skeletal Growth of Aquatic Organisms: Biological Records of Environmental Change*, ed. D. C. Rhoads & R. A. Lutz. New York: Plenum Press, pp. 169–202.

Schefuss, E., Schouten, S., Jansen, J. H. F., & Damste, J. S. S. (2003). African vegetation controlled by tropical sea surface temperatures in the mid-Pleistocene. *Nature*, **422**, 418–21.

Scholz, C. A. (2000). The Malawi Drilling Project: calibrating the record of climatic change from the continental tropics. *GSA Abstracts with Programs*, **32**, A-388.

Sepkoski, J. J. (1992). Ten years in the library: new data confirm paleontological patterns. *Paleobiology*, **19**, 43–51.

Shackleton, N. J. (1995). New data on the evolution of Pliocene climatic variability. In *Paleoclimate and Evolution with Emphasis on Human Origins*, ed. E. S. Vrba, G. H. Denton, T. C. Partridge, & L. H. Burckle. New Haven: Yale University Press, pp. 242–8.

Sharp, J. D. & Cerling, T. E. (1998). Fossil isotope records of seasonal climate and ecology: straight from the horse's mouth. *Geology*, **26**, 219–22.

Skinner, M. F. & Hopwood, D. (2004). Hypothesis for the causes and periodicity of repetitive linear enamel hypoplasia in large, wild African (*Pan troglodytes* and *Gorilla gorilla*) and Asian (*Pongo pygmaeus*) apes. *American Journal of Physical Anthropology*, **123**, 216–35.

Sponheimer, M. & Lee-Thorp, J. A. (1999). Oxygen isotopes in enamel carbonate and their ecological significance. *Journal of Archaeological Science*, **26**, 723–8.

Stager, J. C., Mayewski, P. A., & Meeker, L. D. (2002). Cooling cycles, Heinrich event 1, and the desiccation of Lake Victoria. *Palaeogeography, Palaeoclimatology, Palaeoecology*, **183**, 169–78.

Street-Perrott, F. A. & Harrison, S. A. (1984). Temporal variations in lake levels since 30,000 yr BP: an index of the global hydrological cycle. In *Climate Processes and Sensitivity*, ed. J. E. Hansen & T. Takahashi. Washington, DC: AGU, pp. 118–29.

Thompson, L. G., Mosley-Thompson, E., Davis, M., *et al.* (2002). Kilimanjaro ice core records: evidence of Holocene climate change in tropical Africa. *Science*, **298**, 589–93.

Tiedemann, R., Sarnthein, M., & Shackleton, N. J. (1994). Astronomic timescale for the Pliocene Atlantic delta 18O and dust flux record of Ocean Drilling Program Site 659. *Paleoceanography*, **9**, 619–38.

Tojo, B. & Ohno, T. (1999). Continuous growth-line sequences in gastropod shells. *Palaeogeography, Palaeoclimatology, Palaeoecology*, **145**, 183–91.
Trauth, M. H., Deino, A. L., Bergner, A. G. N., & Strecker, M. R. (2003). East African climate change and orbital forcing during the last 175 kyr BP. *Earth and Planetary Science*, **206**, 297–313.
Valentine, J. W. (1984). Neogene marine climate trends: implications for biogeography and evolution of the shallow-sea biota. *Geology*, **12**, 647–59.
Van Valen, L. M. (1985). How constant is extinction? *Evolutionary Theory*, **2**, 37–64.
Verardo, D. J. & Ruddiman, W. G. (1996). Late Pleistocene charcoal in tropical Atlantic deep-sea sediments: climatic and geochemical significance. *Geology*, **24**, 855–7.
Verschuren, D., Laird, K., & Cumming, B. F. (2000). Rainfall and drought in equatorial east Africa during the past 1,100 years. *Nature*, **403**, 410–14.
Vrba, E. (1985). Environment and evolution: alternative causes of the temporal distribution of evolutionary events. *South African Journal of Science*, **81**, 229–36.
Vrba, E. S. (1992). Mammals as key to evolutionary theory. *Journal of Mammalogy*, **73**, 1–28.
— (1995). On the connections between paleoclimate and evolution. In *Paleoclimate and Evolution with Emphasis on Human Origins*, ed. E. S. Vrba, G. H. Denton, T. C. Partridge, & L. H. Burckle. New Haven: Yale University Press, pp. 24–45.
Vrba, E. S., Denton, G. H., & Prentice, M. L. (1989). Climatic influences on early hominid behavior. *Ossa*, **14**, 127–56.
Vrba, E. S., Denton, G. H., Partridge, T. C., & Burckle, L. H. (1995). *Paleoclimate and Evolution with Emphasis on Human Origins*. New Haven: Yale University Press.

19 *What do studies of seasonality in primates tell us about human evolution?*

DIANE K. BROCKMAN
Department of Sociology and Anthropology, University of North Carolina at Charlotte, Charlotte NC 28223, USA

Introduction

Seasonality in Primates is about the impact of seasonality on the lives of primates. Because the vast majority of primates live in the tropics, most of the chapters in this book focus on aspects of seasonality in primates, including humans, and early hominins living in the tropics. Contributors to this book explore the various responses of primates to environmental seasonality, and their implications for reconstructing our own journey toward becoming human, beginning some seven million years ago. In this chapter, I synthesize the results of these contributions (Table 19.1) and examine the degree, if any, to which responses of primates to seasonality can be extrapolated to the behavior of extinct humans inhabiting mosaic habitats of the tropics.

It will become readily apparent to the reader that this endeavor yields as many questions as it provides insights, a consequence of an imperfect fossil record, the difficulties inherent in documenting seasonality in the distant past (e.g. see Chapter 18), and the paucity of empirical data available directly linking behavioral responses of hominins to environmental flux. Many of the contributions do, nevertheless, provide predictions of how hominins may have responded to increasing seasonality based upon studies of living primate and human populations, and while evaluations of these predictions are limited, they do provide new directions for future research. An important result that emerges from this synthesis is the proposition that hominins may not have responded in the same way to seasonality as did the other primates. In fact, it appears that hominins, or at least some of them, took advantage of increasing seasonality to thrive at the expense of other primates, and that the novel way of dealing with seasonality was a critical part of the hominin lineage.

Seasonality in Primates: Studies of Living and Extinct Human and Non-Human Primates, ed. Diane K. Brockman and Carel P. van Schaik. Published by Cambridge University Press.

What studies of primates may tell us about human evolution

As shown in Table 19.1, 14 of the chapters in this book focus on the responses of living primates and humans to seasonality and 3 concern the impact of seasonality on the habitats occupied by extinct primates and humans. Of the contributions focusing on living primates, seven yield potential insights (e.g. predictions) into how hominins may have responded to similar environmental regimes, three of which have some degree of paleontological support. In the discussion that follows, I focus on the seven chapters that are relevant for assessing hominin responses to increased seasonality, beginning with how habitats respond to seasonal climates, and the responses of primates to seasonal food scarcity (Chapter 2).

Primate responses to seasonal food scarcity: the importance of behavioral flexibility

The proposition that habitats vary in the degree to which food is seasonally abundant is evaluated in van Schaik and Pfannes' (Chapter 2) meta-analysis of 106 studies of the phenology of plant communities in tropical forests and woodlands (Table 19.1). Habitats with more seasonal climates (e.g. woodlands and savannas) are shown to have greater seasonality in food abundance than those in the tropics, a finding that is consistent with earlier studies utilizing smaller data sets (van Schaik *et al.* 1993). As a consequence, primates experience predictable, and occasionally less predictable (e.g. masting), periods of food scarcity, to which they respond with altered foraging strategies, including diet switching to more abundant, but less nutritious and more costly (in term of foraging time), fall-back foods, utilizing highly nutritious but difficult-to-acquire novel foods (e.g. meat), and expanding foraging range, among others.

Indeed, as Hemingway and Bynum (in Chapter 3) indicate (Table 19.1), primates exhibit remarkable behavioral flexibility in their response to food scarcity. These authors analyzed qualitative data on primate responses from 234 studies, covering 119 taxa at 109 sites, and found that most primates, especially Old World primates, respond to food scarcity by changing their diets, rather than tracking preferred foods over larger ranges or altering energetic requirements. They also found that the degree of seasonality does not predict the nature of the response. In this most comprehensive study to date on dietary flexibility, the results further strengthen the widely held view (Harding 1981; Garber 1987; Janson & Chapman 1999) that primates, including humans, are dietary generalists

Table 19.1 *Do studies of seasonality in primates provide insights into human evolution, and, if so, how? Is there fossil evidence to support predictions for hominins?*

Chapter	Summary of conclusions	Implications for human evolution?	Fossil evidence?
2	Habitats with more seasonal climates, such as woodlands and savannas, have greater seasonality in food abundance. In more seasonal habitats, switching to young leaves is less of an option (especially in much of the Neotropics).	Yes. Hominins living in more seasonal habitats faced greater challenges than their ancestors. No predictions for hominin responses to seasonality.	None currently
3	Solutions to the problem of seasonal food scarcity are diverse, including diet switching and modifying ranging patterns. Behavioral flexibility is the key to primate success under changing climatological regimes.	Yes. Hominins experiencing a lean season would be predicted to decrease day range and switch to a more vegetative diet (e.g. USOs), or migrate to a habitat where more preferred food is available.	Yes
4	Predation risk in arboreal primates is associated with the presence/absence of canopy cover. Lemurids living in seasonal habitats employ cathemerality as an antipredator strategy, whereas forest-dwelling monkeys increase vigilance in exposed canopies, lending support to the "better detection in the open" hypothesis (Di Fiore 2002).	No. But the "better detection in the open" hypothesis may apply equally to hominins residing in seasonal habitats where more open (e.g. exposed) savannas predominate.	–
5	Responses in small-bodied primates to seasonal environments include reduction of basal heat, torpor and hibernation, the latter contingent upon the ability to store fat. Benefits include reduced water loss and predation risk, but at a cost, notably the inability to respond to sensory stimuli during the dry season.	No. But early hominins would be predicted to reduce metabolic rates during the dry season and schedule reproduction (e.g. conception) during periods of food abundance. Latter supported in modern agriculturalists (Chapter 13).	–

Table 19.1 (*cont.*)

Chapter	Summary of conclusions	Implications for human evolution?	Fossil evidence?
6	Baboons are remarkably successful in the face of dramatic environmental change, as suggested by (i) flexible social structures that accommodate new ways of interacting when servicing relationships becomes too costly, as during periods of food scarcity; (ii) the ability to mediate effects of food scarcity through careful tracking and exploitation of foods as they became available, (e.g. USOs); and (iii) non-seasonal reproduction.	Yes. Early hominin behavior may have been equally flexible, incorporating handoff foraging, the use of alternative foods during food scarcity, a well-buffered social system, and the ability to alter their environment by moving to more suitable habitats.	As for Chapter 3
7	Seasonality determines the extent to which animals respond to the thermal environment, day length (i.e. latitude) being the primary thermal constraint on time budget allocations. Among temperate-living populations (e.g. DeHoop baboons), long summer day lengths permit more flexible thermoregulatory strategies, which subsequently diminish during the short winter months, when scheduling of such activities as resting becomes more problematic. Populations inhabiting marginal Equatorial habitats experience the greatest thermoregulatory constraints when little time is left over from foraging to seek shade.	Yes. Thermoregulatory responses predicted for Equatorial hominins include habitual bipedality, loss of body hair, and increased body size and physique. Colder winter nights encountered at more seasonal latitudes may have limited hominin dispersal beyond the tropics.	No
8	Chimps, baboons, and capuchins exhibit temporal variation in their tendencies to hunt, but hunting seasonality depends on the type of prey exploited. Seasonal effects can be demonstrated consistently only for chimps that hunt red colobus monkeys. Of the hypotheses advanced to explain hunting seasonality, the ecological constraints hypothesis	Yes. But hunting in extant primates *yields only limited insight* into the evolution of meat-eating in hominins due to fundamental primate–hominin differences, including (i) the location of hunts (arboreal hunts/no scavenging versus terrestrial scavenging/hunting), (ii) hunting methods (opportunistic versus planned and cooperative), and	No

	(ecological factors, i.e. fruit availability, affects social characteristics, including party size, hunting decisions) is supported most strongly at Ngogo, but it has yet to be tested elsewhere.	(iii) the importance of meat in the diet (not essential for survival versus obligatory).	
9	Among the Mardu of Australia, meat-acquisition patterns vary seasonally by sex. Women hunt with digging sticks (wana) year round, and men alternate between wana and gun hunting according to season. Sex differences in hunting strategies are a function of sex differences in social and reproductive tradeoffs, particularly those indicative of skill display. Women gain hunting status through provisioning capacity, while men gain status through superior tracking abilities, although individual differences exist within sexes.	No. The degree of variation observed in human foraging strategies derives from individual differences in foraging goals and reproductive tradeoffs. Thus, we must be cautious in applying generalities from primate responses to seasonality to other taxa, including early hominins.	–
10	Seasonal variation in birth distributions is fairly universal among primates, the consequence of females employing widely divergent resource acquisition and allocation tactics in support of reproductive effort. Tests of the income–capital continuum model show that the reproductive responses of primates to seasonality range from strict income breeders (conception is cued exogenously), to relaxed income breeders (ovarian activity cued exogenously/endogenously), to capital breeders (conception cued endogenously).	No. But early hominins would be predicted to be capital breeders like extant apes (See Chapter 12) and humans (See Chapter 13), creating a situation in which mothers with infants would face periods of very low food abundance in more seasonal habitats.	–
11	Two-thirds of the variation in birth seasonality (as measured by mean vector length) is explained by the combined effects of differences in diet, body mass, latitude, and continent. Mean birth dates in the Equatorial tropics are predicted by dates of peak food availability, whereas in non-Equatorial regions, they occur shortly before the first solstice, coincident with peak plant productivity. Small-bodied (<3 kg) income-I breeders time weaning to food abundance, whereas larger (>3 kg) income-II breeders time birth to food abundance, and capital breeders initiate reproduction 5–6 months after peak food availability.	No. But because early hominins were large-bodied Equatorial omnivores that relied on seasonally available fruits, birth seasonality would have been linked to similar energy constraints as those experienced by extant African hominoids such as chimps. Diet-switching and increased reliance on meat may have relaxed birth seasonality in *Homo*.	No

Table 19.1 (*cont.*)

Chapter	Summary of conclusions	Implications for human evolution?	Fossil evidence?
12	The energetic and reproductive responses of great apes are linked tightly to temporal-spatial variation in fruit availability and fallback foods. Energetic responses of orangs having access to lower-quality fallback foods than chimps include reduced ranging patterns and catabolizing fat stored during mast fruiting to make up for insufficient energy intake during fruit-poor periods. Conception in orangs and chimps (but not in gorillas) is timed to periods of positive energy balance associated with seasonal fruit abundance.	Yes. Great apes yield insights into possible responses of hominins to seasonal fluctuations in food availability. Importance of fallback foods and modifying ranging behavior–*Australopithecus:* (post-canine megadontia suggest low-quality fallback foods, but Conklin-Brittain *et al.* [2002] suggest increased reliance on USOs; evolution of obligate bipedality). *Homo*: processing meat using tools means that limitations are no longer seasonal.	As for Chapter 3
13	Birth seasonality in humans is universal, the consequence of the interaction of social, climatological, and energetic factors. Of these, energetic factors affecting female fecundity is supported most strongly by human data, leading to the coordination of conception with periods of increased energy availability. Ovarian function is particularly sensitive to energy balance, such that ecological constraints (e.g. food availability) on energy available for reproduction can affect probability of conception. Thus, seasonally variable ecological constraints on energy availability yield seasonal patterns of conception and births, as shown by the seasonality of ovarian function in the Lese, where a dearth and a peak in conceptions occur in the "hunger" and new harvest seasons, respectively.	Yes. Unpredictable high-amplitude variation in energy available to early hominins may have resulted in selection for an increased ability to store fat in order to survive, leading to enhanced sensitivity of female reproductive function to environmental energy constraints.	No

14	More seasonal or less productive environments are associated with reduction in body size in primates, but usually without affecting sexual dimorphism. Breeding seasonality is correlated with reduced sexual dimorphism only in multi-male groups, and female group size is not related to seasonality.	Yes. The general primate pattern does not predict the evolution of increased body size and reduced sexual dimorphism in *Homo* from its move into more seasonal environments. Alternative explanations include (i) innovative ways to exploit higher-energy fallback foods, (ii) shift in social organization toward pair-bonding/male care, and (iii) dramatic increase in group sizes, leading to scramble competition in males.	Yes (i)
15	The absence of any effect of seasonality on the presence of primates in the Neotropics and Madagascar suggests the evolution of effective coping strategies in periods of food scarcity (e.g. diet switching and metabolic adjustments). Seasonality is, however, associated with increased folivore biomass, coincident with food peaks and replenishment of energy reserves. Seasonality also affects features of primate communities, high folivore and frugivore biomass being associated with deciduousness and the ease with which frugivores can switch to abundant fallback flush in fruit-poor periods.	No. But see Chapter16 . Analyses of the impact of seasonality on primate communities in mainland Africa might yield more relevant data for hominins.	–
16	The origins and evolutionary trajectories of tarsiers, monkeys, apes, and humans are linked to global changes in climate, variation in life history, diet, and locomotion, influencing the degree to which these radiations were able to cope successfully (or not) with changing environmental conditions, including increased seasonality. Major extinctions in primate evolution are linked to loss of connected canopies, resulting from increasing seasonality and aridity.	Yes. The origin and diversification of the hominin lineage *c.* 7 Ma coincides with increasing, albeit fluctuating, environmental seasonality. Unlike other hominoids that became extinct during the late Miocene Epoch, hominins were able to overcome the constraints of slow life history through social mechanisms, particularly collective childcare. Hominin success is also attributed to reliance on, and innovative ways of processing, meat and obligate bipedality, an exaptation that resulted in global range expansion in *Homo sapiens*.	Yes

Table 19.1 (*cont.*)

Chapter	Summary of conclusions	Implications for human evolution?	Fossil evidence?
17	The impact of glacial cycles on climate, habitat structure, and resource availability are reviewed, focusing particularly on how hominins responded to increasing aridification and seasonality in Africa and Eurasia 3.2 Ma–650 Ka.	Yes. Seasonal fluctuations in resource availability resulted in major dietary and behavioral shifts in tropical hominins, ranging from increased reliance on USOs in australopithecines to confrontational scavenging in early *Homo erectus*. Dietary innovations among temperate species experiencing greater seasonal stress include tactical hunting in Neanderthals and dietary diversification in AMHS, leading to global hegemony.	Yes
18	Changes in the geometry of Earth's orbit, occurring at cycles ranging from 20 to 200 Ka (Milankovitch cycles), affect climate by altering the amount of solar radiation by season and latitude. Accumulating empirical evidence indicates that Equatorial African ecosystems are highly sensitive to orbitally forced climate change and that these perturbations have dramatic effects on seasonal variation in precipitation and temperature. The magnitude of these short-term changes in seasonality matches or exceeds changes invoked over intervals spanning millions of years.	Yes. Although long-term global and regional climate trends have influenced hominin evolution, short-term astronomically driven climatic change coupled with intrinsic feedback loops, threshold effects, and local or regional factors may represent a more significant factor in forcing evolutionary change. Correlating and establishing causal links between shifts in seasonality patterns and specific hominin evolutionary events must be considered within temporal and dynamic framework of Milankovitch cycling.	Yes

AMHS, anatomically modern *Homo sapiens*; *USO*, underground storage organ.

par excellence that cope successfully with seasonality. Faced with similar lean seasons, larger-bodied generalist hominins (e.g. australopithecines, *Homo*) occupying woodland savannah habitats would also be expected to modify their dietary patterns, including perhaps switching to more diverse vegetative diets (see Chapter 3). Of critical importance, of course, is maintaining dietary quality during periods of scarcity. For relatively large-brained primates (and hominins), the development of innovative ways of tracking (e.g. *Cebus* – spatial memory [Janson 1998]) and acquiring (e.g. extractive foraging and tool use: *Pan* – [McGrew 1992], *Pongo* [Fox *et al.*, in press]) alternative or novel foods (e.g. underground storage organs [USOs], meat) may have been a key behavioral innovation that contributed to the success of the hominin lineage.

The finding that primates, and generalist baboons in particular, can succeed and even thrive in highly seasonal savanna environments (Table 19.1, Chapter 6) offers additional evidence that behavioral flexibility is a key adaptive strategy for primates, one that presumably was embraced fully by hominins as well. In their 16-year study of savanna baboons in the Amboseli basin in East Africa, Alberts and colleagues (Chapter 6) documented the behavioral responses of females to dry-season food scarcity, specifically the impact of seasonality on activity budgets (e.g. time spent foraging, time spent in social activities, etc.). Females decrease the time they spend resting and dramatically increase their foraging times on fall-back grass corms during the dry-season; more importantly, they employ a "handoff" foraging strategy in which dietary stability is maintained through careful tracking and exploitation of a greater diversity of foods as they become available (see Chapter 6). As a consequence, savanna baboons appear to be able to maintain positive energy balance, and thus year-round reproduction, in stark contrast to their savanna neighbors, the specialist vervet monkeys. Seasonally breeding vervets at Amboseli eschewed the ubiquitous savanna grasses, preferring a more narrow range of fallback foods (e.g. *Acacia* products) during the dry season, a strategy that has had devastating consequences for populations during Acacia die-offs, when vervets have gone locally extinct (see Chapter 6).

Alberts and colleagues (Table 19.1) argue that the particular attributes shared by humans and baboons – wide geographic distribution, success in (but not restriction to) savanna habitats, and non-seasonal reproduction (but see Chapter 13) – provide a rationale for predicting similar behavioral flexibility in hominins occupying the Plio-Pleistocene savannas of Africa, including perhaps handoff foraging, well-buffered social systems, and the ability to alter their environment by moving to more suitable habitats (see below).

The challenges facing temperate-living baboons, and humans, living on the savannas of southern Africa can be especially daunting. This is because populations living there experience longer day lengths and higher thermal loads than those living closer to the Equator, resulting in individuals having fewer behavioral options available to schedule activities such as resting and feeding (Table 19.1, Chapter 7). In one of the first studies of its kind, Hill (Chapter 7) examined the thermoregulatory responses of chacma baboons to seasonal variation in temperature and day length in the Western Cape Province, South Africa, at 34 °S. Results show that while high midday thermal loads pose substantial challenges for baboons at midday, individuals can successfully mediate their effects by spending more time in the shade while resting and increasing drinking to facilitate evaporative cooling. Day length, on the other hand, was found to be the primary thermal constraint on baboon activity budgets, imposing limits on the numbers of hours per day that individuals could spend feeding, moving, grooming, and resting. In contrast to baboons living in less seasonal Equatorial habitats, those living at temperate latitudes experience a scheduling crunch when it comes to allocating time spent in certain maintenance activities (e.g. resting, foraging).

The importance of seasonal variation in day length resides in the "bottleneck" that it produces in the short winter months, when scheduling of such activities as grooming becomes problematic due to competing foraging requirements (see Chapter 7). These costs are offset, however, by the benefits afforded by long summer day lengths, affording those baboons residing in more open habitats greater flexibility in their thermoregulatory responses, including increased resting time. Those inhabiting more marginal (e.g. less productive) seasonal habitats are not so fortunate, however, because they experience the greatest thermoregulatory constraints when increased foraging time means less time available to seek shade during long hot summer days (see Chapter 7).

We might reasonably predict that hominins, like baboons, would have been equally adept at mediating the effects of increased thermal loads, such that heat stress and dehydration were of little consequence for hominins residing in more open habitats. Nonetheless, various hominin morphological innovations have been linked to avoidance of heat stress, including habitual bipedalism (Wheeler 1991), loss of functional body hair (Wheeler 1992a), body size (Wheeler 1992b), and body physique (Wheeler 1993). The higher evaporative cooling mechanisms exhibited by humans may very well have derived from hominins having to solve an acute thermoregulatory problem occasioned by the move into more open Equatorial habitats, where food was seasonally restricted. Of the morphological innovations linked to thermal stress, obligate bipedality appears to have evolved well before hominins were habitual savanna animals (e.g. 6 Ma [Galik *et al.* 2004]), suggesting that this innovation arose for reasons other than as a response to the thermal environment.

The scheduling bottleneck experienced by chacma baboons during the short winter days has implications for hominins as well, particularly if this season is accompanied by food scarcity. For hominin foragers, this would mean that they would not have enough hours of light to acquire positive energy balance, especially given the colder nights. Thus, by the time hominins entered highly seasonal regions beyond the edge of the tropics, they must have possessed highly effective foraging strategies (Chapter 17, Table 19.1), which when combined with regular use of shelters at night, freed them from the constraints of seasonally shortened day lengths.

Seasonal food scarcity and reproduction: energy balance and conception

The responses of large-bodied primates such as great apes and humans to increased seasonality are especially relevant to understanding the potential responses of hominins to environmental flux. Of particular importance is how seasonal food scarcity impacts energy balance and reproduction. Given that primates in general respond to seasonal food scarcity by resorting to fallback foods (Chapter 3, Table 19.1), it is not surprising that the quality of fallback foods may also impact the reproductive responses of great apes. As with many primates, the energetic and reproductive responses of great apes and humans appear to be linked tightly to seasonal food abundance (Table 19.1, Chapters 12 and 13).

For great apes, and for orangutans and chimpanzees in particular, temporal-spatial variation in fruit availability and the quality of fallback foods are the key ecological constraints impacting reproduction. As Knott shows in Chapter 12, Bornean orangutans and chimpanzees consume a greater diversity of fallback foods during periods of food scarcity, but those utilized by orangutans are of much lower quality, resulting in an energy deficit that must be compensated for. As a consequence, orangutans reduce their range use and catabolize fat stored during mast fruiting to get through the fruit-poor period. For both orangutans and chimpanzees, conceptions occur during periods of positive energy balance associated with seasonal fruit abundance (see Chapter 12).

The implications for human evolution that can be drawn from studies of great apes are that hominins may have been as seasonally constrained in their dietary choices as apes are today, requiring similar responses to seasonal food scarcity. Knott (Chapter 12) derives support for this idea in fossil evidence showing sharp climatically induced seasonal fluctuations in food availability 2.5, 1.7, and 1.0 Ma (deMenocal 1995; deMenocal & Bloemendal 1995). Knott proposes that early hominins may have responded to seasonal food scarcity as apes do today by utilizing fallback foods of varying quality and by reducing daily travel time (see Chapter 12). In particular, Knott cites dental evidence of post-canine megadontia (Teaford *et al.* 2002) as being indicative of the utilization of low-quality fallback foods in *Australopithecus* and the evolution of obligate bipedality as an energy-conserving behavior for exploiting patchy food resources. *Homo*, on the other hand, may have had access to higher-quality fallback foods such as meat as a buffer against the lean season, tool use and meat processing being the key innovations releasing hominins from the constraints of seasonal food scarcity. Knott acknowledges that there is little, if any, evidence from the fossil record to support the link between meat consumption and seasonal food scarcity; she does, however, speculate that the evolution of large brains may have signaled that period in hominin evolution when energy availability ceased to be a limiting factor in reproduction. One might argue, however, that energy availability continues to be a limiting factor in hominoid reproduction (see Chapter 13), particularly for those taxa whose shortened interbirth intervals reach a threshold point and energy ceases to be available to support reproduction.

Energy availability appears to be a key reproductive constraint for humans as well (Table 19.1, Chapter 13). Of the three interacting factors influencing birth seasonality – social, climatological, and energetic – the impact of energetic factors on female fecundity is supported most strongly by the human data, leading to the coordination of conception with periods of high energy availability (see Chapter 13). This appears to be the consequence of the sensitivity of ovarian function to energy balance, such that ecological constraints (e.g. food availability) on energy available for reproduction can affect the probability of conception.

Evidence of a link between seasonally available food and seasonal patterns of conception and births in humans derives from studies of seasonality of ovarian function in the Lese, a population of slash-and-burn horticulturalists living in the Ituri Forest in the Democratic Republic of the Congo (Ellison *et al.* 1989; Bailey *et al.* 1992) (Table 19.1). Results of a decade-long study showed that indices of ovarian function (e.g. estradiol and progesterone levels, frequency of ovulation, duration of menstruation) decline in parallel with weight, indicative of negative energy balance, during the January–June "hunger" season, while the average intermenstrual interval increases. With the onset of the new harvest season in June, women gain weight, thus restoring positive energy balance, and indices of ovarian function normalize (Ellison *et al.* 1989). Not surprisingly, the seasonal patterning of conception in the Lese mirrors seasonality in food abundance, with conception being virtually absent during the hunger season and peaking during the period of positive energy balance following the new harvest season (see Chapter 13). Ellison *et al.* in Chapter 13 document similar links between increased food abundance and conceptions in chimpanzees at Kibale, Uganda (Sherry 2002). Although chimpanzees are able to give birth year-round, Sherry (2002) found that conceptions peaked during periods when high-quality food was both more abundant and more frequently recorded in feeding bouts, a finding that is consistent with that reported by Knott (see Chapter 12) at other sites.

The finding of a shared pattern of linkage between seasonally abundant high-quality food and conception in such large-bodied primates as apes and humans suggests that the adaptive significance of these reproductive responses may be rooted in deep time. Studies of the impact of seasonality on reproductive function (Table 19.1, Chapter 10) and birth seasonality in primates (Table 19.1, Chapter 11) confirm this. As Brockman and van Schaik show in Chapter 10, the reproductive responses of primates to seasonality are diverse, ranging from strict income breeders, to relaxed income breeders, to capital breeders, the latter representing a "special case" that is typically found in great apes and humans (and probably hominins) where conception is cued endogenously. In fact, the only primates exhibiting this capital breeding response are the medium- to large-bodied taxa that have slow life histories, and those living in Southeast Asia (see also Chapter 11). In capital breeders, high-amplitude variation in available energy may have resulted in selection for an increased ability to store fat in order to survive, leading to enhanced sensitivity of female reproductive function to ecological energy constraints, just as seen in hominins (see Chapter 13.)

Most primates, however, rely on increased food intake during the peak lactation period, because it coincides with reliably occurring peaks in food (fruit) abundance. Thus, the seasonal patterning of reproduction in most primates is not directly relevant to understanding human evolution, because they are invariably strict or relaxed income breeders. It is instructive to note, however, that *Papio* was already large enough to be a capital breeder. Hence, it is reasonable to assume that smaller hominins living in seasonal habitats might have been capital breeders as well (their ancestors almost certainly were). Perhaps because they are capital breeders, great apes do not usually occupy highly seasonal habitats. This makes the presence of *Homo* in these habitats all the more interesting, the implications of this being that they must have successfully reduced the effects of seasonality in these habitats.

The impact of seasonality on social organization and sexual dimorphism

This discussion has thus far focused on the impact of seasonality on behavioral ecology and reproduction in primates and humans, and its implications for hominins, but as Plavcan *et al.*'s (Table 19.1, Chapter 14) comparative analysis demonstrates, environmental seasonality also appears to affect body size (but not sexual dimorphism) in primates. Morphological (e.g. canine size, craniometric data), behavioral (mating/breeding systems, group size), and ecological (e.g. annual rainfall, temperature, etc.) data were compiled from the literature, and direct comparisons between dimorphism and seasonality estimates (e.g. rainfall, temperature, estrus overlap, birth seasonality) were carried out using species values and phylogenetic contrast analysis (see Chapter 14). Results showed that seasonal or less productive environments were associated with reduced body size, particularly in females (although species living further from the Equator were also larger than those living at the Equator). Seasonality had no effect on canine or body mass dimorphism, but skull-size dimorphism was correlated directly with breeding seasonality in multi-male groups. Reduced sexual dimorphism was found to be linked tightly with increased breeding seasonality, consistent with a decrease in monopolization potential for dominant males as a result of increased breeding synchrony. Interestingly, seasonality had no impact on female group size, either between or within species (see Chapter 14).

Implications of these results for hominin evolution are that the general pattern found in primates indicates that the evolution of increased body size and reduced sexual dimorphism in *Homo* was not a response to moving into more seasonal environments (*H. erectus*, *H. ergaster* [McHenry 1992]). Instead, Plavcan *et al.* (Chapter 14) propose that increased body size in *Homo* may be associated with greater mobility and dietary flexibility, a consequence of the development of novel ways (e.g. tool use) of obtaining higher-energy fallback foods such as meat, tubers, and nuts, thus buffering them from seasonal food scarcity in more open savanna habitats. As to the reduction in sexual dimorphism observed in *Homo*, Plavcan *et al.* suggest (Table 19.1, Chapter 14) that it reflects a shift in social organization indicative of reduced male–male competition for females, deriving from a shift toward pair–bonds and male care and/or a dramatic increase in groups sizes, yielding new opportunities for scramble competition for mates.

Paleontological evaluations of the proposition that reduced sexual dimorphism in *Homo* is linked directly to shifts in lifestyle are not yet possible, in large measure because adequate proxies for mating and care-taking behavior have yet to be developed. As well, few sites yield large enough sample sizes of taphonomically associated individuals of all ages and both sexes to assess intrasexual variation and sexual dimorphism in hominins, and they are notably absent for *Homo ergaster*.

The question of the relationship between meat eating and increased body size and stature in *Homo* (e.g. *H. ergaster*) is a contentious one, reflecting differing opinions regarding mode of meat acquisition (e.g. hunting or scavenging) and consumption requirements needed to maintain adequate nutrition. Marean (1989) and Potts (1996) contend that Plio-Pleistocene climate change increased the probability that hominins engaged in confrontational scavenging 2.5–1.5 Ma, thus favoring a substantial increase in body size and the ability to exploit a broader range of habitats. Assumptions underpinning the argument that meat-eating is linked to increased body size include that meat obtained from large game was a crucial part of hominin diets and that it was regularly encountered in large enough quantities to maintain a larger body size (O'Connell *et al.* 2002a: 858). O'Connell and colleagues (2002a) argue, however, that while the archeological evidence does support aggressive scavenging in early hominins, evidence from Koobi Fora suggests that encounter rates would have been too low (e.g. one to two per month) to sustain adequate nutrition. The dry season would have posed an even greater nutritional challenge for hominins living on the savanna, the consequence of having to scavenge the meager remains of herbivores, which were themselves experiencing seasonal nutritional deficits (O'Connell *et al.* 2002a), so that without access to other high-quality foods (e.g. USOs, seeds, nuts, etc.), survival would have been dubious at best.

Today, approximately 5% of Africa's flora is edible, providing a broad array of edible plant parts, including nut-like oil seeds, fruit pulp, flowers, leaves, shoots, stems, pith, exudates, and rootstocks, among others (O'Brien & Peters 1999). Rootstocks, including USOs, are a fairly ubiquitous feature of the east African savanna during the dry season and are an important fallback food for primates living there today (see Chapter 6); it is likely that tool-using hominins accessed this fallback food as well during the dry season. Tubers are notoriously difficult to obtain, however, and often have chemical defenses that require pre-preparation (e.g. cooking) before consumption; once accessed, however, the nutritional payoff can be substantial, providing Hadza women today 1000–35 000 kcal/hour, more than enough to meet the nutritional requirements of two individuals per day (O'Connell *et al.* 2002b). One might predict that the evolution of larger body size in *Homo* may have been due to systematic exploitation of USOs in the dry season, the reduction in body-size dimorphism in this genus resulting from having regular access to high-quality fallback foods and the maintenance of positive energy balance.

Most paleontologists agree that maintaining dietary quality would have been crucial to the survival of *Homo ergaster* living in marginal habitats, and that tools were the key cultural innovation that contributed to the dietary shift toward utilization of high-quality novel foods 2.5–1.5 Ma. Beyond that, however, there is little consensus regarding the dietary niche of *Homo ergaster*, and opinions are divided sharply regarding the relative importance of meat and/or USOs in these hominins' diets (e.g. see Chapter 17) (Foley 2001; Lee-Thorp 2002; O'Connell *et al.* 2002a). Recent chemical and isotopic evidence (Lee-Thorp *et al.* 2000; Lee-Thorp 2002) suggests, however, that like their cousins the extant primates, *Homo ergaster* was a generalist, having a diet heavy in C_3 plants (e.g. trees, bushes, shrubs, forbs, herbs, and their edible components – seeds, fruits, nuts, leaves, corms), including animal sources of C_3 (e.g. browsing prey species), and supplemented with C_4 plants (e.g. grass blades, rhizomes, and seeds). This ability to maintain dietary breadth in marginal savanna habitats was arguably a consequence of a unique behavioral innovation that had biological consequences and revolutionized the hominin lineage, which, when combined with increased body size, striding bipedal locomotion, and increased cognitive ability, set our ancestors on the path toward becoming human – the use of increasingly sophisticated tools to exploit high-quality resources during seasonal food scarcity.

The paleontological evidence linking increased body size/stature in *Homo ergaster* to life history, dietary, and cultural innovations is indirect but nonetheless intriguing: it is in *Homo ergaster* that we may be seeing the first indications of a non-pongid life history evolving coincidentally with archeological evidence of meat eating, large body size, and stature, markedly increased encephalization, the development of the sophisticated Acheulean toolkit, and, perhaps, controlled use of fire (Foley 2001; O'Connell *et al.* 2002a, 2002b). The proposition that hominins may be elaborating upon what is essentially a hominoid response to seasonality is supported by the finding that that chimpanzees at Bossou, Guinea, employ tools to obtain otherwise inaccessible high-quality fallback foods (e.g. oil palm nuts and pith) during the period of fruit scarcity (Yamakoshi 2004). Mitani and Watts (Table 19.1, Chapter 6) point out, however, that accessing high-quality fallback foods through hunting is not a direct response to increased seasonality among primates generally, the implications of this being that primate hunting is a poor model for hominin hunting . Thus, although hominin behavioral flexibility certainly contributed to coping with the more seasonal habitats that they found themselves in, a systematic increase in hunting was not part of that coping, at least not during the first three or four million years.

We are only just beginning to understand the neurological correlates of tool-using behavior, but recent evidence suggests that when we use tools, our brains are assimilating the properties of those tools into neuronal space, so that as we develop increasingly sophisticated tools, our brains are reshaped and reorganized (Carmena *et al.* 2003). Arguably, this represents a watershed moment in the hominin lineage, signaling the appearance of "*cumulative* culture ... in which cultural advances were built upon progressively in a way not seen in the social tradition of other animals. Human cognition would henceforth become increasingly complex and differentiated, eventually achieving modern levels of sophistication without the need for further biological change" (Aunger 2002: 305). It is this culturally mediated response to environmental change, supported and maintained by cognitive innovation, that differentiated the hominin lineage from that of other primates ~1.7–1.5 Ma, marking that moment in our evolutionary history when hominins became emancipated (or nearly so) from the constraints of seasonal food scarcity.

Conclusions

Several important conclusions emerge from this review of the impact of seasonality on living primates, including that primates are most diverse in the humid tropics, where they are permanent tree dwellers. They cope well with seasonality, but they are for the most part limited to the Equatorial tropical forests. Far fewer species inhabit more seasonal habitats, the numbers of taxa residing in those habitats exhibiting a declining gradient, depending upon whether they occupy the dry forests, woodlands, or savannas. Yet, the primates living in those habitats cope with increasing seasonality remarkably well. Indeed, as van Schaik *et al.* (Table 19.1, Chapter 15) show, seasonality is correlated strongly with increased folivore biomass, coincident with the season of plenty when energy reserves are replenished. Much of this success is due to the ability to switch diets and exploit foods not used by competitors (other species in different habitats).

The reproductive responses of primates to seasonality also yield intriguing new insights. Capital breeding appears to be a strategy employed by the larger and slower (and Southeast Asian) primates to deal with seasonality, but this would predispose them to living in evergreen forests with limited seasonality, as the great apes do today. Here, again, behavioral flexibility appears to be the solution to the problem of obtaining high-quality fallback foods during the lean season, a strategy that would include diet switching, extractive foraging and tool use, consumption of novel foods (USOs, meat), and, perhaps, decreasing day ranges.

The impact of seasonality on extinct primates and humans

In contrast to studies of living species, it is difficult to assess from the fossil record how primates that are now extinct responded to seasonality, requiring instead an examination of how increasing seasonality may have permanently altered the habitats that primates occupied, and thus the evolutionary trajectory of primates, including humans. In the discussion that follows, I synthesize the paleontological evidence linking the responses of various primate lineages to habitat seasonality.

Jablonski (Table 19.1, Chapter 16) examines the factors that influenced primate diversity through time and at different periods of environmental seasonality, focusing on the origins and evolutionary history of four primate groups – the tarsiers, Old World Monkeys, apes, and humans. As Jablonski notes (Table 19.1), changes in environmental seasonality through geological time invariably created changes in the physical structure of habitats and the availability of food resources that primates depended upon. These changes, in turn, impacted the evolutionary trajectory of primate lineages, particularly of those primates occupying the dense tropical forests where high-quality food was abundant. Jablonski argues that the origins and evolution of tarsiers, monkeys, apes, and humans are linked directly to changes in global climate, resulting in increasing seasonality in temperature and rainfall and less stable environmental regimes. The ability (or inability) of these primate radiations to cope with changing environmental conditions depended in large measure upon the specific life history patterns, diet, and mode of locomotion characterizing the individual primate lineages.

The origin and diversification of Early and Middle Eocene prosimians and Early Miocene hominoids around 50 and 20 Ma, respectively, occurred during periods of global climate equability, when tropical forests and subtropical woodlands were widespread. The earliest primates thrived in these habitats, acquiring the characteristics that defined the order, including the distinctive slow life history pattern that was to mark all, or most, succeeding lineages. Jablonski is quick to point out, though, that this slow life history could be both a blessing and a curse. The frugivorous, large-bodied (5–20 kg), agile tree-climbing hominoids that flourished in the warm tropical forests of Africa and Eurasia experienced major extinction events beginning 10 Ma, when cooler, drier conditions prevailed and forests shrank, giving way to more mosaic habitats, including grassy woodlands and savannas.

We do not yet have the spatial and temporal resolution in the fossil record to link increasing habitat seasonality directly to hominoid extinction events, but Jablonski presents circumstantial evidence from other mammals showing that the Late Miocene Epoch experienced a high turnover event. Paleontological evidence suggests that the increasing aridification that occurred in Africa at the terminal Miocene Epoch resulted in the diversification of ruminants, elephants, and carnivore lineages (Behrensmeyer *et al.* 1992), while that which occurred in Asia produced a steady decline in mammalian diversity, in which large-mammal guilds became dominated initially by equids, followed by a diversity of other grazing ruminants (Barry *et al.* 2002). We do not know the specific cause(es) of the decline in hominoid diversity, but Jablonski believes that in all likelihood it was the very traits that made hominoids especially well-adapted to living in less seasonal forested habitats – their slow life history, their dependence upon high-quality foods (e.g. fruits), and their unique bridging postures – that precluded them from making a living in more open habitats where high-quality fallback foods were less available and the loss of connected canopies made the "curse" of being an obligate tree-dweller a prescription for extinction.

The increasing aridification and habitat seasonality that led to the extinction of many hominoids in the Late Miocene Epoch provided opportunities for other primate lineages to thrive and diversify well into the Pliocene and Pleistocene epochs, including the Old World monkeys and hominins. Jablonski posits that this success was also a consequence of species possessing life history, dietary, and locomotor attributes different from those of apes. For Old World monkeys, it was their smaller body size, and hence much higher intrinsic rate of increase of population, their timing of births to seasonal food abundance, their ability to subsist on lower-quality food, and their generalized quadrapedalism that allowed them to move easily between forest patches. Behavioral flexibility appeared to be the adaptive strategy that led to the success of this radiation well into the Holocene.

The success of hominins is, likewise, attributed to a level of behavioral innovation, but one that is unparalleled in primate evolutionary history and linked inextricably to the challenges presented by living in seasonal-mosaic habitats. The origin and diversification of hominins, *c.* 7 Ma, coincided with increasing, albeit fluctuating, environmental seasonality, but unlike the other hominoids that perished, the hominins flourished in spite of having ape-like patterns of life history. According to Jablonski, the key factor that distinguished hominins from their less successful hominoid cousins appeared to be their ability to overcome the constraints of slow life history through social means, particularly collective childcare. It was this innovation that, when combined with obligate bipedality, omnivory, and the development of innovative ways of acquiring and processing high-quality foods, led to the remarkable success of this lineage, culminating in the global range expansion that characterizes *Homo sapiens* today.

Reed and Fish (Table 19.1, Chapter 17) examine in depth the responses of hominins to increasing seasonality, focusing on the impact of global climate change on habitat structure and resource availability, and its effect on the diet and behavior of extinct humans occupying tropical and temperate regions. Using ecological, morphological, dental microwear, carbon isotope, and archeological data gathered from the literature, the authors examine how seasonal changes over time might have contributed to differences among hominin behavioral adaptations, including diet and foraging strategies. They acknowledge, however, that it is very difficult, and maybe impossible, to link hominin dental morphology with *seasonal* shifts in diet, the consequence of a woefully imperfect fossil record.

With this caveat, Reed and Fish (Table 19.1) marshal evidence suggesting that glacial cycles had major impacts on climate, habitat structure, and resource availability 3.2–1.0 Ma and that this resulted in major dietary and behavior shifts in tropical hominins, including increased reliance on USOs in australopithecines and confrontational scavenging in *Homo ergaster*. Periodic glacially driven cooling periods, represented by lower sea-surface temperatures, are documented at 3.2–2.2 Ma, 2.1–1.9 Ma, and 1.0 Ma–650 Ka, resulting in pulses of aridification and increased seasonality that changed habitat structure and altered the hominin resource base. Evidence from seasonal growth changes in mollusks recovered from Hadar, Ethiopia, dated at 3.4–3.2 Ma indicate that *A. afarensis* living there experienced yearly dry seasons of approximately three months (Hailemichael 1999), although other Pliocene hominin habitats apparently experienced dry seasons of four to six months' duration (see Chapter 17). The implications of this are that hominins most likely faced regular seasons of food scarcity, when access to high-quality fallback foods would be essential for survival. As indicated previously, there is no consensus regarding the precise nature of those fallback foods, but Reed and Fish argue that without tools, australopithecines (e.g. *A. africanus*) would have been unable to access tough foods such as meat or marrow by relying solely on their dentition. This assessment is supported further by evidence from studies of masticatory morphology (Teaford & Unger 2000) and microwear analyses (Ryan & Johanson 1989) showing that australopithecines consumed soft and hard foods, including small hard fruits and perhaps USOs, the latter representing important fallback foods during the lean season (see Chapter 17).

The dietary flexibility exhibited by the australopithecines may very well have contributed to the success of this lineage 4.2–2.0 Ma and allowed them to both adapt to longer dry seasons and thrive in a variety of habitats, including riverine gallery forests, woodlands, scrublands, and open savannas. The success of this lineage contrasts starkly with that of their sister group, the hard-object-feeding paranthropines, whose tenure in the fossil record lasted a mere 1.5 million years (i.e. ~2.5–1.0 Ma). Opinions vary as to the dietary niche of paranthropines (Wood & Strait 2004), but Reed and Fish argue that craniodental evidence and dental microwear (e.g. huge molars, extremely thick enamel, heavy pitting of cheek teeth) suggest that paranthropines ate hard, low-quality food items more consistently than did australopiths, and that USOs were utilized intensively during the lean season. As Reed and Fish note, this narrow dietary breadth did not bode well for this species during the extreme climate shift that occurred ~1.0 Ma, as the disappearance of this taxon from the fossil record coincides with this period of substantially increased aridity. We do not know why, of the two species utilizing USOs during the dry season, the paranthropines, but not the hominins, perished, but it may be that hominins were able to maintain greater dietary breadth during the lean season, thereby forestalling the paranthropines' fate (see below).

The responses of early *Homo* (e.g. *H. habilis* and *H. rudolfensis*) to seasonal food scarcity was markedly different from those observed in their predecessors, involving increased dietary behavioral flexibility, including opportunistic scavenging and the use of stone tools to process meat and marrow. Early *Homo* often is associated with riparian woodlands, where they had access to high-quality, regularly available fallback resources, particularly meat (but see O'Connell [2002a]). With the appearance of *Homo erectus senso lato* at 1.8 Ma at Koobi Fora, Africa, there was a shift from opportunistic to confrontational scavenging, the increased reliance on animal protein buffering them from seasonal stress. Reed and Fish in Chapter 17 argue that this shift correlated with the evolution of new anatomical traits (e.g. increased stature, modern limb proportions, reduction in gut size, more efficient cardiovascular system), representing an adaptation to more open, arid habitats. Evidence for confrontational scavenging derives from the first appearance of the Acheulean tool industry 1.5 Ma, featuring heavy-duty butchering tools associated with the remains of large mammals at Olduvai Gorge (see Chapter 17, but see also O'Connell [2002a]).

The evolution of hunting coincides with the first appearance of *Homo heidelbergensis* in Africa and in Europe ~800–500 ka, and while the authors acknowledge that there is a paucity of archeological evidence directly linking this species with hunting, they nevertheless argue that this hominin's substantially larger brain size and ability to survive in harsh northern latitudes of Europe after 500 ka suggests that they could regularly access high-quality animal resources. It was not until the appearance of Neanderthals, however, that hunting became the key adaptive strategy for surviving in the temperate regions of Europe ~150–28 ka, when glacial activity was at its height. The evidence supporting intensive hunting of large animals is, however, indirect, including an exceptionally robust postcranial skeleton, a high incidence of trauma in this taxon, faunal accumulations dominated by a single ungulate species showing cut marks, and age profiles indicative of specialized hunting strategies (see Chapter 17). The authors cite additional evidence from buccal microwear patterns (Lalueza Fox & Perez-Perez 1993; Lalueza *et al.* 1996) and isotopic analyses (Bocherens *et al.* 1999; Richards *et al.* 2000) to support these conclusions. Thus, it appears that tactical hunting was the principal adaptive mechanism for Neanderthals inhabiting the cold grasslands of Europe, one that had become so integrated into their lifestyle that it was even practiced by populations inhabiting warmer habitats such as those found in the Levant (see Chapter 17).

In contrast, the hunting strategies of anatomically modern *Homo sapiens* (AMHS) incorporated a greater diversity of species. Reed and Fish argue that the first appearance of AMHS 160 ka in Ethiopia and at 125–110 ka in South Africa coincided with a period of dietary expansion, the consequence of their having evolved in the more permissive habitats of the tropics where resources were abundant. This allowed them to hunt a wider range of vertebrates, including marine and avian fauna, such as those found at Klasies River Mouth and Die Kelder, South Africa.

The specific behavioral innovations that facilitated greater flexibility in foraging behavior in this species are not elucidated, but one might well imagine that the global hegemony that ensued was linked inextricably to new cognitive innovations, not least of which was language and its potential for shaping human experience. Among the selective mechanisms advanced to explain the evolution of language, the one proposed by Deacon (1997) is particularly relevant: symbolic communication was a way for hominins to establish social contracts that facilitated family and group structures. Maintenance of these social structures consequently was dependent upon hominins having access to high-quality fallback foods during lean periods, hunting being the most efficient way to acquire meat on a regular basis. The paleontological evidence, however, linking the evolution of language to grouping patterns and seasonal hunting strategies remains elusive.

It is clear from the preceding review that assessing the impact of seasonality on hominin evolutionary trajectories hinges on our ability to document paleoseasonality across a broad range of spatial and temporal scales. The contributions above have elucidated the degree to which long-term global and regional climate trends have influenced hominin evolution in equatorial Africa, but as Kingston (Table 19.1, Chapter 18) points out, short-term astronomically driven shifts in climate can produce equally dramatic impacts on mammalian, including hominin, evolutionary events. Indeed, it appears that the role of climate control in mammalian evolution is exceedingly complex and that climate patterns in Equatorial Africa have undergone almost continuous flux, driven principally by orbital forcing.

Kingston (Chapter 18) proposes that short-term changes in the geometry of Earth's orbit, occurring at cycles ranging from 20 to 200 ka (Milankovitch cycles), affect climate by altering the amount of solar radiation by season and latitude. These orbitally forced climatic changes, in turn, affect systems controlling climate in Equatorial Africa, including African-Asian monsoonal circulation and seasonal shifts of the Intertropical Convergence Zone, so that the tropical and subtropical ecosystems experiencing these regional changes exhibit dramatic shifts in seasonality, including the length, intensity, and duration of wet–dry seasons. Kingston provides evidence from Quaternary paleovegetation studies showing that Equatorial African ecosystems are subject to a number of climatically induced inputs, including orbitally forced climate change, atmospheric CO_2 shifts, and vegetation–soil feedbacks, resulting in rapid shifts in pollen assemblage indices and relative proportions of C_3 and C_4 biomarkers, among others (see Chapter 18). These data suggest that Milankovitch cycles are especially pertinent for understanding how variation in climate and seasonality impacts evolutionary events in Equatorial Africa. Consequently, hominin evolutionary trajectories in Equatorial

Africa should be evaluated in the context of continuous environmental flux, and paleoseasonality assessed at a temporal scale of 10 ka to 1 Ma (Table 19.1).

Short-term climatic change has been documented in Plio-Pleistocene hominin sites of the East African Rift, but as Kingston notes in Chapter 18, few of these events have been linked unequivocally to astronomically mediated insolation (i.e. solar radiation received at the top of the atmosphere) and seasonality patterns. However, studies at Baringo hominin sites in the Central Kenyan Rift Valley have yielded evidence of lake cycling 2.7 and 2.5 Ma consistent with 23-ka Milankovitch precessional periodicity (Kingston *et al.* 2000), suggesting that short-term seasonal change is a more pervasive feature of the African landscape than the fossil record reveals. Our ability to assess the impact of seasonality on hominin evolutionary events, however, requires the development of adequate empirical proxies for paleoseasonality, and while several have been developed, the lack of paleoseasonality records from low-altitude regions represents a significant impediment for understanding the role of seasonality in hominin evolution in Equatorial Africa. The challenge nevertheless remains to establish and correlate causal links between shifts in seasonality patterns and specific hominin evolutionary events, and Kingston (Table 19.1) argues that Milankovitch cycling represents the most appropriate temporal context for such an endeavor. One might conclude from this review that there is a conflict between long-term climate changes and shorter-term changes that are based on Milankovitch cycles, but this is certainly not the case from the consumer's point of view. Seasonality is seasonality for a consumer, no matter what causes it. It can, therefore, be argued that the shorter cycles simply exacerbate the impact of the longer cycles (or dampen them) and that these dynamic interactions might be reasonably expected.

Conclusions

In my effort to answer the question "What do studies of seasonality in primates tell us about human evolution?," several insights have emerged. It appears that the earliest hominins (e.g. australopithecines) exhibited small female body size, occupied seasonal habitats (woodland or savannas), and were most likely capital breeders. Their origin may well have been due to increased seasonality, and they coped with it through behavioral flexibility, devising innovative ways of tracking and exploiting high-quality fallback foods during seasonal food scarcity. Otherwise, as capital breeders, they could not have survived. With the advent of *Homo*, we see the evolution of larger body sizes and reduced body-size dimorphism and movement into highly seasonal habitats. A fundamentally new lifestyle emerged that allowed them to make a living on the savanna. They successfully avoided, indeed *opposed*, the typical primate response to increasing seasonality, the increased female body size arguably resulting from developing innovative foraging strategies that facilitated access to high-quality fallback foods (and, thus, positive energy balance), including the use of tools and controlled use of fire to acquire and process USOs and meat. In conclusion, one might reasonably argue that Primates, the order that does best in non-seasonal tropical habitats, produced a lineage that somehow managed to do very well in the highly seasonal habitats that gradually emerged in Africa during the Pliocene Epoch and that the key to their success was their behavioral flexibility. Rooted in our primate heritage, it was the elaboration of this remarkable attribute in hominins that allowed them to come up with behavioral rather than morphological and physiological adaptations to extreme seasonality, setting in motion the necessary preconditions for the evolution of culturally mediated responses to environmental flux in *Homo*.

Acknowledgments

This chapter has been substantially improved by Carel van Schaik's helpful and thought-provoking comments, for which I am most appreciative. More importantly, I am exceedingly grateful to Carel for his willingness to be the co-editor of this volume and for his encouragement, patience, and dedication to the task.

References

Aunger, R. (2002). *The Electric Meme: A New Theory of How We Think*. New York: Free Press.

Bailey, R. C., Jenike, M. R., Ellison, P. T., *et al.* (1992). The ecology of birth seasonality among agriculturalists in central Africa. *Journal of Biosocial Science*, **24**, 393–412.

Barry, J. C., Morgan, M. E., Flynn, L. J., *et al.* (2002). *Faunal and Environmental Change in the Late Miocene Siwaliks of Northern Pakistan*. Lawrence, KS: The Paleontological Society.

Behrensmeyer, A. K., Damuth, J. D., DiMichele, W. A., *et al.* (1992). *Terrestrial Ecosystems Through Time*. Chicago: University of Chicago Press.

Bocherens, H., Billiou, D., Mariotti, A., *et al.* (1999). Paleoenvironmental and paleodietary implications of isotopic biogeochemistry of last interglacial Neanderthal and mammal bones in Scladina Cave (Belgium). *Journal of Archaeological Science*, **26**, 599–607.

Carmena, J. M., Lebedev, M. A., Crist, R. E., *et al.* (2003). Learning to control a brain–machine interface for reaching and grasping by primates. *Public Library of Science Biology*, **1**, 193–208.

Conklin-Brittain, N. L., Wrangham, R. W., & Smith, C. C. (2002). A two-stage model of increased dietary quality in early hominid evolution: the role of fiber. In *Human Diet: Its Origins and Evolution*, ed. P. S. Unger & M. F. Teaford. Westport, CT: Bergin and Garvey, pp. 61–76.

Deacon, T. (1997). *The Symbolic Species: The Co-Evolution of Language and the Brain.* New York: Norton & Company.

De Menocal, P. B. (1995). Plio-Pleistocene African climate. *Science*, **270**, 53–9.

De Menocal, P. B. & Bloemendal, J. (1995). Plio-Pleistocene climate variability in subtropical Africa and the paleoenvironment of hominid evolution. In *Paleoclimate and Evolution with Emphasis on Human Origins*, ed. E. A. Vrba, G. H. Denton, T. C. Partridge, & L. H. Burckle. New Haven: Yale University Press, pp. 262–88.

DiFiore, A. (2002). Predator sensitive foraging in ateline primates. In *Eat or Be Eaten*, ed. L. E. Miller. Cambridge: Cambridge University Press, pp. 242–67.

Ellison, P. T., Peacock, N. R., & Larger, C. (1989). Ecology and ovarian function among Lese females of the Ituri Forest, Zaire. *American Journal of Physical Anthropology*, **78**, 519–26.

Foley, R. A. (2001). The evolutionary consequences of increased carnivory in hominids. In *Meat-Eating and Human Evolution*, ed. C. B. Stanford & H. T. Bunn. New York: Oxford University Press, pp. 305–31.

Fox, E. A., van Schaik, C. P., & Wright, D. N. (2005). Intra- and interpopulation differences in orangutan (*Pongo pygmaeus*) activity and diet: implications for the intention of tool use. *American Journal of Physical Anthropology*, in press.

Galik, K., Senut, B., Pickford, D., *et al.* (2004). External and internal morphology of the BAR 1002'00 *Orrorin turgenensis* femur. *Science*, **305**, 1450–3.

Garber, P. A. (1987). Foraging strategies of living primates. *Annual Review of Anthropology*, **16**, 339–64.

Hailemichael, M. (1999). The Pliocene environment of Hadar, Ethiopia: a comparative isotopic study of paleosol carbonates and lacustrine mollusk shells of the Hadar Formation and of modern analog. Ph. D. thesis, Case Western Reserve University.

Harding, R. S. O. (1981). An order of ominivores: nonhuman primates in the wild. In *Omnivorous Primates: Gathering and Hunting in Human Evolution*, ed. R. S. O. Harding & G. Teleki. New York: Columbia University Press, pp. 191–214.

Janson, C. H. (1998). Experimental evidence for spatial memory in wild brown capuchin monkeys (*Cebus apella*). *Animal Behaviour*, **55**, 1129–43.

Janson, C. H. & Chapman, C. A. (1999). Resources and primate community structure. In *Primate Communities*, ed. J. G. Fleagle, C. H. Janson, & K. E. Reed. Cambridge: Cambridge University Press, pp. 237–67.

Kingston, J. D., Hill, A., & Deino, A. L. (2000). Pliocene hominid evolution and astronomically forced climate change: evidence from the rift valley of Kenya. *Journal of Human Evolution*, **38**, A15.

Lalueza Fox, C. & Perez-Perez, A. (1993). The diet of the Neanderthal child Gibraltar 2 (Devil's Tower) through the study of the vestibular striation pattern. *Journal of Human Evolution*, **24**, 29–41.

Lalueza, C., Perez-Perez, A., & Turbon, D. (1996). Dietary inferences through buccal microwear analysis of Middle and Upper Pleistocene human fossils. *American Journal of Physical Anthropology*, **100**, 367–87.

Lee-Thorp, J. A. (2002). Hominid dietary niches from proxy chemical indicators in fossils: the Swartkrans example. In *Human Diet: Its Origin and Evolution*, ed. P. S. Ungar & M. F. Teaford. Westport, CT: Bergin and Garvey, pp. 123–41.

Lee-Thorp, J. A., Thackeray, J. F., & van der Merwe, N. J. (2000). The hunters or the hunted revisited. *Journal of Human Evolution*, **39**, 565–76.

Marean, C. W. (1989). Sabertooth cats and their relevance to early hominid diet and evolution. *Journal of Human Evolution*, **18**, 559–82.

McGrew, W. C. (1992). *Chimpanzee Material Culture: Implications for Human Evolution*. Cambridge: Cambridge University Press.

McHenry, H. M. (1992). Body size and proportions in early hominids. *American Journal of Physical Anthropology*, **87**, 407–31.

O'Brien, E. M., & Peters, C. R. (1999). Climactic perspectives for neogene environmental reconstructions. In *Hominoid Evolution and Climactic and Environmental Change in the Neogene of Europe*, ed. J. Agusti & P. J. Andrews. Cambridge: Cambridge University Press, pp. 53–78.

O'Connell, J. F., Hawkes, K., Lupo, K. D., & Blurton Jones, N. G. (2002a). Male strategies and Plio-Pleistocene archeology. *Journal of Human Evolution*, **43**, 831–72.

O'Connell, J. F., Hawkes, K., & Blurton Jones, N. (2002b). Meat-eating, grandmothering, and the evolution of early human diets. In *Human Diet: Its Origin and Evolution*, ed. P. S. Ungar & M. F. Teaford. Westport, CT: Bergin and Garvey, pp. 49–60.

Potts, R. (1996). *Humanity's Descent: The Consequences of Ecological Instability*. New York: William Morrow.

Richards, M., Pettitt, P., Trinkaus, E., *et al.* (2000). Neanderthal diet at Vindija and Neanderthal predation: the evidence from stable isotopes. *Proceedings of the National Academy of Sciences, USA*, **97**, 7663–6.

Ryan, A. C. & Johanson, D. C. (1989). Anterior dental microwear in australopithecus-afarensis: comparisons with humans and nonhuman primates. *Journal of Human Evolution*, **18**, 235–68.

Sherry, D. S. (2002). Reproductive seasonality in chimpanzees and humans: ultimate and proximate factors. Ph. D. thesis, Harvard University.

Teaford, M. F. & Unger, P. S. (2000). Diet and the evolution of the earliest human ancestors. *Proceedings of the National Academy of Sciences, USA*, **97**, 13506–11.

Teaford, M. A., Unger, P. S., & Grine, F. E. (2002). Paleontological evidence for the diets of African Plio-Pleistocene hominins with special reference to early

Homo. In *Human Diet: Its Origins and Evolution*, ed. P. S. Ungar & M. F. Teaford. Westport, CT: Bergin and Garvey, pp. 143–66.

Van Schaik, C. P., Terborgh, J. W., & Wright, S. J. (1993). The phenology of tropical forests: adaptive significance and consequences for primary consumers. *Annual Review of Ecology and Systematics*, **24**, 353–77.

Wheeler, P. E. (1991). The thermoregulatory advantages of hominid bipedalism in open equatorial environments: the contribution of increased convective heat loss and cutaneous evaporative cooling. *Journal of Human Evolution*, **21**, 107–15.

— (1992a). The influence of the loss of functional body hair on the energy and water budgets of the early hominids. *Journal of Human Evolution*, **223**, 379–88.

— (1992b). The thermoregulatory advantages of large body size for hominids foraging in savannah environments. *Journal of Human Evolution*, **223**, 351–62.

— (1993). The influence of stature and body form on hominid energy and water budgets: a comparison of *Australopithecus* and early *Homo* physiques. *Journal of Human Evolution*, **24**, 13–28.

Wood, B., & Strait, D. (2004). Patterns of resource use in early *Homo* and *Paranthropus*. *Journal of Human Evolution*, **46**, 119–62.

Yamakoshi, G. (2004). Evolution of complex feeding techniques in primates: is this the origin of great ape intelligence? In *The Evolution of Thought*, ed. A. E. Russon & D. R. Begun. Cambridge: Cambridge University Press, pp. 140–71.

Index